The Moon's Acceleration
and Its Physical Origins

Other books by Robert R. Newton

*Ancient Astronomical Observations and the Accelerations of
   the Earth and Moon*
Medieval Chronicles and the Rotation of the Earth
Ancient Planetary Observations and the Validity of Ephemeris Time
The Crime of Claudius Ptolemy

ROBERT R. NEWTON

The Moon's Acceleration and Its Physical Origins

Volume 1

As Deduced from Solar Eclipses

The Johns Hopkins University Press
Baltimore and London

Manufactured in the United States of America

The Johns Hopkins University Press, Baltimore, Maryland 21218
The Johns Hopkins Press Ltd., London

Library of Congress Catalog Number 78-20529
ISBN 0-8018-2216-5

Library of Congress Cataloging in Publication data
will be found on the last printed page of this book.

CONTENTS

CONTENTS (Continued)

CONTENTS (Continued)

FIGURES

TABLES

TABLES (Continued)

The purpose of this book is to infer the acceleration of the moon with respect to the sun at any time in history from the earliest usable astronomical observations up to the present. When I began it, I hoped to use all available observations that involve the moon and that can be used to infer its acceleration. It soon became clear, however, that this could not be done within the scope of a single book of reasonable size. The current book, therefore, except for the modern data, uses only observations in which a writer states that he saw a solar eclipse in a known place or region on a known date. I call this book Volume I of the study, and I shall use other types of lunar observation in a later volume or volumes.

I should explain the historical periods that are used in this work, since they differ somewhat from the customary ones. I use "ancient" to refer to any time before about 400, I use "medieval" to refer to any time from then until the introduction of the telescope and pendulum into positional astronomy, and I use "modern" to refer to any time since. There is no Renaissance in this system of periods. So far as positional astronomy is concerned, a Renaissance would begin in the 12th century if we used one.

I cite a reference by giving the name of the author underlined (or the name of the work if it is anonymous, as many medieval works are) and the year it was written or published. If the year is not known exactly, I give a reasonable estimate of the year, preceded by the symbol "ca.". Either the date, or both the author and the date, are enclosed in square brackets, depending upon the context. If a work is cited several times within a short space, I usually omit the year after the first citation.

If detailed information about a reference is needed, such as a page number, I put this also within the square brackets, placing it after the year if the year occurs within the brackets.

In earlier works, I tried to adopt consistent methods of rendering foreign proper names into English, without outstanding success. I have given up most attempts at such consistency in this work, and have merely tried to use English spellings that will be intelligible and useful to the reader.

Dates in this work will be given in astronomical style, as contrasted with historical style. In astronomical style, the year is followed by the month which is followed by the day, without the use of commas to separate the components of the date. The astronomical year and the historical year are the same for years of the common era. The year before the year 1 of the common era is written as 0 in astronomical style, the preceding year is written as -1, and so on. The same years are written as 1 B.C.E. and 2 B.C.E., respectively, in historical style.

In an earlier study of medieval chronicles [Newton, 1972], I found extensive records of such matters as earthquakes, comets, plagues, and the like, and I made extensive tabulations of them. I once intended to do the same here, but there are not enough such records in the sources that I am using here for the first time to make the enterprise worthwhile. However, I have found two compilations on interesting topics that cover a considerable time span and that are valuable for this reason. Accordingly, I present a record of British agricultural success over about a century in Appendix IV and a record of the flooding of the Nile over about half a century in Appendix V.

I thank my colleagues B. B. Holland, R. E. Jenkins, and A. D. Goldfinger, of the Applied Physics Laboratory, for many helpful discussions, and I particularly thank the first two for a detailed reading of the work. I gratefully acknowledge the support of this research by the Department of the Navy through its contract with The Johns Hopkins University, and I thank the Director of the Applied Physics Laboratory of The Johns Hopkins University for encouraging me in this study.

I particularly thank Mr. J. W. Howe and Mrs. M. J. O'Neill of this Laboratory for their dedicated work in preparing this volume for the press.

The Moon's Acceleration
and Its Physical Origins

CHAPTER I

BACKGROUND AND PURPOSES OF THIS WORK

1. Solar Time and Ephemeris Time

It is convenient to begin by reminding the reader of
the system used for measuring time in ordinary life, and to
contrast it with a system used by astronomers for many
purposes.

Ordinary life is regulated by solar time. When the sun
is due south, for people who live in the North Temperate
Zone, we say that the time is noon, or 12 hours. We divide
the interval between successive noons into two equal parts
and say that the midpoint is midnight, which we call 00 or
12 hours if we use a 12-hour clock, and which we call 00 or
24 hours if we use a 24-hour clock. We number other times
of the day accordingly, and we say that a day begins at midnight.

The time system just described is properly called apparent solar time. Because of the details of the sun's
apparent motion through the heavens, the interval between
successive noons is not exactly constant. In practice, we
speak of an imaginary point called the mean sun, which moves
uniformly and which, on the average, is just as far in front
of the real sun as it is behind it. We then say that mean
noon occurs when the mean sun is due south, and so on. The
time system based upon the mean sun is called mean solar
time. For simplicity, I shall use "solar time" for "mean
solar time". The greatest difference between mean and apparent solar time is about 15 minutes.

When solar time is defined in this way, the time system
is different for every different longitude on earth. Solar
time that is based upon using the local longitude is called
"local solar time". Until less than a century ago, every
place, or at least every major city, used its own local
solar time. With the growth of rapid travel and communication over long distances, however, the use of local time
became awkward. In 1884 the system of "zone time" was
introduced at the same time as the convention of saying
that the longitude of Greenwich† is 0°. With only a few
exceptions, the system of zone time has been adopted by all
the earth.

Astronomers and some other scientists find it necessary
to use a single time base rather than the multiplicity of
zone times. All astronomers now use Greenwich mean time;
that is, they use solar time for the meridian of Greenwich

---

†When we speak of the longitude of Greenwich, we mean the
longitude of the axis of a particular telescope in the
Royal Observatory at Greenwich. The working site of the
Royal Observatory has now been shifted from Greenwich to
Herstmonceux, but the first observatory still furnishes
the standard from which longitude is measured.

as their standard solar time, regardless of their longitude.
Many organizations, such as international airlines, that
need communications extending over more than one continent,
often use Greenwich mean solar time also.

In the rest of this work, I shall use "solar time" for
brevity in place of "Greenwich mean solar time", unless it
is necessary to use local time for some reason. The context
will usually make these occasions clear; if it does not, I
shall explicitly say "local solar time".

With the increasing accuracy of mechanical clocks, and
particularly with the introduction of crystal clocks and
cesium clocks, astronomers in this century began to suspect
that solar time does not flow uniformly. Since the recurring phenomenon of noon is produced by the spin of the earth
on its axis, this is equivalent to saying that the earth's
spin is not constant. We now know that this is indeed so,
and that there are two major kinds of variation in the
earth's spin:

a. The earth's spin is subject to changes in angular
acceleration that occur at fairly short intervals of time
and that seem to be random. These sporadic changes occur
non-uniformly, but at an average interval of about 4 years.
The average change in acceleration would be enough to change
the earth's spin by about 350 parts in 10^9 in 1 century if
the change persisted that long.

b. The earth's spin is subject to a slow, secular
change. The secular acceleration changes the earth's
spin by about 20 or 30 parts in 10^9 in 1 century. The
secular acceleration is not constant.

There are other kinds of variation that are relatively small
and that do not concern us in this work. Variations in the
earth's spin have been surveyed in several places (Munk and
MacDonald [1960], Jeffreys [1970], Newton [1973a], and
Newton [1976], for examples).

Since the earth's spin no longer furnishes an adequate
time base, we turn to some phenomenon that furnishes a better
one. Two different time bases are now used by astronomers
and by others who need a measure of time that is more stable
than solar time.

The first of these is atomic time (AT). Atomic time is
provided by a set of cesium clocks that can be compared with
each other by means of radio transmissions or by the physical
transportation of clocks from one place to another. There
are other types of clock that depend upon atomic or molecular
phenomena which may displace cesium clocks as the source of
atomic time in the future.

The second is ephemeris time (ET). The idea back of
ephemeris time is that it should be the independent variable
in the equations of motion of the solar system. That is, it

should be derived from the orbital motion of the bodies that make up the solar system. There are problems in obtaining a practical measure of time in this way, however.

The first problem concerns the fact that the planets move slowly in their orbits, so that we need highly precise measurements in order to obtain a time measure of reasonable precision. In practice, this means that we must use measurements extending over a long time span, and this in turn means that we cannot measure ephemeris time accurately on a current basis. We can only say what relation it has had to some other time base, such as atomic or solar time, at times in the past.

The second problem concerns the motion of the moon. Since the moon moves more rapidly than the planets, our thought is to use it rather than the planets to furnish an operational measure of ephemeris time. By an operational measure, we mean one that can be kept reasonably current. When we try to use the moon, however, we find that it is subject to a secular acceleration that is not of gravitational origin† and that we cannot find from theoretical principles in the present state of knowledge.

An attempt has been made to meet this problem by adopting the results of Spencer Jones [1939]. Spencer Jones studied the positions of the sun, moon, and planets from about 1700 to about 1935. He concluded that the same time base could be used to describe all of the motions provided that the non-gravitational acceleration of the moon was taken to be -22″.44 per century per century.‡ In practice, ephemeris time is now determined from the motion of the moon, when we assume that it has this secular acceleration.

Before we proceed further, it is useful to mention one other time base that is often used in astronomy. This is the base called universal time (UT), which is derived from sidereal time. Suppose that we follow the motion of the imaginary point in the heavens called the vernal equinox, or the "first line of Aries",‡ and that we note the instants when this point passes through our local meridian. The interval between two successive passages is called a sidereal day, which is divided into 24 hours, just as a solar day is. From the apparent motion of the equinox, then, we get a time

---

†The moon does have accelerations due to the gravitational action of the sun and planets. These accelerations can be calculated by theoretical methods to any required accuracy, and we are not concerned with them in this work. We are concerned only with the accelerations that are not of gravitational origin.

‡Spencer Jones did not word his results in this way, because he wrote in terms of solar time as the time base. The description given above is obtained when we translate his results into terms of ephemeris time.

‡The definition of this point is given in most texts on descriptive astronomy. It is the fundamental point from which the positions of the sun, moon, planets, and stars are measured.

measure called sidereal time. If we calculate the solar
time from sidereal time, using a theory of motion of the sun
in order to do so, the result is called universal time.

Some astronomers consider universal time and solar time
to be synonymous. However, they do differ in principle.
Universal time is based upon a theory of where the sun should
be, while solar time, as I have defined it, is based upon
direct observation of the sun. The two times would agree
exactly only if the theory of the sun were exact, which, of
course, it is not. In practice, universal time and solar
time have probably not disagreed by as much as a minute in
the past 2000 years.

The use of ephemeris time rests upon two assumptions.
One is that a single time base is sufficient for the motion
of the sun and planets. The other is that the same time base
suffices for the moon if we determine the secular acceleration of the moon accurately. There have been serious grounds
for suspecting that neither assumption is valid.

2.   Some Earlier Related Work

I have studied some of the problems connected with the
use of ephemeris time in three earlier major works, as well
as some shorter ones. The present work grows directly out
of the earlier ones.

In the first work [Newton, 1970],† I collected a body
of astronomical observations with dates from -762 June 15 to
1241 October 6. When I wrote that work, it had not occurred
to me to doubt the validity of ephemeris time. Accordingly,
I used the collected data to estimate the accelerations of
the moon's orbital motion and of the earth's spin with respect to ephemeris time.‡ More specifically, I assumed that
the theories used to calculate the tables in the American
Ephemeris and Nautical Almanac are correct for the sun and
planets when the time variable is taken to be ephemeris time,
and that the tables for the moon's orbital motion and for
the earth's spin are similarly correct except for small accelerations of non-gravitational origin. Thus I attributed
any discrepancies between calculated and observed positions
of the moon and of the earth's orientation in space to these
accelerations.

In describing the results, let n_M be the orbital angular
velocity of the moon and let $\dot{n}_M$ be its acceleration with
respect to ephemeris time as just defined; the unit of angle
used in giving values of n_M and $\dot{n}_M$ will be the second of arc
and the unit of time will be the Julian century, which

----

† The abbreviation AAO will henceforth be used in reference
to this work. The abbreviation comes from the first words
of the title (Ancient Astronomical Observations).

‡ The value assigned to an acceleration, like the value assigned to a velocity, depends upon the time base used.
Hence it is necessary to specify the time base when one is
talking about an acceleration.

consists of 36525 days. Similarly, let ω_e and $\dot\omega_e$ be the
velocity and acceleration of the earth's spin. Instead of
using $\dot\omega_e$, let us introduce a parameter y defined by

$$y = 10^9 (\dot\omega_e / \omega_e) \qquad (I.1)$$

in order to get numbers of a more convenient size. (The
notation used in Equation I.1 and in the other parts of this
work is summarized in Section I.7.) Thus y gives us the acceleration of the earth's spin in terms of parts in 10^9 per
century. I shall usually omit a statement of the units of
$\dot n_M$ and y.

I divided the data in AAO chronologically into two parts.
From the data before the year 500, I estimated [page 272]

$$\dot n_M = -41.6 \pm 4.3, \qquad y = -27.7 \pm 3.4. \qquad (I.2)$$

The mean epoch of the data used to obtain Equations I.2 was
-200. The data after the year 500 gave [page 272]

$$\dot n_M = -42.3 \pm 6.1, \qquad y = -22.5 \pm 3.6, \qquad (I.3)$$

with a mean epoch of 1000. There are several important points
about the results of AAO:

a. Both values of $\dot n_M$ are fairly close to -40, while the
value used to obtain ephemeris time from the motion of the
moon is -22.44. If the results of AAO are correct, either
the value of $\dot n_M$ has changed by a large amount within historic
times, or the value of $\dot n_M$ at the present time is not close to
-22.44. In either case, the present method of obtaining
ephemeris time from the motion would not be correct. For
some time, it appeared that the second alternative just mentioned was the correct one. Morrison [1973] and Oesterwinter
and Cohen [1972] both found, on the basis of data within the
past two centuries or so, that $\dot n_M$ is close to -40. However,
Morrison and Ward [1975] have recently re-examined some of
the data used by Spencer Jones, and they find $\dot n_M = -26 \pm 2$,
in reasonably close agreement with Spencer Jones's original
result. As I shall describe in Section I.4, I have also made
an independent analysis of the data, without making some of
the assumptions that Morrison and Ward make, and I find
-28 ± 5. Thus it now seems that $\dot n_M$ is in the middle twenties,
and that it cannot be as large (in magnitude) as -40. There
now seem to be two possibilities: Either the results of AAO
are incorrect, or $\dot n_M$ has changed by a large amount within
historic times.

b. In Section VIII.3 of AAO, I analyzed 26 conjunctions
of Venus with stars or other planets, and obtained the following estimate of y from these data alone [page 209]:

$$y = +20 \pm 24 \qquad (I.4)$$

for an epoch near 1000. This estimate is considerably at
variance with the estimates in Equations I.2 and I.3. At
the time, I considered this to be a statistical accident.
However, later work indicates that the discrepancy may be

-5-

significant, as I shall show in a moment.

c. Most of the observations used in AAO involved the
moon. Some were direct measurements of lunar position while
others were measurements that concerned eclipses. Under
these circumstances, it turns out that $\dot{n}_M$ and y are not
strongly determined. However, the parameter called D'' is
strongly determined. D is the lunar elongation, which is
the difference in longitude between the mean moon and the
mean sun, and D'' is its second derivative with respect to
solar time. D'' is related to $\dot{n}_M$ and y by:

$$D'' = \dot{n}_M - 1.6073y, \qquad\qquad (I.5)$$

in the standard units that have been defined. Table XIV.4
of AAO gives estimates of D'' at fourteen epochs between -700
and +1050. [†]

d. The results of AAO give strong evidence that D'' has
changed by a large amount within the past 2000 years. The
values of D'' for epochs before the year +700 are all positive
while those for epochs after 700 are all negative. However,
the distribution of the epochs involved does not let us find
the detailed time history of D''. Clearly, if D'' has changed
by a large amount, either $\dot{n}_M$ or y or both have changed.

The conclusion that D'' has changed by a large amount
within historic times is potentially important for astronomy
and geophysics, and testing this conclusion by means of additional data is important. In the next work [Newton, 1972],[‡]
I collected a large body (about 370) of observations of solar
eclipses taken from medieval annals and chronicles. The
emphasis was on the period from 400 to 1200, although I used
a few observations lying outside that period.

The results confirmed that D'' has changed within historic times. Further, the observations in MCRE, when combined with those already used in AAO, were well distributed
in time, so that it was possible to decide upon the nature
of the variation of D''. D'' rose slowly from about +3 at the
epoch -700 to about +6 at the epoch +700, and then fell
rather quickly to about -20 at the epoch 1300.

Next, I turned to the work of Brouwer [1952] and of
Martin [1969], which allows us to estimate ν_M', the acceleration of the moon with respect to solar time, at a time near
the present. Brouwer assumed that the concept of ephemeris
time is valid, and that $\dot{n}_M$ is equal to Spencer Jones's value

---

[†]When I wrote AAO I had discovered that a certain linear
combination of $\dot{n}_M$ and y is well-determined, but I had not
realized its physical interpretation. If D'' is well-
determined, any constant multiple of it is also well-
determined. As it happened, the combination that I used
in AAO equals $-D''/1.6073$.

[‡]The abbreviation MCRE will henceforth be used in reference
to this work. The abbreviation comes from the title
(Medieval Chronicles and the Rotation of the Earth).

of -22.44. He used a variety of lunar observations from 1621 to the present,[†] and found values of the difference between ephemeris and solar times over that time interval. That is, he calculated the ephemeris time of each lunar observation by assuming that $\dot{n}_M = -22.44$, and subtracted the observed solar time from the calculated ephemeris time. Brouwer felt that his values for epochs before 1820 were tentative because of the questionable quality of the data available to him for earlier years.

Martin tested Brouwer's values for the period before 1820 by using telescopic observations of occultations of stars by the moon. Altogether, Martin used about 2000 observations with dates well distributed from 1627 to 1860, and he analyzed them by using the same basic assumptions that Brouwer used. His results give good confirmation to Brouwer's tentative values.

In effect, the combined work of Brouwer and Martin gives us well-determined values of the longitude of the moon at known values of solar time from 1627 to 1960. Thus we can infer a value of ν_M', the acceleration of the moon with respect to solar time, from these longitudes, and the value inferred does not depend upon the specific assumption that $\dot{n}_M$ equals -22.44. The value is [Newton, 1972c, 1973a]:

$$\nu_M' = -12.1 \pm 0.7. \tag{I.6}$$

ν_M' is related to $\dot{n}_M$ and y by

$$\nu_M' = \dot{n}_M - 1.7373y. \tag{I.7}$$

This is not the same as D'' (Equation I.5), but it is fairly close to it. This means that we can calculate a value of D'' from the value of ν_M' in Equation I.6 by using a tentative value of y, such as either value from Equations I.2 or I.3. If we use the value from Equation I.3, for example, the resulting value of D'' is

$$D'' = -15.0 \pm 1.0, \tag{I.8}$$

approximately. The error estimate in Equation I.8 depends upon the error estimate in Equation I.6 and the uncertainty in y, and it is somewhat arbitrary. The value in Equation I.8 applies at the epoch 1800, approximately.

We can obtain a similar result in several ways. Let G_M denote the mean longitude of the moon as calculated from a strictly gravitational theory. Brown [1919], following a precedent that had been set by Newcomb, added the "great empirical term"

$$10''.71 \sin (140°T + 240°.7)$$

to G_M in order to obtain reasonable agreement with observation. In this, T is solar time in centuries from 1900. The correction was intended to apply only for a century or so from the epoch 1800, and it clearly represents the effect of the lunar acceleration with respect to solar time. Just as clearly, it shows that the acceleration is negative.

Spencer Jones used a quantity B which he called the lunar fluctuation; it is tabulated as a function of time in his Table I and plotted in his Figure I. He defined B as the observed mean longitude of the moon minus a theoretical value. His theoretical value was not G_M, however. In order to find G_M, he removed Brown's empirical term, which would have given him G_M, but he added the quantity

$$GET = 4''.65 + 12''.96T + 5''.22T^2$$

in its place; the coefficient of T^2 corresponds to using $+10.44$ $''/cy^2$ for ν_M'. Thus G_M equals Spencer Jones's theoretical value minus GET. Let us define the lunar discrepancy Δ_M as the observed mean longitude of the moon minus G_M. Then Δ_M equals B + GET.

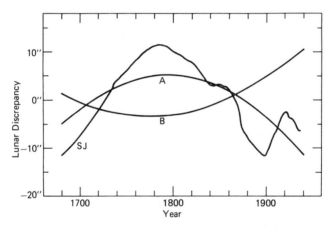

Figure I.1 The discrepancy between the observed longitude of the moon and the longitude calculated from a strictly gravitational theory. The data are those used by <u>Spencer Jones</u> [1939]. The curve marked SJ shows the discrepancy, in seconds of arc, plotted as a function of time. The curve marked A is the best-fitting parabola and the curve marked B is the parabola that Spencer Jones used to represent the data.

The results are shown in Figure I.1. The curve marked
SJ is the lunar discrepancy Δ_M calculated from Spencer Jones's
data in the manner just described. It is again clear from
the figure that the moon has a negative acceleration in modern times. The curve marked A in the figure is the best-
fitting parabola, whose form is

$$-3''.62 - 16''.66T - 7''.86T^2.$$

That is, Spencer Jones should have used an acceleration of
-15.72 rather than +10.44. Finally, the curve marked B is
Spencer Jones's GET.

The fact that Spencer Jones used a value of ν_M' that is
seriously wrong had an interesting effect upon his results.
I shall take this matter up in Section I.4.

The value of D'' in Equation I.8 differs greatly from the
value that we have inferred for the epoch 700, but it is
fairly close to the value inferred for 1300. By combining
this value with the values found in the earlier studies, I
proposed [Newton, 1972c] the following formula for D'' at
times since -700†:

$$D'' = \begin{cases} -14 + T, & -6 \le T \le 0, \\ -44 - 4T, & -12\frac{1}{2} \le T \le -6, \\ 9\frac{1}{8} + \frac{1}{4}T, & -26 \le T \le -12\frac{1}{2}. \end{cases} \qquad (I.9)$$

Since there are no data between 1300 and modern times, the
part of this that applies for times after $T = -6$ is mostly
a guess, although it is a plausible one.

At this point, there were two obvious tasks. One was
to find values of D'' for times between 1300 and the present.
The other was to attempt to isolate the physical cause for
the variation of D''. This task needs some explanation.

Slightly more than two-thirds of the amplitude of the
tides is raised by the moon, with the rest being raised by
the sun. It has long been recognized that friction in the
lunar tide tends to accelerate the moon and, in an ephemeris
time scale, it is the only known effect that does so by an
observable amount. Friction in the solar tide also tends to
accelerate the sun. However, the angular momentum involved,
which is actually the angular momentum of the earth in its
orbit, is so large that the resulting acceleration is trivial

---

†This is a good place to point out that the values of ν_M'
and D'' that have been given, except for those in Equations
I.6 and I.8, are not the actual values at the epochs stated.
Instead, they are the average values between the epochs
stated and the epoch 1900. Similarly, the formula in
Equation I.9 gives the average value between any epoch T
and the epoch 1900. T is time measured in centuries from
1900, so T is negative for times before 1900. The values
in Equations I.6 and I.8 are the best estimates that we can
make for the actual values at 1800; the values at 1800
hardly differ from the average values between then and 1900.

at the present level of accuracy. Friction in both the lunar and solar tides tends to change the earth's spin; that is, tidal friction makes an important contribution to the value of y. However, many geophysical effects other than tidal friction can contribute to y.† Time variations in either tidal friction or in non-tidal effects can thus give rise to time variations in D''. It is valuable for geophysical theory to discover whether the variations in D'' arise from variations in tidal friction, in non-tidal effects, or in both.

In order to answer this question, we need to discover the time dependence of an acceleration parameter that is independent of D''. The independent parameter that is easiest to study is y itself, which we can estimate by analyzing solar and planetary observations. We first calculate the ephemeris time when the sun or a planet had an observed position, and we subtract the observed position, which is a value of solar time, from the calculated ephemeris time. The difference tells us the average acceleration of the earth's spin between the time of the observation and the present.

Accordingly, I next turned [Newton, 1976]‡ to a study of solar and planetary observations. In order to have a work of reasonable length, I confined this work to a study of data from Babylonian, Greek, and early Islamic astronomy. The hope was that I could find values of y from about -500 to about +1000, and that these values, taken in conjunction with the earlier knowledge of D'', would allow us to find the time variation of tidal friction and of non-tidal effects independently.

At the same time that I was collecting the data for this study, I made a careful study of the history of the ephemeris time concept in order to write a useful introduction to the analysis. The study uncovered an interesting situation. Several papers, of which the one by Spencer Jones [1939] was the latest, were written in the first several decades of this century for the purpose of testing the validity of the concept. All the papers indicated that the data are not consistent with the concept. Spencer Jones did find that the data are consistent with it, but only after he discarded an important body of data, namely the measurements of the right ascension of the sun. When the right ascension of the sun was included, he found that the data disagree with the concept by an amount that is statistically significant.

Our first thought is to conclude that the validity of the concept has been disproved by these papers. However, on further study, it turns out that all the investigations suffer from theoretical defects. For example, Spencer Jones assumed that $\nu_M' = +10$, approximately, in modern times, while Equation I.6 shows us that the correct value is about -12. To first order, the conclusions of Spencer Jones are independent of errors in the value used for ν_M', but we may not

---

† Many of these are discussed at length by Jeffreys [1970, Chapter VIII] and by Munk and MacDonald [1960, pp. 227-246]. Newton [1972c] gives a brief summary.

‡ The abbreviation APO will henceforth be used in reference to this work. The abbreviation comes from the first words of the title (Ancient Planetary Observations).

assume that they are independent of errors this large.†
Thus it turns out that ephemeris time was adopted in astronomy before the concept had ever been tested.

Accordingly, I revised the goal of APO after I began
the work. The new goal was to test the validity of the concept of ephemeris time with regard to the secular accelerations of the sun and planets. The method of testing was the
following: First, I estimated the value of $\nu_S{}'$, the acceleration of the sun with respect to solar time. I then analyzed the planetary data by assuming the hypothesis. If the
concept of ephemeris time is valid, the accelerations of the
sun and planets arise solely from changes in the earth's
spin. This means that the accelerations of the sun and
planets with respect to solar time have the same ratio as
their mean motions.‡ That is, the accelerations of Mercury,
Venus, and Mars, for example, obey the relations

$$\nu_{\mercury}{}' = 4.1520\nu_S{}', \quad \nu_{\venus}{}' = 1.6255\nu_S{}', \quad \nu_{\mars}{}' = 0.5317\nu_S{}'.$$

(I.10)

From the observations of Mercury, then, we can make an estimate of $\nu_S{}'$ by using Equations I.10, and we can similarly
estimate $\nu_S{}'$ from observations of Venus and Mars. I did not
use observations of Jupiter and Saturn because of their small
mean motions.

The results of the test were interesting. The value of
$\nu_S{}'$ that was derived from the observations of Mars agrees
well with the value derived from the solar observations.
This means that the same scale can be used for the orbital
motions of the sun (or the earth) and Mars, within the accuracy of the observations. This result does not hold for
Mercury and Venus, however. There are three independent
samples of data for each of Mercury and Venus, making six
samples altogether. The estimate of $\nu_S{}'$ formed from each
sample is smaller than the value found from the solar data.
In other words, with a statistical confidence level that is
estimated at about 10^8 to 1, Mercury and Venus have negative
accelerations in a time scale in which the sun is unaccele-
rated. Thus the result obtained from the Venus data in AAO
(Equation I.4 above) is not an isolated accident.

The Greek data used in APO come to us only through the
writing of Ptolemy [ca. 142]. I knew when I wrote APO that
the observations which Ptolemy claims to have made himself
are fraudulent. I have discovered since that many of the
planetary observations which Ptolemy claims were made by
other astronomers were also fabricated by him, and all of
them may have been.‡ Thus it is not safe to use any Greek

---

†I shall repeat the analysis of Spencer Jones's data, but
with an accurate value of $\nu_M{}'$, in Section I.4.
‡This is proved formally in Chapter I of APO.
‡However, some (but not all) of the solar and lunar observations which Ptolemy attributes to other astronomers are
genuine. See Section XIII.2 of Newton [1977] for details.

data on the planets that we now know of. When we eliminate
the Greek data used in APO, the confidence level of the
conclusion about the accelerations of Mercury and Venus
falls to about 10^6 to 1.

The conclusion reached in APO needs further discussion
which will be deferred to Section I.4. Here, though, it is
desirable to mention two possible explanations of the results, and one consequence of them.

The ephemeris programs used in APO were carefully
tested, and I believe that we can eliminate a program error
as a possible explanation of the results. If we do so,
there seem to be only two remaining possibilities:

a. There are errors in the basic theories of Mercury,
Venus, and the sun that were used in the analysis.

b. There are forces of unknown origin acting on
Mercury and Venus which tend to increase their angular
momenta.

If either possibility is correct, it means that the concept
of ephemeris time, as it is currently formulated, is not
valid. I shall discuss the first possibility in the next
section.

In the meantime, we may not use the concept of ephemeris
time in its general form. That is, we may not assume the
existence of a time scale in which all the planets obey a
strictly gravitational theory of motion. In the rest of this
work, ephemeris time will be used only in a limited sense as
the time scale in which the sun has no secular acceleration.
The moon and planets, however, may have non-zero accelerations
in this time scale.

The work of van Flandern [1974] is quite interesting.
van Flandern has measured the acceleration of the moon with
respect to atomic time by using lunar occultations from 1955
to 1973. He finds that the acceleration is -83 ± 10, although the acceleration with respect to ephemeris time is
between about -25 and -40. The most probable explanation of
this situation, in van Flandern's opinion, is that the gravitational constant is changing at a rate of about -12 parts
in 10^9 per century.†

Finally, the recent works that should be mentioned include the writings of Muller and Stephenson [1975] and of
Muller [1975]. These works raise so many questions that we
must devote most of Chapters III and IV to their discussion.

---

†Since I wrote this, van Flandern (private communication)
has changed his estimate of the lunar acceleration to
-38 ± 5. This makes the change in the gravitational constant about -3 parts in 10^9 per century. See Section XIV.9
for more details.

3.   An Error in Present Orbital Theories

The results of APO led to the realization that there is an error that is common to the present orbital theories of the moon, the sun, and the planets, although the presence of the error should have been clear long before. The error is a consequence of two circumstances: (a) the orbital theories have been based upon incorrect values of the secular accelerations, and (b) the reference epochs used in the theories have been at the extreme end of the span of the observed data instead of being at the average epoch of the observed data. Neither circumstance by itself would cause a serious error; it takes the combination to do so.

As an example, let us suppose that the secular acceleration of some body is a constant, and that observed positions have been fitted to a quadratic function, f_e, say, over the time interval from -1 to +1. For f_e, we may write

$$f_e = a + bt + ct^2. \qquad (I.11)$$

Let us further suppose, for simplicity in discussion, that the observed positions fit f_e with negligible error.

Suppose we think, for some reason such as an inadequate theory, that the coefficient of t^2 is C, which is different from c. We then use a quadratic, f_t, say:

$$f_t = A + Bt + Ct^2, \qquad (I.12)$$

and we find the coefficients A and B by fitting to the data from -1 to +1. The results are

$$A = a + (c - C)/3, \qquad B = b. \qquad (I.13)$$

We do not make any error in the coefficient of t. The error comes in the constant A, and this error is negligible for times that lie at a considerable interval from the range -1 to +1; this error is not important because it does not grow with time.

Suppose, however, that we choose to take the epoch +1 as our reference epoch instead of the epoch 0. The values of $f_t - f_e$ and of its first time derivative at the epoch +1 are

$$f_t - f_e = 2(C - c)/3, \qquad \dot{f}_t - \dot{f}_e = 2(C - c). \quad (I.14)$$

The error at +1 is twice the error at 0, but this is not highly important. The worst problem is that there is now an error in the first derivative, and this will cause an error in position that grows linearly with time.

The reader may object that an error in C causes an error in the term Ct^2 directly, and that this is worse than an error in the first derivative, because it grows quadratically with time. This would be so if we intended to use Equation I.12 to calculate a position. In the work that interests us, however, we are studying the acceleration. We do this by dropping the term Ct^2 from Equation I.12. Then, if we have an observation at time t_1, say, we first calculate a position at time t_1 from the constant and linear terms, obtaining $A + Bt_1$. We then attribute the discrepancy between $A + Bt_1$ and the observed position to the acceleration.

When we do this using $t = 0$ as the reference epoch, we make only the small error in the coefficient A that is shown in Equation I.13, and the resulting error in the acceleration is $2(c - C)/3t_1^2$. When we do this using $t = 1$ as the reference epoch, the error in the calculated position (aside from the term in Ct^2) is $(2/3)(C - c) + 2(C - c)(t_1 - 1)$, as we see from Equations I.14. The error in the acceleration is $2/(t_1 - 1)^2$ times this, or

$$\text{acceleration error} = 4(C - c)/3(t_1 - 1)^2$$
$$+ 4(C - c)/(t_1 - 1). \qquad (I.15)$$

The second term is the larger one if $|t_1 - 1| > 1/3$; it comes from the error in the first derivative in Equations I.14.

In all studies of the accelerations in the solar system that have been made in the past half-century, the orbital theories used have been that of Brown [1919] for the moon and those of Newcomb [1895, 1895a, 1895b] for the sun, Mercury, and Venus, respectively. These theories have several features in common: The empirical constants in them were evaluated using data from about 1700 to about 1890 or 1900, the theories use 1900 as the reference epoch, and the empirical constants were evaluated by using wrong values of the accelerations. Note that the reference epoch is at one end of the data span, which is the other condition required to produce an error.

In his original theory, as I have already mentioned, Brown used a "great empirical term" equal to $10''.71 \sin (140°T + 240°.7)$. This function actually gives a good fit to the fluctuation plotted in Figure I.1. If it had been replaced by a quadratic, we would still have had a good representation of the data, although it would have been better if the epoch had been taken as 1800 rather than 1900. However, when Spencer Jones replaced the great empirical term by a quadratic, he used +10.44 as the lunar acceleration (curve B in Figure I.1), and this is seriously in error. Thus Spencer Jones's form of a "modified Brown theory" using solar time had a seriously wrong value of the coefficient of T in the mean longitude. Again, when ephemeris time replaced solar time, the acceleration was taken as -22.44, but the correct value may well be about -40, particularly for medieval or earlier times. Thus whether we use solar or ephemeris time, the acceleration used in current theories may be wrong by about 20 $''/\text{cy}^2$, and it certainly is wrong when solar time is used.

When Newcomb determined the empirical constants in his theories of the sun, Mercury, and Venus, he worked in terms of solar time. Nonetheless, he assumed, as was natural when he did his work, that the sun and the planets had no accelerations of non-gravitational origin, and we know that this assumption is in error. When ephemeris time replaced

solar time, Newcomb's theories were retained unaltered; it
was merely considered that the theories now supplied the
fundamental definition of ephemeris time. If we base the
definition of ephemeris time upon the sun only, it is correct that we should use a strictly gravitational theory,
but we must take care to use the correct empirical constants.
Since we took over the erroneous constants when we made the
transition to ephemeris time, the constants are still wrong.
Thus with the sun, Mercury, Venus, and the moon the empirical constants are wrong whether we use ephemeris time or
solar time[†]; a possible exception is the lunar theory based
upon ephemeris time, because the value used for the lunar
acceleration with respect to ephemeris time may be accurate.

One of the purposes of this work is to make a new analysis of the data that I have used in earlier work, after
correcting the orbital theories used.[‡] Another purpose is
to obtain additional data to fill in the gap between about
1200 and the data plotted in Figure I.1. Still another purpose will be described in Section IV.6. Since the present
definition of ephemeris time may not be valid, the new analysis will be based upon solar time, using 1800 rather than
1900 as the fundamental epoch of the ephemerides.

Because the quantity $t_1 - 1$ appears in the denominators
of both terms in Equation I.15, and because $t_1 - 1$ is fairly
large for the ancient and medieval data that have been used,
we do not expect a large change in the earlier conclusions.
In particular, it must still be the case that ν_M' and D''
have changed by large amounts, and that they have even
changed sign, in the past 2000 years. However, Equation I.9
will probably require modification.

4.   The Work of Spencer Jones

In Figure I.1 I have already plotted the lunar data
that Spencer Jones [1939] used in his important paper that
led to the introduction of ephemeris time. In Section I.2 I
pointed out that he used curve B in the figure to represent
the secular behavior of the data, corresponding to an acceleration of +10.44, but he should have used curve A, which
corresponds to an acceleration of -15.72.

A number of people in correspondence or in conversation
have objected to this conclusion. They point out that the
data show large fluctuations in the lunar acceleration and
that these fluctuations keep the average acceleration over
the time span of the data in Figure I.1 from being close to
the true secular acceleration. Muller [1975, p. 6.7] says
specifically that the short term fluctuations dominate the
situation over an averaging interval as short as three centuries. Further, in a private communication (1975) he says

---

[†]This statement is also correct for the outer planets,
although the historical details are different.

[‡]I shall also correct a few minor errors in the data used.
See Section II.1.

that omitting the data after 1900 destroys my conclusions
about a change in ν_M' with time.

Brouwer [1952] and Markowitz [1970] have studied the
statistical properties of the fluctuations in the acceleration by independent methods and find results that are in
reasonable agreement. By using their studies, I have found
[Newton, 1973a] the mathematical expectation of the contribution which the fluctuations make to the average acceleration, as a function of the length of the averaging interval,
and I have shown that the expectation is negligible if the
averaging interval is three centuries or more. However, the
actual effect for a particular interval may be much larger
than the expectation. Thus it is more conclusive to compute
directly the average for several intervals and see what
happens.

TABLE I.1

VALUES OF THE LUNAR ACCELERATION USING
DIFFERENT AVERAGING INTERVALS

Interval	$C_T{}^a$ sec/cy^2	$\nu_M{}'$ $''/cy^2$
1627–1960	8.94	-12.62
1710–1960	10.08	-11.34
1810–1960	43.93	+25.93
1627–1900	2.95	-19.19

[a] The coefficient of T^2 in the quadratic
that best fits ΔT, which is ephemeris
time minus solar time.

The data of Brouwer [1952] and Martin [1969], extended
a few years as described in Section I.2, run over a longer
time span than the data in Figure I.1, while agreeing well
with the figure over the time span that they have in common.
Thus I shall use these data, which I shall call the Brouwer-
Martin data, in this discussion. Table I.1 shows the effect
of using various time spans in calculating the average lunar
acceleration ν_M' with respect to solar time.

As I have already mentioned, Brouwer and Martin assume
that $\dot{n}_M$ is -22.44. Using this value, they calculate the
ephemeris time ET when the moon had a particular longitude.
An observation gives the solar time UT when the moon actually
had this longitude, within observational error. Let ΔT denote the difference ET - UT. For a given time interval, we
then find the quadratic that best fits ΔT.

The first column in Table I.1 shows the time interval
used and the second column gives the corresponding value of
C_T, which is the coefficient of T^2 in the quadratic that
best fits the values of ΔT within the specified interval.

C_T and the acceleration parameter y are related by [Newton, 1973a]:

$$y = -0.6338C_T. \qquad (I.16)$$

We calculate y from C_T and then calculate ν_M' from y by using Equation I.7 above. When we use this equation, we should remember that the value $\dot{n}_M = -22.44$ was used in deriving the values of y. This calculation gives us the last column in Table I.1.

The first three lines show the effect of shortening the time span while keeping the same terminal date of 1960. Omitting the initial century, approximately, hardly changes the value of ν_M' but omitting the first two centuries gives a positive rather than a negative value of ν_M', which constitutes a drastic change. We can see from Figure I.1 that ν_M' was indeed positive over almost any interval from about 1800 to about 1920, and we could get a still larger value of ν_M' if we used only data from about 1860 to about 1920. Nonetheless, the excellent agreement of the first two lines in Table I.1 shows that the effect of the fluctuations is almost eliminated when we use the interval from 1710 to 1960, an interval of $2\frac{1}{2}$ centuries, so that the fluctuations certainly do not dominate the average over this particular interval of about 3 centuries.†

The last line in Table I.1 shows what happens if we follow Muller's suggestion of omitting the data since 1900. Instead of destroying my basic conclusion, as Muller says, the omission actually strengthens the conclusion that ν_M' has changed sign within historic times.

Thus it is firmly established that the value of ν_M' for the past $3\frac{1}{2}$ centuries has been about -12 rather than the value +10.44 that Spencer Jones uses in his analysis, and that the change in ν_M' is not the result of the fluctuations. The value over the past $3\frac{1}{2}$ centuries agrees well with the values found in MCRE [page 636, Table XVIII.11] for epochs around 1200. It is this agreement that leads to the first part of Equation I.9 as a representation of the time behavior of D'' for the late medieval period and the modern period. The data that lead to Equation I.9 have a gap between about 1200 and the 17th century. One purpose of this work is to fill in this gap, and, when this is done, we can speak more firmly about the relation between the older data and the Brouwer-Martin data.

So far I have been using terms like discrepancy and fluctuation rather loosely. Now I need to adopt precise definitions for them which will be used in the rest of this

---

†The value -12.62 in the first line of Table I.1 disagrees slightly with the value -12.10 in Equation I.6. More data were used in finding the value in the table, so it is more reliable.

section, but which are not necessarily retained in other parts of the work. Discrepancy will now denote a mean longitude deduced from observation at a measured value of solar time minus a mean longitude calculated for that value of time from a strictly gravitational theory. We shall have to deal with discrepancies for the sun, the moon, and the planets.

The discrepancy for any body is considered to have three components. The first component is an error in observation.[†] The second component will be called the secular variation; it has a long and perhaps infinite period. The third component, which will be called a fluctuation, is distinguished by having only short periods in its Fourier spectrum. In this section, I shall use the following notation:

Δ: a discrepancy,

ε: an error in observation,

δL: a secular variation,

β: a fluctuation.

These quantities for the sun and moon will be denoted by using subscripts S and M, respectively. These quantities for the planets will be denoted by using the classical symbols for the planets. These symbols are given in Table I.6.

The definitions of the secular variation and the fluctuation are still not precise because the spectrum of a discrepancy contains periods ranging from months to millenia or longer. I shall now make the definitions precise by saying that the secular variation is the quadratic function that gives the best fit to the discrepancy in a least-squares sense. The secular variation δL will be written as

$$\delta L = a + bT + cT^2, \qquad (I.17)$$

in which T is solar time measured from the epoch 1900 January 0.5. The definition of δL depends upon the time interval used.

We then complete the definition of β by writing

$$\Delta = \varepsilon + \delta L + \beta. \qquad (I.18)$$

ε and β are fairly well separated in concept, but separating them in an operational sense by using observational data poses a problem.

Spencer Jones used a quantity that he called a lunar fluctuation B, but it is not the same as any of the quantities Δ, ε, δL, or β just introduced. To find B, he first removed Brown's great empirical term from Brown's final expression for the lunar mean longitude and then he added a "great empirical term" of his own, as I explained in Section I.2. Spencer

---

[†]For simplicity, we consider that the measured time is exact and that all the measurement error is in the longitude.

Jones's term is the quantity that I called GET in that section. Thus we have Δ_M = B + GET. Although I have referred to the quantity plotted in Figure I.1 as Spencer Jones's data, it is really the quantity Δ_M derived from his data rather than his "fluctuation" B.

In my discussion of the Brouwer-Martin data, the quantity I have used is strictly equal to the lunar discrepancy Δ_M. The value ν_M' = -12.62 in Table I.1 is the secular acceleration that corresponds to the secular variation δL of the moon. In full, δL is:

$$\delta L = -3''.79 - 14''.48T - 6''.31T^2. \qquad (I.19)$$

As we see from Table I.1, the secular variation does not depend in an important way upon the time interval used provided that the interval is about $2\frac{1}{2}$ centuries or longer. On the other hand, Equation I.9 shows that D'', and hence probably ν_M' as well, can change by a significant amount if the interval is many centuries. Overall, it seems that the best averaging interval for finding the secular variation δL is around 3 to 4 centuries. The interval used in finding Equation I.19 is about 330 years.

The idea back of the present use of ephemeris time is that, aside from the effect of tidal friction upon the moon's orbit, all observed secular variations and fluctuations result from the secular variation and the fluctuation in the earth's spin. The way to test this idea is to answer the following questions:

a. Are the accelerations derived from the secular variations of the sun and planets proportional to their mean motions?

b. Are the fluctuations of the moon, the sun, and the planets proportional to their mean motions?

Note that the moon appears in question b but not in a, because of the effect of tidal friction which acts on the lunar orbit but not on the other orbits.

For the sun and planets, Spencer Jones defines the fluctuations in just the way I have. He also deals with a quadratic like Equation I.17 for the sun and planets without giving it a name. For simplicity, I shall say that he also defines the secular variation in just the way I do. Thus we can use his tabulations of solar and planetary data without change.

There is a complication in dealing with the planetary data. The observed positions are necessarily geocentric, but we want to deal with the secular variations and fluctuations in the heliocentric mean longitudes of the planets. Thus, when I write the secular variation $\delta L_{\m&}$ and the fluctuation $\beta_{\&}$ of Mercury, for example, these quantities refer to its heliocentric mean longitude.

The eccentricities of Venus and the sun are small, and the data used to find the discrepancies are well distributed

around the orbit. Thus, to high accuracy, the observed
difference between an observed position and one calculated
from gravitational theory reduces to $\Delta_{\female} - \Delta_S$, the discrepancy
for Venus minus the discrepancy for the sun. Spencer Jones
tabulates $\Delta_{\female} - \Delta_S$ as a function of time from 1839 to 1935.5
in his Table V.

Matters are more complicated for Mercury. Spencer Jones
uses only measured values of the times when transits of Mercury across the sun's disk occur. Some transits occur in May,
when the heliocentric position of the earth is near the descending node of Mercury, and others occur in November when
the earth is near the ascending node. When a transit occurs,
the heliocentric longitudes of the earth and Mercury are the
same. We can calculate the difference in longitudes from a
gravitational theory for the measured time of transit, and
comparison of this difference with the measured difference†
of 0° gives the disagreement between observation and theory.
A numerical value of the disagreement can be written in the
form $X\Delta_{\mercury} - Y\Delta_S$, in which X and Y are numerical coefficients
to be found from theory.

For Venus, we can take $X = Y = 1$, but we cannot do so
for Mercury. Mercury has a large eccentricity and it is
always about 30° from perihelion at a November transit and
about 30° from aphelion at a May transit. The sun also has
specific values of its mean anomaly at the times of the
transits. Thus, if we let $V_{\mercury}$ denote a value of the disagreement for a November transit and let $W_{\mercury}$ denote a value for a
May transit, we can only write:

$$V_{\mercury} = 1.487\Delta_{\mercury} - 1.01\Delta_S \quad \text{for November,}$$
$$W_{\mercury} = 0.716\Delta_{\mercury} - 0.97\Delta_S \quad \text{for May.}$$

$$(I.20)$$

The numerical values for, say, November transits are not all
exactly the same, but the variation is small enough to neglect here. Newcomb [1882] determined the numerical coefficients in Equations I.20, and Spencer Jones has tabulated
the $V_{\mercury}$'s and $W_{\mercury}$'s from 1677.9 to 1927.9 in his Table IV.

Finally, Spencer Jones gives two collections of solar
data. His Table II gives values of Δ_S determined from
measurements of the solar declination from 1761.3 to 1934.2,
and his Table III gives values of Δ_S determined from measurements of its right ascension from 1839.0 to 1935.3.

Errors in observation should be about the same size for
the sun, the moon, and the planets, and it happens that we
expect them to be about the size of the fluctuations for the
sun and the planets. However, we expect the fluctuations of
the moon to be an order of magnitude larger than the fluctuations of the sun and planets, so we can assume that the observational errors of the moon are negligible, to good accuracy. Thus, for the moon, we write Equation I.18 in the form

---

† I am ignoring the complexities of the actual measurements
in this conceptual discussion.

TABLE I.2

THE LUNAR DISCREPANCY Δ_M OBTAINED

FROM THE BROUWER-MARTIN DATA

Year	Δ_M "	Year	Δ_M "	Year	Δ_M "
1627.0	- 0.24	1764.5	+ 7.72	1862.5	+ 0.73
1637.5	- 4.78	1768.0	+ 8.04	1867.5	- 1.33
1643.3	-16.98	1773.0	+ 8.34	1872.5	- 5.21
1657.0	+ 4.37	1777.5	+ 8.87	1877.5	- 8.04
1663.5	- 9.35	1784.0	+ 8.65	1882.5	- 9.50
1672.5	- 6.82	1788.0	+ 8.44	1887.5	-10.31
1678.2	- 8.65	1793.0	+ 7.90	1892.5	-11.58
1683.5	- 8.42	1798.0	+ 8.80	1897.5	-12.09
1686.0	- 8.80	1803.0	+ 7.15	1902.5	-10.44
1699.0	- 3.23	1808.0	+ 7.58	1907.5	- 8.07
1703.0	- 3.01	1812.5	+ 7.38	1912.5	- 5.60
1708.0	- 1.83	1818.0	+ 7.99	1917.5	- 4.34
1713.0	- 0.90	1822.5	+ 6.34	1922.5	- 4.02
1718.5	- 0.89	1827.5	+ 5.77	1927.5	- 4.88
1728.0	+ 2.26	1832.5	+ 4.32	1932.5	- 6.21
1733.0	+ 1.62	1837.5	+ 3.42	1937.5	- 7.79
1737.5	+ 2.38	1842.5	+ 3.06	1942.5	- 8.74
1746.5	+ 3.92	1847.5	+ 2.59	1947.5	- 9.27
1753.5	+ 5.69	1852.5	+ 2.28	1952.5	- 9.68
1757.5	+ 4.30	1857.5	+ 1.84	1957.5	-10.57

$$\Delta_M = a_M + b_M T + c_M T^2 + \beta_M. \qquad (I.21)$$

Equation I.19 gives the values of a_M, b_M, and c_M, so we can find the value of β_M at any epoch for which we have an observed Δ_M by using Equation I.21. The observed values of Δ_M are listed in Table I.2. When I need a value for an epoch not in the table, I find it by linear interpolation.

In studying question b above, it is convenient to write the fluctuation for the sun or a planet in the form $Q(n/n_M)\beta_M$, in which Q is a coefficient to be found for the sun or a planet by analysis of the data. That is, we now write Equation I.18 for the sun or a planet in the form

$$\Delta = a + bT + cT^2 + Q(n/n_M)\beta_M + \varepsilon. \qquad (I.22)$$

We still have to find the values of a, b, c, and Q for the sun, Mercury, and Venus, for a total of 12 parameters. The Q's will all be unity if the answer to question b is "yes".

Since the Venus observations give values of $\Delta_\varphi - \Delta_S$, there is no interaction between the Venus data and the solar data. If we analyze them simultaneously, the solar data alone determine the solar parameters. These parameters determine the values of Δ_S in the difference $\Delta_\varphi - \Delta_S$, and the Venus data then give the parameters for Venus.

Because of Equations I.20, this separation does not happen for the sun and Mercury. Thus we must analyze the data for the sun and Mercury simultaneously. We then use the values of the solar parameters found from this analysis in the analysis of the Venus data.

There are four points in which I feel that the analysis of Spencer Jones is not correct. They are:

    a. Spencer Jones assigned weights to individual observations that were based upon estimates of their accuracy.

    b. Spencer Jones took the values of c for the sun and planets to be proportional to their mean motions. Specifically, he took c_S to be an independent parameter but he then took c_φ, for example, to be $c_S(n_\varphi/n_S)$.

    c. Instead of finding c_M from the lunar data, he took it to be $+5.22$ "/cy^2. This value comes from a study of ancient data made by de Sitter [1927].†

    d. Spencer Jones found incompatible results from the measurements of the sun's right ascension and its declination, and he eliminated the right ascensions from his final results.

With regard to point a, recent measurements tend to be more accurate than early ones, and they received considerably more weight in Spencer Jones's analysis. However, a main purpose of the analysis is to study the time dependence of the fluctuations β, and how the β's vary from one body to another. For this purpose, an early fluctuation is as important as a recent one and we should try to give equal weights to equal intervals of time.

---

†Unfortunately this study is not reliable. de Sitter gave heavy weight to an allegedly precise measurement of the times of the lunar eclipse of -424 October 9 made in Babylon, but the times were actually calculated rather than observed. See APO, Section IV.9. He also put heavy weight on the so-called eclipse of Hipparchus, which he says was "evidently observed according to a prearranged (and remarkably well arranged!) plan . ." (The exclamation point is de Sitter's.) There is no known basis for this statement, and we do not even know the century in which the observation was made. See the discussion under the heading "-128 November 20?" in Section VI.1 below. There is a more extensive discussion of de Sitter's work in Section VI.3.

Let me anticipate the final results at this time. The best estimate we can make of the errors ε of observation is to equate them to the residuals r defined by

$$r = \Delta - \delta L - \beta,$$

in which we calculate δL and β using the inferred values of the parameters a, b, c, and Q for each body. For the sun and planets, the β's turn out to be nearly twice the r's, on a root-mean-square basis, so that the errors of observation do not interfere particularly with finding the fluctuations and the secular variations. The residuals for the moon are quite small compared with the fluctuations. Thus we should try to weight equal time intervals equally.

For the sun and planets, there are altogether five classes of observation, namely solar declinations, solar right ascensions, longitudes of Venus, May transits of Mercury, and November transits of Mercury. Within each class we should assign uniform weights to each observation regardless of the epoch. We are free, however, to assign different weights to different classes of observation, and we shall in fact find that November transits should receive very small weights.

In point b, Spencer Jones postulates that "yes" is the answer to question a, although finding the answer should have been part of his study.

Spencer Jones was clearly wrong in the value he used for ν_M', and the question now is whether this error affected his results. It turns out that his values of Q are invalid because of this error, but the value we get for $\dot{n}_M$ is not seriously affected.

TABLE I.3

THE ANALYSIS OF SPENCER JONES'S SOLAR DATA

Data sample	ν_S' $''/cy^2$	Q_S	$\dot{n}_M$ $''/cy^2$
δ	+1.08 ± 0.38	1.00 ± 0.16	−27.0 ± 5.1
α	−1.25 ± 1.95	1.16 ± 0.23	+ 4.0 ± 26.0
$\delta + \alpha$	+0.98 ± 0.38	0.90 ± 0.14	−25.7 ± 5.1

In the final inference, the solar and Mercury data must be used together, but it is desirable to make a preliminary investigation using only the solar data in order to compare the right ascensions and declinations. The results are summarized in Table I.3. The first column of the table identifies the data sample. The declinations δ were used alone in the first line, the right ascensions α were used alone in the second line, and both sets were used simultaneously in the third line. The error estimates for the parameters are

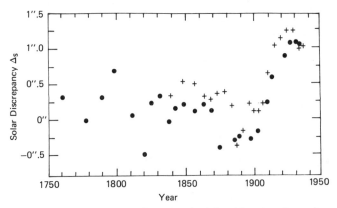

Figure I.2 Values of the solar discrepancy Δs deduced from two classes of observation. Points plotted as circles were obtained from measurements of declination and points plotted as +'s were obtained from measurements of right ascension.

standard deviations.

There is no significant difference in the values of the Q's, although Spencer Jones said that Q_S obtained from the right ascensions is incompatible with the remaining Q's from declinations and from planetary data. Figure I.2 helps us see how he reached this conclusion.

A circle in Figure I.2 is a value of the solar discrepancy Δ_S obtained by using declinations and a plus is a value of Δ_S obtained by using right ascensions. Over the time span covered by both kinds of data, there is almost no difference between the results, except that the right ascensions have a slight positive bias with respect to the declinations. This affects only the constant a_S in the secular variation of the sun, and it does not affect either the acceleration or the value of Q. The main difference between the data sample lies in the time span, which begins in about 1760 for the declinations and in about 1840 for the right ascensions.

We can see by eye that the quadratic which best fits the data has a much larger coefficient of T^2 for the right ascensions than for the declinations, because of the different time spans. Now let us return to Figure I.1. Spencer Jones thought that he was using the short-term fluctuations for the moon, but he was really using the differences between the observed values and the erroneous curve marked B. He should have used the differences between the observations and curve A. At most epochs in the figure, the difference between the curves is bigger than the actual short-term fluctuation. Thus Spencer Jones was really dealing with secular effects when he thought he was dealing with the short-term fluctuations. When he fitted the parameter Q, his fitting process tried to force the apparent secular accelerations of the sun and planets into conformity with the wrong secular acceleration of the moon, and the fitting process was hardly affected by the actual fluctuations. Because of the differing time spans, the process could not force both the declinations and the right ascensions to agree with this wrong ν_M'. Because the planetary data had a long time span, they agreed with the solar value obtained from declinations, leaving the value from right ascensions apparently inconsistent with the other values.

In fact, the residuals have a standard deviation of $0''.22$ for the declinations but only $0''.19$ for the right ascensions. (The difference, however, comes mostly from the early declinations, which clearly have a larger scatter than the values after about 1830.) Thus there is no basis for discarding the right ascensions.

In the remaining analysis, I shall use both right ascensions and declinations, but I shall first subtract $0''.4$ from the discrepancies obtained from right ascensions; this removes the bias shown in the figure. However, simply using both bodies of data gives more weight to times since 1840 than to earlier times, contrary to the principle of weighting equal time intervals alike. Hence I halve the weight of the circles after 1840 in spite of the fact that they show less scatter. The result is the third line in Table I.3, which does not differ significantly from the first line.

We should notice that we cannot find the value of ν_S' by fitting to the discrepancies. Fitting directly to the +'s in Figure I.2 gives a positive value of ν_S' while the value in Table I.3 is negative. The reason is that we subtract the fluctuations from the discrepancies before we find the secular variation. If we were to subtract the fluctuations from the discrepancies and plot the resulting values, we would find that they show a negative acceleration for the time interval from 1839 to 1935, whether we use the right ascensions or declinations.† This should not surprise us. As we see from Table I.1., we also get the wrong sign for ν_M' if we use the time interval from 1810 to 1960.

In analyzing the solar data and the Mercury data simultaneously, I first tried giving the same weight to each Mercury transit as to each solar observation before 1839, with half weight being given to each solar observation after 1839 for the reason explained above. The standard deviation of the residuals for the May transits (the W_ξ) is about $0''.2$, just as it is for the solar data, but it is about $1''.2$ for the November transits (the V_ξ). R. L. Duncombe and T. C. van Flandern (private discussion) point out that this is to be expected. Because of the eccentricity of Mercury's orbit and the position of its perihelion, Mercury is much

<hr>

†We must remember that the fluctuation β_S for the sun is now defined as $\beta_S = (n_S/n_M)\beta_M$, as a consequence of taking $Q = 1$ in Equation I.22.

TABLE I.4

PARAMETERS FOR THE SUN, MERCURY, AND VENUS

Parameter	Sun	Mercury	Venus
a, $''$	0.38 ± 0.05	2.54 ± 0.16	1.43 ± 0.13
b, $''/\text{cy}$	0.99 ± 0.17	4.45 ± 0.40	1.14 ± 0.40
c, $''/\text{cy}^2$	0.59 ± 0.18	1.41 ± 0.26	-3.30 ± 1.16
ν', $''/\text{cy}^2$	1.18 ± 0.36	2.83 ± 0.51	-6.60 ± 2.32
Q	0.86 ± 0.14	0.94 ± 0.08	1.68 ± 0.19

closer to the sun during a November transit than during a
May one and its small dark disk is harder to see against the
sun.

In consequence, I repeated the analysis giving each
November transit a weight equal to 1/36 of that given a May
transit, leaving the relation between the May transits and
the solar observations unchanged. It is not clear that this
is the best set of weights, but I use it for the sake of
definiteness, and in fact all reasonable sets of weights
give results that do not differ significantly. After finding the parameters of the sun from this analysis, I then
used the Venus data to find the parameters of Venus. The
results of the complete analysis for the solar and planetary
parameters are shown in Table I.4.

We are primarily interested in the solar and planetary
accelerations, the values of Q, and the value of $\dot{n}_M$ that
results from the analysis. The solar acceleration ν_S' in
Table I.4 is 1.18 ± 0.36 and the lunar acceleration ν_M'
already found is -12.62 ± 3.10.† These give us

$$\dot{n}_M = -28.38 \pm 5.72 \qquad ''/\text{cy}^2. \qquad (I.23)$$

In spite of everything, this does not differ seriously from
the value found by Spencer Jones, but it does differ, at a
level that looks moderately significant, with the values
found from ancient and medieval data in Equations I.2 and
I.3.

A few closing comments are needed. I have subtracted
the secular variation of Venus, as given by Table I.4, from
the tabulated discrepancy and examined the result. There
is a systematic discrepancy which resembles a sinusoid with

---

†For the error estimate of ν_M', I am taking the difference
between the accelerations found from the Brouwer-Martin
data and the data used by Spencer Jones. I believe that
this is more realistic than an estimate based only upon
the residuals of the data.

a period of about 70 years and an amplitude of about 3″.
Duncombe [1972] has made a new reduction of the Venus data
while extending the time span to run from 1750 to 1949, and
the results show no such effect. Thus it seems that the
anomalous parameters of Venus in Table I.4 are the result
of faulty data reduction and hence they are meaningless. We
should look only at the results for Mercury and the sun.

First let us look at the values of Q, which give the
answer to question b above. Both values are within one
standard deviation of unity. Thus it seems that the fluctuations of the sun, the moon, and Mercury result primarily
from fluctuations in the earth's spin. At most there is
room for about 10 per cent of the fluctuations to arise from
some other source. It is plausible to assume that the same
conclusion holds for the other planets, and the work of
Duncombe indicates strongly that this is the case for Venus.

Morrison and Ward [1975] have made a new analysis of
the Mercury data and have added six transits of Mercury observed between 1940 and 1973. They find

$$\dot{n}_M = -26 \pm 2 \qquad ''/cy^2. \qquad (I.24)$$

Thus it seems to be well established that $\dot{n}_M$ is somewhere
around -25 rather than -40 at the present time.

We should now return to question a above, which is
equivalent to asking whether the secular accelerations of
the sun and planets are the result of a secular acceleration
in the earth's spin. If this is correct, the ratio of the
secular accelerations of Mercury and the sun should be 4.152.
However, the ratio of the secular accelerations shown in
Table I.4 is 2.40 ± 0.85, which differs from 4.152 by about
two standard deviations. Further, the difference is in the
same direction as the difference I find in APO from ancient
and medieval data.

Duncombe [1972], in his study of Venus, assumes that
the secular accelerations of Venus and the sun are in the
same ratio as their mean motions, instead of deriving the
ratio from the data. Morrison and Ward [1975] do a similar
thing for Mercury, and we should remember that Spencer Jones
does the same for both planets. Thus, although the test
provided by Table I.4 has been made with somewhat old data,
it remains the only theoretically valid test of the ratio
that has been made with telescopic data, to the best of my
knowledge.

In conclusion, it is ironic that the paper of Spencer
Jones does not provide a basis for the use of ephemeris time.
There are two aspects to the use of ephemeris time, the long-
term and the short-term. Spencer Jones postulated the equivalent of saying that ephemeris time is valid for the long
term, instead of testing the matter against the data. His
paper does not deal with the short-term fluctuations at all,
although this was his intention in writing it. Instead, it
turns out that the paper deals with matching the secular
accelerations of the sun and planets to an erroneous value
of the secular acceleration of the moon. Thus Spencer Jones

did not provide a test of ephemeris time in either the short-term or the long-term aspect. It is fascinating to speculate what would have happened if he had performed the analysis correctly, or if astronomers had realized sooner that his analysis is in error.

5. The Coordinate η

In order to infer parameters, x and y, say, from a body of observations, we need some property of the observations that satisfies two conditions. Let X denote the property in question. We must be able to assign a value of X to each observation and we must be able to calculate X theoretically for any values of x and y. It helps if X is a sensitive function of x and y. Further, we must not be able to get a given value of X for different values of x and y, at least not within a reasonable range of the variables.

Our first thought is to use the magnitude of the eclipse as the property X, but this does not meet the last condition. Suppose that an eclipse is total, for example. Then the magnitude is 1 at a point on the northern limit of the zone of totality and it is also 1 at a point on the southern limit, and we can shift the point of observation from one limit to the other by changing the accelerations.

In Section XVI.3 of MCRE I introduced a coordinate η which I used there for the property X and which I shall use again here. To find η, we calculate the distance from the observer to the axis of the moon's shadow at the time when the eclipse is a maximum for the observer. We measure the distance in units of the earth's radius. To make the distance definite, we make η positive when the observer is in a northerly direction from the path and negative when he is in a southerly direction. The cited section of MCRE contains a more rigorous definition.

It is clear that the absolute value of η is closely related to the magnitude. The magnitude is large when η is small and vice versa, and the magnitude is obviously a maximum for a particular eclipse when η is zero.

6. The Designation of Eclipse Records, Provenances

In assigning a designation to an eclipse record so that I can refer to it readily, I use two main elements. One is the date and the other is the provenance. I designate the provenance by a set of letters that I shall describe in a moment. To designate a record, then, I first write the date, in the astronomical style that has been described, followed by the letters that designate the provenance.

It often happens that I have more than one record of the same eclipse from the same provenance. In this case, I distinguish the records by adding a lower case letter to the date. The order assigned to the records by means of the lower case letters is merely the order in which I discovered the records, and it has no importance.

To give some examples, I use the letter M to designate the Mediterranean provenance that I shall define in a moment. I have found only one record of the eclipse of -423 March 21 from the Mediterranean provenance but I have found two records of the eclipse of -430 August 3. The single record is designated -423 March 21 M while the two records of the earlier eclipse are designated -430 August 3a M and -430 August 3b M.

In the text I write the names of the months in full. In tables, however, where spacing is important, I use only the first three letters of the English name of the month. Thus the records just mentioned will be designated -423 Mar 21 M, -430 Aug 3a M, and -430 Aug 3b M in tables.

In AAO, where I had relatively few records, I had a fairly simple scheme for designating the provenances. I used B for the British Isles, BA for Babylonia and Assyria, C for China, E for Europe (excluding Greece), Is for Islamic countries, and M for the area around the eastern end of the Mediterranean, including Greece.

In MCRE, where I had many more records, I went to the use of two elements separated by commas to designate the provenance. The first element was the letter (or pair of letters in the case of Is) that had been used in AAO. This was followed by another letter, or perhaps a pair of letters, to designate a smaller region within the larger provenance. For example, I now divided the British Isles into England, Ireland, Scotland, Wales, and the Isle of Man, but I put the Isle of Man together with Scotland because there were few records from Man. In this scheme, the Irish provenance becomes B,I and Egypt will become Is,E.

When I use the name of a country to identify a region, I generally use the de facto boundaries of that country at the present time, and no political judgments should be read into my use of political names. However, I use Germany to designate the totality of the regions usually called East Germany and West Germany in current writing.

TABLE I.5

PROVENANCES AND THEIR DESIGNATIONS

Designation	Description
BA	Babylonia and Assyria
B,E	England
B,I	Ireland
B,SM	Scotland and the Isle of Man
B,W	Wales
C	China
E	Europe (used only for records before 100)

TABLE I.5 (Continued)

Designation	Description
E,BN	Belgium and the Netherlands
E,CE	Central Europe, including Austria, Czechoslovakia, and Switzerland
E,F	France
E,G	Germany
E,I	Italy
E,PR	Poland and Russia
E,Sc	the Scandinavian countries, including Iceland
E,SP	Spain and Portugal
Is,E	Egypt
M	countries around the eastern Mediterranean, including Greece
M,B	Byzantine Empire
M,HL	the medieval country called the Holy Land, sometimes called the Kingdom of Jerusalem, used only for records in the 12th century

In this work, I use the scheme of AAO for all records before the year +100 and for all Chinese records regardless of date. For all other records, I use the scheme of MCRE. The provenances that appear in this work, and their designations, are listed in Table I.5.

TABLE I.6

NOTATION

$\talloblong$: subscript used to identify the planet Mercury

$\female$: subscript used to identify the planet Venus

$\male$: subscript used to identify the planet Mars

A : the constant term in a power series

B : the coefficient in the linear term in a power series

C : the coefficient in the quadratic term in a power series

C : when not used as a coefficient, C is time measured in centuries from the year 600

D : the lunar elongation

D'' : the second derivative of the lunar elongation with respect to solar time, in $''/cy^2$

TABLE I.6 (Continued)

E : an error in observation

ET : ephemeris time

G : when accompanied by a subscript, G is the mean longitude of a body as calculated from a strictly gravitational theory

G : when standing alone, G is the gravitational constant

$\dot{G}$: the rate of change of G

L : a mean longitude, either as observed or as calculated from a theory including forces other than gravitation

M : a subscript used to identify the moon

m : the magnetic dipole moment of the earth

n : the mean motion of a body with respect to ephemeris time, in units of $''$/cy

$\dot{n}$: the acceleration of a body with respect to ephemeris time, in units of $''$/cy^2

S : a subscript used to identify the sun

T : standing alone, T denotes time measured in centuries from 1900, either solar time or ephemeris time

T : as a subscript, T refers to the difference ET - UT

t : time in a general discussion

UT : universal time or solar time measured in centuries from 1900

W : a perturbation in the lunar node minus a perturbation in the sun's mean longitude

y : the acceleration of the earth's spin, expressed as parts in 10^9 per century

β : the fluctuation of the moon, the sun, or a planet

Δ : preceding a symbol, Δ indicates an increment or a perturbation of the corresponding quantity

Δ : standing alone or with a subscript, Δ means the discrepancy in longitude for the moon, the sun, or a planet

δ : a small increment or a perturbation

ε : an error in observation

Λ : longitude of a point on the earth

ν : the mean motion of a body with respect to solar time, in units of $''$/cy

ν$'$: the acceleration of a body with respect to solar time, in units of $''$/cy^2

Ω : the longitude of the lunar node

$\dot{\Omega}$: the rate of precession of the lunar node

TABLE I.6 (Continued)

ω_e : the earth's spin with respect to ephemeris time, in units of $''/cy$

$\dot{\omega}_e$: the acceleration of the earth's spin with respect to ephemeris time, in units of $''/cy^2$

7. Notation; Some Conventions Used in Writing

I have already explained, in a footnote in Section I.2, the convention that will be used in writing dates in this work. A few other conventions concern the use of combinations of letters. I refer to certain works often enough to make short forms of reference useful. These references, and their short forms, are the following:

AAO = Newton [1970].

APO = Newton [1976].

M = Muller [1975].

MCRE = Newton [1972].

MS = Muller and Stephenson [1975].

I use "S." as the abbreviation for the word "Saint", regardless of the language, gender, or number. When I arrange personal or place names alphabetically, I place the names that begin with "S." in a separate listing that precedes all entries beginning with "S".

The notation is listed in Table I.6. The table gives first the symbols that are not letters in any language, then the letters in the Latin alphabet, and finally the letters in the Greek alphabet. If a symbol contains a subscript as well as a letter written on the line, one must look up the elements independently in the table.

The parameters C_T, D, $\dot{n}_M$, y, ν_M', and ν_S' are connected by a number of linear relations that will be used frequently. Three of them have already appeared in Equations I.5, I.7, and I.16. I repeat these here for the convenience of the reader, along with two other relations that have not appeared yet but that will be used frequently:

$$D'' = \dot{n}_M - 1.6073y. \tag{I.25}$$

$$\nu_M' = \dot{n}_M - 1.7373y. \tag{I.26}$$

$$\nu_S' = \quad - 0.1300y. \tag{I.27}$$

$$y = \quad - 0.6338C_T. \tag{I.28}$$

$$\dot{n}_M = D'' - 12.3683\nu_S'. \tag{I.29}$$

The numerical coefficients in Equations I.25 through I.29 are valid only if the parameters have the units defined in Table I.6. The parameter C_T must be in units of seconds of time per century per century.

SOME CORRECTIONS TO EARLIER WORK

1. Some Errors in *MCRE*

In this chapter, I shall take the opportunity to correct some errors that have appeared in recent work, starting with some errors in MCRE [Newton, 1972]. The errors to be dis-cussed here deal with observations and their interpretation; a basic error that appears in ephemeris theories has already been discussed.

In MCRE [page 601], I wrote that "there is no clear reference to the corona in any ancient or medieval record that I have found."† Muller and Stephenson [1975, p. 483] point out that this is not correct, and they cite the record of the eclipse of 968 December 22 written by Leo Diaconus [ca. 990]. I used this record in MCRE, but I stopped read-ing the record too soon. After the part of the record that I quoted [MCRE, p. 549], Leo goes on to say: "Everyone could see the disc of the Sun without brightness, deprived of light, and some dull and feeble glow, like a narrow band shining round the extreme edge of the disc. Gradually the Sun going past the Moon ... sent out its original rays ..."‡ Further, Leo says explicitly that he was in Constantinople, as the city was then known, at the time of the eclipse.

For these reasons, Muller and Stephenson go on to say: "There is no question as to magnitude, identification or place of observation as all are explicitly stated.." I agree that Leo describes the corona and that he is there-fore implying a total eclipse, but it does not follow either that the eclipse was really total as Leo saw it or that the eclipse was total in Constantinople. Ancient and medieval records are full of statements that an eclipse of known date was total at a known place, but in about a fourth of the cases totality at the stated place was impossible. In many cases, records that imply totality simply reflect the natur-al human tendency toward exaggeration rather than the actual astronomical situation. See the discussion of the eclipses of 1133 August 2 and 1140 March 20 in Section XIV.8 for an interesting illustration.

To be more specific, when a record says that an eclipse was total (magnitude equal to 1), let the difference between 1 and the actual magnitude be called the deviation of the magnitude. Let us take the root-mean-square of the devia-tions for records which assert that an eclipse was total; this will be called the standard deviation of the magnitude

---

†Many writers have cited a passage from Plutarch [ca. 90] as the earliest known reference to the corona. However, the meaning of the passage is not clear, at least in my opinion, and we cannot say with certainty whether Plutarch is referring to the corona or to something else. See Section XIV.8 for another possible interpretation.

‡This translation is quoted from Muller and Stephenson.

for such records. We shall see in Section IV.1 that this
standard deviation is 0.034 for records written in places
other than China. Hence I shall use Leo's record as a record which says explicitly that the eclipse was total, and I
shall assume that the standard deviation of the magnitude
for his record is 0.034, which is the value that applies to
this class of record.

Another error in MCRE concerns the simultaneous lunar
eclipse and occultation on 755 November 23. The relevant
record [MCRE, p. 91] describes the occultation of a star
that occurred while the moon was eclipsed, but it does not
identify the star. By calculation, I identified the star,
but I overlooked the possibility that the star might have
been a planet. I corrected this error later [Newton, 1972a].
The "star" was Jupiter, and I thank Mr. Joseph Ashbrook of
Sky and Telescope for calling this error to my attention.

Another error concerns the eclipse that I identified
[MCRE, p. 533] as 291 May 15. Since the year stated in the
record furnishes the only means of identifying the eclipse,
I should have allowed for the possibility of an error in the
year, and 292 May 4 is also a possible identification. The
error has no effect on the results of MCRE; I did not use
the record because the observation could have been made anywhere in the Roman Empire.

The only other errors in MCRE that I am aware of concern the locations of certain medieval monasteries. On page
285, I referred to a monastery at a place called Murbach,
which I did not attempt to locate since I found no eclipse
records from there. A map given by Riché [1962, p. 486]
shows that it was about 75 kilometers northwest of Basel.

Several eclipse records are found in the annals of a
monastery located in Blandin or Blandigny [MCRE, pp. 216-
217], which I took to be the modern Blandain, near Tournai.
Grierson [1937] has edited these annals. The place is Mont
Blandin, within the modern city of Ghent, about 50 kilometers
north of Tournai. Since none of the eclipses involved was
taken to be very large, changing the position has a negligible effect upon the quantitative results. I thank Professor
Grierson for calling the error to my attention.

I also thank Professor A. Fletcher, of the University,
Liverpool, for a letter calling my attention to errors in
the locations of S. Benet Holme, near Oxnead [pages 136-137],
and of Weingarten [page 377]. I took Oxnead as being in
Oxfordshire but it is really in Norfolk, at about 52°.7N
latitude, 1°.5E longitude. There is a Weingarten at the
place I said, but it is not the origin of the annals used.
The correct Weingarten is a few kilometers from Ravensburg,
at about 47°.7N, 9°.1E. Both places can be found on maps
of large scale. I thank Mr. Kaye Weedon of Oslo for also
calling my attention to the error about Weingarten. Neither
of these errors has an appreciable effect upon the results.

On page 216 of MCRE I incidentally referred to a place
near Liège called Trajectus or Trajectum, but without using
it as a place where observations were made. There are

many places called Trajectus, and Professor Fletcher points out that this particular one is the modern Maastricht. I thank him for also calling my attention to the error about Blandin.

William of Malmesbury has left us eye-witness accounts of the eclipses of 1133 August 2 and 1140 March 20 [MCRE, pp. 161 and 163].† I thank Professor Kenneth Harrison of Charles Street, London, for a letter saying that William mostly lived not at Malmesbury, as I assumed, but at Glaston-bury, about 50 or 60 kilometers to the southwest. According to the editor of the cited edition of William's writing,‡ William does refer to himself as a monk of Glastonbury in a passage that cannot be dated accurately. However, most in-dications that I can find show him to be at Malmesbury, where he seems to have attained the position of librarian. I shall respectfully disagree with Professor Harrison and continue to use Malmesbury as the site of the eclipse obser-vations.

On pages 80-81 of MCRE I decided to use 0.01 as the standard deviation of the magnitude when a record says that stars were visible‡ during an eclipse; the planets are to be counted among the stars in this context. People too numerous to acknowledge, whom I thank for their interest, have written giving me counter examples, in which stars were seen when the magnitude was less than 0.99. In fact, it seems to be possible to see Venus with the naked eye on a clear day under favorable circumstances even when there is no eclipse. Further, the ancient and medieval records them-selves show a number of counter examples. Three of these are the three independent records of the eclipse of 891 August 8 from Constantinople [MCRE, pp. 547-548]. All three records say that stars were seen during the eclipse, but the eclipse was annular with a maximum magnitude at Constantino-ple of only about 0.94.

As a statistical matter, I believe that it is reasonably accurate to use 0.01 for the standard deviation of the mag-nitude when stars (in the plural, but counting planets as stars in this connection) are seen during an eclipse. That is, in records made by people with astronomical training, or at least with astronomical interest, and which record that stars were seen, I believe that the magnitude exceeds 0.99 more times than not.

However, this is not the consideration that is impor-tant to us. We want to know what happens in records made by typical annalists or chroniclers who usually have no interest in astronomical matters but who record an eclipse because it is a fairly striking phenomenon. In order to answer this question, I have calculated the standard devia-

---

†See a further discussion of these records in Section XIV.8.

‡This edition is found in Rerum Britannicarum Medii Aevi Scriptores, Number 90, William Stubbs, editor, H.M. Sta-tionery Office, London, 1887.

‡That is, I assumed that the chances are about 2 out of 3 that the magnitude is 0.99 or greater.

tion of the magnitude for the records from Europe and the Near East which say that stars were seen during an eclipse. I did the calculations using provisional values of the accelerations. It is not likely that use of the finally inferred values will make a change that is significant so far as present interests are concerned, but I shall repeat the calculations with the final values in Section XIV.5 as a matter of safety.

The answer is surprising. There are 45 records in the sample used, and the standard deviation of the magnitude for this sample is 0.054, corresponding to a magnitude of 0.948. At this magnitude, I doubt that any star except Venus can be seen, and I doubt that it would be seen often.

We should remember that the standard deviation of the magnitude for eclipses which are explicitly described as total is 0.034. The combination of this value and 0.054 when stars were seen is hard to explain. It is possible that an error in assigning the place of observation accounts for the small magnitude in a few cases, but this cannot explain the numerous cases in which "stars were seen" or "there was complete darkness" when the eclipse was actually annular by a considerable margin. I can only suppose that the annalists, like most people, had a strong tendency to exaggerate their experiences.

I shall use 0.054 as the standard deviation of the magnitude for records which report that stars were seen but which do not explicitly assert totality. This value is limited to records from Europe and the Near East. Records from China will require different treatment, as we shall see in Chapter V.

2.   A "Record" of the Eclipse of -1130 September 30

Since there are some passages in the Hebrew and Christian scriptures that refer to an unusual behavior of the sun, it is inevitable that some scholars of these scriptures will try to interpret such passages as records of solar eclipses and try to date them. J.F.A. Sawyer has done this with regard to two scriptural passages.

These passages involve matters of high religious and emotional significance, and I feel reluctant to discuss them in a work like the present one. However, other writers have injected the conclusions of Sawyer into recent astronomical literature and have championed his conclusions in doing so. Since his conclusions are supported by neither textual nor astronomical evidence, in my opinion, it is necessary to present an analysis of the evidence.

The first passage [Sawyer, 1972a] is from Joshua 10:12-13: †

---

†The translation and the arrangement of the text are taken from the Revised Standard Version [1952].

Then spoke Joshua to the LORD in the day when the
LORD gave the Amorites over to the men of Israel; and
he said in the sight of Israel,

"Sun, stand thou still at Gibeon,
and thou Moon in the valley of Aijalon."
And the sun stood still, and the moon stayed,
until the nation took vengeance on their enemies.

Is this not written in the Book of Jashar? The sun
stayed in the midst of heaven, and did not hasten to
go down for about a whole day.

This passage has a simple and straightforward interpretation. The Hebrews needed time to "take vengeance on their
enemies." If darkness had come before they had done so, the
enemies might have escaped. Therefore, according to the
story, the LORD by a miracle prevented the arrival of darkness until the vengeance had been accomplished. When there
is a simple interpretation, it seems to me that the simple
interpretation is most likely to be the correct one. However, Sawyer interprets this as the record of a solar eclipse,
and he uses an interesting argument in doing so.

Sawyer starts by citing a number of medieval records of
eclipses from monastic annals in which the duration of totality is apparently exaggerated, sometimes being measured in
hours. He suggests that the dramatic impact of an eclipse
causes this exaggeration, and that an experience which lasts
at most for a few minutes can seem to last for a substantial
part of a day.† Thus, when the spectators were suddenly deprived of the sun, they did not see it again for what seemed
like a large part of the day. When they could see it again,
it was still in essentially the same place, since the time
was really only a few minutes. That is, it seemed to them
that the sun had stood still for a long time.

It seems to me that this interpretation, in addition to
being complicated and conjectural rather than simple and
straightforward, goes contrary to the text in several ways:

---

†My experience suggests a different explanation. The aspect of an eclipse that first struck me was not the coming
darkness but the cold. The total time interval during
which either cold or darkness is impressive lasts for half
an hour or longer. Thus, if the time interval is given as
2 hours, which is a common medieval figure, the exaggeration is by less than a factor of four; it is not a matter
of magnifying a few minutes into a large part of a day.
Further, the exaggeration is only apparent in many
cases. For example, the record 1133 August 2 E,I that will
be given in Section X.3 below says that the sun was "almost
totally eclipsed from the 6th hour to the 9th hour ..."
This seems to say that near-totality lasted 3 hours, but I
do not believe that this was the intent of the writer. I
think he meant to state that the eclipse was nearly total
and that the entire eclipse lasted about 3 hours, but his
conciseness gives rise to an apparent exaggeration.

(a) According to the passage, the sun "stayed in the midst
of heaven"; it did not vanish from the midst of heaven as it
would in an eclipse. (b) In Sawyer's interpretation, the
writer focussed on perhaps the most subtle aspect of an
eclipse, while ignoring its most striking features such as
cold, darkness, and appearance of the stars. (c) The
Hebrews took advantage of the event, whatever it was, to
complete their vengeance. They could not have done this
during a few minutes of total darkness, when they could not
have seen their enemies to strike them.† (d) The sun and
moon, according to the text, were in different parts of the
sky and an eclipse was therefore impossible. Since Aijalon
is due west of Gibeon, the passage suggests that the battle
was between the two, that the time of the sun's stopping
was in the morning, and that the moon was near the last
quarter.

Even if we should conclude that the event is an eclipse,
we cannot date it. Traditionally, the battle described took
place at some uncertain time interval after the Exodus,
whose time is itself uncertain. Further, some scholars believe, on the basis of archaeological and other evidence,
that many of the actions attributed to Joshua in the traditional account actually took place long before the Exodus,
perhaps as much as two centuries before. Thus we must allow
any time from about -1400 to -1100, say, as a possibility.

Sawyer [page 142] writes that astronomical calculations
have "shown that there was no eclipse at all several centuries earlier, .." By this, I assume he means that there
had been no total eclipse visible in Palestine for several
centuries before -1130. Unfortunately, this is the kind of
thing that it is impossible to know. I have estimated
[Newton, 1974] that the uncertainty in calculating the magnitude of an eclipse around the year -1300 is about 0.18 on
a standard deviation basis. Therefore, for many eclipses
between -1400 and -1100, we are simply unable to say whether
the eclipse was total near Gibeon or not.‡

The reader may argue that the passage in its present
form was suggested by the occurrence of an eclipse that took

---

†The sense of the Hebrew (Masoretic) text, particularly on
this important point, is the same as that of the translation given above. I thank my colleague A.D. Goldfinger
for verifying this point. A person who wishes to maintain
the eclipse interpretation might try to argue that the
eclipse terrified the enemies to the point that they fled,
and hence that vengeance was accomplished in this way.
This interpretation is not supported by any evidence, and
it is contrary to a straightforward interpretation of the
text.

‡If we anticipate the final results of this work, the magnitude of the eclipse of -1130 September 30 at Gibeon could
have had any value from 0.74 to 0.97, but it is unlikely
that the eclipse was total there. See Section XIV.8.

place some time after the event. It is almost certain that the book of Joshua, in its present form, was not written until centuries later, say not before about -500, and that any text that may have existed in -1000, say, may have been extensively altered before the present form became canonical. Thus, if the passage were suggested by a later eclipse, it could have been any eclipse between -1400 and -500 that was total in Palestine.

On pages 143-144, Sawyer writes: "There was not another eclipse observable in Palestine anything like this one for more than 700 years, ..." Again, we do not know enough about astronomy to make such a statement. To give only a few examples, the eclipses of -1123 May 18, -1083 March 27, -1062 July 31, and -1040 November 23 were all large in Palestine, and any one of them could have been total so far as we can say on the basis of present astronomical knowledge.

In summary, to interpret the passage from Joshua as the record of a total eclipse of the sun goes against the whole meaning of the text. Even if we assume that the event was a total eclipse, we cannot date it. There must be a dozen possible dates, of which I have enumerated four besides the date -1130 September 30 that Sawyer selected.

3.   A "Record" of the Eclipse of 29 November 24

The second Biblical passage that Sawyer interprets as a report of an eclipse is <u>Luke</u> 23:44-45. This is the passage which describes the death of Jesus; there are similar passages in <u>Matthew</u> and <u>Mark</u>. In <u>Mark</u> 15:33-38, there was "darkness over the whole land" from the sixth to the ninth hour,† and the "curtain of the temple was torn in two." In <u>Matthew</u> 27:45-53, we find these events and we find also that "the earth shook, and the rocks were split, the tombs also were opened" and many spirits of the dead appeared to the living. In <u>Luke</u>, we do not find the events that are added in <u>Matthew</u>, but we do find the statement that the darkness happened because the sun was eclipsed. We do not find any of these events in <u>John</u>.

These quotations are from the <u>Revised Standard Version</u> [1952]. I thank my colleague S. M. Krimigis for looking up the passage from <u>Luke</u> in the Greek. The phrase του 'ηλιου 'εκλιποντος occurs in the Greek. I shall grant Sawyer's point that this means an eclipse for the sake of argument, although other interpretations are possible.

Before I take up Sawyer's arguments about the passage, it is useful to mention an important point in Biblical criticism. There are persons named Matthew, Mark, and Luke who figure in the Biblical narrative. As I understand the matter, there is no evidence that they are responsible for the Gospels now attributed to them; there are only rather late traditions. The Gospels were probably not written down until

---

†In our terminology, this is from noon to mid-afternoon.

some time after the event, perhaps not for 75 years in some cases. This makes it unlikely that the characters in the narrative also wrote the Gospels, although they might have been responsible for earlier Gospel forms, either written or oral, that are now lost.

Sawyer [1972b] asks: Why does the author of Luke, alone among the Gospel authors, say that the darkness was caused by an eclipse? We should note that the event could not have been a natural solar eclipse because the Crucifixion occurred within a day or so of the full moon. To me, a simple explanation also seems like a plausible one: The author of Luke had some education in such matters and knew that a solar eclipse could produce a great darkness. However, he did not know, or perhaps did not care, that there could not have been a natural eclipse at the time stated. He may simply have used "eclipse" (εκλειποντος) as a synonym for darkness, as many other writers have done.

Sawyer suggests the following explanation: Luke was probably from Antioch while Matthew and Mark were connected with Jerusalem or the surrounding area. The eclipse of 29 November 24 was the only eclipse near the correct time that was large in Palestine, and it was total in Antioch but not in Jerusalem. Luke saw this eclipse and this caused him to insert the reference to an eclipse.†

This is the kind of conclusion that cannot be disproved, but I see little reason to accept it. In astronomical work, we must study this record with no attention to its religious significance. When we do so, we must immediately put it in the class that I have called the magical eclipse [AAO, pp. 44-45]. As I wrote there, ancient peoples showed a remarkable tendency to fight important battles during total eclipses, and I should have added that they also showed a remarkable tendency to die during eclipses. The death of a prominent person is often accompanied by miracles, including a miraculous darkness. The deaths of Julius Caesar, Roland, and S. Olaf are examples that come immediately to mind. The fact that a miraculous darkness is associated with the death of Jesus therefore does not need to call for any comment, nor does the fact that one author out of three extended the reference from a darkness to an eclipse.

Even if we grant that the reference to an eclipse was occasioned by the occurrence of one, the eclipse of 29 November 24 is not the only possibility. The eclipse of 59 April 30 was probably total in Antioch and it is thus a possibility even if we accept the restriction that the eclipse had to be observed by the Biblical character called Luke. If we remove this restriction, the eclipse could have been one that was large anywhere around the eastern half of the Mediterranean at any time until at least the year 100. Thus the eclipses of 71 March 20, 75 January 5, 80 March 10, and 83 December 27 are all possibilities.

---

†There were two ancient cities named Antioch. The one connected with Luke is the present Antakya in Turkey. Note Sawyer's implicit assumption that the Biblical characters were also the authors of the Gospels.

4. A Reappraisal of Ptolemy

The most grievous error in the literature about astronomical accelerations is that of accepting Ptolemy's work as genuine. We must now accept as fact [Newton, 1977] that much of the information contained in the Syntaxis† is not merely erroneous but fraudulent. It is probable that some of the information in the Syntaxis is genuine, but Ptolemy's writing by itself gives us no basis for separating the genuine from the fraudulent. Thus we cannot accept any information from the Syntaxis unless it is confirmed by sources which are known to be independent of Ptolemy.

The realization that the Syntaxis is unreliable has come in four main stages. ibn Yunis [1008, p. 142] is the earliest writer I have found who recognized that Ptolemy's observations of the times of equinoxes and solstices are seriously in error. We know now that these times are all too late by more than a day, although a reasonable error with the methods and instruments available to Ptolemy would have been a few hours. We can be sure of this because we find that similar observations made centuries before Ptolemy agree with results from modern calculations within a few hours.

Not all medieval astronomers agreed with ibn Yunis, and we find major astronomers even as late as Copernicus [1543] trying to devise theories which would reconcile Ptolemy's solar observations with those of other astronomers, both ancient and more recent. However, by about 1800, astronomers had realized that almost all of Ptolemy's observations were in error by unlikely amounts and Delambre [1817, volume 2, p. xxv], because of these errors, could ask: "Did Ptolemy do any observing? Are not the observations that he claims to have made merely computations from his tables, and examples to help in explaining his theories?"

In other words, on the basis of the size of Ptolemy's errors, Delambre argued that Ptolemy did not make the claimed observations at all, and that he instead simply calculated them from his theories. However, this argument has a weak point. It rests upon preconceptions about the observing methods that Ptolemy might have employed. If we can show plausible methods that would lead to errors of the size that we find in Ptolemy's observations, we have answered Delambre's argument.

Scholars since Delambre's time have spent much effort in trying to find methods of measurement that Ptolemy might reasonably have used and that lead to errors of the size

---

†This is the short term that I often use in referring to Ptolemy [ca. 142]. Most literature on the subject refers to this work as the Almagest, but this name, which means The Greatest, has an implication that is now unacceptable. For this reason, I have substituted the neutral term Syntaxis, which is a word that occurs in the Greek title. At least it occurs in the title of modern Greek editions, and it may well have been used by Ptolemy in the original.

that we find. In some cases, these efforts have been moderately successful while in others they have not. The solar
observations are outstanding in the way that they have re-
sisted explanation. The trouble is that all likely sources
of error have different effects upon equinoxes and solstices.
An error that tends to make an autumnal equinox come too
late tends to make a vernal equinox come too early by about
the same amount, and it tends not to affect a solstice at
all. However, we find that Ptolemy's times are all too late
by more than a day, whether they refer to vernal or autumnal
equinoxes or to a solstice. Even though the solar data have
so far defied explanation on the assumption that they were
actually observed, however, we must still recognize the possibility that some future scholar may succeed where all
earlier ones have failed. In other words, an argument based
upon the sizes of the errors cannot be conclusive.

The second main stage begins with a later work of
Delambre. He received many comments on his history of
ancient astronomy [Delambre, 1817], and he used the preface
to his history of medieval astronomy [Delambre, 1819] to
reply to some of them. Here he found a conclusive argument
about Ptolemy's solar observations. Ptolemy claims two
things about these observations: (a) he made them with
great care and (b) they confirm that Hipparchus was accurate
in deciding that the length of the tropical year is 365 +
(1/4) - (1/300) days. Delambre now showed [Delambre, 1819,
p. lxviii] that Ptolemy fabricated the observation of a summer solstice that he claims to have made, as well as one of
the equinoxes, and he showed how the fabrication was done.
I shall illustrate the fabrication by using the summer solstice.

Ptolemy says that Meton measured the summer solstice in
Athens in the year that we call -431, and that he found it
to come at the time we call 06 hours on June 27. The Julian
day number at noon on -431 June 27 was 1 563 813, so the
Julian day number at the time of the observation was
1 563 812.75 in terms of Athens time. Ptolemy claims that
he measured the summer solstice in the year 140, just 571
years later. Now 571 years equals 208 555.847 days, if the
year has the length that Hipparchus inferred. When we add
this to the Julian day number at 06 hours on -431 June 27,
we get 1 772 368.597. This is the day number at 02.3 hours
on 140 June 25. Since Ptolemy gave his times only to the
nearest hour, we should round this to 02 hours on 140 June
25. This is exactly the time which Ptolemy claims that he
found by measurement.

Britton [1967], in a doctoral dissertation that seems
to be still unpublished, did calculations of the same sort
for all of Ptolemy's equinox times but not for the solstice.
He found again that the calculated values, when rounded to
the hour, agree exactly with the times that Ptolemy claims
to have measured.

I did the same calculations somewhat later [AAO, Section II.2], except that I did them for all of Ptolemy's
claimed solar observations instead of for just a sample as

Delambre and Britton had done. Because Delambre's calculations are found in his work on medieval astronomy, I did
not find them when I was doing my first work with ancient
astronomy. Because Britton's dissertation is unpublished,
I was also unacquainted with his work at that time. Thus
I did the calculations independently of Delambre and Britton, although I do not have the priority. My findings were
the same: The calculations made in the way that has been
described give exactly the values which Ptolemy claims to
have measured, without exception.

It is important to realize that this argument is quite
different from earlier arguments based upon the fact that
Ptolemy's errors seem to be unreasonably large. Those arguments depend upon assumptions about the specific methods of
measurement that Ptolemy might have used, and thus they are
not rigorous. The present argument depends upon the exact
agreement, within the level of rounding used, of Ptolemy's
alleged measurements with values that we can calculate in
advance from older theories that Ptolemy "confirmed". We
must recognize that there is a finite probability that the
agreement happened because of the chance occurrence of observational errors, but the probability turns out [Newton,
1973] to be something like 10^{-100} for the solar data. I
believe that any reader is willing to neglect a probability
like this and say that there is no possibility that the
measured results happened by chance.

We must also realize that the conclusion is independent
of any assumption about the way the measurements might have
been made, and that the argument is thus mathematically rigorous. The reason is that the probability of the agreement
happening by chance is enormously small for any possible
method of measurement, whether it is one with basically high
accuracy or basically low accuracy. See Newton [1973] for
the explanation of this fact.

It is interesting to watch the reaction of scholars to
a development of this sort. I have never seen a published
reference to the work under question, either by Delambre or
Britton, except in my own writing.[†] Further, even though I
was convinced that Ptolemy's solar observations had been
fabricated, it did not occur to me for a long time to suspect that any of his other alleged observations had been.
The exploration of this suspicion, when it did develop,
marks the third stage in our understanding of Ptolemy's
fraud.

Late in 1972, I began a study of ancient planetary
observations. As part of the process of gathering data
for this study, I read straight through the Syntaxis in
order to find all of the observations in it. As I did so,
I was intrigued by some of the obvious errors in such matters as the lunar parallax or the latitude of Alexandria.
I first studied these errors from the standpoint of trying
to find sources of error that would explain them. After
several unsuccessful attempts, I again tested the hypothesis

[†]This is written in final form in December, 1977.

that they had been fabricated, and I published the results in three papers [Newton, 1973, 1974a, and 1974b] and in Chapter XIII of APO. In these works I analyzed all the observations that Ptolemy claims to have made except the star catalogue.

The results are unequivocal. All of Ptolemy's claimed observations that he uses and that can be tested are fabricated. Two qualifications are used in stating this conclusion, but this does not mean that there are exceptions. Ptolemy derives some of his planetary parameters by using only observations that he claims to have made himself. There is no way to test the fabrication of these observations, although the errors present in them make them look suspicious; these observations number about a third of Ptolemy's claimed planetary observations. There are also measurements of the declinations of 12 stars that Ptolemy quotes but does not use at all. Surprisingly, these turn out to be genuine.

The fourth stage came about a year later. I studied all of the observations in the Syntaxis, including the ones that Ptolemy attributes to other astronomers as well as the ones that he claims to have made himself; this review is published as part of an extensive survey of the Syntaxis [Newton, 1977]. In spite of my earlier experiences, I was still astonished to find that many of the observations attributed to other astronomers are also fabricated.

Altogether we can establish that about a third of the observations which Ptolemy attributes to others have been fabricated. This does not necessarily mean that the remaining two thirds of the observations are genuine, however; it may mean only that I have not been clever enough to find the procedures by which Ptolemy fabricated them.

The lesson of experience is now clear, it seems to me: We cannot accept any statement in the Syntaxis on the basis of Ptolemy's evidence. We can accept only those statements that have independent confirmation. It will take much time and thorough study to establish which of Ptolemy's statements do have independent confirmation. In the meantime, we must abstain from using any statement from the Syntaxis unless we have found independent confirmation of it.

In the limited study that I have been able to conduct to date, I have established confirmation for a few observations [Newton, 1977, Section XIII.2]. These are the observations of the summer solstice and of the equinoxes made by Aristarchus and Hipparchus, and the measurements of the declinations of 18 stars attributed to Timocharis, Aristyllus, and Hipparchus. Ptolemy also claims to have measured the declinations of these same 18 stars, but he used only 6 of them in studying the precession of the equinoxes. Interestingly, the 12 declinations that he does not use are genuine but the 6 that he does use are fabricated.

Almost all of the observations involving the moon that are found in the Syntaxis, including those that Ptolemy attributes to others as well as those that he claims to have made himself, are definitely fabricated. In particular, all the occultations of stars by the moon, all direct measurements

of the lunar position, and most of the lunar eclipses are
fabricated. Of the three oldest lunar eclipses in the Syntaxis, the record of the eclipse of -719 March 8 is definitely fabricated. I could not reach a firm conclusion about
the records of -720 March 19 and -719 September 1, but the
preponderance of the evidence is that they are fabricated.

It will take a thorough search of the relevant literature to establish whether any observations other than those
mentioned above† have independent confirmation. In the meantime, we cannot safely use any other observations in the
Syntaxis. This wipes out three categories of lunar observations that were used in AAO. Luckily, their loss is somewhat balanced by Babylonian lunar observations that have
been found in the meantime [APO, Chapter IV].

The elimination of the Syntaxis as a source of astronomical observations also affects the work of APO. There I
used observations of the planets from Babylonian and Islamic
astronomy, along with the observations found in the Syntaxis.
Each body of observations indicates that Mercury and Venus
are gaining orbital energy. With the removal of the Syntaxis
from consideration, there are only two independent bodies of
observations, and the conclusions of APO are somewhat weakened. However, there is a large body of Babylonian observations that is currently being edited, and it should more
than compensate for the loss of the observations from the
Syntaxis when it becomes available. This is fortunate, because the conclusions of APO are disconcerting, and it is
important to see whether additional data will confirm or
deny those conclusions.

5.   The Eponym Canon Eclipse

Ptolemy's fabrication of data has an effect upon Babylonian chronology as well as upon Greek astronomy. This
effect stems from Ptolemy's "king list" and from his use of
the list in dating observations.

The king list starts with a list of Babylonian kings,
followed by their Persian successors, then the Greek rulers
of Egypt, and finally the Roman emperors through Antoninus
Pius, who was emperor at the time when Ptolemy probably
wrote the Syntaxis. For each ruler on the list, Ptolemy
gives the number of Egyptian years to be attributed to his
reign.

It was customary for many ancient peoples to use kings'
reigns as the method of specifying the years. In doing so,
it is necessary to have a convention about the way in which
accessions are to be counted. As Ptolemy uses the list,
the year in which a king comes to power is counted as the
last year of his predecessor. In other words, the first
year of a particular king begins on the first day of the
Egyptian year that next follows his accession, in Ptolemy's
usage.

---

†That is, certain solar observations and star declinations.

The first king in Ptolemy's list is Nabonassar, and the beginning of his first year, in the sense just used, is the basic epoch to which Ptolemy refers all his astronomical ephemerides. This epoch is -746 February 26.[†] By counting the years assigned to the various kings in Ptolemy's list, we can find the first year of any king in the list and hence we can find the nth year of any king.

In Chapter IV.6 of the <u>Syntaxis</u>, for example, Ptolemy says that there was a lunar eclipse during the night between the 18th and 19th of the Egyptian month Thoth, in the second year of the king we designate as Merodachbaladan. The middle of the eclipse came at midnight at Babylon, and only the southern fourth of the moon was eclipsed. When we use the king list and the known properties of the Egyptian calendar, we conclude that the eclipse came at the midnight between -719 March 8 and 9, Babylon time. Since <u>Oppolzer</u> [1887] lists a small eclipse that came within an hour or so of this time, we seem to have simultaneous confirmation of the record and of the king list.

This confirmation is illusory, however. There is little doubt [<u>Newton</u>, 1977, Section VI.7] that Ptolemy fabricated this eclipse. The fact that the circumstances stated by Ptolemy agree fairly well with those listed by Oppolzer merely means that the calculations upon which Ptolemy based his fabrication were fairly accurate. When he calculated the time of the eclipse, Ptolemy in effect calculated that the eclipse came a certain number of years before his own time. When he fabricated the year in the Babylonian calendar, he merely counted the years in his king list and said that the eclipse came in the second year of Merodachbaladan.

Suppose that Ptolemy's king list is not accurate. Instead of saying the second year of Merodachbaladan, let us suppose that his list said that March in the year -719 came in the eighth year of king K. Ptolemy would then have stated the eighth year of K as the year of the eclipse. Later astronomers would use Ptolemy's king list, they would equate the eighth year of K to -719, and they would arrive at the date -719 March 8/9 for the eclipse. Thus astronomical calculations would "confirm" Ptolemy's king list whether it is accurate or not, if the eclipse is fabricated.

In the study just cited, I concluded that all of the early observations which Ptolemy dates by means of Babylonian kings are fabricated.[‡] Therefore it follows that there

---

[†] Contrary to many statements in the historical literature, there is no reason to assume that any ancient people used this date as a reference epoch for history.

[‡] The evidence that leads to this conclusion is quite strong except for two lunar eclipses. For these two eclipses, the evidence is not strong, but the preponderance of the evidence leads to the conclusion that they are also fabricated. In research studies, whether of astronomy or chronology, it is clearly not safe to use any of the observations that Ptolemy cites, unless they have confirmation.

is no astronomical confirmation of the early kings, and
it is not legitimate to use the king list in chronological
studies except where it has independent confirmation.

There is such confirmation for the later part of the
king list. Neugebauer and Weidner [1915] have published an
extensive astronomical diary from the 37th year of Nebuchad-
nezzar II† of Biblical fame, and they have established that
the year in question began on -567 April 23. Nebuchadrezzar's first year therefore began in -603, and this agrees
with Ptolemy's list.

Further, extensive historical evidence links the reign
of Nebuchadrezzar with that of his predecessor Nabopolassar,
and it seems safe to conclude that the first year of Nabopolassar began in the spring of -624. This date also agrees
with Ptolemy's king list.

This does not automatically confirm the king list from
Nabopolassar on. However, the dates of the Persian kings
also have strong independent confirmation, and there is not
much room for error between Nebuchadrezzar and the Persians.
Thus we may accept Ptolemy's king list from -624 on, with at
most a small error.

The reign of Nabopolassar, however, followed an extensive period of political confusion, and there is room for
appreciable error about kings before Nabopolassar and hence
for years before -624. This brings us to the "eponym canon"
eclipse.

In the Assyrian kingdom over a period of many centuries
there was an annually chosen officer called the limmu whose
name was used to designate the year. In this respect,
Assyrian practice was the same as that of the Romans, who
often designated a year by using the names of the consuls.‡
More than a century ago, Assyriologists discovered a long
continuous list, often called the eponym canon, of these
eponymous officials. Along with the name of the limmu for
the year, the list sometimes records a notable event that
happened during the year, such as the accession of a new
king. Thus the eponym canon forms a kind of annals. One
event [Fotheringham, 1920] is an eclipse of the sun which
took place in the month called Simanu. If we can identify
this eclipse, we can fix the chronology of a long period of
Assyrian history.

For more than a century, it has been accepted that this
eclipse is the eclipse of -762 June 15. Trying to find the
basis of this identification is an interesting exercise in
frustration. Fotheringham [1920] says that the identification was first made by George Smith in 1867, and he cites

---

†Modern scholars seem to prefer the spelling Nebuchadrezzar.

‡There were two consuls who were chosen annually, and almost
all historical writers under the Republic identified a year
by stating the names of the consuls that year. Many writers
under the Empire continued this practice, while others used
the number of years that the current emperor had been ruling.

Smith [1875] for further information. However, Smith does
not give any further information. He merely says that he
has identified the eclipse, and he does not give any information that could form the basis for the identification. He
gives no clue to any source in which additional information
can be found.

Fotheringham writes; "The date is beyond question,
since the Assyrian eponym canon overlaps Ptolemy's Canon of
Kings, and the year -762 is obtained by counting backwards
to the year of Ishdi-Sagale."† As I understand the matter,
this method of identification is impossible rather than certain. Ptolemy's king list applies only to Babylonia while
the eponym canon applies only to Assyria, I believe. However, there are some synchronisms between Babylonian and
Assyrian chronology which may allow connecting the eponym
canon with the kings in Ptolemy's list, and Fotheringham's
statement may tacitly assume the knowledge of these synchronisms, along with the unproved assumption that the synchronisms are established beyond question.

An extensive study of the Assyriological literature for
the period around 1867 might reveal the basis on which the
eponym canon eclipse was identified as that of -762 June 15.
For present purposes, however, the reward of such a study
would not warrant the effort involved.

If Fotheringham is correct that the eclipse was dated
by connecting it with Ptolemy's king list, the identification immediately becomes uncertain. We cannot rely upon the
accuracy of the early part of the king list, which would be
the part involved in identifying the eponym canon eclipse.
Hence we must see if we can identify the eclipse without
using the king list.

In identifying the eclipse, we have two items of information to help us. First, we know the general historical
period, so the date of the eclipse cannot be too far from
-762. In testing the identification, I shall assume that
the eclipse must be within 40 years of -762. Second, we
have the statement that the month was Simanu.

The Assyrian calendar of this period was presumably the
same as the Babylonian calendar in its general structure,
although it may have differed in some details. If this is
correct, a month was a lunar month, and a month began with
the sunset at which the new moon could first be seen. The
average length of the year was kept close to the correct
value by having 13 rather than 12 months in about 7 years
out of 19.‡ The need for an added month was presumably
determined by some kind of astronomical observation, but
we must assume some variability in the accuracy with which

_____

†This is the name of the limmu in the year of the eclipse.

‡I do not mean to imply that the Assyrians had discovered
 this rule by the time of the eponym canon.

the rule was applied at this early stage in history.†

Since a month began when the new moon was visible, an eclipse could only occur a day or two before the beginning of a month. Thus an eclipse in the month Simanu had to be one that came near the end of the month, which is the third month of the year. Parker and Dubberstein [1956] give the dates when each month began in the Babylonian calendar at a time when the rules of the calendar were being followed rather accurately. By inspecting the dates over an interval of a century, I find that the end of Simanu could have come on any date between June 2 and July 26 in the Julian calendar. This tells us the variation in the end of Simanu that is imposed by strict astronomical considerations. However, we must allow the possibility that there could have been an error of a month in either direction because of errors in applying the strict astronomical rules. Hence we must allow any date between about May 3 and August 25 for an eclipse that came in Simanu.

In such an important matter, I do not believe that we can assume that the scribe who wrote the original record was being astronomically precise. Instead, I believe that we must admit the possibility that an eclipse which came just before the beginning of Simanu was described as coming in that month. This means that the earliest possible date must be extended to about April 3.

TABLE II.1

POSSIBLE DATES FOR THE EPONYM CANON ECLIPSE

Date	Magnitude at Nineveh
-802 Aug 6	0.56
-790 Jun 24	0.49[a]
-777 Apr 4	0.67
-769 May 5	0.66
-762 Jun 15	0.95
-754 Jul 16	0.53
-750 May 5	0.67[b]
-736 Jul 26	0.40
-722 Apr 25	0.18

[a]Maximum eclipse occurred at sunset.
[b]Maximum eclipse occurred at sunrise.

---

†The rule in the Jewish calendar, for example, at least at a somewhat later period, was that the 14th day of the first month had to come after the vernal equinox.

In Table II.1 I give the magnitudes of nine eclipses that were visible at Nineveh between the years -802 and -722, for dates between April 3 and August 25. I exerted reasonable care in listing the possible dates, but I do not guarantee that the list in Table II.1 is exhaustive. The reader may wish to assume that the scribe was rigorously precise in saying that the eclipse was in Simanu. If so, he may eliminate the eclipses of -777 April 4 and -722 April 25.

Fotheringham [1920] writes: "As the eclipse is the only eclipse mentioned in this Chronicle, which covers an interval of 155 years, there can be no reasonable doubt that it had been reported as a total eclipse." Unfortunately there is no basis in fact for the assumption back of this statement. We have too many examples in which an eclipse was the only one reported in a chronicle over an interval of $1\frac{1}{2}$ centuries and which was nonetheless far from total.[†] The most we can assume is that the eclipse was large enough to be observed with a reasonable probability. This does not require a very large eclipse. As we shall see from Table XIV.3 in Section XIV.6, the eclipse of 1181 July 13 was recorded at Mont S. Michel although the magnitude was only 0.29. In Section XIV.7 we shall see that the eclipse of -15 November 1 was recorded in Charng-an at a magnitude of only 0.056.

The eclipse of -762 June 15 did indeed have the largest magnitude at Nineveh of any eclipse in Table II.1, but it is not the only striking eclipse. In fact, any eclipse in the table would have a good probability of being recorded except perhaps the eclipse of -722 April 25. Even a rather small eclipse that occurs at sunrise or sunset is readily visible. On -750 May 5, two thirds of the sun was eclipsed in Nineveh at sunrise, and on -790 June 24 about half of it was eclipsed at sunset. Either of these eclipses would have been striking.

Thus, as matters now stand, any of the dates -790 June 24, -762 June 15, and -750 May 5 must be accepted as a possible date for the eponym canon eclipse, and still other dates cannot be excluded. We can assign the date of the eclipse only by using chronological evidence that is independent of Ptolemy.

It will be a formidable task to try to determine the date by evidence that is strictly independent of Ptolemy, and I have made no attempt to do so. The identification of the eclipse as that of -762 June 15 was made more than a century ago. During the past century, this date and others that are derived from it have become firmly embedded in the Assyriological literature. Thus a date that seems to be independent of Ptolemy's king list may in fact be derived from it. It will be necessary to pursue every Assyriological or Babylonian date in the literature back to its origin to find whether it ultimately depended upon Ptolemy's king list or not.

[†]See AAO, page 60, for an example.

The eponym canon eclipse has been considered to be the oldest astronomical observation that can be dated. The next oldest observations have been considered to be the lunar eclipses of -720 March 19, -719 March 8, and -719 September 1, which occur in Ptolemy's Syntaxis. These eclipses must also be eliminated from the list of dated observations. Thus the oldest datable astronomical observation that appears in the current astronomical literature is a Chinese observation [AAO, p. 62] of the solar eclipse of -708 July 17.†

6.   Multiple Identifications

For several records in AAO, I was unable to decide upon a unique identification, but I could decide that the date had to be one out of a few possibilities. In the analysis of these records, I took the total statistical weight that would normally be assigned to a single record and divided it among the possible dates. If all possibilities seemed to be equally likely, I divided the total weight uniformly. If one possibility seemed more likely than the others, for some reason, I correspondingly gave it more of the total.

I later decided that this procedure is incorrect, for a reason that I shall describe in Section III.7. Thus I did not use the procedure in MCRE nor shall I use it here.

---

†However, one earlier observation will be presented in Section V.2.

THE WORK OF MULLER AND STEPHENSON

1. Preliminaries

Recent publications by Muller and Stephenson [1975] and by Muller [1975] raise so many questions that we must devote two separate chapters to their discussion. For brevity, I shall designate these publications by MS and M, respectively, in the rest of this work. Before we proceed directly to discussing MS and M, it is desirable to outline a few preliminary considerations.

First let us look at the mean longitude L_M of the moon, after we remove the contributions to its secular acceleration that come from solar and planetary perturbations. Over a limited time interval, we can write L_M as a quadratic function of time:

$$L_M = A_M + B_M T + \tfrac{1}{2}\dot{n}_M T^2. \qquad (III.1)$$

In discussing MS and M, T will represent ephemeris time, since those writers work with ephemeris rather than solar time. T is measured in Julian centuries from 1900 January 0.5 ET, and $\dot{n}_M$ denotes the acceleration of the moon with respect to ephemeris time at that epoch. A_M and B_M are empirical constants that are determined from observations. I shall call them the constant and linear coefficients in L_M.

Instead of using an angle to give the orientation of the earth in space, MS and M use the difference ΔT between ephemeris and solar times. They write ΔT in the form

$$\Delta T = A_T + B_T T + C_T T^2. \qquad (III.2)$$

We may call A_T and B_T the constant and linear coefficients in ΔT; they are also determined from observations. C_T is directly related to the acceleration parameter y that was defined in Table I.6 in Section I.7, and also to ν_S', the acceleration of the sun with respect to solar time. In fact (Equations I.28 and I.27):

$$y = -0.6338 C_T, \qquad \nu_S' = -0.1300y. \qquad (III.3)$$

Equations I.25 and I.26 in Section I.7 relate the parameters ν_M' and D'' to $\dot{n}_M$ and y. If we put these relations in terms of $\dot{n}_M$ and C_T, they become

$$\nu_M' = \dot{n}_M + 1.1011 C_T, \qquad D'' = \dot{n}_M + 1.0187 C_T.$$
$$(III.4)$$

2. Determining the Constant and Linear Coefficients

$\underline{MS}$ use† a number of observations of solar eclipses in order to derive the acceleration parameters. Some of the eclipses are ones that have been used in earlier work and some are new to the astronomical literature. $\underline{M}$ also uses some solar observations; with one exception, all these have been used in earlier studies of the solar acceleration.

In addition to the acceleration parameters $\dot{n}_M$ and C_T (or y), $\underline{MS}$ and $\underline{M}$ determine the constant and linear coefficients A_T and B_T from the ancient and medieval data. They do not determine A_M and B_M in this way, however, for reasons that I do not understand. If we must make new determinations of A_T and B_T, it seems to me that we must also make new determinations of the corresponding quantities for the moon. The only clue to their reasons that I have found comes on page 3.3 of $\underline{M}$, where Muller writes with regard to the lunar coefficients that "there cannot be a significant error there so far as the purposes of the present analysis are concerned." If the discussion of Section I.3 is correct, this statement must be in error. The constant and linear coefficients in L_M have been found by fitting to data from about 1700 to 1900. If the fitting process had been done using the finally adopted value of $\dot{n}_M$, or if the reference epoch had been the center of the time interval, the values of A_M and B_M would be accurate enough for the present analysis. However, neither condition holds, and a revision of both coefficients seems necessary, along with the revision of A_T and B_T.

Before we pursue this point further, let us take up the other main statement that $\underline{MS}$ and $\underline{M}$ make about A_T and B_T. The earlier studies that I have made show a time dependence of the parameter D'' that is represented rather well by the formula [$\underline{Newton}$, 1973a, Equation 8.2]:‡

$$D'' = \begin{cases} -14 + T, & -6 \leq T \leq 0, \\ -44 - 4T, & -12\frac{1}{2} \leq T \leq -6, \\ 9\frac{1}{8} + \frac{1}{4}T, & -26 \leq T \leq -12\frac{1}{2}. \end{cases} \qquad \text{(III.5)}$$

$\underline{MS}$ and $\underline{M}$ say that this time dependence is a consequence of the fact that I keep A_T and B_T unchanged when I estimate the parameters $\dot{n}_M$ and y from which D'' is derived. When A_T and B_T are adjusted at the same time that $\dot{n}_M$ and y (or C) are adjusted, they say, the time dependence disappears. ´ From the study of ancient and medieval eclipses in $\underline{MS}$, for example, they find,‡ in their Section III.7:

---

†Since the symbol $\underline{MS}$ denotes two people, I use the plural verb.

‡We should remember that $D'' = \nu_M{}' - \nu_S{}' = \dot{n}_M - 1.6073y$.

‡These are the values that they find by using only the data in $\underline{MS}$, before they combine their own results with those found in other analyses.

$$\dot{n}_M = -37.5 \pm 5.0 \quad ''/cy^2,$$

$$\Delta T = 66.0 + 120.38T + 45.78T^2 \quad \text{seconds.} \tag{III.6}$$

They do not give error estimates for the individual coefficients in ΔT, but they say that the expression is accurate
to within ± 100 seconds or $\pm 0.5T^2$ seconds, whichever is
larger. The values in Equations III.6, they say, fit all
ancient, medieval, and modern data.

I am not able to verify this statement. To see this,
let us derive the values of D'' and ν_M' from the values of
$\underline{MS}$ in Equations III.6:

$$\nu_M' = -37.5 + (1.1011)(45.78) = 12.91,$$

$$D'' = -37.5 + (1.0187)(45.78) = 9.14, \tag{III.7}$$

if we omit the statements of the units and the uncertainties.
For comparison, Equation III.5 gives $D'' = 6.0$ when $T = -12\frac{1}{2}$
centuries. This is somewhat smaller than the value in Equation III.7, but the difference is within the combined uncertainties. Thus there is no significant difference be-
tween my results and those of $\underline{MS}$ and $\underline{M}$ for times before
about the 7th century.

Now look at the dependence of L_M upon solar time that
is shown in Figure I.1. This is the dependence that comes
from the data used by $\underline{\text{Spencer Jones}}$ [1939], and all independent studies show almost identical results. As we saw in
connection with that figure, the average value of ν_M' from
about 1680 to 1935 is $-12.10 \ ''/cy^2$, but the value of ν_M' in
Equations III.7 is $+12.91 \ ''/cy^2$. Thus the results of $\underline{MS}$ do
not even give the correct sign of ν_M' for modern times.

Since Figure I.1 shows considerable short-term fluctuation in addition to the overall parabolic dependence, the
reader may object that the difference between $+12.91$ and
-12.10 is no more than an accidental excursion of the fluctuations. I have already studied this possibility in Section I.4, where I show that the fluctuations do not affect
the value of ν_M' (or of D'') by more than about $1 \ ''/cy^2$ over
the time span of the data.

We find a large change in ν_M' and D'' from Muller's and
Stephenson's own results. In Figure 11 of $\underline{MS}$ [p. 513], they
plot a quantity they call $\delta\Delta T$. To find this quantity, they
use the value of $\dot{n}_M$ from Equations III.6 and find the corresponding value of ΔT for each observation. $\delta\Delta T$ is this
value of ΔT minus the value of ΔT calculated from Equations
III.6. They then plot the value of $\delta\Delta T$ from each observation as a function of time.

If Equations III.6 were correct, the values of $\delta\Delta T$
would scatter randomly about the horizontal axis in Figure
11 of $\underline{MS}$. A glance at their figure shows that this does
not happen. Instead, it is clear from their figure that
the values of $\delta\Delta T$ from about 1200 or even earlier fit rather

well to a parabola which has a maximum of about 110 seconds at about the year 1560, when T = -3.4. By eye, I estimate that the following parabola fits rather well to the values of $\delta\Delta T$ from MS:

$$\delta\Delta T = 110 - 15(T + 3.4)^2$$

$$= -63.4 - 102.0T - 15.0T^2 \quad \text{seconds.}$$

If we add this to the right member of the second of Equations III.6, we get

$$\Delta T = 2.6 + 18.38T + 30.78T^2 \quad \text{seconds} \qquad \text{(III.8)}$$

This, rather than the second of Equations III.6, represents the results of MS from about the year 1200 to the present.

In other words, the value of C_T for the past seven centuries or longer is about 30.78 rather than 45.78, according to the data used by MS. This value was calculated using $\dot{n}_M = -37.5$. The combination of these values yields $\nu_M' = -3.5$. This is algebraically somewhat larger than the value that I find for recent times (since about 1200), just as their value for ancient times is somewhat larger than mine. However, we both find comparable changes in ν_M' (or in D'') between ancient times and the present.

Thus the values in Equations III.6 do not fit the data, nor do any other relations that imply constant accelerations. This result is shown both by my results and those of MS. The main difference between us is a matter of words, not of quantitative results. I look at a certain time history of ν_M' and conclude that it has decreased by about 25 $''/\text{cy}^2$ since ancient times. They look at essentially the same time history and say that ν_M' has remained constant. I believe that my description of the phenomenon is somewhat more accurate.

If we knew beyond question that C_T were a constant, we could use the ancient and medieval data in estimating A_T and B_T, as MS do. If our goals include deciding whether or not C_T is a constant, on the other hand, using the ancient and medieval data in estimating A_T and B_T is contrary to correct kinematic principles. We can see this from a simple example. Suppose that a coordinate x obeys the differential equation

$$\ddot{x} = a(t).$$

The solution is

$$x = x_0 + v_0 t + \iint a(t)dt.$$

The double integral is the definite integral, with the lower limit being zero in both stages of integration. The function a(t) is the acceleration at any instant of time t. We can define an average acceleration A(t) between 0 and t by writing

$$x = x_0 + v_0 t + \tfrac{1}{2}A(t)t^2.$$

Clearly $A(t) = (2/t^2)\iint a(t)dt$. The numerical values of D'' that I have quoted in Equation III.5 are averages defined in this way. If $a(t)$ varies with time, so does $A(t)$, and vice versa.

If we have a priori knowledge of the function $a(t)$ or $A(t)$, we can find the empirical constants from observed values of x at different times by first evaluating $x - \tfrac{1}{2}A(t)t^2$ at each observation. We then find x_0 and v_0 by a suitable inference process, such as the method of least squares, from the values of $x - \tfrac{1}{2}A(t)t^2$. In this case, we can use values of x at any time t.

If we do not know $A(t)$ in advance, however, and if our goal is to find the time dependence of $A(t)$, which it is in studies of the astronomical accelerations, we are more restricted in our procedure. We assume that $a(t)$ can be approximated by a constant $a(0)$ in the immediate vicinity of $t = 0$, and we estimate x_0, v_0, and $a(0)$ by fitting to values of x for times near 0. We use only values of t as near 0 as possible; the range of t that we must use depends upon the availability of data and the "noise" level of the data.

After we evaluate x_0 and v_0, we study the time dependence of $a(t)$ or $A(t)$ by using data for times well removed from 0. For each observation of x, we first evaluate $x - x_0 - v_0t$. We then equate this quantity to $\tfrac{1}{2}A(t)t^2$, and this gives us $A(t)$ as a function of time. It is clear that we do not adjust x_0 and v_0 in this process; the parameters x_0 and v_0 are fixed only by observations near $t = 0$. We cannot infer x_0 and v_0 by using data well removed from $t = 0$ because we do not know the integral $\iint a(t)dt$ if we do not know the function $a(t)$ itself in advance. When we use this procedure, we will find constant values of $A(t)$ and $a(t)$, within allowable errors of observation, if $a(t)$ and $A(t)$ are in fact constant. That is, we may use this procedure, which is designed to cope with a variable $a(t)$, even if $a(t)$ is a constant.

To be sure, we could also start by assuming that $a(t)$ is a constant, a_0 say. We would then evaluate x_0, v_0, and a_0 by using observations from all ranges of t that are available. Finally, we would test the hypothesis that $a(t)$ is constant by seeing whether the form $x_0 + v_0t + \tfrac{1}{2}a_0t^2$ agrees with the data within allowable errors of observation. If it does, we conclude that $a(t)$ is probably constant. Otherwise, we must conclude that $a(t)$ is variable and we must start the analysis all over again, using the procedure outlined earlier. MS and M used the procedure just outlined, that of assuming a constant $a(t)$, and they carried it to the point of plotting the residuals (the values of $\delta\Delta T$). Having done so, they failed to observe that the residuals differ from zero in a systematic fashion that is outside the range of allowable errors of observation. Thus they failed to observe that the accelerations have varied with time by a large amount, even within historic times.

In Section I.3, I concluded that the constant and linear coefficients now used in the standard ephemeris theories need to be changed. Since Muller and Stephenson also say that these coefficients, or at least those in the rate of rotation of the earth, need to be changed, the reader might suppose that they and I are saying the same thing. Actually, I believe that we are saying almost opposite things. I have concluded that an error has been made, because of a complex set of circumstances, in the process of finding these coefficients from modern data. When this error has been corrected, my conclusion is that the coefficients should not again be altered because of results from ancient or medieval data.

In contrast, M[p. 3.13] writes: "This approach permits, and indeed, (*sic*) demands the simultaneous use of all ΔT data, ancient through modern. The modern data will provide coefficient A in {3.10}, the intermediate data (*i.e.* medieval) will give B, and the entire suite will be able to yield C, the desired acceleration of the Earth's rotation."† This seems to say that all data enter into the determination of all coefficients, but that the error levels and hence the weights restrict the practical effects of the data to those stated. If I have understood correctly what M and MS have written, their view is different from mine, and I believe that theirs is theoretically incorrect.

3.   "Population Bias" in Records of Solar Eclipses

In my earlier work, I have used all valid records of solar eclipses that I could find,‡ provided that two conditions are met: (a) it must be possible to date the eclipse unambiguously on the basis of the record itself, and (b) it must be possible to locate the place of observation, either as a specific spot such as a city or as a region of acceptably small size.

MS [p. 472] call this a "statistical approach" to the use of eclipses. An alternate approach is to use only those records from which we may infer, with high confidence, that the eclipse was central‡ at a known place. I shall discuss the basis for such an inference, and its validity, in Sections IV.1 and IV.3.

MS prefer the second approach and they write [MS, p. 472]: "We consider a statistical approach to large solar

---

† "This approach" means the approach of M and of MS. Equation {3.10} in this quotation is the same, except for notation, as Equation III.2 above for ΔT, and A, B, and C are the constant, linear, and quadratic coefficients in that equation. The parenthesis is M's.

‡ There are some minor exceptions. For reasons that will be explained in a moment, I have not used a few records of the eclipses of 1239 June 3 and 1241 October 6 that I have found.

‡ That is, that the sun was totally eclipsed as seen by the observer, or, for an annular eclipse, that the observer was inside the zone in which the moon is entirely surrounded by the solar disk.

eclipses in medieval times as of doubtful validity on account of the non-random distribution of medieval centres of population." It is clearly necessary to decide whether their objection is valid before we can decide upon the eclipse sample to be used in this and future studies. MS use the term "population bias" to describe the effect of the non-random distribution of population centers.

There is a phenomenon that may be called population bias that may arise under unusual circumstances. In order to understand the phenomenon, let us start by noting two possible approaches to follow in amassing a collection of eclipse records: (a) We may select a certain set of eclipse dates and use all the records of those eclipses that we can find. (b) We may select (or let the availability of records select for us) a certain set of places and use all the records from those places that we can find. If we pursue either approach as far as we reasonably can, we clearly end up at the same place, because we will end up with all the records that we can find. If we stop far short of this point, however, the first approach is more likely to produce a bias† than is the second, for equal numbers of records.

For example, let us consider the eclipse of 1239 June 3. Celoria [1877a] has collected all of the records (about 35) of this eclipse that he could find, and it happens that his collection was one of the secondary sources of eclipse records that I discovered while writing AAO. It also happens that the path of totality of this eclipse in Europe travelled almost west-to-east through northern Spain, along the Mediterranean coast of France, through northern Italy, and on through the Balkans. In western Europe, then, it is clear that there were many more possible places of observation to the north of the path than to the south, and observations of this particular eclipse are inherently biased by an accident of geography. If I had had a large number of eclipse records that were free of bias, the bias in this set would not have been important. As it happened, however, the number of records that Celoria collected for 1239 June 3 was about equal to the total number of records that I had collected for all earlier eclipses, and the use of Celoria's set could have seriously biased the situation.

As a result, I did not use Celoria's collection for 1239 June 3 in AAO. Also as a result, I explicitly adopted the second approach in MCRE. That is, I did not use any reliable records from a particular source unless I used all reliable ones from that source. This meant, among other things, that I ignored the records in Celoria's collection except those in sources that I studied directly in their primary versions.

Now let us consider the second approach. The center of an eclipse path is just as likely to pass to the north

---

†I shall not consider the question of whether it would be better to call this a statistical fluctuation rather than a bias.

as to the south of a specific point,† and it is just as likely to pass to the east as to the west. Thus, even if we took eclipse reports from only a single place, it is physically impossible for there to be a bias in records that are based upon the second approach. This is the reason that I adopted the second approach in MCRE.

In order to word the matter in another way, let us consider the coordinate η introduced in Section I.5. For eclipses that are visible at a given point, the algebraic average of η is zero, and this is the condition that we need in order to have an unbiased sample.

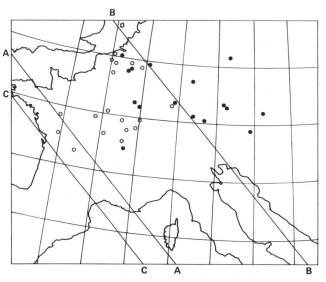

Figure III.1. Some circumstances of the eclipse of 1039 August 22. Line AA is the center line of the eclipse path, as calculated from the results of MS, while line CC is the center line as calculated from the results of MCRE. The solid circles are places where the eclipse was recorded, and line BB is a line through the centroid of the solid circles parallel to lines AA and CC. The open circles are places where known annals were being kept at the time, but where the eclipse was not recorded. Weather may explain the distribution of the records.

---

†This statement requires that we consider both penumbral and umbral eclipses, and that we consider the path in space, not just the path on the surface of the earth. In other words, it applies to all visible eclipses, but it does not apply to umbral eclipses alone.

When I said that the average value of η is zero, I should have said that the average tends to zero as the sample size tends to infinity. With a finite sample, of course, we usually find an average that is different from zero. The average for a finite sample is sometimes called the statistical fluctuation for that sample.

The second approach leads to smaller statistical fluctuations than the first approach, for samples of the same size, as we can see in a simple fashion. Suppose, for example, that the center line of an eclipse passed through northern Africa, as nearly happened for 1239 June 3, but that all known records of the eclipse come from Europe. Suppose further that we have found 25 records of this single eclipse. Then all the records in this sample of 25 have positive values of η, and the sample is severely biased. With the second approach, however, if we have found 25 records of different eclipses from a single place, it is highly unlikely that all of them will have the same sign of η; the probability is only 1 out of 2^{24}.

Thus, with the procedure that I followed in MCRE, it is physically impossible for there to be a bias introduced by the "non-random distribution of medieval centres of population." There would not be a bias even if all of the records I used had come from the same place.

In attempting to demonstrate the existence of population bias in MCRE, MS [p. 473] consider explicitly the eclipse of 1039 August 22. They did not notice that this example in fact proves the absence rather than the presence of population bias in MCRE.

The situation is shown in Figure III.1. The solid line marked AA is the center of the eclipse path, as calculated by MS from their inferred values of the parameters. The solid circles are the locations of places from which MS found records of this eclipse. Line BB is the center of the path that we would infer from these records alone; we should note that none of the records indicates a large eclipse, by the way. As MS note, the records are all biased with respect to the calculated path. Thus we would get a seriously wrong result if we inferred the parameters from these records without using records from other eclipses.

In MCRE, I used only five records of this eclipse, and these five are among the points that MS plot. The center of the path as calculated from the parameters found in MCRE is shown by the line CC, which is about 3° farther west than the path calculated by MS. Thus the maldistribution of the records that I used is even worse than MS indicate.

The figure proves that the "statistical approach" did in fact work, contrary to the claim of MS. The poor distribution of the records for the eclipse of 1039 August 22 is balanced by the distribution of records for other eclipses, and the overall distribution is clearly well balanced. If it were not, my solution for the eclipse path would lie close to the line BB. Instead, it lies far to one side of all the places where this particular eclipse was recorded.

There is still another aspect of the situation. The open circles in Figure III.1 are places where we know that annals were being kept in 1039; we know this from the study of those annals in MCRE. We see that many of these places lie between lines BB and CC. Although the compilers of the corresponding annals were interested in eclipses,† and although the eclipse must have been large and perhaps even central‡ at several of the places, the eclipse was not recorded at any of them although it was recorded in places where the eclipse was fairly small. The only likely explanation I can think of is the weather. The distribution of the records suggests that 1039 August 22 was cloudy over southwestern France but clear to the northeast.

MS [p. 520] write that their analysis shows a distinct bias in the data from MCRE but not in the data from AAO, and they cite this as further evidence of population bias in MCRE. I cannot verify the existence of a bias in the data from MCRE. Certainly the plot of individual values of D'' from AAO and MCRE [MCRE, Figure XVIII.2, p. 641] shows no appreciable difference between the two works. I can think of only one reason for a belief in a bias in the data from MCRE.

If we use the records in MCRE to infer values of $\dot{n}_M$ and y, and if we ignore the fact that at least one of them changed by a large amount during the period covered by the data, we find values of both $\dot{n}_M$ and y that are considerably larger in magnitude than other recent estimates [MCRE, p. 633]. Since MS assume that the accelerations are constant, their analysis of the data presumably shows the same result. However, this result is a consequence of a peculiar situation that exists with eclipse data, and it is not physically significant.

The peculiar situation is that eclipses determine D'' strongly, but they determine the individual accelerations $\dot{n}_M$ and y only weakly. Now let us suppose for the moment that we are dealing with a single parameter which is time-dependent. If we ignore this fact and infer a value as if the parameter were constant, the value that we find is the mean of the time-dependent parameter for the relevant time period. However, if we are dealing with two parameters, the values found by treating them as constant are not necessarily close to the means, particularly if the normal equations are poorly conditioned. In fact, the values found this way may lie outside the physically possible range of values.‡ It is likely that this phenomenon is what has happened with the values of $\dot{n}_M$ and y from MCRE, and the values therefore need indicate

---

†We know this because I used only annals that contain eclipse records in MCRE.

‡The eclipse was total at some points in the central zone and annular at others. I have not tried to determine the points where it was total and those where it was annular.

‡I thank my colleague R.J. McConahy for pointing this out to me.

nothing beyond a statistical accident. A bias in the inferred parameters in this situation does not indicate a bias in
the data sample used.

The reader may convince himself of the existence of
this phenomenon by a simple experiment. On page 272 of AAO
I found

$$\dot{n}_M = -41.6, \qquad y = -27.7, \qquad \text{(III.9)}$$

by using all data before 500. By using all data after 500
I found

$$\dot{n}_M = -42.3, \qquad y = -22.5. \qquad \text{(III.10)}$$

However, if the reader uses all of the data in a single solution for the parameters, ignoring the possibility that the
parameters might change with time, he will find [Stoyko,
1970]:

$$\dot{n}_M = -34.5, \qquad y = -21.6. \qquad \text{(III.11)}$$

In the composite solution, neither $\dot{n}_M$ nor y lies within the
range of the individual solutions.† I am sorry that I was
not aware of this phenomenon when I wrote MCRE and AAO.

We should note briefly the eclipse of 1241 October 6.
Celoria [1877b] also collected all the records of this eclipse
that he could find. The center line of this eclipse passed
across Europe, approximately from Hamburg to Istanbul. The
density of population at a considerable distance to the east
of the path was probably much less than at points near the
path or in Europe to the west of it. Thus, if I had used
all of the eclipse reports that Celoria collected, there
would have been a population bias for this eclipse also. For
this reason, in AAO, I used only the records which said that
stars were seen or that the eclipse was total. Within the
narrow band in which the magnitude is large, there is no
appreciable bias.‡

To sum up the matter of a "population bias", such a
bias is physically impossible provided one simple rule is
followed: If we use a record from a particular source, we

---

†The residuals calculated using the composite solution in
Equations III.11 are significantly larger than those calculated using the individual solutions in Equations III.9
and III.10. This suggests that the composite solution is
not valid. Stoyko did not note this point.

‡Muller and Stephenson did not notice this point. They
"demonstrate" [MS, p. 474] that there is a bias in the reports of this eclipse that I used, but their demonstration
assumes that I used all the records that Celoria published.
Their demonstration does not apply to the set of records
that I used. However, some of my arguments in AAO were
poorly worded and did not express well what I did.

must use all independent records from that source. A non-random distribution of the places cannot introduce a bias provided that we follow this rule. With any actual sample, which is necessarily of finite size, there is a statistical fluctuation, but we cannot avoid this by any sampling procedure.

4. A "Record" of the Eclipse of -1374 May 3

MS [p. 481] and M [pp. 8.11-8.12] assume that a text from the ancient city of Ugarit contains a record of the eclipse of -1374 May 3. This identification comes from earlier studies by Sawyer and Stephenson [1970] and by Stephenson [1970]. The translation of the text is given as:

> Side A: The day of the new Moon in the month of Hiyar was put to shame. The Sun went down (in the daytime) with Rashap in attendance.

> Side B: (This means that) the overlord will be attacked by his vassals.

Rashap, according to the writers, means the planet Mars.

Stephenson [1970] begins the identification by writing: "The absence of any parallel to this text in the Ugaritic literature suggests that the phenomenon was rare and spectacular, and the description can hardly refer to anything other than a total eclipse of the Sun." Fotheringham [1920] makes a similar statement with regard to the eponym canon eclipse (see Section II.5): "As the eclipse is the only eclipse mentioned in this Chronicle, which covers an interval of 155 years, there can be no reasonable doubt that it had been reported as a total eclipse." Unfortunately, there are so many counter-examples that we cannot accept these statements at all. Thus Stephenson and Sawyer and Stephenson base their entire studies of this record upon an invalid assumption. †

After starting with an invalid assumption, the writers base the identification of the eclipse upon the fallacy that I have called the "identification game" [AAO, Section III.2]. Before we take up this matter, however, it is desirable to discuss the significance of the calendar month Hiyar.

The name Hiyar sounds like the name of the second month

---

†See the discussion of the eponym canon eclipse in Section II.5, as well as page 60 of AAO. Further, a study of the magnitudes of eclipses recorded in medieval European annals [Newton, 1974] shows that the number of recorded eclipses is almost independent of the magnitude for magnitudes ranging from totality down to 0.85.

of the year in the Babylonian calendar.† It is plausible
that Hiyar was the second month in the Ugaritic calendar and
that the Ugaritic calendar was constructed on the same general lines as the Babylonian one. If so, the months were
lunar ones, and we must establish the limits within which
the second month could fall.

Inspection of the tables of Parker and Dubberstein
[1956] for years around -600 shows that the first day of the
second month could come anywhere from April 5 to May 29. In
the period of the Ugaritic record, the dates would be somewhat later in the Julian calendar, say April 10 to June 3.
These are not the limits we want, however, for two reasons.

First, if the Ugaritic calendar does indeed resemble
the Babylonian one, a calendar day begins at sunset and a
month begins with the first sunset at which the crescent new
moon is visible. This necessarily happens after the conjunction of the moon with the sun, and the average interval between a conjunction and the first day of a month is probably
about two days. Thus a solar eclipse in the month of Hiyar
must come on the 28th or 29th day of that month under most
conditions; an eclipse cannot come on the first of the month.
Thus we should add, say, 28 days to the limits just found.
This means that the eclipse must have come between May 8 and
July 1, if our understanding of the calendar is correct.

Second, most years in the Babylonian calendar have 12
months, but slightly more than 1 year in 3 has 13 months.
At the period of the Ugaritic document, the choice of which
month would be the first month of the year must have been
made annually on the basis of some kind of astronomical observation, and it was nearly a millenium later before the
Babylonians adopted a fixed numerical scheme for choosing
the first month. In the Jewish calendar, which is patterned
closely after the Babylonian calendar, the observational rule
is that the 14th day of the first month must come after the
vernal equinox. We do not know whether the Babylonians
followed this rule or not, but, whatever rule they did follow, the results were almost identical.

Since the choice was based upon astronomical observation,
and since the observation could not be highly accurate at
this stage of history, we must assume that the rule governing
the choice was not rigorously applied. Further, the choice
was made by officials who had strong political interests, and
we can see many reasons why hastening or delaying the start
of a new year would confer political advantage. Hence we
must allow some latitude in the choice of the first month and
a corresponding latitude in the range of any date in the calendar. If we allow no more than a month's latitude in either
direction, we must still allow a range from about April 8 to
about July 31 for an eclipse that came in Hiyar, at least in
a preliminary search for the identification of the eclipse.

---

†The information that the reader will need to understand
this discussion of calendars is summarized in Section II.2
of APO. See also Section II.5 above.

In "identifying" the eclipse, Stephenson [1970] now proceeds as follows: From archaeological evidence, he says, we know that the eclipse came between -1450 and -1200, and these dates seem reasonable. For this period in history, he further writes, we can calculate the circumstances of a solar eclipse with errors that "do not exceed 1 per cent in the magnitude and 5 minutes in the time of occurrence." He now allows for the possibility that the eclipse may have been total at some point near Ugarit rather than at Ugarit itself, and he finds that only four eclipses which were total somewhere had magnitudes greater than 0.88 at Ugarit; the eclipse must have been one of these. The four dates are -1405 July 14, -1374 May 3, -1339 January 8, and -1222 March 5.

Stephenson [1970] then writes: "It is clear that only the eclipse of 1375 BC May 3 could have occurred in the month of Hiyar." MS [p. 481], however, soften this to read: "If we could be certain that the Ugaritic calendar at this very early period corresponded with the Babylonian and Hebrew, we would have no hesitation in assigning the date of the eclipse as 3 May 1375 B.C. However, until it can be shown that this is definitely the case, the suggested date is somewhat questionable." In contrast, M [p. 8.12] asserts without qualification that this date is correct; we should note that M was written later than MS.

For the sake of argument, let us assume that the eclipse was total in Ugarit or somewhere nearby, and let us further assume that the Ugaritic calendar corresponded exactly with the Babylonian one. Even if we make these assumptions, which are fundamental to Stephenson's conclusion, we see that he has made two errors either of which is serious enough to invalidate his conclusion completely.

The first error concerns the date. We have just seen that the limits on the date are April 8 to July 31. Therefore the eclipses of both -1405 July 14 and -1374 May 3 are possibilities. I agree that we can eliminate -1339 January 8 and -1222 March 5 if we assume that the scribe did not make an accidental error when he wrote Hiyar.† In fact, for the specific years -1405 and -1374, we can now be more definite.

In -1405, if the rule relating the calendar to the vernal equinox was being followed, the first month probably began with the crescent moon that was seen near the middle of April. This means that -1405 July 14 came near the end of the third month, not the second month. In -1374, the first month probably began in early April, which means that -1374 May 3 came in the first month, not in the second one. For either year, if the scribe was correct, we must

---

†This assumption is not particularly strong. In the monastic records from medieval Europe, I have shown [Newton, 1974] that the stated year is wrong about 1 time in 4. There is no reason to assume that this incidence of error is peculiar to Europe, and there is no reason to assume that errors are confined to the statements of the year. We must assume a similarly high probability of error in any detail that is written about an eclipse.

assume that the officials made an error of a month when they decided upon the beginning of the year. That is, no date found by Stephenson fits unless we assume either that the calendar was in error or that the scribe made an error in writing the month.

The second error concerns the accuracy with which we can calculate the circumstances of an eclipse around the year -1400. I have shown [Newton, 1974] that the error in calculating the magnitude is about 0.20 rather than 0.01, and even this estimate is somewhat optimistic.† Even if we grant that the observer saw the eclipse as total,‡ we must still admit any eclipse that was total somewhere and that attained a calculated magnitude of, say, 0.75 at Ugarit. Thus there are many possibilities beyond the ones that Stephenson admitted. As a reasonable estimate, there are three times as many, and we cannot assume that the possible dates are limited to -1405 July 14 and -1374 May 3.

The reference to Mars is interesting. The natural interpretation of the statement is that the observer saw Mars during the eclipse. However, as Stephenson points out, Mars was below the horizon during the eclipse of -1374 May 3 and could not have been visible. He suggests that the observer saw Aldebaran or Capella, which were above the horizon, and that the observer mistook one of them for Mars. Further, although Stephenson does not note the fact, Mars was also below the horizon during the eclipse of -1405 July 14.

We now have an interesting situation. There are only two conditions that allow us to identify the eclipse. One is the month and the other is the sighting of Mars. The dates of -1405 July 14 and -1374 May 3 do not satisfy either condition. Therefore neither date is correct unless there is an error in both specific details that are recorded. However, Mars was above the horizon on both -1339 January 8 and -1222 March 5. In order to accept either of these dates, then, we need to assume only a single error (in the month) and not two independent ones. In other words, either of these dates is more likely than the date that Stephenson adopted, while the date of -1405 July 14 is equally likely.

However, this does not end the matter. For the eclipse of -1374 May 3, Mars was only a short distance east of its conjunction with the sun; it was below the horizon during the eclipse only because the eclipse happened early in the morning. When Mars is in conjunction, it is at its greatest distance from the earth, and M [p. 8.11] believes that it would be too faint to be seen during the eclipse even if it had been above the horizon. He suggests that Mars was in the period of invisibility that it always has when it is near the sun. He further suggests that the recorder knew

---

†In Section XIV.8 I shall show that the uncertainty is much worse than 0.20 for this particular eclipse. We cannot even tell whether the eclipse was total or invisible at Ugarit.

‡We may not reasonably assume even this, as I shall show in a moment.

that Mars was close to the sun, and that this was what he meant by saying that Mars was "in attendance".

If this is correct, it destroys the last prop that supports Stephenson's identification of the eclipse. This means that the reference to Mars was "theoretical" knowledge. Hence it does not imply that Mars or any other "star" was visible during the eclipse; it implies only that Mars was near conjunction and that the scribe knew this. Hence there is no reason to assume that the eclipse was outstandingly large. If our experience with European eclipses is a guide, we must accept any eclipse, total or annular, that could have attained a magnitude of 0.85 at the place of observation. This means that we must accept any eclipse whose calculated magnitude at Ugarit was about 0.6 or greater.† The number of possibilities in $2\frac{1}{2}$ centuries is thus so large that there is no hope of identifying the eclipse from the information given in the record.

5. A "Record" of the Eclipse of -1329 June 14

Under the Shang dynasty of China, information about magical procedures, prognostications, and their consequences was often inscribed on the plastrons of turtles or the scapulae of cattle. Bones inscribed in this way are often called "oracle bones". Some oracle bones refer to astronomical matters.

MS [p. 489] identify one of the oracle bones as a record of the eclipse of -1329 June 14. As we shall see, this "identification", like the one in the preceding section, is reached by playing the identification game. Further, in the case of the oracle bone, the inscription does not seem to refer to a solar eclipse at all.‡

MS give the following translation of the inscription: "On the day ping-shen the oracle was consulted; on the day ting-yu, a sacrifice was made; it was a clear day. On the following day there was a great eclipse of the Sun (ta shih jih). It was a clear day." This translation overlooks two important parts of the text, which have a profound significance for its overall meaning. The text is found on page 83

---

†This is a simplification of what we must actually do. The uncertainty in the magnitude is not the same for all eclipses, even for a given uncertainty in D''. What we must do is to vary D'' within its possible range and see what eclipses attain a magnitude of 0.85 at some place near Ugarit, for some value of D'' within its possible range.

‡Since the language did not contain a technical term for an astronomical eclipse, we cannot be rigorously sure that the event in the Ugaritic text of the preceding section was an eclipse rather than some other kind of darkness. However, interpretation of the event as an eclipse is plausible and even probable. This is not true of the event on the oracle bone, as we shall soon see.

of Jao [1959].†

By the method that I shall describe in a moment, MS identify this inscription as a record of the solar eclipse of -1329 June 14. They then go on to reach the following conclusions [MS, p. 523]: " . . . the Chinese cyclical day calendar was correctly synchronized in -1329, and has pre-sumably been carried forward to the present day without error. The tortoise shell on which the inscription appears, and all the other records preserved there, are thereby dated; the eclipse record to the day. This provides the earliest established date in the Chinese Calendar."

We must begin by explaining what is meant by the Chinese cyclical day calendar. The Chinese, in addition to their mixed solar-lunar calendar, used a 60-day cycle in giving dates. The 60-day cycle is thus analogous to our week. "Ping-shen" is the 33rd day and "ting-yu" is the 34th day of a cycle. Thus the event called ta shih jih, which MS trans-late as a great eclipse of the sun, came on the 35th day of the same cycle, according to the translation of MS.

In times that are much later than the Shang dynasty, we find dates given by means of a 60-day cycle that can be identified with certainty. From these dates, we can formu-late the rule: Add 50 to the Julian day number and divide the sum by 60. The remainder is the number of the day in the 60-day cycle, with one exception: If the remainder is 0, we call the cycle number 60 rather than 0. If the count in the 60-day cycle has not been interrupted since Shang times,‡ the Julian day number of the ta shih jih is 45 plus some multiple of 60.

MS [p. 489] find "that only five solar eclipses were total near Anyang‡ during the Shang dynasty and one of these was the eclipse of BC 1330 June 14 which occurred on the 35th cyclical day." That is, it occurred on the 35th day according to the rule just given. This date, then, is their identification of the eclipse, upon which they base their other conclusions.

---

†Jao in turn is quoting from Fang Fa-Chien, K'u-Fang-Erh-Shih's Collection of Divination Inscriptions on Oracle Bones, published by the author in 1935. This work is usually iden-tified as K'u-Fang in the relevant literature and the in-scription is usually identified as K'u-Fang 209.

‡This is presumably what MS mean by saying that the cyclical day calendar was correctly synchronized in -1329. In spite of what MS conclude, we do not know the rule of synchroniza-tion in Shang times.

‡The Chinese capital at the time, and the place where the inscription was presumably made. China, considered as a political entity, was quite small at the time in comparison with its modern extent. Thus the eclipse, if it were ob-served anywhere in the kingdom of the time, was observed close to Anyang.

As Keightley [in press]† says, these conclusions are wrong for at least five reasons, which I shall now summarize. I refer the reader to Keightley's paper for more details.

a. The inscription K'u-Fang 209 occurs on a shell fragment that contains no other inscriptions. There are no "other records preserved there" to be dated by means of the eclipse.

b. The identification and the conclusion that the cyclical calendar was "correctly synchronized in -1329" are based upon reasoning in a circle. MS do not give the dates of the other four eclipses that were visible in Anyang, but let us suppose for the sake of illustration that one of them came on Julian day number 1 233 053.‡ Let us postulate a rule in which we add 42 rather than 50 to the Julian day number before we divide by 60. This day then has number 35 in the 60-day cycle by the new rule, and thus it agrees with the inscription as translated by MS. By the reasoning of MS, then, this rule of synchronization has been established by the eclipse record, and the eclipse now has this new date. We can "confirm" as many dates as there were visible eclipses by postulating the corresponding rule for synchronizing the 60-day cycle, and we can "prove" that each visible eclipse was the one recorded.

c. MS are wrong in saying that the event called ta shih jih occurred on cycle day 35. As the translation about to be given will show, the event was on day 34. Thus the date -1329 June 14 does not correspond to the record if the rule of synchronization has been unchanged since Shang times.

d. Perhaps the most serious point is that the inscription makes no reference to a solar eclipse or to any other unusual behavior of the sun. Before offering the translation which shows this point, it is useful to discuss briefly the form and purpose of these "oracle bone" divination inscriptions. Their purpose is to record the actions of the king as a magician or seer, his predictions as to what would occur, and what actually happened. We should note that the total body of known inscriptions records only successes on the part of the king; a scribe with the temerity to write that the king had failed probably had a short life expectancy. The two main parts of the inscription are called the charge (or divination) and the verification. The charge records the sacrifices or other actions that the king intends to take, along with his predictions. The verification records what happened as a consequence and, as the name implies, it always verifies what the king predicted. This particular inscription also has two shorter parts, which Keightley calls

---

†Most of the discussion in this section is taken from Keightley's paper. I have elaborated on a few details in an attempt to help the reader.

‡I do not have a table of solar eclipses for the historical period in question, and I have no idea whether there was an eclipse on this date or not. I have chosen this date arbitrarily.

the preface and the postface. The translation of the inscription, including the identifications of its parts, is [Keightley, in press]:

Preface:	Crack-making on ping-shen.
Charge:	On the next day, ting-yu, we will offer wine libation and a beheading sacrifice; it will be clear weather.
Verification:	On the ting-day† in the early morning, it was overcast‡; at the time of the great meal, the day became clear.
Postface:	In the first month.

The translation of MS has a number of errors besides the wrong identification of the day when the verification occurred. They turn the prediction of clear weather into a statement of what had happened on day 34. They translate the phrase ta shih jih, which according to Keightley means "at the time of the great meal", as a great eclipse of the sun. Finally, they omit the postface entirely.

It is correct that ta shih, which means "great eating", was used to denote an eclipse, as well as to denote the great meal of the day or the time when it was eaten. In this case, the role of the phrase in the standard form of these inscriptions tells us what is meant. It would be without precedent for the verification to record a solar eclipse, which was an undesirable event, as the outcome of the king's prediction of clear weather, which was a desirable event; we must remember that these inscriptions always show that the king was right. Actually the verification tells us that the day did become clear, and the only reasonable role for ta shih to play in the verification is to tell us when this happened, namely at the time of the great meal.

e. The postface tells us that these events took place in the first month, a matter that MS omit from their translation. The first month moved around slightly with respect to the seasons, but it always came in the winter. It is not possible for the date -1329 June 14 to come in the first Chinese month, and therefore MS's identification would be wrong even if the inscription did record an eclipse.

---

†Ting-day refers back to the day ting-yu, which means cycle number 34 in the 60-day cycle. Therefore the events in the verification took place on day 34. MS translate this as "the following day" and hence they put these events on day 35.

‡The exact meaning of the word translated as "overcast" is not known, but "overcast" conveys the general sense. Other possibilities are foggy and misty.

These are Keightley's five reasons for deciding that the conclusions of MS are wrong. We can add a sixth to these:

f. Let us assume for the sake of argument that MS's translation of the inscription is correct, and let us further agree to ignore the matter of the month. We still cannot accept the conclusions of MS because they depend upon reasoning in a circle. MS say that only five solar eclipses were total near Anyang during the Shang dynasty. Nothing about the inscription implies a total eclipse even by their translation, but let us also grant them this point. What MS have done is to assume certain values of the astronomical accelerations and to calculate the visible eclipses using these values. This automatically excludes any other values of the accelerations.† They then identify the eclipse as one of the calculated ones and turn around to calculate the accelerations on the basis of this identification. The resulting accelerations necessarily agree with the values assumed in the first place, no matter what these values are.

As I pointed out in the preceding section, the uncertainty in calculating the magnitude of an eclipse this long ago is at least 0.2. Since the eclipse was a great one according to MS, let us assume that the observed magnitude was at least 0.9. Then any eclipse with a calculated magnitude of 0.7, whether total or annular, is a possible candidate. I estimate that the number of possible eclipses is around 60 or 70, rather than being 5 as MS assert. It is clear that MS's identification is unreliable, even if we were to grant all the other steps in their arguments.

I thank my colleague Dr. Chih-Kung Jen of this Laboratory for his help in understanding the literature about the inscription K'u Fang 209.

6. The Places Where Eclipses Were Observed in China

Beginning with the accession of the Former Han dynasty, which ruled from -201 to +9, and continuing with few interruptions until modern times, we have extensive annals of Chinese affairs. The annals are grouped by dynasties, and it was traditional for the rulers of one dynasty to order the compilation in a single source of the annals belonging to the preceding dynasty. Two main types of annals will concern us here. One type is called the imperial or basic annals. This type contains records of political affairs and other matters of imperial concern, and it also contains records of many astronomical matters such as novae, comets, and solar eclipses; the solar eclipses are what interest us here. The other type is called an astronomical treatise, and it usually appears in the same principal compilation as the imperial annals. It typically contains a summary of the astronomical knowledge of the time, often including a treatise on

_____
† More accurately, it excludes equations of condition other than those with certain specific coefficients, but I shall use the simpler language of referring to the accelerations.

the calendar, along with an annalistic collection of astronomical observations.

Few Chinese records of solar eclipses say where the eclipse observation was made. Most students of the subject have concluded that the observations could have been made anywhere in China when the record does not state a specific place. Luckily for our purposes, China in this context does not have the extent of modern China. We may take it to be approximately the region between the Yangtze River on the south and the line of the Wei and Yellow Rivers on the north, stretching from about Charng-an (the modern Sian) on the west to a point somewhat short of Nanking on the east. Cohen and Newton [in preparation] have used a rectangle extending from 31°N to 34°.5N and from 108°.95E (the longitude of Charng-an) to 117°E. To be sure, China today rules and during many times in its past history has ruled areas far beyond this region. Nonetheless this region can be considered as the Chinese "heartland", at least for the moment. I shall return to the question of using this region at a later point.

With regard to the problem of where the Chinese observations of eclipses were made, M [p. 5.28] says that it "was not solved for the Chinese material until Stephenson (1972)† undertook a detailed study of it." M and MS cite a number of points that are involved in the solution of the problem; these points presumably originated in the unpublished work of Stephenson.

a. The term jih‡ can be translated as "complete". In reference to an eclipse, this is taken to be the same as "central". That is, a jih eclipse may be annular or total, but the observer is within the central zone. MS [p. 489] say that the term has this meaning throughout Chinese history.

b. MS [p. 478] write: ". . . there is no evidence of systematic observation in the provinces. Dubs‡ conveys a false impression of the proficiency of provincial observers . . ."

c. Still on page 478, MS refer to the eclipses of -187 July 17, -27 June 19, and 120 January 18, which are reported as "almost complete" in the astronomical treatises, and they say: "In each case the wide zone of totality crossed the entire country over heavily populated areas, but if reports of totality reached the astronomer royal, they have not found

---

†This refers to the unpublished doctoral dissertation of Stephenson that I have not had an opportunity to see.

‡Many writers render this term as chi in an English transcription. This jih is not the same term as jih in the preceding section. In rendering Chinese names and terms, I shall follow Cohen and Newton for terms that appear in their work. Otherwise, I shall follow the immediate source being cited.

‡At this point, MS cite page 338 in volume 1 of the work that I cite as Dubs [1955].

their way into the records from which the astronomical† sections of the histories were compiled."‡ Note the importance of the distinction between the astronomical treatises and the imperial annals. The eclipses of -187 July 17 and -27 June 19 are reported as total in the imperial annals [Dubs, 1938]; I have not seen any statement about the eclipse of 120 January 18 in the annals as opposed to the astronomical treatise.

d. Portents of many kinds, including solar eclipses, were regarded as warnings from Heaven‡ to the ruler that he was not performing his duties satisfactorily in some regard, and there were heavy penalties for fabricating portents. Bielenstein [1956] accepts this as the general idea but he points out that the portents in practice were reported by members of the bureaucracy. He gives evidence for concluding that the portents actually functioned as a means whereby the bureaucrats expressed their criticism of the ruler. They did not often fabricate portents, but they often abstained from reporting portents under a ruler with whom they were pleased. That is, according to Bielenstein, few of the observed portents, including eclipses, were actually recorded under the reign of a ruler who was popular with the bureaucracy, while almost all of them would be reported under a ruler who was unpopular with them.* Bielenstein's study deals only with the Former Han dynasty.

e. M [p. 8.7] writes: "It also seems unlikely that the Emperor would admit observations from anywhere in China as affecting his* administration . . . It is inconceivable that the Emperor would have sat on the 'edge of his throne' for every solar eclipse and meteor seen everywhere in China."

f. From the preceding considerations, M and MS conclude that an observation appearing in an astronomical treatise was made at the capital unless there is a specific statement that the observation was made somewhere else. However, they then

---

†The emphasis is that of MS.

‡My calculations made with provisional accelerations confirm this statement for the eclipses of -27 June 19 and +120 January 18, but they disagree for the eclipse of -187 July 17. According to my calculations, the path of this eclipse went from the northwest to the southeast and missed completely what I shall define as heartland China. It went only through mountainous regions which probably had a small population, and it is not likely that areas in the path of the eclipse would routinely send eclipse reports to the Han capital.

‡Heaven, spelled with a capital letter, is often used in English to denote the principal deity of the Chinese state religion.

*M and MS do not refer to the work of Bielenstein; I have added this reference in order to clarify the meaning of point d. I join many other scholars in not accepting Bielenstein's conclusions. My reasons are given in Appendix VI.

*The emphasis is M's.

attach a special meaning to "capital". They cite the eclipses
of 756 October 28 and 761 August 5, which both occurred during
a prolonged period of revolution when the emperor was not able
to be in his nominal capital. Both eclipses were reported as
jih, yet, as M and MS find from computation, neither was complete at the nominal capital. However, they find that both
were complete at the place where the emperor was temporarily
residing. From this they conclude that the records in the
astronomical treatises were made by the imperial astronomers
who, as M [p. 8.45] picturesquely puts it, "are always with
the Emperor, even when they must pack bag and baggage, and go
into hiding at the remotest edge of the Empire!"† Hence, M
and MS use the term capital to mean the place where the emperor was at any particular moment, and I shall follow their
usage in this immediate discussion. When I need to refer to
the city which was the usual capital, I shall refer to the
nominal capital.

g. M and MS make two further points in emphasizing the
distinction between the imperial annals and the astronomical
treatises. The first is [MS, p. 479]: ". . . the imperial
annals, which always commence a dynastic history, contain
eclipse and occasional cometary records but it is clear that
these are in the main abbreviated from the records of the
court astronomers with additions from elsewhere. Material
from this source is of lower reliability."

h. After they conclude that eclipses in the astronomical treatises almost surely were observed at the capital
unless there is a specific statement to the contrary, MS
[p. 479] write: "However, during the Former Han (202 B.C.-
A.D. 9), Later Han (A.D. 23-220) and T'ang (618-906) dynasties we can be more specific than this. During these
periods, a measurement of the right ascension of the Sun,
usually expressed to the nearest degree, normally accompan-
ies the observation. When this is the case it is a positive
proof that the report comes direct from the records of the
imperial astronomers."

This collection of points and the accompanying evidence
cited in support of them seem to form a structure of great
strength. Unfortunately, I must point out that this structure is built upon sand. Let us start with point b, about
which two comments are needed.

First, it does not take proficiency to see a solar
eclipse. In fact, with the probable exception of a nova,‡
it does not take proficiency to see any of the recorded
portents. I saw a solar eclipse the first time I tried,
even though I was a boy at the time.‡

---

†The emphasis and the exclamation point in this quotation
are M's.

‡It does not take proficiency to see a nova, but in most
cases it takes detailed knowledge to know that it is a
new star and not an old one.

‡However, seeing a partial eclipse requires some method of
darkening the sun's disk.

Second, there is evidence of systematic observation in the provinces, by M's and MS's own arguments. They cite 11 Chinese eclipses, not counting the special cases of -1329 June 14 and 1221 May 23.[†] They are forced to assume that two eclipses, those of -600 September 20 and +65 December 16, were observed well outside the capital, but the records give no indication of this. There is no reason to assume that special arrangements were made to observe these two eclipses, so we must assume the routine existence of observations outside the capital. Further, many other eclipses that appear in the astronomical treatises were observed outside the capital with no mention being made of this fact. Dubs [1938] notes a number of these under the Former Han dynasty while Cohen and Newton [in preparation] have noted others under the Tarng[‡] dynasty; examples will be given later in this section. Finally, there are often conflicts between eclipse circumstances reported in the imperial annals and those in the astronomical treatises. Hence observations must have been made in many different places.

Point e is an assertion with no supporting evidence. It reads like a statement about how its writer would act if he became the emperor of China without changing his background or personality. The contrary conclusion is more plausible. If the emperor believed that a portent was sent to warn him about his relations with his people and territories, it is likely that he would sit "on the edge of his throne" because of a portent seen by any of his people anywhere in his territories. However, there would be some limits to his routine concern. At a time when the Chinese emperor ruled as far as the Caspian, he might not be routinely concerned with eclipses seen only on the shores of the Caspian. By the time the news reached him from there, it would probably be too late to take the appropriate action. Besides, the Chinese may not have regarded these territories as truly Chinese. This is among the reasons that led us [Cohen and Newton, in preparation] to use only "heartland" China in studying Chinese records of eclipses.

During the long course of Chinese history, there was undoubtedly much change in beliefs, practices, and court procedures. Hence we cannot take conclusions drawn from one dynasty and apply them to other dynasties. Further, we cannot use a mixture of evidence, some from one dynasty and some from another, in order to reach a conclusion. We must consider each dynasty, or at least each major historical period, as a separate unit and draw independent conclusions for it. The Tarng dynasty is the dynasty for which we have the firmest evidence, and I shall now take it up. All the information to be used about the Tarng records comes from Cohen and Newton.

[†] This eclipse was observed in Mongolia by a Chinese expedition that was on its way to Samarkand. See Section V.5 below.

[‡] Cohen and Newton use this spelling while MS use T'ang.

In point h, MS say that the presence of a "measured" value of the right ascension is a "positive proof" that the report comes from the records of the court astronomer and hence that the observation was made at the capital. They apply this to both Han dynasties and to the Tarng dynasty. The statement is certainly wrong for the Tarng dynasty. The Tarng astronomical treatises give the right ascension for slightly more than 100 eclipses. Of these eclipses, 27 were not visible in heartland China and most of these 27 were not visible anywhere in Asia, including the major adjacent islands. With only 15 exceptions, nothing distinguishes the records of observable eclipses from those that could not be observed. It is therefore incorrect to refer to the right ascensions as measured values. They were certainly calculated. The presence of a value of right ascension thus proves nothing about how an eclipse was observed; it does not even tell us that the eclipse was observed at all. It merely tells us that someone calculated the right ascension for a certain group of dates and inserted the values in the records; this could have happened at any time between the original date and the compilation of the records in the final form that we have today.

We must conclude that the unobservable eclipses represent predictions, not observations. We must also conclude that many of the other records also represent predictions, but we can separate predictions from actual observations in only a few cases.† Thus as I just said, the presence of a right ascension does not even prove that an eclipse was observed at all, and it certainly cannot prove "positively" that the eclipse was observed at the capital.

The finding that at least 1 eclipse in 4 in the Tarng astronomical treatises was predicted reacts heavily upon point d. If 1 eclipse in 4 is fabricated,‡ fabrication can hardly have been regarded as a serious matter. On the other hand, if an eclipse could affect imperial policy, it is plausible that an eclipse was a serious matter. Hence we seem forced to the conclusion that a solar eclipse was not necessarily a portent or warning under the Tarng rulers. Something that is predicted cannot be a portent.

---

†In nine and only nine cases, there are special remarks which imply that an observation was made. For example, on 808 August 8, the emperor said to the prime minister than an eclipse had been predicted for the day before and that it all happened as predicted. Hence we may conclude that the eclipse was observed, in all probability by the emperor himself, who was in his nominal capital at the time. The recorder made an error of a month in giving the date; the eclipse was on 808 July 7.

‡I do not mean for this term to imply any form of improper behavior. The eclipses were fabricated only in the sense that the dates were calculated. Even though the predicted dates were left in the records, we have no reason to assume that deceit was intended; there is no claim that the eclipses were observed.

Point g says that the astronomical treatises are the important and reliable sources as far as eclipses are concerned, and that the imperial annals are not as valuable. The Tarng records show that this is not so, as Cohen and Newton have demonstrated. Their conclusion is shown by several lines of evidence, of which I shall take the space to give only one here. Eight records say that an eclipse was predicted for a given date but that it was not observed.† These records occur only in the imperial annals, not in the treatises. The evidence shows that the annals often give other astronomical information that is not found in the treatises. In fact, for two eclipses, those of 904 November 10 and 906 April 26, the right ascension appears in the imperial annals but not in the corresponding astronomical treatise. See notes 121 and 122 to Table 3 of Cohen and Newton.

The correct situation seems to be that the compilers of both the astronomical treatises and the annals used the records of the astronomical office, or intermediate compilations made from them, but different compilers had different interests and hence compiled different events. The compiler of the astronomical treatises apparently was not interested in eclipses that did not occur. The compiler of the annals, on the other hand, was apparently interested in the favor shown the emperor by the withholding of a predicted eclipse.‡ There is no reason to assume that either compilation is more reliable than the other.

M and MS use the Tarng eclipses of 756 October 28 and 761 August 5 to prove point f, which says that the eclipses in the treatises were observed only where the emperor was, and not "anywhere in China". The revolution mentioned above began in 756 and was not completely put down for many years. Both these eclipses were total and at both times, according to M and MS, the eclipse was not total at the nominal capital although it was total where the emperor happened to be.

Let us take up 761 August 5 first because it is the easier case to discuss. The zone of totality for this eclipse passes through heartland China although it missed the nominal capital. Therefore it proves nothing relevant to the point of M and MS.

The case of 756 October 28 is complicated by the fact that there were two emperors at this time, not counting the leader of the revolution. Shortly after the emperor fled because of the revolution, his son proclaimed himself as the

---

† In two of these cases, the eclipse was large in heartland China, but clouds are given as the reason for not seeing the eclipse. In the other cases, the eclipse was either very small or invisible.

‡ Almost all the entries about predicted eclipses that did not occur conclude with a statement that the emperor was congratulated by the court officials as a consequence of the fact that the eclipse did not occur as predicted. This leads to interesting conclusions that will be discussed further in Section V.7.

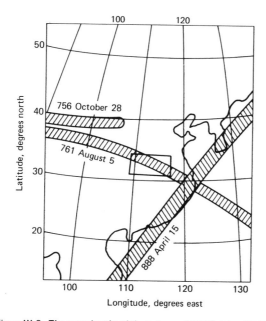

Figure III.2. The central paths of the eclipses of 756 October 28, 761 August 5, and 888 April 15. The rectangle near the center of the figure encloses the area that I call "heartland" China. The figure is reproduced from Cohen and Newton.

emperor and henceforth referred to his father as the "retired" emperor. On 756 October 28 [Cohen and Newton, in preparation], the father was in the modern Cherng-du, at latitude 30°37'N, longitude 104°6'E. The son was in the modern Chinq-yang, at about 36°0'N, 107°42'E.

Figure III.2 shows the central zones for the eclipses of 756 October 28, 761 August 5, and 888 April 15; we shall come to the last of these in a moment. There is a slight uncertainty about the longitude where the zone for 756 October 28 ends, but we need only the latitude, which is firm. We see that both father and son were definitely south of the zone of totality. Thus, contrary to the claim of M and MS, the eclipse was not total where the "emperor" was, no matter which emperor we consider.†

The records specify that two other eclipses were "complete" during the Tarng dynasty. The astronomical treatise says that the eclipse of 879 April 25 was complete, but does not apply the term to the eclipse of 888 April 15. The

---

†M and MS do not note that there were two emperors. M names only the father, but puts him at the location of the son. Incidentally, the usurper was in "heartland" China. If we regard him as a possible emperor, the eclipse was not total where he was either.

imperial annals do just the opposite.[†] Now there was no
eclipse on 879 April 25; it happens to be one of the predicted eclipses that could not have been observed. It is
clear what has happened. The annals contain the reliable
record (contrary to point g above). The compiler of the
treatise accidentally copied the remark about completeness
opposite the date preceding the correct one.

M and MS do not mention the eclipse of 888 April 15,
presumably because the statement about completeness occurs
in the annals rather than the astronomical treatise. M
[p. 8.8] notes that the eclipse of 879 April 25 is described
as jih and, quoting the unpublished work of Stephenson, he
correctly notes that it must be fabricated. He fails to
note that the annals contain the reliable source material
for this eclipse and hence that the occurrence of jih in the
annals, not in the treatise, is the important occurrence.[‡]

The eclipse of 888 April 15 was annular rather than
total, and Figure III.2 shows its central path. The path
passed well to the southeast of heartland China, and we know
that the emperor was in his nominal capital of Charng-an
(the modern Sian, at 34°16′N, 108°57′E) on the day of the
eclipse [Cohen and Newton, in preparation]. Thus this observation could not have been made where the emperor was if
it is accurate, contrary to the conclusions of M and MS.
The reader may say that this eclipse does not contradict
their conclusion since it does not occur in an astronomical
treatise. However, as we saw, the eclipse would occur in
the astronomical treatise if the scribe had not made an error in compiling that treatise. Therefore the record properly belongs in the astronomical treatise and it is relevant.

We now have a problem. If jih does indeed mean central,
the eclipses of 756 October 28 and 888 April 15 show that
observations were made well outside of heartland China. If
observations were made only in heartland China,[#] jih cannot
mean central. Cohen and Newton consider this problem briefly.
They conclude that the more probable solution is that the
observations were made in heartland China and that jih, by
the time of the Tarng dynasty, was not used with its rigorous
meaning of complete. It probably meant only a large eclipse.
In the cases where jih is used with no added detail, the
eclipses have magnitudes of only about 0.9. On the other

[†]There are actually three astronomical treatises for the
Tarng dynasty, two of which usually give identical information about eclipses. Here they do not. The treatise I
have just mentioned is the one that Cohen and Newton call
STS-Astronomy. The other treatise, called WSTK by them,
records both eclipses, along with the right ascension of
the sun, but does not say that either eclipse was complete.

[‡]Evidence that the annals are independent of the treatise occurs throughout most of the Tarng dynasty. For example, the
first "predicted but did not occur" eclipse is dated 714 February 20 (February 19 in Greenwich time). This is well before
the important records of 756 and 761.

[#]Or if they were made only where the emperor was.

-82-

hand, in all cases where there is more detail, the magnitude
is larger. However, Cohen and Newton cannot eliminate the
possibility that jih does mean complete and that observations
were made well outside of heartland China.

In either case, we may eliminate the conclusion of M
and MS for the Tarng dynasty, and we must turn to other dynasties. M and MS assert that their conclusions are correct
for both Han dynasties. In fact, as we shall see in Section
IV.5, their parameter values depend critically upon assuming
that the eclipse of 120 January 18 was observed in Luoh-yang,
the capital at the time. This eclipse occurred under the
Later Han dynasty.

There has been no detailed study of the eclipse records
of the Later Han dynasty, but Dubs [1938 and 1955] has made
a detailed study for the Former Han (-201 to +9). It would
take too much space to investigate the situation in the
Former Han dynasty with the same detail that we have used
for the Tarng dynasty, and I shall only outline the necessary study.

M and MS may have started their line of argument with
a remark that Dubs [1938] makes about the eclipse of -187
July 17. This eclipse was total† according to the imperial
annals but it was almost total according to the astronomical
treatise. By calculation, he finds that the eclipse had a
magnitude of 0.92 at the capital. Hence, he concludes, the
reports in the astronomical treatise come only from the
capital but the reports used in the annals can have come
from other places.‡

Dubs finds a similar situation for the eclipse of -79
September 20. Here, however, the calculated magnitude at
the capital‡ was only 0.77. I think we may question whether
this would be called "almost complete" or "almost total",
but I shall ignore this point. He also finds the same situation for the eclipse of -27 June 19, for which the calculated
magnitude at the capital is 0.96. I think there is some
question about the accuracy with which we can calculate this
magnitude in view of the uncertainties in the astronomical
accelerations, but I shall ignore this point also.

In spite of his statements about these three eclipses,
Dubs is forced to contradict them. Between the eclipses of

---

†I presume that Dubs is using "total" for the term jih
(complete).

‡It is only my guess that M and MS started from this statement by Dubs. Although they cite Dubs extensively, neither
source refers to this statement. M [p. 5.28] explicitly
credits Stephenson with originating the idea in his unpublished work.

‡As it happens, for the eclipses that are involved in testing
the place where the observations were made, the emperor was
at the nominal capital, so far as we can tell. Hence there
is no distinction between the nominal capital and the capital in the sense of MS and M.

-137 November 1 and -133 August 19, only two eclipses were visible in China and two were recorded in China. These are the eclipses of -135 April 15 and -134 April 4.† Both are recorded in the astronomical treatise and both were visible in China but not at the capital. The eclipse of -55 January 3, which is correctly dated, likewise was not visible at the capital, although it was visible elsewhere in China. Finally, the eclipse of +2 November 23 is recorded as total in the astronomical treatise but it could not have been total there.‡

Thus we have three instances in which the records support the conclusion of Dubs and four in which they contradict it. Since M and MS have a condition in their conclusion which Dubs does not have, this statement does not necessarily apply to the conclusion of M and MS, and we must look at the latters' conclusion in more detail.

M and MS restrict their conclusion to records which state the right ascension of the sun and which are found in the astronomical treatise. Most records that give the right ascension say that it was a certain number of degrees in (east of the beginning of) a certain constellation.‡ The record of +2 November 23 gives only the constellation without stating the number of degrees; we cannot say whether this is an accident, a change in policy, or an indication that this record is basically different from those which do state the number of degrees.

We were able to show conclusively that the presence of the right ascension in the Tarng records is no indication that the eclipse was observed by the imperial astronomers at the capital. We cannot be as conclusive as this for the Former Han records, but the available evidence does not support the idea. The evidence indicates the same thing that it did for the Tarng dynasty: The presence of the right ascension merely means that the right ascension was calculated by someone between the time of the eclipse and the preparation of the record in its present form, and this tells us nothing about the conditions of observation.

---

†The Chinese dates given in the records do not correspond to the dates of any actual eclipses, but they do correspond to these dates if we assume certain plausible errors in writing. Hence there may be some question about the dates. Whatever the correct dates may be, the eclipses were not visible at the capital.

‡The record of the eclipse of -14 March 29 in the astronomical treatise says explicitly that the day was overcast in the capital and that the record comes from the provinces. This proves that the treatise did sometimes use records from elsewhere, but it does not bear directly upon the main question. (It might be argued that the treatise used provincial reports only if none were available from the capital.) It also proves that observations were made routinely at places other than the capital.

‡With exceptions that are few enough to be accidental, the Chinese constellations are different from the ones used in Mediterranean regions.

If the presence of the right ascension has the significance that M and MS wish to impute to it, it must have been
inserted into the record at the time when the observation
was made. There is considerable evidence that this was not
the case for the Former Han records. The error in the right
ascension given for the eclipse of -79 September 20, for
example, is more than 40°. It is hard to see how this error
happened if the imperial astronomer inserted this value at
the time of the eclipse.

The record that Dubs [vol. 2, p. 136] tentatively dates
as -140 July 8 is particularly informative. The Chinese
date that appears in the record corresponds to -138 March 21,
but there was no eclipse on that date, nor was there an
eclipse at that time of year in any nearby year. By postulating three errors,† the date can be made to correspond to
-140 July 8, when there was an eclipse visible in China.
Hence Dubs concludes that we may possibly have a garbled
record of this eclipse.

The record occurs in both the annals and the astronomical treatise, and the latter also gives the right ascension.
The right ascension given is approximately 358°. This is
close to the right value for -138 March 21, but it is not
the right value for any possible eclipse. Dubs finds that
this situation is general. When the Chinese date does not
correspond to an eclipse, the right ascension nonetheless
is usually about right for the date given, but it is not
right for the date of any possible eclipse. This can only
mean that the right ascensions were inserted into the records
long after the event, so long after that the compiler did
not realize that the date was not that of a visible eclipse.‡

It is far from clear that we should restrict ourselves
to records that state the right ascension of the sun with
the precision of a degree. However, if we do so, we are
left with the records of -187 July 17, -79 September 20,
-55 January 3, and -27 June 19 out of the records that were
enumerated a few pages ago. Of these four, three records
support the contention of M and MS while one (-55 January 3)
contradicts it. This is not strong support for their contention. The distribution of 3 to 1 in favor is one that could
easily happen by chance with the small sample that is available.

Further, the contention depends upon the assumption that
jih or chi means that an eclipse was central. We shall find
strong evidence in Section IV.1 that the term did not have
this significance at any stage in Chinese history from which
solar eclipse records have survived.

---

†Each error by itself is one that is fairly easy to make,
but the presence of all three errors in one record is
unlikely.

‡I write this on the assumption that the Han eclipse records
originated from observations of eclipses. We shall see in
Section V.3 that this is probably not so, and that many of
the records originated from predictions of eclipses. This
is the same situation that we find for the Tarng dynasty.

There is no available evidence to guide us toward a decision about the places of observation in the Later Han dynasty. The best we can do is to apply the evidence from the Former Han dynasty, but with a lowered confidence level. Since the confidence level in favor of the conclusion of $\underline{M}$ and $\underline{MS}$ is low at best for the Former Han, it is quite low for the Later Han.

As I have noted, $\underline{M}$ and $\underline{MS}$ assume that the observation of 120 January 18 (under the Later Han) was made at Luoh-yang, and their quantitative results are heavily influenced by this assumption. $\underline{M}$ [p. 8.38] assigns a probability of 0.98 (odds of 49 to 1) that this assumption is correct. As I shall show explicitly in Section IV.5, the probability is certainly less than 0.5. I can see no justification for using odds of 49 to 1 for an assumption that almost dominates the results of $\underline{M}$ and $\underline{MS}$.

From specific references in the historical sources, we know the names of several places where eclipse observations were regularly made at some time in the course of Chinese history. All these places are within or near the boundaries of what I have called heartland China. Since bureaucracy has an enormous inertia, the safest assumption is that observations continued to be made at these places in all the times that concern us in this study. For this reason, I shall assume that the Chinese eclipses were observed in heartland China (that is, if they were observed at all), and I shall define this term more specifically in Section V.1. Of course, in the rare cases when we have a specific statement about the place of observation, I shall use the stated place.

7. Some Miscellaneous Matters

In $\underline{AAO}$, I used many kinds of data in inferring the acceleration parameters, and I was concerned with the extent to which various kinds of data agree with each other. In order to do this, I did some experiments in which I tested the stability of the solution against the deletion of groups of data. When I used all the data before the year 500, I found [$\underline{AAO}$, pp. 271-273]

$$\dot{n}_M = -41.6, \qquad y = -27.7, \qquad (III.12)$$

in the usual units. I am omitting the estimates of the standard deviations. When I omitted all data except the places where solar eclipses were observed, I found

$$\dot{n}_M = -41.8, \qquad y = -27.7. \qquad (III.13)$$

I remarked that the data from the observations other than the eclipses carried about two thirds of the total weight of the data. That is, about two thirds of the data, by weight, are omitted in going from the first result to the second. From this, and from one other experiment, I concluded that the data were consistent and stable.

<u>M</u> [p. 4.5], in an attempt to prove that one should use only total solar eclipses and measurements of equinoxes, says that this is not correct, and that the "solar eclipses clearly dominate the entire set" in <u>AAO</u>. Unfortunately, he seems to have said this without testing his statement. When I use only the data other than the solar eclipses, I get

$$\dot{n}_M = -40.8, \qquad y = -27.4. \qquad\qquad (III.14)$$

It is clear that the eclipses do not dominate the data set, since the answer is hardly changed when the eclipses are omitted entirely in deriving Equation III.14.

I did one other experiment in connection with the stability of the solution. I remarked in Section II.6 that I allowed several different dates to be assigned to a few eclipse records in <u>AAO</u> when I was unable to assign a unique date to a particular record, and that I dropped this practice in all subsequent work. I was already suspicious of this practice in <u>AAO</u> and I performed a test to see if it affected the results. When I omitted these records from the inference, I again found only a small change in the parameters. I therefore concluded that the use of multiple identifications, whether it was valid or not, had not affected my results.

<u>M</u> [pp. 4.6-4.7] objects to this procedure, and I agree with his objection even though I do not agree with his reasons for it. For one thing, he has misstated what I did. On page 4.6, he writes: "For example, the observation of Thucydides (-430) is taken to be at <u>either</u>† Athens or Thasos, since both are plausible." By this, he means that both are plausible locations for the location of the observer. However, this is not what I did. I used the distance from Athens to Thasos in order to estimate the uncertainty in the location of the observer, and I decreased the weight assigned to the record in accordance with this uncertainty. This is quite different from what <u>M</u> says.

His other example does represent what I did. It is known that Hipparchus used an eclipse that was total near the Hellespont, but the date of the eclipse is not known. I showed [<u>AAO</u>, p. 262] that the dates -309 August 15, -281 August 6, -189 March 14, and -128 November 20 are all consistent with the facts stated in the record. I then treated the record as if it were four different records, one associated with each date listed. If W represents the weight that I would assign to a record with a unique date, I divided the total weight W among the four dates used.

<u>M</u> [p. 4.7] writes of this procedure: "This is incorrect statistically, in my view, because three-fourths of the data is (sic) <u>known</u> to be <u>wrong</u>".‡ This puts the emphasis in the wrong place. We also know that one-fourth of the data are correct, and this is valuable knowledge that we should not

---

†The emphasis is <u>M</u>'s.

‡The emphasis in this quotation is <u>M</u>'s.

throw away lightly. If the errors in the erroneous data are unbiased, the correct data obviously lead us to correct inferences of the parameters when we use a collection of such records. The correct data lead to correct parameters while the erroneous ones contribute nothing on the average and hence they do not interfere with the process except to make it more laborious. On the other hand, if one arbitrarily selects a single date and throws out the other three, he may "throw the baby out with the bathwater."

We have a similar situation with regard to the place of observation. Sometimes we can tell from textual and historical studies that a record of known date was made in one of several places but we cannot tell which one. For the sake of illustration, let us suppose that there are four such places for a record that would receive a weight W if the place were known uniquely. I then treat the record as if it were four records, dividing the total weight W among the four places used. Again one-fourth of the data (locations) are right and three-fourths are wrong. Again, if the erroneous locations are unbiased, use of these records leads to valid parameters.

However, there is a big distinction between the case of multiple dates (multiple identifications) and the case of multiple locations. In the latter case, we assign the date from the text alone with no assumption about the astronomical accelerations. The locations are also assigned with no reference to the accelerations, and thus they can have no bias with respect to the inference of these accelerations. Therefore the assumptions needed to use records with multiple locations are satisfied.†

This is not so for the case of multiple identifications. Between writing AAO and MCRE I realized that the list of multiple dates is governed by the eclipse maps in Oppolzer's Canon [Oppolzer, 1887], and hence the multiple dates used were biased toward the accelerations that Oppolzer used. In other words, I was playing a concealed form of the identification game. When I dropped the use of multiple identifications in MCRE, I should apparently have explained the reasons more fully.

Now let us get back to the stability of the accelerations found in AAO. M [p. 4.6] performed an interesting experiment that did not occur to me. Pages 262 and 263 in AAO give the weights and the coefficients of condition for the solar ·eclipse records before the year 500. M omitted the records with a

---

†The case of multiple locations is treated differently from the case in which we must take the location of the observation to be anywhere in an extended region. The main reason for this is that the distribution of errors has different properties in the two cases. In the case of an extended region, the uncertainty in the place makes a contribution to the standard deviation of the variable η used in the inference process, and this contribution is combined in a root-mean-square fashion with the contributions from the path geometry and the uncertainty in the magnitude.

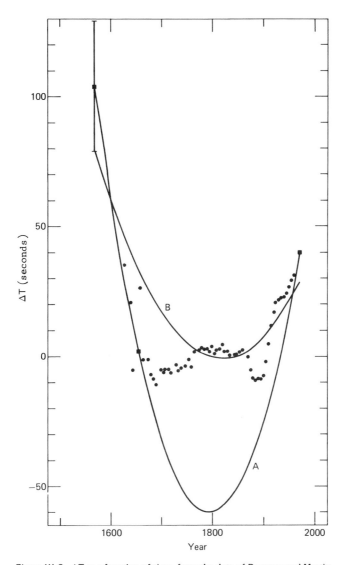

Figure III.3. ΔT as a function of time, from the data of Brouwer and Martin. The plotted points are five-year averages of the data. The vertical bar shows the limits on ΔT that M̲ finds from a record of the eclipse of 1567 April 9. M̲ passes the parabola A through the three points marked with squares and says that this parabola represents the behavior of the data. Curve B is a parabola chosen by a method described in the text.

relative weight of 0.01 or less; this deleted only 0.019 of
the total weight of all the eclipses. The parameters inferred from this slightly deleted set are

$$\dot{n}_M = -72.5, \qquad y = -47.6, \qquad (III.15)$$

which differ by a large amount from the values in Equations
III.13. The corresponding values of D'', however, are almost
the same.

     The conclusion that I would reach from Equations III.13
and III.15 is that the solar eclipse data are well suited for
finding the single parameter D'' but poorly suited for finding
two independent parameters. I am not sure what conclusions
$\underline{M}$ reached. He uses Equations III.15 to lead up to two points.
One is that my selection of data is biased; I have already
discussed this point in Section III.3 and shown that it is
not so. The other is that the method of inference used to
derive Equations III.13 and III.15 is not valid. I shall
deal with this point in the next chapter.

     In his pages 6.4-6.9, $\underline{M}$ considers the modern lunar data
of $\underline{Brouwer}$ [1952] and $\underline{Martin}$ [1969] that I have studied in
two earlier papers [$\underline{Newton}$, 1972c and 1973a].† These data
show the presence of large but short-term fluctuations in
the acceleration parameter ν_M', and these fluctuations interfere with finding the appropriate secular value of ν_M'. In
the references cited, I have shown that the mathematical expectation of their effect on the secular ν_M' is about 1 $''/cy^2$
if the averaging time is 2 or 3 centuries.

     Brouwer and Martin assume that $\dot{n}_M = -22.44$. With this
assumption, they use the lunar data to estimate ΔT, the difference between ephemeris time and universal time. Their
results are plotted in Figure III.3. Each point plotted in
the figure is a five-year average except for a few early
points where the data are not sufficiently dense to let us
calculate such an average.

     $\underline{M}$ supplements the data of Brouwer and Martin with a
record of the eclipse of 1567 April 9. From this record, he
concludes that ΔT in 1567 was between 79 seconds and 129
seconds; these limits are shown by the vertical bar in Fig-
ure III.3. $\underline{M}$'s values of ΔT at this point are also based
upon $\dot{n}_M = -22.44$.

     $\underline{M}$ does not seem to have seen my studies of the effects
of the fluctuations. He says [p. 6.7] that the short-term
effects "dominate" the secular effects "over two or three
centuries." As we have seen, this is not so; the short-term
variations, when averaged over three centuries, should affect
ν_M' only by about 1 $''/cy^2$ out of a total value of more than
10. $\underline{M}$ then contradicts himself in his next sentence, where
he says that fitting a parabola to "400 years of data will

---

†These data cover a slightly longer time span than the data
of $\underline{Spencer\ Jones}$ [1939] that were used in Figure I.1. We
get almost identical quantitative results from the two collections of data.

average out these variations very well." If the short-term variations dominate the average over 3 centuries, they cannot be averaged out "very well" by adding one more century.[†]

$\underline{M}$ then arbitrarily selects the three points in Figure III.3 that are marked with squares and passes the parabola marked A through them. The equation of this parabola is

$$\Delta T = -24 + 68T + 32T^2. \qquad (III.16)$$

When we use Equation III.16 with the first of Equations III.4, we find $\nu_M' = +12.8$, which agrees excellently with the value that applies to ancient times. Thus $\underline{M}$ claims to have proved that the accelerations have not changed within historic times.

This proof has two obvious defects besides the one that has already been mentioned. First, $\underline{M}$ has used the center of the limits for 1567, although the value of ΔT may be anywhere on the vertical bar in Figure III.3. Second, it is clear that the curve A disagrees seriously with the data over almost all of the time span in the figure.

The quadratic that best fits the data of Martin and Brouwer is [Newton, 1972c][‡]

$$\Delta T = 9.0 + 22.3T + 8.94T^2.$$

The value of this in 1567 is 33.7 seconds, which is well below the vertical bar in the figure. I do not have as much confidence in the strictness of the limits for this eclipse as $\underline{M}$ does, for reasons that I will explain when I discuss it in Section X.3. However, for the sake of argument, I am willing to accept $\underline{M}$'s limits for the present.

Ideally, we should assign suitable weights to the eclipse of 1567 and to the data points of Brouwer and Martin and find the quadratic that best fits the expanded data set. However, we do not need to do this to see that Figure III.3 leads to a value of ν_M' which is far different from the value that is correct for ancient times. Values of ΔT of 80 seconds in 1567, 0 seconds in 1800, and +8 seconds in 1900 obviously fit all the data in Figure III.3 fairly well. The parabola that passes through these points is

$$\Delta T = 8.0 + 20.7T + 12.71T^2. \qquad (III.17)$$

This parabola is curve B in Figure III.3; it clearly represents the situation better than curve A. Equation III.17 leads to the value $\nu_M' = -8.4$, which differs greatly from the value 2000 years ago. Thus we find once more that the nongravitational effects on the earth and moon have changed by a

---

[†]It is easy to show [Newton, 1973a, p. 130, for example] that the average effect of the short-term variations is proportional to $t^{-3/2}$, in which t is the averaging interval. Thus the effect over 4 centuries is $(3/4)^{3/2} = 0.65$ times the effect over 3 centuries.

[‡]The reference gives only the coefficient of T^2, but it is not hard to find the constant and linear terms.

large amount within historic times, contrary to the claims of M̲ and M̲S̲.

We have seen in this chapter that M̲ and M̲S̲ have many errors in their fundamental approach to the study of the accelerations, and we shall see in the next chapter that they have serious errors in their formal analysis. Against this, they have made a useful contribution to the field by finding a number of eclipse records that have not been used previously in the literature. Unfortunately, they cancel out much of this contribution by the errors that they make in discussing individual records of eclipses, and by the invalid records that they have introduced into the astronomical literature. I shall deal with these matters in the discussions of individual eclipse records in the relevant chapters.

CHAPTER IV

COLLECTION OF DATA AND METHODS OF INFERENCE

1.   The Weight Attached to a Solar Eclipse Record

If we are to use a record of a solar eclipse in the
inference of the accelerations, four conditions must be met.
First, we must be able to date the eclipse without ambiguity.
Second, we must be able to say that the observer was within
a certain geographical region. The smaller the region is,
the more valuable the record is. However, we do not have to
place the observer in a specific city, for example, in order
for the record to be valuable. Third, the record must tell
us or let us estimate the magnitude of the eclipse as the
observer saw it. Fourth, because different observations have
different value, we must be able to attach a weight to each
observation before we use it.

The weight depends upon several factors, but the only
factor that poses a particular problem concerns the magnitude.
I shall discuss the way I handle the magnitude in this section, while deferring the discussion of the other factors to
Section XIII.1.

In Section I.5 I introduced a coordinate η, which is the
distance from the observer to the axis of the moon's shadow
when the eclipse is a maximum for the observer. η is taken
to be positive if the observer is northward from the shadow
axis at maximum eclipse.

Now let us consider the class of all eclipse records
that give no indication of the magnitude of the eclipse.
Some eclipses in this class were central as the observer saw
them while others had fairly small magnitudes. As I remarked
in Section III.3, the coordinate η is positive for some of
these eclipses and negative for others. The algebraic average value of η for the eclipses in this class is therefore
zero.† The best estimate of η that we can make for any
eclipse in this class, in the absence of detailed information, is therefore zero. We must now make an estimate of
the error that we make in saying $\eta = 0$, and this error estimate will then be used in determining the weight of the
record.

If for simplicity we speak only of eclipses that were
total somewhere, saying $\eta = 0$ is the same as saying that the
eclipse was total for the observer (magnitude ≥ 1). (For

---

†Rigorously, all we can say is that we expect the average to
approach zero as the size of the sample approaches infinity,
but we do not expect the average to be zero for a sample of
finite size. In order to avoid the frequent and cumbersome
statement of this fact, I shall ignore the effect of the
sample size unless it plays a crucial role in some discussion.

annular eclipses, there is an equivalent but more complicated statement.) This is correct for some records but wrong for others, and we must attach a standard deviation to the estimate that the eclipse was total. I shall use the symbol σ_μ for this standard deviation. We must then convert σ_μ into an equivalent standard deviation σ_η for the coordinate η, which I shall do in Section XIII.1.

In earlier work I commonly used $\sigma_\mu = 0.06$ for ancient eclipses and 0.10 for medieval ones. I based these values partly upon observations of the eclipse of 1970 March 7 made by several persons including myself, and partly upon the estimates of various people about the smallest eclipse that would probably be observed by a person who was not expecting an eclipse. I used a larger value for medieval eclipses because we expect people in medieval times to be more aware of eclipses than people in ancient times.[†]

TABLE IV.1

MAGNITUDES OF ECLIPSES RECORDED
IN VARIOUS PROVENANCES

Provenance	Number of records	Standard deviation of the magnitude
China	32[a]	0.36[a]
British Isles	94	0.21
France	87	0.18
Germany	134	0.17
Italy	90	0.16
Other Parts of Europe	127	0.17
Near East	35	0.18
All except China	567	0.18

[a]In stating these values, I have omitted about half the Chinese records, for a reason explained in the text.

These estimates were acceptable for preliminary studies, but we should do better in the present extensive study. Accordingly I shall use a rigorous procedure here. I start by using provisional values of the accelerations, taken from earlier work, calculating the magnitudes of all eclipses that will be used in this work, and finding the standard deviation of the magnitudes from unity. In order to see if σ_μ differs significantly from one provenance to another, I have calculated σ_μ in this way for each provenance from which we have a sufficiently large sample. The results are given in Table IV.1.

[†]I am really comparing people in medieval Europe with people in the ancient Near East in making this statement.

The first seven lines in the table list the number of records, and the standard deviation of the magnitude σ_μ for those records, from seven different provenances, counting the catch-all "other parts of Europe" as a single provenance. The last six provenances have essentially the same value of σ_μ, which is about half of the value for China. In fact, if we put together all provenances except China, the value of σ_μ is 0.18, which is just half of the value 0.36 for China. As a result of Table IV.1, I shall use $\sigma_\mu = 0.36$ for Chinese records and 0.18 for all other records, when nothing is said about the magnitude.

I used only about the first half of the Chinese records in calculating the value in Table IV.1. The reason is that the second half of the records, beginning with a record of the eclipse of -441 March 11, is biased toward records that indicate a large magnitude. That is, the records from -441 March 11 on do not form a random sample with respect to the magnitude. The reason for this will be explained in Chapter V.

I also examined the records from the non-Chinese provenances to see whether the value of σ_μ has a tendency to change with time. For the British Isles, the value of σ_μ does tend to increase uniformly with time, but this is not so for the other provenances. It seems likely that σ_μ is substantially constant in time for the records that will be used in this work, and that the time dependence found for the British Isles is a statistical accident.

There is a myth of long standing which says that an eclipse is likely to go unnoticed and hence unrecorded unless it is nearly total. As we saw in Section III.4, Fotheringham [1920] and Stephenson [1970] explicitly use this assumption in attempting to establish important parts of their arguments. M [p. 5.25] writes explicitly that ". . . few eclipses of lesser magnitude (statistically speaking) will even be noticed." He is referring here to eclipses with a magnitude less than 0.98.

Table IV.1 confirms an earlier conclusion [Newton, 1974] that the assumption made by these various writers is not true, and hence that important parts of their arguments are invalid. In particular, as the cited work shows, about four out of five eclipses recorded in the annals of European monasteries and similar sources had magnitudes less than 0.98, the level below which M asserts that an eclipse is not likely even to be noticed.

We must also consider some special classes of record that do give a clue to the magnitude. The records which say that stars were seen were studied in Section II.1, where we found that the value of σ_μ for such records is 0.054. This is the value of σ_μ that I shall use for these records in the present work, instead of the value of 0.01 that I have used in earlier work. This applies only to records from provenances other than China. The Chinese records that refer to stars pose a special problem that I shall discuss in Section V.5.

Now let us turn to non-Chinese records which say explicitly that an eclipse was total, and to Chinese records which say that an eclipse was chi or jih. We may remember from Section III.6 that this term is always taken in the literature to mean an eclipse that was central to the observer, so that it is applied to both total and annular eclipses. However, if we included annular eclipses in a discussion of the magnitudes of jih eclipses, we would need to know the details about the maximum magnitude of each annular eclipse. For simplicity, then, I shall consider here only jih eclipses that are shown to be total somewhere by calculation.[†]

TABLE IV.2

MAGNITUDES OF ECLIPSES REPORTED AS TOTAL

Provenance	Number of records	Standard deviation of the magnitude
China	15	0.071
All others	40	0.034

Table IV.2 shows the values of σ_μ for the Chinese jih eclipses that are known to have been total somewhere, and for the non-Chinese eclipses that were explicitly described as total. Again the value of σ_μ for the Chinese records is about twice that for the other records. There is no discernible variation of σ_μ with time. This means that jih did not really mean a central eclipse at any stage in history from which we have records, unless observations recorded in China annals were observed at points well outside China.

I can only conclude that the people who recorded the eclipses simply did not use total or jih in the way that I would. Perhaps it is a matter of astronomical training. If a person is seeing a large eclipse for the first time, so that he does not necessarily know what to expect, he may think that an eclipse has gone as far as it can. Hence he may say that it is "total" or "complete" well before a trained astronomer would apply these terms. In this connection, we should remember that an individual has almost no chance of seeing two central eclipses unless he specifically arranges his travel to do so. Hence, an observer who saw a large eclipse almost never had previous experience with large eclipses.

[†]If the accelerations were large enough to affect the question of whether an eclipse was total or annular, they would have many other effects that would be easy to observe. Since we do not observe any of these effects, we can calculate accurately whether an eclipse was annular or total without having accurate values of the accelerations.

Next we must deal with records which state that an eclipse was partial. Some of these simply state that an eclipse was partial rather than total, but make no attempt to indicate the magnitude. Using provisional values of the accelerations, I have calculated the average magnitude of these eclipses at the places of observations, along with the standard deviation of the magnitude.

TABLE IV.3

MAGNITUDES OF ECLIPSES REPORTED AS PARTIAL,
WITH NO INDICATION OF THE MAGNITUDE

Provenance	Number of records	Average magnitude
China	6	0.894
All others	50	0.800
All	56	0.810

The results are shown in Table IV.3, separately for China and for all other provenances. We see that the average magnitude for the Chinese records is appreciably larger than it is for the other provenances, and the difference is enough to have a modest statistical significance. However, in view of the small size of the Chinese sample, I am loath to accept the difference as an established one. Accordingly, I shall use the same value for all provenances. That is, if a record states that an eclipse was partial, but gives no other clue to the magnitude, I shall take the magnitude to be 0.81. The standard deviation of the magnitude with respect to this value is 0.14.

Finally we have the records which state that an eclipse was partial and which give an estimate of the magnitude. Here we can only treat each record as a special case, particularly with regard to assigning a standard deviation to the estimates. There are not enough records relating to a particular estimated value to let us derive standard deviations statistically, so I assign a separate standard deviation to each record in accordance with the accuracy which it seems likely that the observer would achieve under the circumstances. This is a highly subjective process. Nonetheless, by the exercise of caution, I believe that we can assign values in this way that will avoid any serious consequences of the subjectivity. I have followed one general procedure which I believe helps greatly in doing this.

TABLE IV.4

OLD AND NEW VALUES OF THE STANDARD DEVIATION OF THE MAGNITUDE

Old σ_μ	New σ_μ
0	0.034
0.01	0.054
0.02	0.068
0.03	0.082
0.04	0.096
0.05	0.110
0.06	0.124
0.07	0.138
0.08	0.152
0.09	0.166
0.10[a]	0.180

[a]The value σ_μ = 0.06 when used
in AAO is equivalent to 0.10
for medieval records.

The procedure is based upon Table IV.4, which compares
the values of σ_μ used in MCRE (old σ_μ) with those adopted in
this work (new σ_μ). For example, if a record asserts that an
eclipse was total, I assigned it a standard deviation of 0 in
MCRE, but I assign a value of 0.034 in this work, Similarly,
where I used 0.10 in MCRE for a record that says nothing
about the magnitude, I use 0.18 here. In preparing Table
IV.4, I have interpolated linearly between the lines in which
the old σ_μ equals 0.01 and 0.10.

To illustrate the use of the table, suppose that a record gives an estimate of the magnitude which I interpreted
as equalling 0.98 in MCRE.[†] This is 0.02 from totality. In
this work, I shall change the magnitude to 1 - 0.068 = 0.932;
0.068 is the value that appears opposite 0.02 in Table IV.4.
To be conservative, I then usually round 0.932 down to 0.93.
I do this in accordance with a suspicion that ancient and
medieval observers, as well as many modern observers, often
wish to emphasize the dramatic quality of their experiences,
and thus they tend to exaggerate the degree of darkening in
the eclipses they saw.

[†]The records rarely state a specific numerical value in indicating the magnitude. Instead, they usually state it by
using terms which require some interpretation, and thus the
exact magnitude implied by a record is usually somewhat uncertain itself. Thus there is an error in the way that an
observer states the magnitude, as well as an error in the
way that he estimated it.

This deals only with assigning the estimate of the magnitude. In addition, we must assign a standard deviation to this estimate. As I said earlier, I do this separately for each record in accordance with the circumstances, but I shall never use a value less than 0.05.

The reader may object to the values of σ_μ that I have adopted here, and in particular to the values 0.034 and 0.054 for records that state totality and the visibility of stars, respectively. There are two grounds for objection. One is that the values are based upon provisional values of the accelerations and hence that they may be too large. It is true that the finally inferred values of the accelerations will lower the estimated standard deviations, but it is doubtful that the standard deviations will decrease enough to affect the general considerations of this section. Nonetheless, for safety, I shall repeat the estimates after the accelerations have been inferred, and revise any conclusions that are seriously in error. This will be done in Section XIV.5.

Another objection is that I may have made errors in assigning the places where the eclipses were observed. It is possible that this accounts for some of the large deviations from totality, but it cannot account for all of them. Particularly this cannot account for the annular eclipses that were reported as total.

The matter of eclipses that are reported as total brings up an interesting point with regard to records from the British Isles. The point is suggested by the following statement of MS [p. 476]: "On the Continent of Europe, as distinct from the British Isles, the chronicles are to a considerable extent independent of one another and are mainly concerned with local events. Thus, except in rare cases, we can be confident that the place of observation was where the chronicle was compiled. This conclusion is not true for the various annals of the British Isles. Several of these are the work of individuals who are known to have travelled throughout the country. Others bear the names of monasteries, but at no time did they attach special significance to the affairs of the monastery."

This statement is not correct. For one example out of many, consider the annals of the monastery of Bermondsey. In MCRE [p. 133] I quoted the editor of these annals on just this point. After giving the reasons for his conclusion, the editor writes: "Their value is, therefore, chiefly confined to the history of the house of Bermondsey itself."

The British and continental annals seem to me to follow pretty much the same course. In almost all provenances, the chronicles or annals that cover the earliest recorded times tend to deal with the affairs of the entire provenance. They are compilations made at a later time from a variety of sources, and we cannot tell from them where in the provenance an observation was made. Throughout the medieval period, there were indeed individuals who continued to compile annals and histories of the whole provenance, just as MS say. At the same time, however, many and probably most monastic annals dealt with a large region so far as high political and religious matters are concerned, but otherwise they concentrated upon local affairs and in particular upon the affairs of the monastery (or church). This is true in both the British Isles and on the continent. Further, in a number of cases, both British and continental, the writer tells us specifically that he is giving us his own observation of an eclipse, and in most of these cases we know where the writer was.

However, this is not the interesting point. MS go on to say that they do not use British records for the reason given. The interesting point is that the reason is irrelevant to their choice of records. They use only records in which it is specifically stated that an eclipse was total, or that it never became total but reached a magnitude on the verge of totality. There are no British records known to me of either kind. I think it is interesting that no British records make an explicit statement of totality, as I read them, even though the eclipses were total in many instances. Thus it does not matter to MS whether they know the places of observation or not; there are no known British records for which the question even comes up.

Among the records which indicate a large but not central eclipse are those which state that the sun appeared like the moon at a certain age. For example, the record that I labelled 1178 September 13e E,F on page 349 of MCRE says that the sun appeared like the moon on its second or third night (after the crescent new moon appeared). The record also says that Venus could be seen near the sun. A moon at this age has about 0.04 of its surface visible. If it had not been for the remark about Venus, I would have assigned 0.04 as the standard deviation of the magnitude. As it was, I compromised and used 0.02.

M [p. 5.25] disagrees with this procedure because of irradiation in the eye, which causes one to exaggerate the size of the sun. In my experience, irradiation is so severe that one cannot tell with the naked eye that the sun is being eclipsed [MCRE, pp. 81-82] until the magnitude is very large, probably of the order of 0.99.† However, several old sources describe rather accurately the orientation of the solar crescent at advanced stages of the eclipse but still far from totality. They could not have done this unless they had optical aids which largely eliminated the irradiation. For this reason, I assumed that the comparison of the sun with the moon was done with optical aids.

M [p. 5.25] raises a new consideration. He has calculated the magnitude of the eclipses which compare the sun with the moon and finds that the " . . vast majority . . come

---

†I commented there that a person who carefully examined the after-image, which can be seen while one is blinking, might realize that it consisted of two parts. One is the large round blob that comes from irradiation. The other is the true figure of the sun. I doubt that an ancient or medieval observer would reliably interpret the after-image even if he observed the effect.

from magnitudes in excess of .98!"† For the record of 1178 that was just mentioned, he calculates that the magnitude was 0.995. Thus the comparison of the sun with the moon may have been made for eclipses so near totality that the naked eye could see the sun as a crescent. In this case, the apparent size of the sun would be much larger than its correct size and the estimated magnitude would be too small. M suggests that we should use these comparisons only after we have done a series of experiments to calibrate the effect of irradiation.

However, the matter is more complicated than this. For one thing, comparing the sun to the moon at a certain age is just one way of giving an estimate of the magnitude. Numerous records, however, give an estimate of the magnitude by stating directly a fraction of the sun that appeared to be eclipsed. The problem of possible irradiation arises for all these estimates, and it is independent of the way in which the observer stated his estimate. Thus we must consider all such estimates together.

For another thing, many records suggest that observers, even those who were not trained in astronomy, frequently used some kind of aid in viewing an eclipse. We cannot calibrate the effect of irradiation alone; we can calibrate only the effect of irradiation in connection with a specified optical aid, and I do not think of any instance in which we know the nature of the aid.

For the time being, I shall continue to handle estimates of the magnitude, regardless of how they are stated, in the same basic way as in MCRE. However, I shall modify some of the estimates of the magnitude, and the associated estimates of the standard deviation, in the way indicated earlier in this section. After the final inference of the accelerations, I shall then make a study of these records in order to study the effect of irradiation and, perhaps, of other phenomena upon the magnitude estimates. I shall do this in Section XIV.6.‡

This is a good place to point out an important consideration about weights. We can rarely estimate the error level associated with a method of measurement as accurately as 40 percent. Suppose that we assign a standard deviation equal to σ. In most cases, then, the value 1.4σ is just as likely to be right. If the weight depends only upon the value of σ, and if the weight is 1 in the first case, it is $1/1.96$ in the second.

In other words, we rarely know the "correct" weight within a factor of 2. It follows, then, that we can change the weights assigned to individual measurements or classes of measurement by factors of 2 without making any "real" change in the situation. If we change any inferred parameters by an important amount when we make such changes in

---

†The exclamation point is M's.

‡The study in Section XIV.6 shows that the estimates of magnitude were not appreciably affected by irradiation, contrary to M's statement.

the weights, we are probably doing something wrong and we need to study our procedures carefully.

For a final point, let us return to the records that explicitly state totality. Let η_e denote the value of the coordinate η for an observer on one edge of the central zone. If we really know that the eclipse was total for the observer in question, and if we really know where the observer was, we can say with confidence that the value of η at the observer's position lies between $-\eta_e$ and $+\eta_e$. A record of this sort then gives a linear inequality for the coordinate η:

$$-\eta_e \leq \eta \leq +\eta_e. \tag{IV.1}$$

This inequality says that the probability of η lying in some interval $d\eta$ is uniform between $-\eta_e$ and $+\eta_e$ and zero outside that range.

However, we have seen that the standard deviation of the magnitude for these records is 0.034, and the value of η where the magnitude is $1 - 0.034$ is typically about $2\eta_e$. Thus there is a large chance that η lies outside the range from $-\eta_e$ to $+\eta_e$.

The difference between this probability distribution and the one implied by inequality IV.1 underlies one of the most fundamental differences between the method of analysis I use and the one that $\underline{M}$ and $\underline{MS}$ use. Because of the importance of this topic, I shall devote a separate section (Section IV.3) to it.

Before we can carry out the analysis, we must be able to convert a standard deviation σ_μ into an equivalent standard deviation σ_η of the coordinate η. This will be done in Section XIII.1. The value of σ_η will then help determine the weight to be attached to a record.

2.   Kinds of Data to be Used

In studies of the astronomical accelerations, I have used many kinds of data. In $\underline{AAO}$, I used the places where solar eclipses were observed, measured magnitudes of partial solar and lunar eclipses, measured times of lunar and solar eclipses, occultations or conjunctions involving the moon, the times of equinoxes and solstices, occultations or conjunctions involving the planet Venus, and the mean longitudes of the sun and moon taken from tables. In $\underline{MCRE}$, I restricted the study to the places where solar eclipses were observed; I did this in order to restrict the study to one that could be handled within the confines of a single book, and not out of a belief that this is necessarily the best kind of observation to use. In a later paper [$\underline{Newton}$, 1972b], in addition to equinoxes and solstices, I used many measurements of solar coordinates made at known times. Most of the coordinates were meridian elevations, but a few other types of coordinates were also used. In $\underline{APO}$, I restricted the study to solar and planetary observations, because of the particular purpose of that work. There I used all the solar observations

that had been used in earlier work, as well as some additional measurements of solar coordinates, and additional
values of the mean longitude of the sun taken from tables.
I also used occultations or conjunctions involving Mercury,
Venus, and Mars, as well as measurements of the rising or
setting times of the planets.

In this work, I originally intended to use all available
observations that involve the moon; this would include all
data that involve eclipses. It would also include measured
rising and setting times of the moon, in addition to the
kinds of data that have been used earlier. It soon became
clear that a work of this scope was not possible within the
confines of a single book, and it became necessary to adopt
a modest goal. Thus I shall limit this work to the use of
the places where solar eclipses were observed. This means
that the work will contain two main parts, so far as the
collection of data is concerned. One part will be a review
and in some cases a reassessment of solar eclipse records
that have been used in earlier work. The other will contain
records, mostly from years after 1200, that have not been
used before. The latest eclipse that will be used is that
of 1567 April 9, which actually belongs in the Renaissance
as most people date historical periods.

We should note that the study is limited to the places
where solar eclipses were observed, and that it does not include the use of measured times or magnitudes. There is one
exception to this. Some chroniclers made rough estimates of
the magnitude of an eclipse, as I discussed in the preceding
section, and I shall use these estimates in assigning a value to η rather than using $\eta = 0$. I do this in order to decrease the scatter of the η values. However, I do not include measurements of magnitude made by people with astronomical training. These measurements will be used in a later
work.

We can word the discussion of the inference method
either in terms of the accelerations with respect to ephemeris time or in terms of the accelerations with respect to
solar time. Since the observation that a solar eclipse was
seen at a particular place involves an implicit measurement
of solar time, I shall use solar time in this work, contrary
to my practice in most earlier work.

Although we shall use only the places where solar eclipses
were observed, it is interesting to look at the kind of information that we can derive from other kinds of observations.
The magnitude of a lunar eclipse gives an estimate of $\dot{n}_M$, the
acceleration of the moon with respect to ephemeris time.
Occultations or conjunctions involving the moon and a star
give the position of the moon against the star background at
a measured value of solar time. Hence each occultation or
conjunction furnishes an estimate of ν_M'. The mean longitude
of the moon taken from a table also gives an estimate of ν_M'.
The measured times of eclipses, whether lunar or solar, however, tell us the solar time when the elongation D of the
moon from the sun had a particular value, as do the rising
and setting times of the moon. Each such measurement thus
furnishes an estimate of D''. Finally, the magnitude of a

solar eclipse, or the place where an eclipse was observed, gives us a linear relation between ν_M' and ν_S', the accelerations of the moon and sun with respect to solar time.

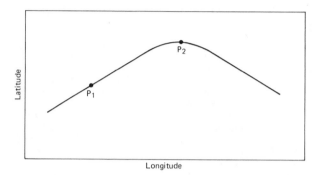

Figure IV.1. The central line of an eclipse path. Almost all central lines, when plotted on scales that are uniform in latitude and longitude, consist of a curved portion and a longer portion that is almost straight. Most paths contain two straight portions joined by a curved section. The extremum in latitude can be either a northern limit, as shown in the figure, or a southern limit. The relation between the accelerations which an eclipse furnishes us depends upon the location of the observer with respect to the curved portion of the path. P_1 and P_2 are two points used in discussion.

We can write this linear relation in the form

$$A\nu_M' + B\nu_S' = Z. \qquad (IV.2)$$

The values of A and B depend upon the geometric relation between the point of observation and the eclipse path. Eclipse paths, almost without exception, contain a curved portion and one or two nearly straight portions; most have two straight portions. The general shape of an eclipse path, plotted with the use of uniform scales in latitude and longitude, is shown in Figure IV.1.

Now consider what happens when we accelerate the sun and moon. If we accelerate the sun (positively), we delay the time of the eclipse. This means that the earth has rotated farther to the east and hence that the path is shifted to the west on the surface of the earth. An acceleration of the moon would have exactly the opposite effect if the lunar orbit had no inclination. Because the lunar orbit does have an inclination, an acceleration of the moon also shifts the entire path in latitude, but the shift in latitude is about 0.1 times the shift in longitude.

Now consider a point that is on the eclipse path for some set of accelerations, and ask what combination of accelerations would leave the point still on the path. If the point, like P_1 in Figure IV.1, is on a straight portion of the path, the required accelerations are nearly equal. Similarly, if a point is at the same latitude as P_1, we do not change its distance from the path if we use equal accelerations of the sun and moon.

Suppose a record says that an eclipse was total at some point not on the calculated path in Figure IV.1. We must then find accelerations which put the point on the path. The change in the distance from the point to the path depends upon how far the point was from the path in the first place. Thus, if the point is at the same latitude as some point on the straight portion, the record gives us the equation

$$\nu_M{}' - \nu_S{}' = Z. \qquad (IV.3)$$

In other words, to high accuracy, most solar eclipses give an estimate of D'', the difference between the accelerations of the moon and the sun.

Matters are quite different for a point like P_2 that is on the curved portion of the path. On the curved portion, a small shift in latitude requires a large shift in longitude, much larger than the longitude shift that is needed on a straight portion. For observations made at the latitudes of such points, then, we leave the relation in the form of Equation IV.2, with the understanding that A and B are no longer equal to +1 and -1, respectively.

The value of Z for an observation on the curved portion is approximately proportional to the latitude shift of the path, which is small compared with a longitude shift. Thus Z in Equation IV.2 tends to be about 0.1 times Z in Equation IV.3. The error in knowing the location of the point is the same in both cases, so the relative error in Z in Equation IV.2 is about ten times what it is in Equation IV.3. Other things being equal, a single instance of Equation IV.2 thus has about 0.01 times the weight of a single instance of Equation IV.3, and there are also fewer instances of Equation IV.2. Thus the magnitudes of solar eclipses, or the places where they were observed, give very little strength in separating the two accelerations $\nu_M{}'$ and $\nu_S{}'$. However, a reasonably large sample of eclipses can give a formal solution for both accelerations.

MS and M both disagree vigorously with my choice about the kinds of data to use. Since M writes more extensively on the matter than MS, I shall refer mostly to M in discussing the question. Since M uses the accelerations with respect to ephemeris time, I shall now word the discussion in terms of them.

Various kinds of observations give the following relations between the acceleration parameters $\dot{n}_M$ and y:

a. The magnitude of a partial lunar eclipse gives

$$\dot{n}_M = Z. \qquad (IV.4)$$

b. Occultations or conjunctions give

$$(\nu_M{}' =) \; \dot{n}_M - 1.7373y = Z. \tag{IV.5}$$

c. Eclipse times, most magnitudes of solar eclipses, and most places where solar eclipses were observed, give

$$(D'' =) \; \dot{n}_M - 1.6073y = Z. \tag{IV.6}$$

d. A few magnitudes or places of solar eclipses give

$$A\dot{n}_M - By = Z, \tag{IV.7}$$

in which it is understood that A and B are not equal to the values in Equation IV.6.

e. Solar observations give

$$y = Z. \tag{IV.8}$$

In all these, Z is a quantity derived from observation, while the coefficients of $\dot{n}_M$ and y in the left members of the equations are determined by theoretical considerations. The fact that I use the same symbol Z for all the right members does not mean that the quantities represented by Z are all the same.

Since the variables are separated in Equations IV.4 and IV.8, we might think that we should find $\dot{n}_M$ and y by means of the magnitudes of lunar eclipses and of solar observations. Unfortunately, the magnitude of a lunar eclipse is rather insensitive to $\dot{n}_M$ and, conversely, $\dot{n}_M$ cannot be determined accurately from measured magnitudes. Thus we must turn to other quantities.

$\underline{M}$ [p. 4.12] remarks that lunar eclipse magnitudes have insufficient sensitivity to be useful. He ignores lunar conjunctions and occultations (Equation IV.5) completely, so far as I can see. He then repudiates the use of measured eclipse times for a reason that seems odd to me. He notes that they provide the same linear relation (Equation IV.6) between the parameters as the solar eclipses, except for small effects that I am representing here by Equation IV.7. Hence, if we use solar eclipses, he says, there is no point in using measured times because they do not help us separate the parameters. The only observations that $\underline{M}$ admits into his study are the places where solar eclipses were observed and the observations (Equation IV.8) that involve only the sun.†

---

†This applies only to observations before modern times, and there is in fact one exception for ancient times. $\underline{M}$ originally includes in his data sample the times of two solar eclipses, those of -321 September 26 and -135 April 15, that were measured in Babylon. He then deletes the data from -321 because they do not agree with his other results, leaving only the times for -135 as a single exception.

Further, $\underline{M}$ and $\underline{MS}$ do not admit most of the places where solar eclipses were observed. They use only records in which the magnitude and the place of observation are, they believe, accurately known. In practice, I believe, they use only eclipses which were very large at the place of observation.† Thus they exclude most of the eclipses in $\underline{AAO}$ and $\underline{MCRE}$.

For the purposes of this discussion, observations that involve the moon can be divided into four classes:

      (a) Solar eclipses stated to be central at a known place.

      (b) Solar eclipses that were simply observed at a known place, with no statement about the magnitude.

      (c) Either solar or lunar eclipses for which the time or the magnitude (or both) was measured at a known place.

      (d) All other kinds of lunar observations.

In classes (a) and (b), we assume that the time was not explicitly measured, although there may be some vague statement about the time.

Both $\underline{M}$ and $\underline{MS}$ advocate that we use only class (a) in studying the accelerations and that we ignore all other classes, except for two or three unusual observations. They do not recognize in any place I noticed that class (d) exists. So far as I can judge from their writing, they advocate ignoring class (c) because it is useless, but they claim that use of class (b) is actually harmful. In other words, they say, we can get improved results if we deliberately omit most‡ of the relevant data.

This is manifestly not so, but it is surprising how often scholars advocate this position. The following rule is necessarily correct: If data are irrelevant to a problem, ignoring them does not affect the situation. If data are correct, relevant, and unbiased, using them always improves the situation. In writing this, I of course assume that the data are used correctly and in particular that they are suitably weighted. Another way to word the matter is to say that correct, relevant, and unbiased data should always receive a weight greater than zero.

---

† In a few places, notably 120 January 18, they explicitly take the eclipse to be very large but not total. I find no clue in their writing about the way they assign the magnitude in these cases. After they assign the magnitude in some way, I believe that they then do the equivalent of finding the quantity η_e in relation IV.1 and then using only the appropriate single inequality rather than the double inequality shown there.

‡ I believe that there are more surviving observations in class (b) than in all other classes put together.

I do not plan to use observations of classes (c) and (d) in this work for reasons of space, but I have used such observations in earlier work and I plan to use them again in later work. Hence it is useful to determine the relative weights that should be placed upon various kinds of observation.

For example, let us look at the measurements of eclipse times in class (c). These observations, in common with most observations in classes (a) and (b), give us estimates of D'' (Equation IV.6). $\underline{M}$ says the measured times are useless because they do not help us separate $\dot{n}_M$ and y in the equation for D''. This is not the important question, it seems to me. The important question is whether they improve the accuracy with which we know D''. The answer to this is obviously yes.

We now want to estimate the accuracy with which eclipse times could be measured with the naked eye, with ancient or medieval methods of measuring time. The standard deviation of early Islamic measurements of solar eclipse times [Newton, 1974] is 9 minutes and of lunar eclipse times is 15 minutes. The accuracy of Babylonian and Greek time measurements is about 30 minutes, according to the same reference. However, with one exception, all the Babylonian and Greek measurements studied in the references come to us only through Ptolemy [ca. 142], and we know now [Newton, 1977, Chapter XIII] that all these observations were fabricated by him. The errors in fabrication were generally large because Ptolemy's solar and lunar theories were rather inaccurate. Hence the Islamic data furnish the only reliable estimate of the accuracy of measured eclipse times.

Now let us consider records of observations in class (a); these are the records which assert that a particular eclipse was central at a known place. We saw in the preceding section that the standard deviation of the magnitude for such records is 0.071 for Chinese records and 0.034 for records from other provenances. In order to be as favorable toward class (a) records as possible, let us use only the value 0.034. On a standard deviation basis, then, the magnitude was 0.966 or greater for records in class (a).

By plotting a number of eclipse paths, and finding the edges of the zone within which the magnitude is 0.966 or greater,† I find that the root-mean-square width of this zone is about 16° in longitude. It takes the earth 64 minutes to turn through 16° in longitude. In the analysis of observations in class (a), we necessarily assume that the observer was on the center line of the eclipse. The maximum error we make in doing so is thus about 32 minutes, and the standard deviation is almost exactly 20 minutes.

That is, the use of a record in class (a) is equivalent to a time measurement with a standard deviation of about 20 minutes. In contrast, the standard deviation of a measured solar eclipse time is about 9 minutes and that of a measured

---

†I did this only for eclipses that were total somewhere, for reasons explained in the preceding section.

lunar eclipse time is about 15 minutes. Hence an observation
which says that a solar eclipse was total at a known place
(a class (a) observation) should receive about half the weight
of a measured lunar eclipse time and about a fourth the weight
of a measured solar eclipse time. The weight for a class (a)
observation from China is even lower.

Now let us turn to the observations in class (b), which
simply say that a solar eclipse was observed at a particular
point, with no clue to the magnitude. M has two objections
to the use of such records. First, he and MS claim that such
records are biased. We discussed this matter in Section III.3.
We saw there that a bias in these records is physically impossible and that the example which MS use in an attempt to demonstrate the presence of bias in fact demonstrates its absence.
Second, M couples the type of observation (class (a) versus
class (b)) to the method used to infer the acceleration parameters. He argues that the method of weighted least-squares,
which is the method I have used in my analysis, is inaccurate.
In fact, on his page 4.8, M "suggests" that the value of the
observations "is wasted if the data are processed by least
squares." He uses instead the method that he calls "minimum
deletion linear inequalities", which he describes [M, p. 9.4]
as a "mathematically exact" procedure. I shall compare the
accuracy of least squares and linear inequalities in the next
section, but I want to close this section by comparing the
relative weights we should give to observations in classes
(a) and (b).

For the records in class (b) that originate from places
other than China, the standard deviation of the magnitude is
almost exactly 0.18, as we saw in the preceding section. We
found there that the records in class (a) have a standard
deviation of 0.034. In order to assign weights, we need the
corresponding standard deviations σ_η of the coordinate η. We
shall see in Section XIII.1 that these values are about
0.100 for class (b) and 0.032 for class (a). Thus a record
in class (a) should receive almost exactly 10 times the
weight of a record in class (b). For the Chinese records
in classes (a) and (b), the standard deviations are larger,
but their ratio is about the same as for the non-Chinese
records. Thus the conclusion is the same for both Chinese
and non-Chinese records.

The number of records in class (b) is about 15 times
the number of records in class (a), and the ratio of weights
for individual records is about 10. That is, the collection
of records in class (b) has about 1.5 times the weight of the
collection of records in class (a).

In summary, most writers on the subject, including M
and MS, take it as a basic premise that records stating
centrality (usually totality) of a solar eclipse at a known
point (class (a) records) are uniquely valuable among ancient
and medieval astronomical observations. Quantitative investigation shows that this is far from the case. Instead, of all
the classes of naked-eye observations that were enumerated a
moment ago, the class of "eclipses central at a known point"
is perilously close to being the least valuable rather than
the most valuable.

3. Least Squares and Linear Inequalities

When we begin the inference of the accelerations, we have observations that we number from 1 to N. For the present, let us speak in terms of the accelerations with respect to ephemeris time, which are $\dot{n}_M$ and y. The ith observation gives us a relation between the accelerations that has the form

$$A_i \dot{n}_M + B_i y = Z_i. \tag{IV.9}$$

With each observation, we associate a weight W_i. We then define a function $F(\dot{n}_M, y)$ that is related to the entire set of observations:

$$F(\dot{n}_M, y) = \tfrac{1}{2} \sum_{i=1}^{N} W_i (A_i \dot{n}_M + B_i y - Z_i)^2. \tag{IV.10}$$

For brevity, I shall omit the arguments of F in later writing unless it is necessary to call attention to them explicitly. I shall also omit the limits on the sum.

We calculate the values of A_i and B_i theoretically for the circumstances of each observation, and we assume that these values are free of error. We assume that observational errors enter only into the values of Z_i, which result from some kind of observation or measurement. Now let us define some new variables by the relations:

$$a_i = A_i \sqrt{W_i}, \qquad b_i = B_i \sqrt{W_i}, \qquad z_i = Z_i \sqrt{W_i}. \tag{IV.11}$$

In terms of these variables, F becomes

$$F = \tfrac{1}{2} \sum (a_i \dot{n}_M + b_i y - z_i)^2. \tag{IV.12}$$

The new "observational" quantities z_i all have the same standard deviation, namely unity. Thus the use of a_i, b_i, and z_i facilitates mixing observations of different weights or classes in a single inference of the accelerations.

In order to explain the method of least squares, let us make up a set of data by a synthetic method. Let us assume that $\dot{N}_M$ and Y are the "correct" values of $\dot{n}_M$ and y, respectively. We use the values $\dot{N}_M$ and Y to calculate the circumstances of each observation and hence to calculate "theoretical" values of the quantities represented by the z_i, and we use ζ_i to denote the set of values obtained this way. Then we pick numbers E_i at random from a population that is distributed according to the Gaussian law with a standard deviation of unity, one number E_i for each observation. We finish generating the synthetic data by letting z_i in Equation IV.12 equal $\zeta_i + E_i$, and we form the function $F(\dot{n}_M, y)$.[†]

[†] Note that we do not use the values $\dot{N}_M$ and Y directly in forming $F(\dot{n}_M, y)$. We use $\dot{N}_M$ and Y only indirectly in calculating the z_i. When we substitute these values of z_i into Equation IV.12, F then naturally becomes a function of $\dot{n}_M$ and y.

F is a positive function that has a minimum for some
pair of values of $\dot{n}_M$ and y. When F has been generated this
way, we know from texts on statistical methods that the pair
of values $(\dot{n}_M, y)$ which minimizes F is the pair $(\dot{N}_M, Y)$. More
precisely, the pair $(\dot{n}_M, y)$ which minimizes F approaches the
pair $(\dot{N}_M, Y)$ as the number of observations approaches infinity.
For a collection containing N observations, there are methods
of calculating an error estimate associated with each parameter and I shall show in Section XIII.3 that these estimates
are proportional to $N^{-\frac{1}{2}}$. Thus the difference between the
pairs $(\dot{n}_M, y)$ and $(\dot{N}_M, Y)$ approaches zero as N approaches infinity.

Since the function F is a sum of squares that we minimize, the method just described is called the method of least
squares. If the errors in the values of z_i are distributed
according to the Gaussian law, the method of least squares
furnishes the best estimate of the parameters that we can
form. The Gaussian law says that the probability dp of finding an error between E and E + dE equals

$$dp = (2\pi)^{-\frac{1}{2}} e^{-\frac{1}{2}(E/\sigma)^2} (dE/\sigma). \qquad (IV.13)$$

In this, σ is the standard deviation of the distribution.
The integral of dp from $-\infty$ to $+\infty$ is unity for any value of σ.

It is often written that physicists believe that the
Gaussian law of error is a law of statistics while mathema-
ticians and statisticians believe that it is an experimental
law of physics. In a simple situation, there are often reasons for suspecting the accuracy with which the Gaussian law
applies. However, the success of the method of least squares
depends in part upon the following law: In a situation where
there are many independent sources of error, the resultant
error distribution tends toward the Gaussian law of Equation
IV.13, whatever may be the distributions for the individual
error sources. Most physical situations are complex ones in
which there are many independent sources of error, and thus
the Gaussian law usually applies with reasonable accuracy.

If we know that the error distribution does not follow
the Gaussian law, we may be able to devise a method of solution that is better than the method of least squares. M
and MS do not use any observations involving the moon, with
trivial exceptions, except the places where solar eclipses
were total or nearly so. For these observations, M and MS
say that we know the error distribution, and it is not Gaussian. Instead, they say, we have the relation IV.1, which
says that the error distribution is uniform between two limits and that there are no errors outside these limits. For
the present, I shall not question the accuracy with which
this distribution applies. Instead, I shall start by describing the statistical method that M and MS use with this
distribution.

In this method, we can start the solution by assuming
a value of $\dot{n}_M$. For this value of $\dot{n}_M$, we calculate the edges
of the central path of an eclipse and plot the edges using
the coordinates of latitude and ephemeris longitude.

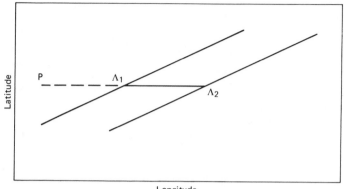

Figure IV.2. Limits on longitude set by a record of a central solar eclipse. If we fix the value of $\dot{n}_M$, we can plot the edges of the zone of totality as functions of latitude and ephemeris longitude. Suppose a record says that the eclipse was total (or annular and central) at point P whose latitude and geographic longitude are known. The ephemeris longitude of P, at the time of the record, was therefore between Λ_1 and Λ_2, if the record is correct.

Figure IV.2 shows the central path in a typical case. Suppose we have a record which says that the eclipse was central at a point P. We know the latitude and geographical longitude of P, and the horizontal coordinate should be interpreted as geographical longitude when we are talking about P, although it should be interpreted as ephemeris longitude when we are talking about the path limits.

The earth's spin acceleration changes the ephemeris longitude of P but not its latitude. If the record is correct, then, the ephemeris longitude of P at the time of the eclipse was between Λ_1 and Λ_2. Let Λ_g denote the geographical longitude of P. If Λ denotes the ephemeris longitude of P at any time, let

$$\Delta\Lambda = \Lambda_g - \Lambda. \tag{IV.14}$$

Then $\Delta\Lambda$ at the time of the eclipse lies between $\Lambda_g - \Lambda_1$ and $\Lambda_g - \Lambda_2$. $\Delta\Lambda$ is related to the parameter y by

$$\Delta\Lambda = -0.0065925yT^2. \tag{IV.15}$$

$\Delta\Lambda_1$ defines a value y_1 of y and $\Delta\Lambda_2$ defines a value y_2 of y. The eclipse then tells us that y lies between y_1 and y_2 if the record is correct and if the assumed value of $\dot{n}_M$ is correct.†

The easiest way to describe the method used by $\underline{M}$ and $\underline{MS}$, which we shall call the method of linear inequalities, is by example. For reasons that will become clear as we go along, the solution in the method of linear inequalities, for any collection of data derived from eclipses that were recorded as central, usually depends in the end upon only four eclipses, which I shall call the critical eclipses. For the collection of eclipses that $\underline{M}$ admits to his data sample for times before +400, the critical eclipses in his analysis are those of -1130 September 30, -600 September 20, +65 December 16, and +120 January 18.‡

For these four eclipses, I calculate the values y_1 and y_2 for two different values of $\dot{n}_M$, and I assume that the y's are linear functions of $\dot{n}_M$. Instead of stating the values of y_1 and y_2 directly, I shall use the form $y \pm \Delta y$, in which y is half the sum of y_1 and y_2 while Δy is half their difference. For $\underline{M}$'s four critical eclipses, I thus get the following equations of condition:‡

$$y = 0.65506\dot{n}_M - 4.0015 \pm 0.3380,$$

$$y = 0.71658\dot{n}_M - 2.5472 \pm 0.4946,$$

$$y = 0.75070\dot{n}_M - 0.4212 \pm 0.5253,$$

$$y = 0.69034\dot{n}_M - 3.9287 \pm 0.9737.$$

(IV.16)

---

†The method I am about to describe is what the method of $\underline{M}$ and $\underline{MS}$ would reduce to if they inferred only the two parameters $\dot{n}_M$ and y. As we saw in Section III.2, they infer two parameters in addition to $\dot{n}_M$ and y, but their method of doing so is incorrect in principle. The fact that their work involves four rather than two parameters means that their method is more complicated in detail than the description that follows, but it does not differ in essential features.

‡As we saw in Section II.2, the so-called record of the eclipse of -1130 September 30 probably does not refer to an eclipse at all, and there is no basis for choosing this date even if it does. The other records are genuine records of eclipses, but there is no reason to assume that they were observed at the places that $\underline{M}$ assumes, for the reasons discussed in Section III.6. In this immediate discussion, we are interested in the way that $\underline{M}$ analyzes the data, and hence I use his data even though they are not correct.

‡In order to reproduce $\underline{M}$'s model with four parameters as closely as possible by a model with only two parameters, I have replaced T^2 in Equation IV.15 by $\Delta T/45.78$ from Equations III.6. $\underline{M}$ actually takes the eclipse represented by the last of Equations IV.16 to be almost but not quite total, and thus he uses only one limit in this equation where I have shown two. This does not affect the general nature of the argument.

The equations are listed in the chronological order of the corresponding eclipses.

When $\dot{n}_M = 0$, the limits on y imposed by the first of Equations IV.16 are -3.6635 and -4.3395, while the limits from the second equation are -2.0526 and -3.0418. Since there is no overlap between the two ranges of y, no solution of Equations IV.16 is possible if $\dot{n}_M = 0$. As $\dot{n}_M$ becomes negative, the ranges of y approach each other and a solution first becomes possible when $\dot{n}_M = -10.106$. For this value of $\dot{n}_M$, the ranges of y allowed by the first and second equations, respectively, are -10.2836 to -10.9596 and -9.2944 to -10.2836.

As $\dot{n}_M$ becomes still more negative, a solution remains possible for a long time but finally becomes impossible again. The last value for which a solution is possible is $\dot{n}_M = -37.174$. For this value of $\dot{n}_M$, the ranges of y are -28.0147 to -28.6907 and -28.6907 to -29.6799.

Hence the first pair of equations in Equations IV.16 restricts $\dot{n}_M$ to the range from -10.106 to -37.174, and it restricts y to the range from -10.284 to -28.691, if we round the y values to three decimal places. Similarly, the other pair of equations restricts $\dot{n}_M$ to lie between -33.275 and -82.944, while it restricts y to lie between -25.926 and -62.162. The set of all four equations thus provides the "solution"

$$-37.174 \le \dot{n}_M \le -33.275,$$

$$-28.691 \le y \le -25.926. \quad \text{(IV.17)}$$

The solution in relations IV.17 is given by means of inequalities. It is customary to give a solution by means of a central value, which is what we usually mean when we refer to an estimate of a parameter, and an uncertainty that is attached to the estimate. In this style, relations IV.17 become

$$\dot{n}_M = -35.224 \pm 1.950,$$

$$y = -27.308 \pm 1.382. \quad \text{(IV.18)}$$

Even Equations IV.18 do not give the solution in the form we finally want. It is customary to express uncertainties by means of standard deviations, but the uncertainties in Equations IV.18 give the extreme limits if it is indeed true that the eclipses in question were central at the places we assumed when we derived the equations of condition (Equations IV.16). Hence we should divide the uncertainties in Equations IV.18 by $\sqrt{3}$, obtaining finally:

$$\dot{n}_M = -35.224 \pm 1.126,$$

$$y = -27.308 \pm 0.792. \quad \text{(IV.19)}$$

We can also solve Equations IV.16 by the method of least squares. We start by replacing the uncertainties quoted in Equations IV.16 by the corresponding standard deviations, and we do this by dividing each uncertainty in Equations IV.16 by $\sqrt{3}$. In Equation IV.10, which defines the "least-squares" function $F(\dot{n}_M, y)$, we then assign the weights W_i to be inversely proportional to the standard deviations just found. I believe that the rest of the process is obvious, and the least-squares solution of Equations IV.16 is

$$\dot{n}_M = -34.223 \pm 3.520,$$

$$y = -26.576 \pm 2.437.$$
(IV.20)

The standard deviations shown in Equations IV.20 are obtained from the standard deviations attached to the individual observations by the method described in Section XIII.3.

The central values in Equations IV.19 and IV.20 do not differ significantly, and thus I do not understand the already quoted contention of $\underline{M}$ and $\underline{MS}$ that the use of least squares destroys the value of the data. The method of linear inequalities (Equations IV.19) seems to give standard deviations that are only about a third of those given by the method of least squares, but I shall show in the next section that this is not correct. The standard deviations shown in Equations IV.20 are seriously over-estimated, for reasons that I shall describe in detail.

It is now desirable to define carefully what we mean by erroneous observations (or equations of condition), and what we do when we delete observations.

Suppose that we have records of the eclipses of 1133 August 2 and 1140 March 20 observed at point A, and that there is no question about the point of observation. Suppose further that the original records say that the first eclipse was total but say nothing about the magnitude of the second. What happens if a copier accidentally transfers the statement of totality from 1133 August 2 to 1140 March 20?

The original record of 1133 August 2 belongs in the class of records that asserts a central eclipse, and the original record of 1140 belongs in the class that says nothing about the magnitude. The mistake in copying puts the record of 1140 March 20 into the class that asserts centrality while putting the record of 1133 August 2 into the class that says nothing about the magnitude. Thus the copier has created two records that do not belong in the classes to which we assign them.

We express this situation by saying that the class of records that assert centrality, for example, contains two different populations. One population contains the records that asserted centrality in their original forms; we often say that a population of this sort is the parent population. The other population contains records that have been altered from their original form, and we say that such records do not belong to the parent population.

When we refer to erroneous records, we usually mean records that do not belong to the parent population. In every situation I can think of, we want to remove such erroneous records or observations before we infer the parameters involved in the problem. There are two basic methods of doing so.

As an example of the first method, suppose we can show from historical evidence that a particular eclipse record was written in place A and that it was later copied in place B. The first record belongs to the parent population while the second does not. Hence we remove the second record from the data sample. Removing an "observation" from the data sample in this way differs from the process called deletion that I shall define in a moment.

In the second method, we use all the observations that have not been removed by the first method, and we infer values of the relevant parameters from the remaining observations. We then calculate the circumstances of each observation theoretically using these parameters, subtract the calculated value from the observed value, and call the difference the residual. We next examine the residuals in detail, paying particular attention to the frequency with which residuals of a particular size occur. If we have done our work correctly and if all the data belong to the parent population, we expect to find that the residuals form a cluster. If we find residuals that lie far outside the cluster, we suspect that the corresponding observations do not belong to the parent population. We then remove these observations and repeat the inference of the parameters. I use "deletion" to mean the removal of observations accomplished this way.

The crucial distinction between deletion and other types of removal is the following: The decision to delete is based upon the size of the residual and hence the decision is a function of the observed value itself. Other kinds of removal are accomplished before the residuals are known and hence the decision to remove is not a function of the observed value.

The mere fact that we find a few residuals lying far outside the cluster is not in itself a justification for deletion. We are justified in deleting observations only if we have strong grounds, based upon considerations other than the size of the residuals, for believing that our sample may contain observations that do not belong to the parent population. Further, we should never delete more than a small fraction of the total sample.

As it turns out, I did not delete any data in MCRE, using the technical definition of deletion that was just given, and I shall not delete any data in this work. The matter of deletion is important nonetheless because it has been necessary to do some deletion in all of my work that uses data other than the places where solar eclipses were observed.

I shall close this section by referring to some further advantages which M claims for the method of linear inequalities over the method of least squares. In every case I noticed, the advantage claimed has nothing to do with the issue, and I shall give two examples.

On his pages 9.10-9.11, $\underline{M}$ writes: ". . . the linear inequalities filter is virtually unaffected by a limited number of erroneous equations!† In particular, the worse the equation is (if in error), the less chance it has of affecting the solution at all. This is in marked contrast to least squares, where even a weakly weighted equation will markedly affect the solutions if its conditions are grossly in error."

I cannot find a basis for this statement. The method of linear inequalities is, if anything, more sensitive to erroneous observations than the method of least squares, and we shall see a striking example of the sensitivity of linear inequalities in Section IV.5. Here we can look at a simpler example of what $\underline{M}$ apparently has in mind.

Since we are merely looking for a numerical illustration of the problem, let us assume that Equations IV.16 represent valid observations. Let us suppose further that the "observational" quantity -4.0015 in the first of Equations IV.16 has somehow been changed to -2.9875, a change of 3 times the limiting uncertainty shown. This is more than 5 standard deviations. For brevity, I shall identify the equation in question as Equation A.

Equation A and the equation following it can be solved only if $\dot{n}_M$ is in the range from +6.3768 to -20.6908, and if y is in the range from +1.5277 to -16.8792. We remember that the second pair of equations restricts $\dot{n}_M$ to the range from -33.275 to -82.944 and that it restricts y to the range from -25.926 to -62.162. Since there is no overlap between the two ranges, the equations have no solution in the method of linear inequalities.

However, this is not exactly what would happen in dealing with the data sample. The four equations in Equations IV.16 are only the critical equations from the data sample, but there are, say, 15 equations in the sample altogether. When we start to solve the problem with all 15 equations, we do not know which 4 equations are the critical ones, nor do we know that Equation A is in error. We soon discover that the set of equations has no solution.

Without writing down the other 11 equations, we cannot say exactly what will happen, but the probability is that events will unfold in the following way: After we discover that the set of 15 equations has no solution, we try the effect of deleting each equation in turn, and we discover that we can find a solution if we delete Equation A. In all probability, we cannot find a solution by deleting any other single equation. In order to find a different solution, we would have to delete at least two equations with most samples of data formed this way. We are more likely to have a single serious error than to have two or more serious errors, so the solution with the maximum likelihood is the one that deletes Equation A.

---

†The exclamation point is $\underline{M}$'s.

Hence we delete this equation and solve the remaining 14 equations in the usual way by the method of linear in-equalities. If the remaining equations are indeed correct, as we have assumed, the new solution is probably close to the original one, which was given in Equations IV.19. Thus, as $\underline{M}$ says, a serious error has little effect on the solution.

Events go in almost exactly the same way with the method of least squares. We start with all 15 equations and form the function F in Equation IV.12. The fact that Equation A is seriously in error does not keep us from finding the val-ues of $\dot{n}_M$ and y that minimize F. Using these values of $\dot{n}_M$ and y, we calculate the residuals and examine them. Fourteen of the residuals form a tight cluster, since they come from correct observations, but the residual from Equation A lies outside the cluster by about 5 standard deviations. Since the probability of finding an error of 5 standard deviations in a sample size of 15 is about 1 in 10^5, the odds are high that Equation A is in error. We delete it and solve again with the remaining equations. The new solution differs but little from the solution in Equations IV.20.

We see that neither method is affected particularly by the presence of a large error, provided that we solve the problem by using well-known and accepted methods of deleting an individual observation. Thus $\underline{M}$'s point has nothing to do with the relative merits of linear inequalities and least squares. Instead, it deals with the merits of deleting a small number of observations that disagree seriously with the main body of data.

On page 9.7, $\underline{M}$ writes: "Newton (1970)† notes that certain eclipse geometries cannot be framed in terms of least squares equations when the geometry is unfavorable." He then goes on to explain how these same geometries are handled by the method of linear inequalities, thus giving another example of the superiority of the latter method.

$\underline{M}$'s statement about eclipse geometries is not correct. Figure IV.1 shows the situation that he is referring to. If the point of observation has the same latitude as P_1, we get two possible equations of condition from the observation. One comes from using the western portion of the eclipse path and the other comes from using the point with the same lati-tude on the eastern portion. One equation is correct and the other is erroneous, and the task is to eliminate the wrong one.

When the point of observation has the latitude of P_1 in Figure IV.1, the two equations differ by a large amount and there is no difficulty in eliminating the wrong equation. However, as the latitude approaches that of P_2, the equations approach each other, and we need accurate knowledge before we know which equation to delete.

In ten cases in $\underline{AAO}$, the point in question was so close to P_2 that I could not choose between the two equations with

---

†This is a citation of $\underline{AAO}$ [$\underline{Newton}$, 1970]. The passage in question is on pages 258-259 of $\underline{AAO}$.

the limited knowledge of the situation that I had when I wrote AAO. Since I could not decide which equation to eliminate, I was forced to drop these cases from the inference. These are the cases that M is referring to; he is saying that they could have been included if I had used the method of linear inequalities.

Again we see that the matter has nothing to do with linear inequalities versus least squares. If we have enough information, we can eliminate the wrong equation whichever method we use. M failed to note that I used all the medieval cases in question in MCRE† with no difficulty, even though I used the method of least squares; the reason was that the information from AAO allowed me to make the necessary decisions when I wrote MCRE. In the present work, I shall use all the cases in question, whether ancient or medieval, that represent genuine observations.

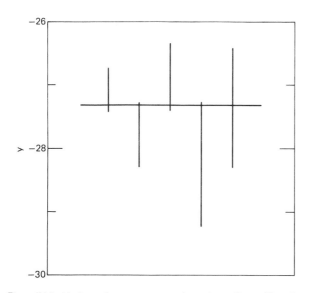

Figure IV.3. Limits on the parameter y set by various eclipses. The values of y are calculated on the assumption that $n_M = -35.224$. The eclipses represented by the first four bars set narrow limits to the value of y in the method of linear inequalities. The fifth eclipse straddles the limits set by the first four eclipses and thus it does not contribute to the solution for y. Most eclipses have the characteristic of the fifth eclipse plotted, and thus they do not contribute to the solution in the method of linear inequalities.

---

†MCRE dealt only with medieval observations.

4. The Probability Distribution for an Eclipse Recorded as Central

Before we take up the probability distribution expected for records which assert that an eclipse was central, let us look again at the solution of Equations IV.16. If we use $\dot{n}_M = -35.224$ (Equation IV.19), we find that y must lie between -26.74 and -27.41 for the first eclipse, and so on. The limits of y imposed by the four eclipses, for this value of $\dot{n}_M$, are shown in Figure IV.3.

The vertical bar farthest to the left in Figure IV.3 shows the limits imposed by the first of Equations IV.16, which, in $\underline{M}$'s erroneous assumption, represents a record of the eclipse of -1130 September 30. The middle three bars in the figure show the limits imposed by the other equations. I shall explain the bar farthest to the right in a moment. The horizontal axis in Figure IV.3 has no significance; I have simply separated the various vertical bars by a convenient amount.

The horizontal line in the figure is the value y = -27.308 from Equations IV.19. We see that the vertical bars corresponding to the four eclipses are not consistent with a value of y that differs much from this; this is the reason that these four eclipses are important in $\underline{M}$'s analysis. However, there are many eclipses in $\underline{M}$'s sample besides these four. The bar on the right shows the limits on y imposed by a typical eclipse in $\underline{M}$'s sample. The limits imposed by this eclipse lie well to either side of the limits allowed by the first four, and such an eclipse does not play a direct role in $\underline{M}$'s analysis.

Although the last eclipse in Figure IV.3 does not play a direct role in the method of linear inequalities, it is as valuable as any other eclipse in the method of least squares. Altogether, the method of linear inequalities typically uses perhaps a fourth of the total sample that would be used in the method of least squares.

I mentioned in the preceding section that the standard deviations quoted in Equations IV.20, which come from applying the method of least squares to Equations IV.16, are overestimated by a large amount. We can now see why this is so. Each standard deviation in Equations IV 20 is the product of two factors. One is proportional to $\sqrt{N}$, in which N is the number of observations, and the other is the standard deviation of the residuals. By a residual, I mean the difference between an observational quantity and the value of the same quantity calculated from theory using the inferred values of the parameters.

We have seen that N in the method of least squares is about four times the number of critical equations, which is all that were used in deriving Equations IV.20. Thus the standard deviations in Equations IV.20 should be divided by 2 for this reason. Further, we see from Figure IV.3 that the observations which are not among the critical ones in the method of linear inequalities tend to have smaller residuals than the critical observations have. Thus the standard

deviations in Equations IV.20 should be divided by a further
number that we cannot specify exactly but that is substantially larger than unity.

Altogether, then, we divide the standard deviations in
Equations IV.20 by a number that is probably not far from 3.
Hence there is no important difference between the standard
deviations obtained by the two methods. Since this result
may be an accident arising from the particular example used,
we need to make a more general inquiry. Therefore we ask:
Under what circumstances, if any, do we expect the method of
linear inequalities to be better than the method of least
squares?

When the error distribution follows the Gaussian law,
we know that the method of least squares is the best method
of estimating the parameters. The method of linear inequalities can be better only if the probability distribution is
not Gaussian. As an extreme example, let us confine our attention for the moment to records which assert that an eclipse
of known date was central at a known point. Further, let us
suppose for the moment that this assertion is correct, with
no possibility of error.

In this example, let x be the distance from the point
where the observation was made to the center of the eclipse
path, and let 2ℓ be the width of the path. Then the parameter x/ℓ equals -1 at one edge of the path and +1 at the
other. The value of x/ℓ for the point of observation, under
the assumptions just made, certainly lies between -1 and +1,
and it may be anywhere between these limits with equal probability. Its probability of being outside these limits is 0.

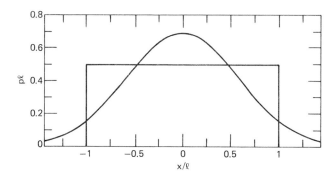

Figure IV.4. Two probability density functions. If we know that an eclipse
was central at a specified point, the variable x must lie between the limits
$-\ell$ and $+\ell$, and the probability density p is uniform between those limits.
Hence p is $0.5/\ell$ within the limits and 0 outside them. In the corresponding
Gaussian curve of error, the standard deviation is $\ell/\sqrt{3}$, and the maximum
probability density is $0.6909/\ell$. The figure shows the product $p\ell$ as a function
of x/ℓ.

The corresponding probability density p is shown by the rectangular distribution in Figure IV.4. Between $x/\ell = -1$ and $x/\ell = +1$, p equals $1/2\ell$. The quantity that is plotted, namely $p\ell$, equals 0.5 between those limits and 0 outside those limits.

The standard deviation of x/ℓ for the rectangular distribution is $1/\sqrt{3}$. The Gaussian probability distribution for this standard deviation is shown by the bell-shaped curve in Figure IV.4. The maximum value of p is $0.6909/\ell$, and p falls to about $0.16/\ell$ at the points where $x/\ell = \pm 1$. Within the range for which the rectangular distribution is uniform, the Gaussian distribution varies by a factor of about four. This means that the method of least squares "tries harder" to put the point of observation at $x = 0$ than the method of linear inequalities does.

M [p. 9.6] says that this effect "introduces bias into the equations of condition themselves, . . ." This is clearly incorrect. The equations of condition are the equations that have the form and function of Equations IV.16. When we change the probability density from the rectangle to the Gaussian curve in Figure IV.4, we do not change the coefficient of y, nor the coefficient of $\dot{n}_M$, nor the term that is independent of y and $\dot{n}_M$. (This term is the "observed" value.) We change only the uncertainty quoted, and this does not constitute a bias.

We can put the matter another way. When the method of least squares "tries" to put the point of observation on the center line of the eclipse path, it is just as likely to try to move the point to the east as to the west, to the north as to the south. Thus the "attempt", which is clearly symmetrical, cannot produce a bias in the distance from the point to the center of the path. In fact, if we use any probability distribution whatsoever, no matter how unreasonable it looks, we do not introduce a bias, provided only that the mean value of x for the distribution is zero.

Up to this point in the immediate discussion, I have accepted the supposition that an eclipse of known date was central at a known point if the record says so. As we saw in Section IV.1, we are not entitled to make this supposition. With the non-Chinese records which state explicitly that an eclipse was total, we saw that many of the eclipses were not total, and that the standard deviation of the magnitude for such records is 0.034. This means that the actual error distribution falls off even more slowly than the bell-shaped curve in Figure IV.4.

M himself assumes that the rectangular distribution is not correct, although he does not make this explicit statement in any place that I noticed. To each record, M assigns a number that he calls the probability of truth. This is his estimate of the probability that the date, magnitude, and place of observation that he uses in the analysis are all "true", and thus M's probability of truth is the same thing as the number that I call [AAO, MCRE] the "reliability". For the four eclipses used in the example of Section IV.3, M assigns probabilities of truth of 0.50, 0.90, 0.80, and

0.98. M̲ does not use a record to which he assigns a value less thān 0.50. For details about how the probability of truth is used, I refer the reader to Chapter X of M̲.

What does it mean to assign a probability of truth of, say, 0.80? It means that the total probability that x lies between -ℓ and +ℓ is not unity, as it is in Figure IV.4. Instead, the probability is only 0.80, and the remaining probability of 0.20 lies outside the limits ±ℓ. It is safe to assume in almost every case that the date is correct† and hence to assume that the probability of error concerns either the magnitude or the place assigned to the observation.

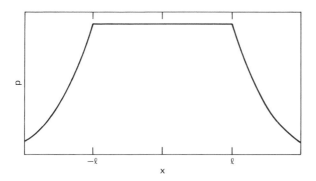

Figure IV.5. An alternative probability density function. In a real case, we cannot be certain that an eclipse was total at a specified spot. Thus the probability density, though constant between −ℓ and +ℓ, does not vanish outside those limits. Instead, it falls gradually to 0.

I think it is clear that the probability of committing an error is greater for small errors than for large errors, whether we are dealing with the magnitude or the place. Thus the correct probability distribution seems to be like the distribution shown in Figure IV.5. Between the values ±ℓ, the probability density is uniform and equal to the probability of truth multiplied by 0.5/ℓ. Outside the values ±ℓ, the probability gradually falls to zero.

Now what can we say about the relative merits of the methods of least squares and linear inequalities? In order to answer this question carefully, it is convenient to introduce some terminology that I have not used yet. To start with, I shall use the term "estimator" to designate a particular method of inferring a set of parameters from a set of observations. Let there be n parameters to be inferred, and let the set of parameters be represented by a column

---

†However, this assumption is not correct for the "record" of the eclipse of −1130 September 30 that M̲ uses.

vector $\underline{x}$ that has n components. Let there be N observations, and let the set of observations be represented by a column vector $\underline{Z}$ that has N components.

I now wish to divide estimators into linear estimators and non-linear ones. To do so, we note that $\underline{x}$ can always be written as the product of a matrix $\underline{\underline{M}}$ and the vector $\underline{Z}$:

$$\underline{x} = \underline{\underline{M}} \cdot \underline{Z}.$$

$\underline{\underline{M}}$ is a matrix with n rows and N columns.

The matrix $\underline{\underline{M}}$ in the method of least squares is independent of the observation vector $\underline{Z}$, so that $\underline{x}$ is a linear function of $\underline{Z}$. That is, the least-squares estimator is linear. The least-squares estimator when used with deleted data, however, is not. To see this, suppose that the Kth observation is deleted. This means that we first calculate all the residuals and then change all the coefficients in the Kth column of $\underline{\underline{M}}$ from their original values to 0, and we do this because of the size of the Kth residual. That is, the matrix $\underline{\underline{M}}$ is now a function of the Kth component of $\underline{Z}$ and hence it is a function of $\underline{Z}$. Since $\underline{x}$ is the product of $\underline{Z}$ with a function of $\underline{Z}$, $\underline{x}$ is a non-linear function of $\underline{Z}$ and the deleted least-squares estimator is a non-linear estimator. The linear-inequalities estimator is also non-linear.

From texts on advanced statistical methods, we can learn the following: If the error distribution follows the Gaussian law of error, the least-squares estimator is the best possible estimator. For any error distribution, the least-squares estimator is unbiased and, as we have seen, it is linear. For any unbiased error distribution whatever, the least-squares estimator is the best unbiased linear estimator. If the error distribution is not Gaussian, there may be a non-linear estimator that is better than least-squares.

There is no way known to me to prove whether the method of linear inequalities is better than the method of least squares or not. Even if we could assume that the rectangle in Figure IV.4 gives the correct error distribution, we still do not know which method is better. All the more so, then, we do not know which is better for the actual error distribution in Figure IV.5. There is no basis for the claim of $\underline{M}$ and $\underline{MS}$ that the least-squares estimator is markedly inferior to the linear-inequalities estimator.

I believe that we are entitled to appeal to the principle of continuity in this problem. This principle says that any measurable property of a physical system is a continuous function of its arguments. In other words, a small change in one property of a system cannot cause a large change in some other property. We see from Figure IV.5 that the actual error distribution for the eclipse problem is reasonably close to the Gaussian curve of error. For the Gaussian curve, the least-squares estimator is the best one. For the actual curve of error, then, the least-squares estimator is reasonably close to the best one. Since it is close to the best estimator when we use eclipses, and since it is the natural estimator to use with other types of data, I choose to use it for the eclipse data as well, in order to facilitate combining eclipses with other observations.

In the example of the preceding section, we saw that the method of linear inequalities is about as good as the method of least squares, and it may be slightly better, if we confine ourselves to the use of eclipse records that assert totality or centrality. Thus a person may legitimately use the method of linear inequalities with the eclipse data if he chooses to do so. If he does choose to do so, however, it seems to me that two warnings should be explicitly attached to his results: (a) Since the method involves an incorrect assumption about the error distribution, the uncertainties given formally by the method are too small. The same warning applies, of course, to the method of least squares; the question is which method gives more accurate estimates of the uncertainties. (b) The method of linear inequalities contains a serious danger that is not present in the method of least squares. This danger will be illustrated in the next section.

5.   The Eclipse of 120 January 18

We saw in Figure IV.3 that most records do not directly affect the solution when we use the method of linear inequalities and that the solution depends upon only a few records, perhaps only a fourth or less of the total. In the method of least squares, on the other hand, all valid records affect the solution.

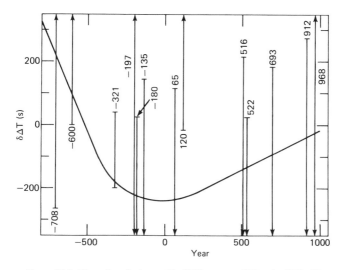

Figure IV.6. The eclipse limits used by MS between −800 and +1000. The eclipse record of 120 January 18 does not seem to be consistent with the other eclipse records. If it is deleted or corrected, the nature of the solution can change drastically. The curve drawn in the figure is only one of many curves for δΔT that become possible when the limits for 120 January 18 are changed. Thus the solution found by M and by MS depends critically upon a single record.

Now let us suppose that there is an error in our interpretation of one of the records. If the error is quite large,
we will probably discover it readily, and we can then delete
it whether we use the method of linear inequalities or the
method of least squares. What happens if the error is not
large enough to be obvious? The record involved is only one
of many used in the method of least squares and thus it does
not affect the solution in an important way. In the method
of linear inequalities, it does not affect the solution at
all unless it is one of the records that happens to set overall limits, in the way shown in Figure IV.3. However, if it
is one of the records that does limit the parameters, the error has a serious effect upon the solution.

Just this may have happened, I believe, with the eclipse
of 120 January 18. We saw in Section IV.3 that this is one
of the critical eclipses in the solutions found by $\underline{M}$ and by
$\underline{MS}$. This eclipse was observed somewhere in China, and $\underline{M}$ and
$\underline{MS}$ take the specific place of observation to be Luoh-yang,
which was the capital at the time. The situation is shown
in Figure IV.6, which is plotted using the calculations that
are summarized on page 509 of $\underline{MS}$.

In these calculations, $\underline{MS}$ use the value $\dot{n}_M = -37.5$,
which is the value in their final solution for the accelerations. With this value of $\dot{n}_M$ they calculate the limiting
values of ΔT that correspond to the path limits for each observation. Then they calculate a value of ΔT from their
formula:

$$\Delta T = 66.0 + 120.38T + 45.78T^2 \text{ seconds;}$$

this is their final solution for ΔT. The quantity $\delta \Delta T$ that
is plotted in Figure IV.6 is the value of ΔT obtained from
an observation minus the value obtained from this formula.
In other words, the values of $\delta \Delta T$ are the residuals in the
analysis of $\underline{MS}$.

The earliest class A record[†] that $\underline{MS}$ use is a Chinese
record of the eclipse of -708 July 17.[‡] Between then and
968 December 22, $\underline{MS}$ use records of 13 eclipses. ($\underline{M}$ also has
a record of 484 January 14 in his data sample, and, as we
have just seen, the "record" of -1130 September 30 has a
major position in his analysis.) The limits on $\delta \Delta T$ for these
13 eclipses are plotted in Figure IV.6.

In most cases, one limit for each eclipse lies outside
the scale of Figure IV.6, and $\underline{MS}$ do not tabulate the missing
limit. A missing limit is indicated by an arrow at one end
of the appropriate vertical bar in the figure. A limit that

[†]$\underline{MS}$ class their records as class A or class B. Class A
records are those in which they feel high confidence. Class
B records are those in which they feel some appreciable uncertainty about either the place of observation or the magnitude. I could not decide from their writing whether they
did not use the class B records at all or whether they used
them but with lower relative weights than the Class A ones.

[‡]Both $\underline{M}$ and $\underline{MS}$ give this date as -708 July 8.

is within the scale is indicated by a short horizontal line
at one end of the vertical bar. In a typical case, the
total length of a vertical bar is 1500 to 2000 seconds, so
that most limits are in fact far off of the scale. The
eclipse record of -321 September 26 is of a different type
from the others. The eclipse was very small at Babylon,
where the observation was made, but the observer measured
the time when the eclipse began. MS assign an error of ± 2
minutes to the time measurement,† for a total bar length of
240 seconds.

The lower limit appears in Figure IV.6 for the eclipse
of 120 January 18, but only upper limits appear for all
other eclipses for many centuries on either side of this
one.‡ Both M and MS allow this fact to govern their solution for the accelerations. I would draw a different conclusion from this fact: Since the eclipse of 120 January 18
disagrees with all other data in the sample over a period of
more than a millenium, it is likely that we have made an error in interpreting the record.

In my opinion, an error in the interpretation is highly
likely on textual grounds alone. As we saw in Section III.6,
MS conclude that Chinese observations of eclipses were made
in the capital‡ unless the record has an explicit statement
to the contrary. The record of the eclipse of 120 January
18 does not contain a statement about where it was observed,
and so MS (and M) take the place to be Luoh-yang, the capital
at the time. I find two serious objections to this conclusion:

(a) We studied relevant evidence in considerable detail
in Section III.6 for the Tarng and Former Han dynasties,
which are the only dynasties for which detailed evidence is
available. We saw that the evidence from the Tarng dynasty
contradicts the conclusion at a confidence level that is near
certainty. The evidence from the Former Han dynasty does not
contradict the conclusion, but it does not support it either.
Since there is no basis for the conclusion in either dynasty
that has been studied in detail, there is no basis for it in
the Later Han dynasty, which has not been studied in detail.

(b) Even if we should grant that the conclusion is correct, there is still a serious difficulty. The conclusion
that the observation was made at the capital does not rest
upon an actual statement that appears in the record. Instead,
it rests upon the absence of a statement, and every historian

---

†MS [p. 481] say that there is also a delay in detecting the
beginning of the eclipse, which they will discuss in their
Part III. I have not been able to find this discussion.
M [p. 8.27] says that the delay should be about 2 minutes,
but that it depends upon several factors. If I understand
his Figure 10.1, he concludes that the delay was actually
somewhat more than 4 minutes.
‡Except for the "timed" record of -321 September 26.
‡We remember that this means, in the terminology of MS, the
place where the emperor was; it is not a fixed position.

knows that a conclusion based upon the absence of information is a risky one.

In assessing the importance of this absence of information, we must remember that we do not have the original records. The information that we have comes from a compilation made [MS, p. 491] between the years 398 and 445; this is about three centuries after the event. So far as I am aware, we do not know the literary history of the records between the event and the compilation, and so we do not know whether we are dealing with secondary, tertiary, or even higher order records.

With regard to this absence of information, M [p. 8.39] says that "there is no significant chance that the lengthy expression regarding a provincial source could have been left off this observation, . . ." I fear that the evidence gives no basis for this statement. To give a counter-example from Chinese records, we saw in Section III.6 that the compiler of the astronomical treatise for the Tarng dynasty accidentally transferred a statement of centrality from the actual eclipse of 888 April 15 to the (inaccurately) calculated date of 879 April 25. I shall give two other counter-examples out of many, both connected with European records of the eclipse of 1039 August 22. The information about both these examples is found on pages 342-343 of MCRE.

The French chronicler Rodulfus of Cluny has left us lengthy records of the eclipses of 1033 June 29, 1039 August 22, and 1044 November 22. It is probable that these records are based upon Rodulfus's personal observations. About 60 years later, Hugo of Flavigny compiled a new chronicle and used the chronicle of Rodulfus as one of his sources. When he came to deal with the eclipses, Hugo started by putting down the date for 1039 August 22 but not the date for 1033 June 29. He then copied, under the date 1039 August 22, the first half of the record of Rodulfus for 1033 June 29, and completed this entry by copying the second half of Rodulfus's record for 1039. He then copied accurately the record of Rodulfus for 1044 November 22, except that he compressed it somewhat by omitting a few details. Thus what started out as three records in Rodulfus's chronicle ended up as two records in Hugo's chronicle.

The chronicler Clarius of Sens, about 20 years after Hugo, then compiled still a different chronicle and used Hugo's chronicle as one of his sources. Amazingly, he did almost the same thing as Hugo. He started by putting down the date 1039 August 22 and copied the first half of Hugo's record for this date; we should remember that this is already a conflation of Rodulfus's records for 1033 June 29 and 1039 August 22. His eye then wandered to Hugo's record for 1044 November 22 and he copied the last half of this record, still under the date of 1039 August 22. He then went on from 1044, so that he did not compile the eclipse of 1044 November 22 into his chronicle at all. Thus what started out as records of three eclipses in Rodulfus's chronicle ended up as a record of a single eclipse in Clarius's chronicle, by the process of transferring "lengthy expressions" from one record to another, accompanied by the omission of still other "lengthy expressions".

The records that precede and follow the eclipse of 120 January 18 are important in assessing the record of 120 January 18. These are the records dated 118 September 3 and 120 August 12.† In both these records, it is stated that the eclipse was not observed in the capital but that it was reported from the provinces [M, p. 8.39 and MS, p. 491]. There is certainly a reasonable probability that one of these remarks was originally attached to the record of 120 January 18 and that the compiler, from inattention or boredom, accidentally transferred it to one of the adjoining records.

In fact it is an even bet that this is just what happened with the record dated 120 August 12. The eclipse of 118 September 3 was total in the regions just south of the modern China and it was readily visible in any of the provinces.‡ However, there was no eclipse on 120 August 12, although this was the date of a new moon. There was an eclipse one month earlier, on 120 July 13, but this eclipse was not visible anywhere in Asia, Europe, or northern Africa.‡

In fact, we cannot explain the record of 120 August 12 by an error in dating. The first eclipse that was visible in China after 120 January 18 was on 124 October 25, and it is duly recorded. Thus any eclipse that appears between these dates seems to have only one explanation: It must have been a calculated or predicted eclipse, and it could not have been observed in the provinces, as the record says. The remark which says that it was so observed must have been transferred by accident from the record of some other eclipse. We may safely assume that the remark was originally attached to one of the records immediately adjacent to 120 August 12 rather than to a record lying at a greater interval.

The preceding eclipse, as we have said, is 120 January 18 and the following eclipse is 124 October 25. Both eclipses were fairly large anywhere in China and the remark could have been attached properly to either of them; weather was presumably the reason why the eclipse was not seen in the capital. It is equally likely that the remark was originally attached to either date. Hence the probability is 0.5 that the remark now dated 120 August 12 was originally attached to the record of 120 January 18.

Further, as we have also said, the record of 118 September 3 says that the eclipse was not observed in the capital. There is a small but finite chance that this remark

---

†The relevant records appear in Gaubil [1732b, p. 277], with the year being incorrectly given as 119 for the eclipse of 120 January 18. Most of the Chinese year in question came during our year 119, and Gaubil failed to note that the eclipse date had passed our 119 December 31 and gone into the year 120.

‡It was also readily visible in the capital, weather permitting.

‡The northernmost point on its path was at latitude 26°S.

was originally attached to 120 January 18 and accidentally transferred to this date. Altogether, then, the probability is more than 0.5 that the record of 120 January 18 originally said that the eclipse was not observed in the capital.

Let us put the matter another way. M interprets the record of 120 January 18 to mean that the eclipse of that date was observed in the capital, and he attaches a probability of 0.98 to the truth of this interpretation. However, as we have just seen, the probability is more than 0.5 that the eclipse was not observed in the capital. This means that the probability of truth for M's interpretation is less than 0.5 instead of being 0.98. By M's rules of analysis, then, he should have omitted this record from his study, and MS should have done the same. This omission would have had a profound effect upon their results.

Now let us return to Figure IV.6. The place where the eclipse of 120 January 18 was observed may easily have been as much as 5° in longitude from the place that M and MS assume. This is the equivalent of 1200 seconds in time, which means that the limits one should assign extend 1200 seconds in both directions beyond the limits that M and MS use. Thus both limits for 120 January 18 are off the scale of Figure IV.6, and the eclipse gives no limits that are useful in the method of linear inequalities. It is still valuable in the method of least squares, however, and I shall use it with a suitable weight.

In the solution that MS find, $\delta\Delta T$ is given by the horizontal axis, and thus $\delta\Delta T$ is a constant, namely zero. Now that the eclipse of 120 January 18 no longer provides a critical limit, this is not necessary. The curve shown in Figure IV.6 is now a possible solution, but it is not the only one. MS do not tabulate the limits indicated by arrows in Figure IV.6, and I have not calculated them. However, we know that the typical length of a vertical bar is around 30 minutes, or 1800 seconds, and we can assume this length in the present approximate discussion. This fact, combined with an extension of the limits for 120 January 18 by 1200 seconds, means that possible solutions can lie almost entirely off the scale in Figure IV.6. The only requirement that keeps the solution within the figure at all is the requirement that the curve for $\delta\Delta T$ should pass close to the bar shown for -321 September 26.†

In fact, when we look at Figure IV.6, I do not see how the method of linear inequalities can yield a precise solution for the accelerations, nor how it can even tell us whether the accelerations are constant or variable, unless

---

†This is not an observation of a large eclipse, but a measurement of the time when a small eclipse began. Although both MS and M play down the value of ancient measurements of time, this record in fact gives the tightest limits of all the valid ones that MS and M consider before the year +1000. It is ironic that M deletes this record on the basis that it is too inaccurate. The true significance of this record is consistent with the conclusion of Section IV.2.

the data sample is many times the size of M's and MS's sample.

In a remark that I quoted near the end of Section IV.3, M says that erroneous equations of condition have virtually no effect in the method of linear inequalities but that they can have a serious effect in the method of least squares. We see now that the situation is just the opposite. Since the method of least squares uses all of the data rather than only a small part of them, a moderate error in an equation of condition is diluted to the point of being unimportant. In the method of linear inequalities, there are two possibilities. If an erroneous equation does not provide one of the critical inequalities, it does not affect the solution at all. However, if it does happen to provide a critical inequality, as the question-able equation that M and MS use for 120 January 18 does, an er-roneous equation will have a serious effect upon the solution.

In my opinion, it is dangerous to use a method that al-lows any single observation to have an important effect upon the solution. This is one of the reasons why I have always chosen to use the method of least squares in the study of the astronomical accelerations.

6. The Lunar Node

In Section IX.2 of AAO, I used measured magnitudes of partial lunar eclipses in an attempt to see whether there is a perturbation in the lunar node beyond that given by the standard theory. However, because the magnitudes were not measured accurately, and because they are not very sensitive to changes in the parameters, it was not possible to reach any conclusions. In the rest of AAO and in subsequent work, I have ignored possible perturbations in the node, after con-vincing myself [AAO, p. 287] that nodal perturbations do not affect the inference of the accelerations.

M notes this conclusion on his page 3.5, where he writes: "Newton . . dismisses node as a possible cause of systematic[†] error in the analysis of ancient observations on the grounds that the effects average zero. His argument is sound for the timed[‡] observations, but as we shall now see, it is incorrect for the untimed data including the large solar eclipses (which dominate his solutions and those of this paper)."

By way of background, we should note that a small change in the position of the lunar node changes the latitude of the moon at a particular time but it does not change the longi-tude of either the mean moon or the true moon. Thus a change in the node does not affect the observations in which a time is measured. This includes the rising and setting times of the moon, lunar conjunctions, and the times of eclipses. How-ever, a change in the node may affect the magnitude of an eclipse, and this includes as a special case the place where a solar eclipse was large or total. The latter case provides

---

[†]The underlining is M's.

[‡]This underlining is also M's.

the data which <u>M</u> calls the untimed data.

We should also note that the value of $\dot{n}_M$ which comes from the work of <u>Spencer Jones</u> [1939] is -22.44† while the value found from the ancient data [<u>AAO</u>, p. 272] is about -40. <u>M</u>, however, introduces a perturbation $\delta\dot{\Omega}$ to the rate of precession of the lunar node and infers the parameter $\delta\dot{\Omega}$ from the data at the same time as the acceleration parameters. When he does this, he finds [line 1 in his Table 12.2, p. 12.9]:

$$\dot{n}_M = -26.6 \pm 2.7 \ ''/cy^2, \qquad \delta\dot{\Omega} = +8.2 \pm 3.6 \ ''/cy. \qquad (IV.21)$$

Since this value of $\dot{n}_M$ is close to the value found from modern data, <u>M</u> says that the nodal perturbation $\delta\dot{\Omega}$ has removed the apparent discrepancy between the ancient and modern values.

This statement is unfortunately mistaken. As I pointed out in Section III.7, the ancient data other than eclipses in <u>AAO</u> carry about two-thirds of the total weight of the data, so that they have about twice the weight of the eclipses. Hence, contrary to <u>M</u>'s statement, my solutions are not dominated by the eclipses. In fact, when the eclipses are omitted entirely, the resulting solution gives (Equations III.14) $\dot{n}_M = -40.8$.

This value of $\dot{n}_M$ hardly differs from the solution obtained by using all the data, but there is an even more important point in this connection. The data which lead to the value -40.8 are not affected by the perturbation $\delta\dot{\Omega}$ in the precession rate of the lunar node. Thus the introduction of $\delta\dot{\Omega}$ into the inference process does not change the value found for $\dot{n}_M$, and thus $\delta\dot{\Omega}$ does not remove the discrepancy between ancient and modern values of $\dot{n}_M$.

I am not claiming that the value -40.8 is necessarily correct; in fact I finally conclude in Chapter XIV that it is seriously in error. I am merely saying that, if it is seriously wrong, the error does not come from omitting $\delta\dot{\Omega}$ in the analysis. Instead, it must come from errors in the data used,‡ or from other problems.

How then do we account for the fact that <u>M</u> found $\dot{n}_M = -26.6$ when he introduced the nodal perturbation $\delta\dot{\Omega}$? The answer is that <u>M</u>'s method of introducing $\delta\dot{\Omega}$ is theoretically incorrect, and thus any value he finds is only accidental. The proof of this statement, and a description of a correct method of introducing $\delta\dot{\Omega}$, will occupy almost all the rest of this section.

---

†In new analyses of the data that Spencer Jones used, <u>Morrison and Ward</u> [1975] find $\dot{n}_M = -26 \ ''/cy^2$ while I find (Section I.4) $\dot{n}_M = -28.4$.

‡We shall see in Section XIV.4 that this is indeed the case, and that eclipse data are not capable of yielding an accurate value of $\dot{n}_M$ or of $\delta\dot{\Omega}$.

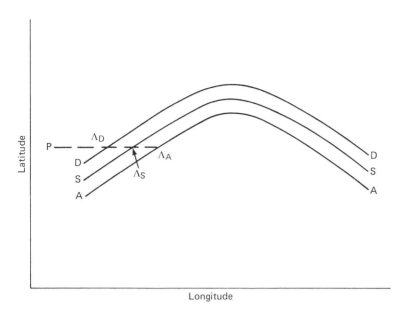

Figure IV.7. The effect of a nodal perturbation upon the place where an eclipse was central. The curve SS is the central path that we calculate with the standard lunar theory. Introducing a perturbation $\delta\Omega$ in the longitude of the lunar node shifts the path to AA if the moon is near the ascending node and to DD if the moon is near the descending node. The point P is the place where the eclipse was observed to be central. The abscissa is ephemeris longitude for the curves but it is geographic longitude for the point P. The perturbation $\delta\Omega$ changes the ephemeris longitudes of points on the central path, but the change is zero when averaged over the curves DD and AA.

First, we should remember that the parameter D'' is the acceleration parameter that is strongly determined by eclipse data.[†] In dealing with eclipse data, then, the parameters we use should be D'' and some independent parameter, say $\nu_S{}'$. D'' has another important property in connection with eclipse data. The value of D'' that we infer from the data is not affected, to first order, by the nodal perturbation $\delta\dot{\Omega}$. We can study the relation between D'' and $\delta\dot{\Omega}$ with the aid of Figure IV.7.

In the figure, the curve SS is a typical center line of an eclipse path, like the curve shown in Figure IV.1; it is to be drawn using the formula for the node given by Brown's theory of the moon [Brown, 1919]. Point P is a point at which the eclipse was observed to be central. The abscissa is ephemeris longitude when we are talking about the curve and it is geographic longitude when we are talking about P.

[†]When I write "eclipse data" in this section, I mean what M̲ calls the untimed data. That is, the eclipse data in this̲ sense consist of the places where solar eclipses were observed, along with the other circumstances, such as the eclipse dates, needed to use this information.

We first derive the equation of condition for this observation as we did in Section IV.3, except that we deal only with the central value and not the limits. If Λ_g is the geographic longitude of P, $\Delta\Lambda = \Lambda_g - \Lambda_S$, and the parameter y is given by Equation IV.15; call this value y_S. We then get a value $D''_S = \dot{n}_M - 1.6073 y_S$.

Now suppose that we add a small amount $\delta\Omega$ to the node Ω of the lunar orbit. If the moon is near its ascending node, increasing the longitude of the node decreases the latitude of the moon and this shifts the path of the eclipse southward, say to the curve AA. $\Delta\Lambda$ is now $\Lambda_g - \Lambda_A$, and this leads to a value D''_A. However, if the moon is near the descending node, increasing the longitude of the node by $\delta\Omega$ increases the latitude of the moon, shifts the path to curve DD, and yields a value D''_D.

It is clear that the average of D''_A and D''_D is equal to D''_S, the value that we calculate without using the perturbation of the node. In other words, the quantity that is well determined by the untimed eclipses, which is the parameter D'', is independent of a perturbation in the position of the node.

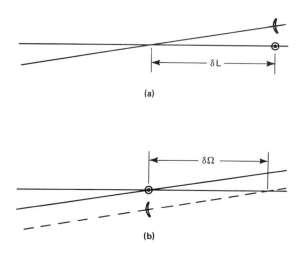

(a)

(b)

Figure IV.8   The effect of two perturbations upon a solar eclipse. In part
(a), the position of the sun is advanced by the amount δL while the
parameter D'' is held constant. Thus the position of the moon is also
advanced by δL. If the sun and moon were both at the node (the intersection
of the two lines) when $\delta L = 0$, they advance to the positions marked ☉ and
☾ by virtue of the perturbation δL. In part (b), the position of the node is
advanced from the intersection of the two solid lines to the intersection of
the horizontal line (the plane of the sun's orbit) with the dashed line. In
consequence, the position of the moon is shifted to the position marked ☾ .

The averaging of D'' to a value that is independent of a perturbation in the node is a very strong tendency. We see that D'' is averaged for eclipses that occur on the western half and the eastern half of the curve in Figure IV.7. Further, the eclipse path is just as likely to have a southern extremum in latitude as a northern extremum, which is what it has in the figure. If the reader will imagine a set of curves that is reversed top to bottom, he will see that averaging occurs for these two types of curve also.

Now let us see what happens when we have an acceleration ν_S' or a nodal perturbation $\delta\Omega$, under circumstances in which we keep D'' constant. Instead of using ν_S' directly, it is simpler in this discussion to use a perturbation δL in the position of the sun.

Since we keep D'' constant when we introduce either δL or $\delta\Omega$, we do not change the solar time of an eclipse. Since we do not change the solar time, we do not change the orientation of the earth with respect to the sun and hence we do not change an eclipse path in longitude. The only change is in latitude.

In part (a) of Figure IV.8, the horizontal line represents the plane of the sun's orbit and the slanting line represents the plane of the moon's orbit. The motion of the sun and moon is to the right in the figure, so the node that is drawn is the ascending node. For simplicity, suppose that the sun and moon are exactly at the node at the center of the eclipse, in the unperturbed condition. The reader should be able to convince himself that this simplification does not affect the conclusions to be reached.

Now advance the longitude of the sun by the amount δL. Since there is no change in the lunar elongation (because D'' is held constant), the longitude of the moon is advanced by the same amount. Hence the sun and moon advance to the positions marked $\odot$ and $\mathbb{C}$, respectively. The latitude of the moon is clearly increased by the amount $\delta L \sin \underline{i}$, in which $\underline{i}$ is the inclination of the moon's orbit. This increases the latitude of the path on the earth's surface.

In part (b) of Figure IV.8, the horizontal line is again the plane of the sun's orbit and the slanting solid line is the plane of the moon's orbit before we introduce the perturbation $\delta\Omega$ in the lunar node. If we advance the node by the amount $\delta\Omega$, the plane of the moon's orbit changes to the dashed line. Since the moon is on the slanting line, but with the same longitude as before, it is at the position marked $\mathbb{C}$ while the sun, marked by the symbol $\odot$, is still at the unperturbed position of the node. We see that the latitude of the moon has decreased by the amount $\delta\Omega \sin \underline{i}$.

If δL and $\delta\Omega$ are equal, the changes in the moon's position are equal and opposite.[†] Thus the path of an eclipse

---

[†] By drawing another figure like Figure IV.8, the reader can readily see that this conclusion is also correct if the eclipse occurs near the descending node of the lunar orbit.

cannot distinguish between perturbations in L and Ω. It can tell us only the difference between δL and $\delta\Omega$. This means that the eclipse data, in addition to the parameter D'', can tell us only the difference, W say, between the perturbations:

$$W = \delta\Omega - \delta L. \qquad (IV.22)$$

Thus $\underline{M}$ is correct in saying that my conclusion about the lunar node is incorrect for the untimed data. When I reached that conclusion, I was considering only the first order effects, and the node does not affect D'' to first order. I failed to appreciate the fact that finding a second parameter involves higher order effects, and that these effects may depend upon the node.

It is interesting to see how the introduction of a nodal perturbation changes the value inferred for $\dot{n}_M$. In looking at this question, let us assume for simplicity that ν_S' is a constant, so that $\delta L = \frac{1}{2}\nu_S'\tau^2$; τ is time measured in centuries from 1800. We further suppose that the main perturbation in the node comes from changing its precession rate by an amount $\delta\dot{\Omega}$ from the value used in the standard ephemeris. Then the quantity W is the following function of time:

$$W = \tau\delta\dot{\Omega} - \frac{1}{2}\nu_S'\tau^2 .$$

If we do not introduce a nodal perturbation, we take $\delta\dot{\Omega} = 0$ and infer ν_S' on this basis. When we introduce the nodal perturbation, we do not change the inferred values of W and thus we necessarily change the estimate of ν_S'. Let us use $\delta\nu_S'$ to denote the change in the estimated value of ν_S'. Then $\delta\nu_S'$ is the coefficient of $\frac{1}{2}\tau^2$ that makes $\frac{1}{2}\delta\nu_S'\tau^2$ the best fit possible to $\tau\delta\dot{\Omega}$ in a least-squares sense. If the data run from $\tau = -\tau_0$, say, to 0, $\delta\nu_S'$ is chosen to make the function

$$\int_{-\tau_0}^{0} (\tfrac{1}{2}\delta\nu_S'\tau^2 - \delta\dot{\Omega}\tau)^2 d\tau$$

a minimum. It is a trivial calculation to find

$$\delta\nu_S' = -5\delta\dot{\Omega}/2\tau_0 .$$

$\underline{M}$ [p. 12.9] finds $\delta\dot{\Omega} = +8.2$ "/cy, and τ_0 is about 25. Thus $\overline{\delta\nu_S}' = -0.82$.

To find the corresponding change in $\dot{n}_M$, we remember (Equation I.25 in Section I.7) that $D'' = \dot{n}_M - 1.6073y$ and (Equation I.27) that $\nu_S' = -0.1300y$. If we eliminate y from these equations, we find

$$\dot{n}_M = D'' - 12.3683\nu_S' . \qquad (IV.23)$$

The coefficient of ν_S' is the number of months in a year. Thus if we change ν_S' by -0.82 without changing D'', we change $\dot{n}_M$ by $+10.1$. This is about the change that $\underline{M}$ finds.

So far as I can see, there is only one way to separate

a solar perturbation from a nodal perturbation by using eclipse data. Suppose we know from theoretical grounds that δL should be of the form $\delta L = \delta L_0 f_1(\tau)$ and that $\delta\Omega$ should be of the form $\delta\Omega = \delta\Omega_0 f_2(\tau)$, with f_1 and f_2 being known distinct functions of time. Then, if we construct the time dependence of the parameter W, we can find the parameters δL_0 and $\delta\Omega_0$ by fitting the sum $\delta L_0 f_1(\tau) + \delta\Omega_0 f_2(\tau)$ to the function that describes W.

$\underline{M}$ assumes that $\delta\Omega$ is dominantly a linear function of time, $\tau\delta\dot\Omega$, say. This is a plausible assumption. He also assumes that ν_S' is constant, so that $\delta L = \frac{1}{2}\nu_S'\tau^2$. Since the functions of time are quite different, he can then in principle separate them and hence he can separate ν_S' and the perturbation $\delta\dot\Omega$.

The difficulty in this procedure comes from the fact that D'' varies strongly with time, as we have seen; it changes from about +10 or so in the year 0 to about -10 or so for times since about 1200 or 1300. Since $D'' = \nu_M' - \nu_S'$, either ν_M' or ν_S' has changed, and it is probable on physical grounds that both have changed if either has.† Thus there is no basis for assuming that ν_S' is constant, and thus there is no basis for $\underline{M}$'s solution. In fact, since ν_S' probably has a complex time dependence, $\underline{M}$'s procedure is inherently incapable of finding the nodal perturbation unless he can postulate the correct time dependence of ν_S' (or of δL) in advance.

In this work, I shall divide the sample of eclipse data into time intervals that are as short as possible while still yielding reasonably precise solutions for the independent parameters D'' and W. The hope will be that the functions $f_1(\tau)$ and $f_2(\tau)$ can be approximated by constants with sufficient accuracy during each time interval. This process, if it succeeds, will exhibit D'' and W as functions of time. We can then look at these functions to see whether we can develop plausible patterns of behavior for the parameters ν_S', ν_M', and $\delta\dot\Omega$. Study of the node by this procedure will constitute a goal of this work beyond those outlined in Chapter I.

I have now completed a review of the work of $\underline{M}$ and $\underline{MS}$. They have contributed to the field by introducing some records of solar eclipses that have not been used in the astronomical literature before. However, we cannot accept their study of the methods of analysis that should be used, nor can we accept their conclusions. The most we can do is to accept a few comments they have made on minor points. Even in the limited matter of the new eclipse records, as I said at the

---

†According to Equations I.26 and I.27, a change in the acceleration y of the earth's spin changes both ν_M' and ν_S'. Such a change can come from magnetic effects, for example, which are known to have changed by a significant amount within historic times. A change in tidal friction seems less likely, although it is certainly possible on the basis of current knowledge. See Section XIV.9 for further discussion.

end of Section III.7, we should approach their work with caution; they have introduced about as many incorrect records as correct ones. We saw three incorrect records in Sections II.2, III.4, and III.5, and we shall see still others when we come to a detailed discussion of the solar eclipse records in the following chapters.

CHAPTER V

CHINESE RECORDS OF SOLAR ECLIPSES

1. Some Preliminary Matters

The method used to designate eclipses, including the symbols used to identify the various provenances, is described in Section I.6. The symbols themselves are tabulated in Table I.5. I remind the reader that the Chinese provenance is designated by the single letter C.

In some Chinese records, the source identifies explicitly the place where the observation was made. In Section III.6 I discussed what to assume for the place when there is no explicit statement of place or region. While all earlier literature known to me took the place to be anywhere in China when there is no explicit statement, MS and M reach a different conclusion. They say that the place for such records is always the place where the emperor was, which, for brevity, they describe as the capital. Thus the place where the emperor was at any moment, which is called the capital, must be distinguished from a nominal capital. They give several reasons for this conclusion, one of which is the alleged low proficiency of observers outside the capital.†

In the same section, following Cohen and Newton [in preparation], I decided to take the place for such records to be heartland China. This is a suitable place to define accurately the meaning that "heartland China" will have in this work.

Both Charng-an and Luoh-yang were capitals of China‡ over extended periods, and we know from specific statements in some records that some solar eclipses were observed in each. In addition, Needham [1974] identifies four other cities as the sites of astronomical observatories. These are Yang-cherng, Khaifeng, Nanking, and Peking, although it seems likely that Peking was the site of an observatory only in times too late to concern us in this work. These six places, and their geographical coordinates, are listed in Table V.1. The coordinates are taken from the Times Atlas [1955] except for Yang-cherng, which I cannot identify with any point listed in the atlas. The coordinates of Yang-cherng are taken from Cohen and Newton [in preparation].‡

---

† The proficient observers in the thesis of MS and M were the court astronomers who always accompanied the emperor wherever he went.

‡ In the rest of this work, I shall use "capital" to mean a specific city, in the usual meaning of the word, rather than giving it the meaning used by MS and M. If any exceptions are needed, they will be specifically noted.

‡ Needham [1974, caption of Figure 9] says that Yang-cherng is the modern Kao-ch'eng. Times Atlas lists two cities in China with this name, but neither is close to Yang-cherng.

TABLE V.1

KNOWN OBSERVING SITES IN CHINA

Site	Latitude, degrees	Longitude, degrees
Charng-an	34.25	108.97
Luoh-yang	34.78	112.43
Yang-cherng	34.43	113.03
Khaifeng	34.78	114.33
Nanking	32.08	118.80
Peking	39.92	116.42

Five of the six places listed in Table V.1 have served as capitals of China at different times in its history. Thus the reader might try to say that the observatory used at any particular time was in the capital and we have so many observing sites only because there were so many capitals. However, Yang-Cherng was never the capital, so far as I can discover, but Needham [1974] describes it as China's central observatory for many centuries. Khaifeng was apparently the capital only for part of the 10th century, yet it is known to have received an important addition to its observing equipment [Needham, 1974] in 1088, a century after it ceased to be the capital. Hence we know that there were important observing sites in China outside the capital, contrary to the contention of M and MS, and we do not have to rely upon untrained observers in order to have provincial observations.

The Chinese eclipse observations that will concern us span a time interval of more than 1900 years. During this interval, we should expect to find some change in the observing sites. However, so far as we can tell, the observing sites seem to be about the same for the entire interval, with an exception that will be noted shortly. Hence, still with the exception to be noted, I shall assume that the eclipses could have been observed at any of the main observing sites. Thus we want the centroid of the sites, we want an estimate of the uncertainty in the place of observation, and we want a way to introduce the uncertainty into the analysis.

The centroid of the places listed in Table V.1 is at latitude $35°.14$ and longitude $114°.00$. However, as I remarked a moment ago, Peking was probably not an observatory site during the historical periods that concern us. If I omit Peking, the centroid shifts to latitude $34°.06$ and longitude $113°.51$.

The Chinese eclipses that we shall use occurred about 20 centuries ago, on the average. If we use an age of 20 centuries and a change in longitude of $1°$ in Equation IV.14 in Section IV.3, we find that an error of $1°$ in longitude gives an error of 0.38 in y; this corresponds to an error of 0.61 in D''. Both errors are considerably less than the accuracy with which we can hope to find the corresponding parameters. The effect of an error in latitude is harder to assess, but it should be considerably less. Hence it makes little difference whether we include Peking among the observatories or not.

Since it seems unlikely that Peking was a principal observatory during the periods that concern us, I shall take the centroid of the Chinese observations to be:

$$\text{latitude} = 34°.06, \quad \text{longitude} = 113°.51 \quad (\text{V.1})$$

The standard deviations of the coordinates about the centroid are 1°.15 in latitude and 3°.57 in longitude.

I shall base the analysis of the observations upon the coordinate η (Section I.5), which tells us how far and in what direction the observer was from the center line of an eclipse when the eclipse was a maximum for him. In the analysis, we need a central value of η and a standard deviation. Hence, in the analysis, I shall calculate the value of η at the four points whose coordinates are at latitude 34°.06 ± 1°.15 and at longitude 113°.51 ± 3°.57.[†] I shall then take the central value of η to be the mean of the four calculated values. For the standard deviation σ_η of η, I shall use the largest difference between the mean and any one of the four individual values.[‡]

Another contribution to σ_η comes from the uncertainty in the magnitude, which is expressed by a quantity σ_μ that is called the standard deviation of the magnitude. This was discussed at length in Section IV.1. There we decided to divide the Chinese records into three main classes: (a) If a record says that an eclipse was central or jih, $\sigma_\mu = 0.071$; this applies also if a record says that stars were seen or if it gives other specific indicators of a great eclipse. The reason for this will appear in Section V.5. (b) If a record says that an eclipse was partial but does not state a specific magnitude, we take the magnitude to be 0.81, with a standard deviation of 0.14 attached to this estimate. (c) If a record gives no clue to the magnitude, we take the magnitude to be 1, with a standard deviation of 0.36 attached to this estimate. There are a few records that do not fit into these classes which will have to be handled individually as they arise.

[†] The rectangle with the corners just stated is the region that I call heartland China.

[‡] More accurately, this will be taken as the contribution to σ_η made by the uncertainty in the place of observation; there will be other contributions from other sources. See Section XIII.1 for a complete discussion of σ_η. Use of the largest difference does not accord with the definition of a standard deviation, but I use it here in order to be conservative in assigning a weight to a record that does not indicate the place of observation.

When we reached a decision about the places of observation a moment ago, I mentioned that there would be an exception. The exception concerns the records dated -480 or earlier. These records come entirely from the annals of the Chinese state of Lu (also spelled Loo or Lou). Choosing a place for these observations poses a problem that we can discuss more conveniently in Section V.6, after we have discussed all the Chinese records that we shall use.

In addition to giving a calendar date, the Chinese records usually specify a day by giving it a number within a cycle of 60 days, which goes on without interruption even when the months and years change. Thus the 60-day cycle resembles our week, except for being much longer. Use of the 60-day cycle goes back at least to Shang times, and we encountered it in Section III.5 in the discussion of the alleged eclipse record of -1329 June 14. We shall also encounter it in this chapter. I shall commonly use the term "cycle number" in speaking of the number assigned to a day within the 60-day cycle. Appendix II describes the method by which the cycle number is stated in the records.

Gaubil [1732b], Hoang [1925], and Wylie [1897] have made extensive tabulations of Chinese eclipse records down to the modern period; their tabulations include about 1000 records. However, in order to use any records safely, we must study carefully all the records from the same historical period, in order to be sure that we understand the records and the circumstances under which they were written. Altogether only 58 records can be used in the present work, but 32 of these are dated -480 or earlier.†

2.   Records Earlier Than the Han Dynasties

Gaubil [1732a, p. 140] quotes an extensive passage about two mythical astronomers named Hsi and Ho during the reign of Emperor Chung K'ang, about -2000. These astronomers, being drunk with wine, neglected their duties and allowed the sun and moon to be out of agreement in their conjunction. Because of their gross neglect of duty, they were executed.‡

Gaubil assumed that this is a record of a solar eclipse and he decided that the eclipse happened on -2154 October 12. Other scholars have followed Gaubil in assuming that this is a record of an eclipse and they have assigned dates covering a span of more than two centuries. Fotheringham says that the record "is taken by all Chinese scholars, ancient and

---

†There are more than 58 valid records, but a few records pose special problems that keep them from being used readily in the analysis. See Section XIII.4.

‡This passage is discussed at length in AAO [pp. 62-66].

modern, to mean an eclipse."† It would be remarkable if the scholars in any subject showed such unanimity, and they fail to show it here by a large margin.‡ To me, the record seems to deal with the difficulties of keeping a lunar calendar in step with the seasons, and it does not seem to have anything to do with an eclipse.

I believe that anyone who reads the record with a judicial rather than a romantic eye will conclude that this passage does not record an eclipse, and recent writers on the subject of ancient eclipses tend to ignore it, so far as I have seen. Let us hope that it will not be necessary to mention this passage again in a serious discussion of eclipses.

$\underline{M}$ and $\underline{MS}$ both use the "eclipse" of -1329 June 14, and $\underline{M}$ assigns it a reliability (which he calls the probability of truth) of 0.80, which is rather high. As we saw in Section III.5, the inscription in question probably does not deal with an eclipse at all, and we cannot date it even if we assume that it does deal with an eclipse. Since the event, whatever it was, came during a winter month, a date in June is wrong unless there is an error in the record. Hence I take the reliability of this "eclipse record" to be zero.

We can now turn to passages in Chinese literature that actually refer to solar eclipses. The earliest known body of such passages comes from the annals of the state of Lu, which is near the northeastern corner of the region that I have called heartland China. We must start by describing the nature of the annals, which come to us through two main sources. One of these sources may have been written by the philosopher Confucius.‡

Confucius was born, in either -550 or -549, in or near Kufow, which was the capital of Lu.* The month and day of his birth are known, but the oldest records differ by a year in the matter of his birth. He devoted himself to study and to the service of the state until he was about 55 years old. At this time, he left Lu and spent about 12 or 13 years in travel. Finally he returned to Lu, where he stayed until his death in -478.

---

† In the Encyclopaedia Britannica, in the form that was being published in 1958, J.K. Fotheringham wrote a section called "Ancient Eclipses" for the general article called "Eclipse". The statement just quoted appears in this section. The article has been revised in the present form of the encyclopaedia, but the quoted statement still appears.

‡ It would also be remarkable if anyone in the year -2000 connected an eclipse with a conjunction of the sun and moon.

‡ I use this form of his name because it has become traditional in English. Confucius belonged to the family or clan named K'ung, and a better Romanization of his name is K'ung-fu-tze, which means something like "the philospher K'ung".

* I use the spelling Kufow because it is the spelling used in the Times Atlas [1955]. The spellings Chu-fu and Chu-fou are also used in English writing. Webster's Geographical Dictionary, in the form that was published in 1972, says that known descendants of Confucius still live in Kufow.

After his return to Lu, Confucius according to tradition wrote a compilation of the annals of the state of Lu
that I shall cite as Confucius [-479].[†] In this compilation, he grouped events by the four seasons.[‡] Even if he
had no event to record in a particular season, he still made
an entry for that season. For example, no event is recorded
for autumn in the year -716, but an entry reads "Autumn, the
seventh month." The seventh month is specified because that
month marks the beginning of autumn in the Chinese usage of
the time.

Because of this method of grouping events, the compilation is often called Ch'un-Ch'iu, which means Spring-Autumn
in this context; two seasons are allowed to stand for all
four in this title. However, we may doubt that the author
himself chose this title, and I shall refer to the compilation as the Annals of Lu, which describes the contents better.

Some critics in the modern period have been less than
enthusiastic about the Annals of Lu. Legge [1911] in fact
writes: "Ignoring, concealing, and misrepresenting are the
characteristics of the Spring and Autumn." I have not seen
any other writer who expresses himself as vigorously, but all
writers on the subject imply that there are difficulties in
understanding why events are presented as they are.

If Confucius is misleading or obscure in what he says,
we should like to think that this applies only to such matters
as personalities or politics, and not to simple non-contro-
versial events such as solar eclipses. However, eclipses may
have been controversial. It is often said that the ancient
Chinese regarded eclipses as warnings sent by Heaven to the
ruler to express dissatisfaction with his performance of his
duties. If this is so, eclipses are political and they may
also have been distorted in the circumstances of their occurrence. Thus we may need to extend our scepticism to them as
well as to other events. I shall return to this question in
Section V.7 after we have discussed all the Chinese records
of eclipses. There I shall decide that the eclipses in the
Annals of Lu are genuine.

I shall assign a reliability of 0.5 to most of the
eclipses from the Annals of Lu; I generally use this reliability for compilations made long after the events they describe. This reliability does not usually express doubt
about the occurrence of the eclipses recorded. The difficulty

---

[†]As well as I can determine, earlier scholars tended to accept these annals as the unquestioned work of Confucius.
Later writers tend to be more cautious and to say that the
annals are traditionally ascribed to Confucius. I believe
that the present scholarly consensus is that Confucius had
some connection with the annals, but the exact nature of
this connection is obscure. For the sake of definiteness
and brevity, I shall write as if Confucius were the author,
but the reader should remember that the matter is not certain.

[‡]However, he was following a custom that was already old. See
Legge [1872, p. 8].

is that we usually cannot tell whether the eclipses were observed in the place where the annals were kept, or whether reports of eclipses coming from elsewhere were inserted in the annals with no record of the origin of the report. In a few cases, a record from Lu tells us where the observation was made, and I give a reliability of 1 to these records.

The annals were continued for a few years after -479, and several writers also wrote commentaries on them almost immediately. The most important of these is the commentary by Tso Ch'iu-Ming [ca. -450].† Tso does not confine himself to comments on the Annals of Lu. Instead, he often gives a large amount of information that is not found in them. Further, his account of events often conflicts with that of Confucius, although Tso rarely if ever points this out explicitly. For these reasons, it might be fairer to call this an independent history rather than a commentary.

Other sources that have survived allow scholars to compare Tso and Confucius. As I understand the matter, these sources allow us to decide which is the more accurate and unbiased, and Tso comes out ahead in almost all instances; the exceptions, if any, may be accidents of writing. Further, the continuation of the main body of annals is found as a part of Tso's commentary, as I understand the matter. If these statements are so, we owe a double debt to Tso.

Let us make an amusing digression at this point. As I pointed out in Section III.2 of MCRE, several recent medievalists have seriously tried to maintain the position that annals were a medieval Christian invention, and that their invention is closely connected with Easter tables; that is, with lists of the date on which Easter occurs in various years. In Section III.3 of MCRE I pointed out several sets of pre-Christian annals, including a set from Assyria that goes back to about -800. The Annals of Lu go back almost as far. However, the set of annals that holds the record, so far as I know, comes from Egypt [Gardiner, 1961, pp. 62ff]. These annals are inscribed on both faces of a stele now called the Palermo Stone because it is preserved in Palermo. It originally gave the annals of the first five dynasties of Egypt, with at least one event for every year, although most of the entries have been lost because the stone has been badly broken. The years involved are roughly -3100 to -2350. This is safely before the medieval period.

Returning to the main subject, we need to note some points about the chronology given by Legge, the cited editor and translator of the Annals of Lu. A year in the Annals is identified as the year Y of a certain ruler. The rule about successions is that year 1 of a ruler begins on the first day of the year after he comes to power.‡ Legge [1872] starts

---

† I shall use only the name Tso in later references to this writer.

‡ I do not know what would happen if the succession occurred on the first day of the year.

his work by giving a table for converting the years stated this way into our calendar. He makes two serious errors in doing this. On page 88, he says that he will use the astronomical manner of stating the year, rather than the historical manner, in deference to a request made by a certain astronomer. It is clear, however, that he has not realized that the number differs in the two styles. He says in his table that the years are years B.C. (in spite of his statement that he will use the astronomical style), but they are actually years in the astronomical style with the minus sign left off. Thus, where he writes 721, for example, we should read -721, and so on.

Legge then accidentally repeats the year that he calls 497 B.C. (which is really -497). Thus the second year that he calls this is really -496, which happens to be 497 B.C. By the cancellation of two errors, then, his table is right after the second appearance of this year in his table.

Most scholars whom I know use the Julian calendar for such times in history. Legge, however, says that he will use the Gregorian calendar throughout, and the reader should keep this in mind when he uses Legge's work.[†]

The contents of the Annals of Lu are interesting. Most of the events, naturally enough, are political in nature, and they refer to events in both Lu and many other states. They include deaths of rulers, wars and alliances between states, state marriages, and so on. The annals cover about 250 years, and a cursory inspection suggests that there are about 6 events per year, so that there are about 1500 events in all.

Of these 1500 events, 37 are solar eclipses.[‡] Early Chinese literature is popularly supposed to contain records of many comets, novae, and other astronomical matters. However, I found only 4 comets and 2 meteor showers, no lunar eclipses, and no other astronomical events in the Annals of Lu. When we leave out political events, the overwhelming majority of other entries deals with the weather.

We may now turn to the records of solar eclipses. Most records contain identical words, differing only in the relevant numbers required. This is perhaps best shown by example. The eclipse of -654 August 19 is recorded in words that may be translated this way: "In the ninth month, on cycle number

---

[†] Both M and MS use the Gregorian calendar sometimes and the Julian calendar sometimes, even in talking about the same eclipse, and without stating which calendar they are using in a given context.

[‡] Three records contain recording errors that we cannot unsnarl, so only 34 eclipse records are usable from the textual viewpoint. Some of these present problems in analysis, so that only 32 of the total of 37 will be used in this work. See Section XIII.4.

45, the first day of the moon, the sun was eclipsed." Occasionally some of this information is omitted, probably by accident.† If the eclipse is the first event in a season, the name of the season precedes the statement of the month. If it is also the first event in a year, the number of the year precedes the name of the season; the season is always spring when this happens.

In the usage of the time, spring contained the first three months of the year, summer contained the next three months, and so on. Since the Chinese year began at about the same time as our year, spring in this usage was approximately the same as our winter, and so on.‡ This explains why the annalist can note as an unusual event [Confucius, -479, p. 336] that there was no ice during the spring of a certain year.

In all the instances when the day of the month is stated, the solar eclipse came on the first day of the month. This indicates strongly that the lunar month in China at this time began on the day of the true conjunction of the sun and moon. This fact is relevant to the study of the Athenian calendar of a period somewhat later than the Annals of Lu. Allowing the Athenian month to begin on the day of the true conjunction would simplify many problems connected with trying to reconstruct the Athenian calendar, but most students of the subject have said that this was impossible before the development of a sophisticated astronomy. Yet we see that the Chinese, at an earlier stage, were apparently beginning the month with the true conjunction. In APO [Section II.4] I show how this could be done with only simple observations that were well within the abilities of the times, but I did not know when I wrote APO that it had apparently been done. Of course, we do not know whether the Chinese used the method described in APO or some other method.

We need individual discussions only for records that present special features.

-719 February 22. Unless the eponym canon eclipse (Section II.5) can be identified by using only information that is strictly independent of Ptolemy, this is the oldest solar eclipse record that can be dated. In fact, only two older records than this are known to exist, whether they can be dated or not.‡ One is the record of the eponym canon eclipse. The other is a record from Ugarit that cannot be dated (Section III.4) but that is from the period between -1450 and -1200.

---

† Since Confucius's name is connected with the annals, many critics have tried to read profound moral and ethical judgments into these omissions. Legge's notes usually bring the reader back sharply to reality on these occasions.

‡ This fact makes me wonder if the translation I have used is justified in the way it translates the names of the seasons.

‡ However, there are several older records of lunar eclipses, but they cannot be dated confidently. See Keightley [1977].

-708 July 17. The record says that this eclipse was central or complete.

-668 May 27. The record reads: "In the sixth month, on cycle number 8, the first day of the moon, the sun was eclipsed, when we beat drums, and offered victims at the altar of the land." We must note two things about the performance of these rites. First, they do not seem to be associated necessarily with central eclipses, because there is no mention of them on -708 July 17 when the eclipse, if central, was actually total.† The question of when the rites were performed will be discussed further in connection with the eclipse of -524 August 21. Second, the record tells us that the eclipse was observed in the capital Kufow. At least, I think it is safe to assume that the "altar of the land" was in the capital. Certainly the eclipse was observed in Lu.

-663 August 28. The rites performed on -668 May 27 were performed again, so this eclipse was also observed in Lu and probably in Kufow. Legge, in a comment in the cited edition of Confucius [-479, p. 109], thinks that these rites were observed at all eclipses. If so, it seems odd to me that they are mentioned in 3 and only 3 records out of 37. See -668 May 27, -611 April 27, and -524 August 21.

-644 August 28 ? The record [Confucius, -479, p. 166] says: "In summer, in the fifth month, the sun was eclipsed." Legge uses the date -644 February 3 in referring to this eclipse but, as he points out, the eclipse of that date was not visible in China. The fifth month could correspond to any month from April to June. The eclipse of -647 April 6 is recorded for the third month. Except for it, I can find no eclipse visible in China in an appropriate month for many years on either side of -644. The record under discussion cannot be an accidental repetition of the eclipse of -647 April 6 because the months are different. It is possible that the eclipse of -644 August 28 was visible in China and that the month is accidentally wrong; the matter cannot be settled without detailed calculation. It is unfortunate that the cycle number was omitted from this record, so there is no satisfactory way to test an identification. I shall calculate the circumstances in Section XIII.4 to see whether the eclipse of -644 August 28 was visible, but I shall not use the record in any case.

-611 April 27. Again the drums were beaten and victims were sacrificed, so this eclipse was presumably observed in Kufow. The date in China was -611 April 28.

-601 May 7. The record says that this eclipse was observed in the 17th year of the ruler, in the 6th month, on cycle number 40. The year stated is -591, and there was no

---

† Calculation with the final values of the accelerations gives the eclipse of -668 May 27 a magnitude of 0.83 at Kufow.

possible eclipse in that year. However, as Legge points out
in the cited edition, everything fits perfectly if we assume
that the 17th year was written by mistake for the 7th year.
Hence the eclipse is -601 May 7, but it happened on May 8 in
China.

-600 September 20. The record says: "In autumn, in
the seventh month, on cycle number 1, the sun was totally
eclipsed." "Totally" is the word the translator used; it is
probably the word jih that I have discussed earlier. The
eclipse of -600 September 20 was total in parts of China and
it came on a day with cycle number 1. However, it could not
have been in the 7th month. Further, it is the only event
recorded for autumn in this year. Legge in his notes has
suggested the most probable solution: The eclipse was originally recorded "in winter, in the tenth month"; this would
be correct. However, after the compiler or a copyist wrote
"in autumn, in the seventh month," he accidentally inserted
the record of the eclipse in the wrong place.

-598 March 5 and -557 May 30. These eclipses actually
came on the following days in China.

-551 August 20. Successive entries in the annals read:
"In the ninth month, on cycle number 47, the first day of the
month, the sun was eclipsed. In winter, in the tenth month,
on cycle number 17, the first day of the moon, the sun was
eclipsed." The first entry is correct and the second entry
is clearly an error. Legge, in his comments about these entries, says that the critics have tried in vain to find an
explanation of this error. It seems to me that there is a
simple and plausible explanation. The annalist first recorded
the eclipse correctly. He then wrote "In winter, in the tenth
month, on cycle number 17" with the intention of recording
some other event. However, he accidentally copied the statement about the eclipse again before he went on to the other
events. This explanation, though plausible, is not necessarily correct, of course.

-548 June 19. The record of this eclipse reads: "In
autumn, in the seventh month, on cycle number 1, the first
day of the moon, the sun was completely eclipsed." Three
entries later, we find: "In the eight month, on cycle number
30, the first day of the moon, the sun was eclipsed." This
entry is clearly wrong. The explanation that I proposed for
-551 August 20 does not work here, because the word "completely" does not occur in the second entry. It is probable
that the entry was transferred by accident to this point from
somewhere else, but I cannot find any other eclipse that it
might have applied to. This is the only eclipse entry in the
Annals of Lu that is incorrect and for which the error lacks
a simple and plausible explanation.

-524 August 21. This record has two errors. It says
that the eclipse came "in summer, in the sixth month" but it

should have said "in autumn, in the ninth month". Again we seem to have an accidental transfer from one season to another. The record also says that the cycle number was 11, but it should have been 10. In spite of these errors, there is no reason to question the identification.

The annals do not say that any rites were performed in connection with this eclipse, as they were on three earlier dates. However, Tso [ca. -450, p. 667] comments that there was much discussion about the matter in the court on this date. One group held that drum beating and sacrifice should be performed for all eclipses, while another group held that these rites applied only if the eclipse came on the first day of the first month. Since it seems that the rites were not performed, the second group must have won the argument. Incidentally, this example goes against Legge's conclusion (see -663 August 28) that the rites were observed for all eclipses. Further, since the matter was being debated in court, the eclipse must have been observed in the capital Kufow.

-520 June 10. The record of this eclipse in the annals is accurate and it follows the standard formula. However, it gives Tso [ca. -450, p. 688] an opportunity to comment further on eclipses. When this eclipse happened, the ruler asked one of his advisors: "What is this for? What calamity does it indicate, or what blessing?" The answer was: "At the solstices and equinoxes, an eclipse of the sun does not indicate calamity. The sun and the moon, in their travelling, are, at the equinoxes, in the same path; and at the solstices, they pass each other. On other months, an eclipse indicates calamity." This remarkable passage tells us two things.

First, if this is a real event and not something made up for didactic reasons, the ruler must have seen the eclipse himself. The annals give no indication that he was travelling, so he presumably saw the eclipse in Kufow.

Second, whether the event happened or not, this passage was presumably written by -450 or earlier.† It shows a knowledge that solar eclipses were connected with the moon and it shows a groping toward an understanding of the geometrical relations that govern eclipses. The fact that the geometry is not accurate does not matter for present purposes.‡ The passage shows an understanding that some eclipses happen because of the natural motions of the sun and moon, and that these eclipses do not mean anything. An eclipse is a portent only if it does not result from the natural working out of the solar and lunar motions.

---

† It is possible that this passage was not written until about -200. See Section V.7.

‡ With regard to the geometry, I wonder if the translation is accurate. Is it possible that the word translated as equinox really meant conjunction?

There are several later eclipses in the Annals of Lu, but we do not need to comment upon any except the last. Most commentators say that the main body of the annals† ends with the first entry for the year -480; this entry is an interesting one. It says that hunters in the west (presumably in the west of Lu) captured a lin. Legge in his notes to the cited translation discusses the nature of the animal called a lin at some length. It is possible that there was once a real creature called a lin. However, it finally became, and probably already was in -480, a fabulous beast, and there are suggestions that it was a beast with one horn. In other words, it may be the Chinese version of the unicorn. Legge doubts that this entry is really connected with Confucius.

The annals that are preserved as part of Tso' commentary continue for a few years, and most authorities, as I understand the matter, believe that they were prepared by disciples of Confucius. The continuation contains an accurate record of the eclipse of -480 April 19, which follows the standard formula. The only oddity about it is that Legge, in his summary of the solar eclipses, does not identify it.

It is interesting to summarize the chronological accuracy of these records briefly. There are altogether 37 records of solar eclipses in the Annals of Lu. Three of them undoubtedly represent scribal errors of some sort, and one of these three is probably mere dittography. The others probably result from the accidental transference of a record from its correct position to a wrong position in the annals. Thus we are left with 34 records out of 37 that are usable from the textual viewpoint. This is a remarkably high fraction, especially for records so old.

Within the 34 usable records, one has the year wrong (-601 May 7) and one has the cycle number wrong (-524 August 21). The other chronological item that occurs in the records is the month, and we find two clear cases in which the month is enough in error (-600 September 20 and -524 August 21) to put the eclipse in the wrong season. Legge [1872, pp. 93-97] gives a table of the number of months in each Chinese year, and a date when each year began. On the basis of this table, apparently, he says [p. 89] that the months are wrong in 14 of the records. It seems quite unlikely that recorders who were so accurate with regard to the year and cycle number would be so inaccurate with regard to the month.

It is more likely, I think, that we do not know in most cases how the numbers of months were allocated to individual years. Further, so far as I can determine, we do not know the principle on which the first month was chosen, except that it was fairly close to the winter solstice. Also, when there was an intercalary month, we do not always know where it was placed in the year; it was not necessarily the last month. Thus I think we do not know how many errors in the month are in fact present. I cannot understand what Legge says on this question and I am not sure whether he was aware of this point or not.

---

†That is, the part of the annals connected with Confucius.

TABLE V.2

CHINESE ECLIPSES BEFORE THE HAN DYNASTIES

Designation	Standard deviation of the magnitude	Reliability
- 719 Feb 22 C	0.36	0.5
- 708 Jul 17 C	0.071	0.5
- 694 Oct 10 C	0.36	0.5
- 675 Apr 15 C	0.36	0.5
- 668 May 27 C	0.36	1
- 667 Nov 10 C	0.36	0.5
- 663 Aug 28 C	0.36	1
- 654 Aug 19 C	0.36	0.5
- 647 Apr 6 C	0.36	0.5
- 644 Aug 28 C[a]	0.36	0
- 625 Feb 3 C	0.36	0.5
- 611 Apr 27 C[b]	0.36	1
- 601 May 7 C[b]	0.36	0.5
- 600 Sep 20 C	0.071	0.5
- 598 Mar 5 C[b]	0.36	0.5
- 574 May 9 C	0.36	0.5
- 573 Oct 22 C	0.36	0.5
- 558 Jan 14 C	0.36	0.5
- 557 May 30 C[b]	0.36	0.5
- 552 Aug 31 C	0.36	0.5
- 551 Aug 20 C	0.36	0.5
- 549 Jan 5 C	0.36	0.5
- 548 Jun 19 C	0.071	0.5
- 545 Oct 13 C	0.36	0.5
- 534 Mar 18 C	0.36	0.5

[a]This date is questionable, and this record will not be used in the analysis.

[b]The eclipse happened on the next day in China.

TABLE V.2 (Continued)

Designation	Standard deviation of the magnitude	Reliability
- 526 Apr 18 C	0.36	0.5
- 524 Aug 21 C	0.36	1
- 520 Jun 10 C	0.36	1
- 519 Nov 23 C	0.36	0.5
- 517 Apr 9 C	0.36	0.5
- 510 Nov 14 C	0.36	0.5
- 504 Feb 16 C	0.36	0.5
- 497 Sep 22 C	0.36	0.5
- 494 Jul 22 C	0.36	0.5
- 480 Apr 19 C	0.36	0.5
- 441 Mar 11 C	0.071	0.1
- 381 Jul 3 C	0.071	0.1
- 299 Jul 26 C	0.071	0.1

Table V.2 summarizes the eclipse records from the Annals of Lu. It gives the designation of each usable eclipse, plus one unusable eclipse, along with the standard deviation of the magnitude and the reliability. The table also lists the eclipse of -480 April 19 from Tso and three later eclipses which need a brief discussion.

According to Needham [1959, p. 420], annals say that stars were seen during the eclipses of -441 March 11, -381 July 3, and -299 July 26, but I have not found references to these eclipses in any other source. Further, the statement is just a passing remark made to illustrate another point, rather than part of a detailed study of the annals in question. For these reasons, with no intention to disparage anyone's scholarship, I shall use a reliability of 0.1 for these eclipses; this is what I did in AAO. The observations could have been made anywhere in heartland China, as this term is defined in Section V.1. Since stars were seen, I shall use 0.071 for the standard deviation of the magnitude.

3.   Records from the Han Dynasties

The dynasty called the Former Han or Western Han ruled from -201 to +9. After a short interval, the dynasty called the Later Han or Eastern Han ruled from 25 to 220. Dubs [1938, 1955] counts the period from 9 to 25, which some writers class as an interregnum, as part of the Former Han period, and I shall follow him for convenience, without implying any opinion about the historical question.

Dubs has made extensive studies of the eclipses recorded

-153-

in the annals and the astronomical treatises of the Former
Han, and I used his studies in AAO. There I used only the
eclipses which, on his reading, were described as central or
nearly so. Here I shall use all eclipses from Dub's study
which are dated correctly and without ambiguity, if the record indicates that the eclipse was actually observed.

In about a fourth of the records from the Former Han
dynasty, the date listed is not that of an eclipse visible
in China. In a few of these cases, postulating a single
plausible error in writing converts the date into the date
of a visible eclipse, and we may justifiably assume that the
wrong date is simply a recording or copying error; I shall
not use these cases, however. In many other cases, the date
can be converted into a possible date only if we assume several copying errors, and these are implausible. Cases of
this sort may occur in the annals, in the astronomical treatises, or in both.

In many of these "impossible" cases, the astronomical
treatise gives the right ascension of the sun. When this happens, the right ascension is reasonably accurate for the "impossible" date but it is grossly in error for any possible
eclipse date.

For example [Dubs, 1938, pp. 501-502], both the annals
and the astronomical treatise of the Former Han dynasty list
an eclipse on the 7th year, 1st month, 1st day, cycle number
38, of the Emperor Hui, and the treatise gives the right ascension as 312°, approximately.† A study of this record is
instructive, because both ancient and modern errors appear in
its study.

In order to change the recorded date into the date of
an eclipse that was visible in China, we must change every
item in the record except the name of the emperor and the
day of the month. Dubs concludes that the record actually
refers to the eclipse of -191 September 29, which came on
the 3rd year, 9th month, 1st day, cycle number 57, under the
same emperor.

The 1st day of the 1st month of the 7th year came on
-187 February 21. This is the date of a new moon and it has
cycle number 38, just as the record says. Further, there
was an eclipse exactly 1 month away, on -187 January 22, although it was not visible in China. Thus -187 February 21
is a date that could easily have been calculated for an
eclipse by a person who did not yet have accurate theories of
eclipses.

There is one discrepancy between this date and the record. The right ascension on -187 February 21 was about 331°

---

†The Chinese records give the right ascension by saying that
the sun was a certain number of degrees east of a particular
star, and the Chinese degree was $1/365\frac{1}{4}$ of a circle, not
$1/360$. Thus a Chinese statement of right ascension to the
nearest Chinese degree does not translate into an integral
number of our degrees.

while the record says 312°. Thirteen lunar months earlier, however, on -188 February 2, the right ascension was 312°. Thus it seems likely that the person who calculated the right ascension of the sun used the table of values for the wrong Chinese year.† A mistake of this sort is easy to make. Many medieval annalists in Europe used both a lunar and a solar calendar and they often gave equivalent dates in both calendars. Sometimes they made a mistake in doing so. In all such cases that I have noticed, the mistake came from using the wrong year in relating the two calendars.‡

Thus, if we assume that this record gives a predicted rather than an observed date for an eclipse, we need to assume only one simple error (in calculating the right ascension) instead of many. Further, the error concerns only a secondary calculated quantity and not a quantity that is important for the record. For this reason, I believe that there was no error in the recorded date.

There are three errors in Dubs's discussion of this record. One is a simple typographical error. He says that the cycle number given in this record is the same as that for the recorded eclipse of -1 February 15, but there was no eclipse on that date. He meant -1 February 5; an eclipse is recorded for that date and the cycle number was 38.

The other errors also concern the cycle number. Dubs says that the cycle number on -191 September 29, which he concludes is the date meant in the record, is 58, although it is actually 57. This causes Dubs to make a further error. A Chinese source which Dubs calls Han-chi, which was written about a century later than the annals, gives 58 rather than 38 for the cycle number on the recorded date. Since the Han-chi compiler usually copied his data from the annals, Dubs concludes that the annals originally read 58 and that someone, after the Han-chi was compiled, changed the cycle number in the annals to correspond to the rest of the recorded date. Thus he concludes that the annals originally gave the correct cycle number for the actual eclipse date even though it was wrong about the rest of the items in the date. Hence, he says, there is "good reason for considering that the eclipse of 192 B.C. was actually recorded."

We realize now that the cycle number on -191 (192 B.C.) September 29 was not 58, and this removes the basis for Dubs's conclusion. Since the Chinese forms for cycle numbers 38 and 58 differ in only one character (see Appendix II) it is more likely that the compiler of the Han-chi simply misread the cycle number when he was copying from the record.

----

†Thirteen lunar months would be exactly 1 Chinese year if there were an intercalary month between the two dates, or if the calculator thought there were.

‡See MCRE [p. 414] for an example. Such a mistake does not mean that the annalist did not know what year it was. It merely means that he let his eye wander to the wrong line or column of a table.

In summary, we have to postulate some errors in the record whether we assume that it refers to a predicted eclipse or to an observed one. If we assume that the eclipse was predicted, all the items in the date are correct, and we need to postulate an error in only a single secondary quantity, namely the calculated right ascension of the sun. Further, there is a simple explanation of the error in this case. If we assume that the eclipse was observed, we have to assume three errors in the date, which is an item that would have been recorded at the time of the observation, and we still need to assume an error in the right ascension of the sun. Further, in this case, the error in the right ascension is a large and implausible one with no simple explanation. Thus, so far as the text of the record is concerned, it is far more likely that the eclipse was predicted than that it was observed and recorded.

I have checked several of the other "impossible" dates, but not all of them. All the ones I have checked are the dates of new moons, or are within a day of a new moon, with one exception. Some of them are dates of actual eclipses (but eclipses that were not visible in China) while some are not. Even when they are not, they are only 1 month away from an actual eclipse, and thus they are less than a month from an eclipse season.†

As I have discussed in Section III.6, we have a similar situation in the Tarng dynasty. There, we find that about 1 date in 4 is not the date of an eclipse visible in China, that many of these dates occur in the astronomical treatises with a statement of the right ascension of the sun attached, and that the right ascension is reasonably accurate for the date given. The main difference is that the date in the Tarng records is nearly always the date of an eclipse that actually occurred but was not visible in China, while this is less often true in the Former Han records.

Scholars of the subject have usually assumed that the "impossible" dates in the Former Han records were copying errors, with at most one or two dates that were calculated or fabricated. The findings with respect to the Tarng records tell us, it seems to me, that we must consider this question again, and ask whether the "impossible" Han dates were the result of inaccurate prediction. A detailed study of this question is beyond the scope of this work, but we may explore it tentatively.

---

†We saw in the preceding section that the solar eclipses from the Lu period all came on the first day of the month. This indicates that the beginning of the month was based upon the true conjunction and that it was probably determined by observation. In the Han records, the eclipses are sometimes on the first day and sometimes on the last day. This suggests that the month was now determined by calculation based upon the mean moon rather than the true moon. In the Tarng dynasty, however, eclipses could be on the last day, the first day, or the second day. This situation is harder to understand, and I have not looked for an explanation.

Whether the "impossible" dates come from eclipse calculations or not, they tell us an important fact. If the
values of right ascension were recorded when the eclipse
observations were made, the responsible astronomer would
know the right ascension of the sun fairly accurately simply
from knowing the time of year. He would rarely make an error such as writing that the right ascension was 312°, which
corresponds to a day in February, if he were writing this in
September. Yet such apparent errors occur frequently. If
the records imply actual observations, this could happen only
if the values of right ascension were put into the records
long after the event, when no one knew any longer that the
recorded date was impossible for an actual eclipse. Hence
the occurrence of the right ascension in the record tells us
nothing about the conditions of observation.

Dubs tried to see whether certain "impossible" dates
were the results of eclipse calculations. In doing so, however, he assumed that the Chinese in Han times had no methods
of calculating eclipses except the use of cycles.† This is
by no means the only elementary way of predicting eclipses.

To start with, all the eclipse dates in the Han records
are dates of new moons or close to it, and thus we may assume
that the Han Chinese knew that solar eclipses can come only
at the new moon. If they had also discovered that eclipses
can come only during certain seasons, they were well on the
way to eclipse calculation. Eclipses can come only during
seasons that last about 34 days and that recur at intervals
of about 173 days between their centers. Once the seasons
are discovered, any new moon that comes during a season is a
possible candidate for an eclipse.

In fact, every new moon that comes during an eclipse
season is the date of a solar eclipse that is visible somewhere. The only question is whether the eclipse was visible
at a particular place. There are elementary ways of deciding
this with moderate accuracy, of which I shall mention only
two. If we observe the angle between the moon and the sun
during the last few days of a lunar month, or if we equival-
ently measure the difference between their rising times, we
can predict the time of the true conjunction with an error
of only an hour or so.‡ Thus we can easily eliminate eclipses
that occur during the night and which are therefore invisible.
We then need to eliminate those eclipses which will occur in
the wrong range of lunar latitude, and we can do this by observing the difference between the rising azimuths of the sun

---

†For example, if a solar eclipse is observed on a particular
day, there will almost surely be another eclipse 18 years
plus 11 days later, although there is only about 1 chance
in 6 that the later eclipse will be visible at the place
where the first eclipse was seen. This is the cycle that
is usually miscalled the saros. There are many other cycles,
all of which depend upon approximate coincidences between important astronomical periods, and all of which are only approximate.

‡This is one of the elementary ways of predicting the true
conjunctions. See APO [Section II.4] or the discussion in
Section V.2 above.

and moon on the last few days of the month.

I am not claiming to have proved that Chinese astronomers in the Former Han dynasty used these or other equivalent methods. I merely say that methods such as those described give reasonably accurate predictions of eclipses on a short-term basis and that they were within the technical capacity of Chinese astronomers in Former Han times.

In fact, a Chinese writer of the -4th century gave instructions [Needham, 1959, p. 411] for predicting solar eclipses based upon the relative positions of the sun and moon at times near the new moon. It is believed that this writer did not yet realize that the physical body of the moon was interposed between us and the sun. Instead, it is thought, he believed that an influence radiating from the moon affected the sun.† However there is ample time between this writer and the Former Han dynasty for the development of more substantive theories and methods of prediction.

In studying the methods of prediction that Han astronomers might have used, it would be valuable to know how far in advance they made their predictions. I have not attempted to study this question. The only evidence I have noted that may be relevant to this question comes from the Tarng dynasty. For example, the Tarng records say explicitly [Cohen and Newton, in preparation, note 100 to Table 3] that the eclipse of 794 May 4 was predicted by the astronomers and that the prediction was submitted to the court on the day before, namely 794 May 3. This is suggestive but not conclusive. It is possible that the prediction had been made long in advance but that the astronomer waited until the day before to send it to the court.

The possibility that the "impossible" eclipses came from prediction or calculation raises still another disconcerting possibility: The records in and of themselves give us no basis for distinguishing between the "impossible" eclipses and most of the ones that were visible in China. Thus it is possible that most of the records of observable eclipses also represent calculation and not actual observation. If this is so, the only difference between the "impossible" eclipses and most of the others is the accuracy of the prediction.

For this reason, I shall use an eclipse record from the Former Han dynasty only if it contains a statement that explicitly indicates actual observation as opposed to calculation. Thus the policy of using all valid observations, regardless of the indicated magnitude, yields only two usable records beyond those used in AAO.

---

†In some respects, this writer does not seem to be as well informed as the writer who commented on the eclipse of -520 June 10 (Section V.2), although he is presumably two centuries later.

TABLE V.3

ECLIPSE RECORDS FROM THE HAN DYNASTIES
THAT INDICATE ACTUAL OBSERVATION

Designation	Statement Indicating Observation
- 197 Aug 7 C	Central.
- 187 Jul 17 C	Central according to annals, almost central according to treatise.
- 180 Mar 4 C	Central according to both sources.
- 146 Nov 10 C	Almost central according to treatise.
- 88 Sep 29 C	The treatise says: "It was partial, like a hook. In the late afternoon the lower part of the sun was eclipsed from the northwest. In the late afternoon the eclipse was over."
- 79 Sep 20 C	Central according to annals, almost central according to treatise.
- 34 Nov 1 C ?ᵃ	The treatise says: "It was partial like a hook; then it set."
- 27 Jun 19 C	Central according to annals, "like a hook" according to treatise.
- 15 Nov 1 C	The treatise says this could be observed only in the capital.
- 14 Mar 29 C	The treatise says that the four quarters all observed this eclipse, but it was cloudy in the capital.
- 1 Feb 5 C	The treatise says it was like a hook.
+ 2 Nov 23 C	The treatise says it was total, but Dubs says it was not total in the capital.
65 Dec 16 C	The treatise says it was central.
120 Jan 18 C	The treatise says it was almost complete, and it was "like evening" on the earth. The annals say it was complete, that stars appeared, and that it caused fear to the people.

ᵃThe correct date may be -48 August 9; see the main text.

The records that I shall use are listed in Table V.3. This table gives the designations of the eclipses and the nature of the statements which indicate actual observation. Before I comment on the table, however, let me comment on another implication of the possibility that some of the eclipses were calculations and not observations.

The idea that the eclipses in the Former Han records were observed has always posed a problem. Many of the

-159-

observable eclipses in the records were quite small, with
magnitudes of 0.1 or less everywhere in heartland China.†
It is certainly possible to observe eclipses this small,
but it is not likely that they would be observed unless they
had been predicted and that the observers had thus been
alerted to their occurrence. Thus the idea that they were
observed perhaps implies some degree of prediction in connection with them. Further, when Dubs tries to find a possible date for an "impossible" record, he is often forced to
an eclipse that was very small. For example, there is an
eclipse that he tentatively dates as -144 March 26, but which
the record actually dates about 4 or 5 months later [Dubs,
1938, pp. 505-506]. There was no eclipse on the recorded
date, and he thinks that -144 March 25‡ is the only eclipse
date which might have been mistaken for the recorded date in
copying. However, the only place in China where the eclipse
of -144 March 25 was visible, according to Dubs's calculations, was the eastern tip of the Shantung peninsula, well
outside of heartland China, and the eclipse was quite small
there. Thus he concludes that regular watches for even tiny
eclipses may have been maintained even in such remote spots.
It is more likely that the record refers to a predicted
eclipse.

If we assume that many of the records reflect calculation rather than observation, we remove all of the problems
that concern small magnitudes except one. The record of the
eclipse of -15 November 1 says, in the translation of Dubs
[1938, p. 512]: "Heaven caused the capital alone to know
of it; the kingdoms in the four directions did not perceive
it."‡ Dubs calculates that the eclipse had a magnitude of
only 0.08 at the capital (Charng-an). I shall check this
calculation, but there seems to be little question that the
eclipse was small at Charng-an. Thus at least one quite
small eclipse was actually observed, unless this remark was
attached to the wrong date by accident. However, it is
reasonable to suppose that a very small eclipse was observed
occasionally, especially if it had been predicted, and the
existence of this single record is not disturbing.*

---

†The smallest eclipse that has been found in the European
annals had a magnitude of 0.29. See Section II.5 above.

‡This is the date in Greenwich time, but the eclipse was on
the next day in China. Hence Dubs uses the date -144 March
26.

‡Incidentally, this implies an expectation that the provinces
("the kingdoms in the four directions") would usually observe and record an eclipse.

*Dubs says that an eclipse this small could be observed only
by special means, such as looking at the reflection of the
sun in a mirror or the surface of a stream. I believe this
gives an erroneous impression. Any effect that dims the
sunlight and thus removes most of the irradiation (see Section IV.1) allows us to see the true outline of the sun,
and this allows us to see eclipses even smaller than 0.08.
Haze or thin clouds create the right conditions on a few
occasions, and any material that is fairly dark, such as
colored glass, can also do so. I have tried to observe

Before we turn to specific records of eclipses, we need to consider a final point. I have rejected most Han records of eclipses on the basis that many of them were calculated and that all of them may have been, except for those that contain special remarks. However, Sivin [1969] and Bielenstein and Sivin [1977] explicitly deny the possibility that astronomers under the Former Han dynasty could calculate the occurrence of solar eclipses. We must consider what weight to give their arguments before we can decide whether my rejection of most Han records is justified.

Bielenstein and Sivin [1977, p. 185] say that there is "a simple and well-known historical fact: the Han astronomers did not believe that solar eclipses can occur only at the new moon." They go on to say that this point "has been documented in the English language in Sivin's Cosmos and Computation in Early Chinese Mathematical Astronomy (Leiden, 1969),† pp. 6 et passim." However, the cited work by Sivin does not even mention the point in question, on page 6 or any other page I can find, and it certainly does not demonstrate the point.

Further, it is about as certain as a matter of this sort can be that this "simple and well-known historical fact" is not a fact at all. To see this, we note first that a lunar eclipse always comes about the middle of the month, and a solar eclipse always comes about the first of the month, when a lunar calendar is used to record the eclipse dates. Recognition of this fact is a necessary but not sufficient condition for being able to calculate or predict eclipses.

Solar eclipses are recorded in the appropriate annals and treatises of the Han dynasties, but lunar eclipses are not. Sivin [1969, p. 6], after noting this fact, writes that "for lunar eclipses, which could be predicted, it was sufficient to publish the method of their calculation . ." Later he discusses the Han methods of predicting lunar eclipses in considerable detail.‡ This means that Chinese astronomers, in or before the Han dynasties, had learned from studying records of eclipses that lunar eclipses can come only near the middle of a month. (They had of course learned much more than this about lunar eclipses since they predicted them.) It does not seem possible to me that the astronomers, from

---

the sun's disk by reflection in water many times and have found it almost impossible. Even a very small breeze ruffles the surface enough to prevent observation by reflection, and I have never seen a stream that was smooth enough.

†This is a reference to the work I cite as Sivin [1969].

‡We should note that the Annals of Lu (Section V.2 above), five centuries before the Former Han dynasty, already record solar but not lunar eclipses. If Sivin's explanation for the absence of lunar eclipse records is correct, it implies that the Lu astronomers could already predict them. If they could not, Sivin's explanation is not adequate. I mildly doubt that astronomers from Lu in −700 could already predict lunar eclipses. However, there is no doubt that Han astronomers did so, because their methods of prediction have been preserved and discussed in the modern literature.

the same set of records, would not have noticed the parallel fact that solar eclipses come only near the first of the month, particularly since a search for regularity in the occurrence of solar eclipses seems to have been a preoccupation of early Chinese astronomers.

Another argument seems even more compelling. At least 13 out of 55 reports of solar eclipses in the Former Han dynasty give dates on which there was no eclipse visible in China. In every case, the date given is the first day or the last day of a lunar month; this is always the date of a new moon or it is at most one day away from a new moon. There are two possibilities that I see. If the "impossible" eclipses were predicted, as I believe was the case, astronomers under the Former Han knew that solar eclipses come only at the new moon. If on the other hand the "impossible" eclipses were scribal errors, it seems incredible that the writers or copiers made errors in every other feature of the date but never made a mistake in the day of the lunar month. Thus they must have known automatically that a solar eclipse could come only on the first or last day of a lunar month, which by definition are at the new moon.

Sivin [1969] does write repeatedly that the Han astronomers could not predict solar eclipses, but he does not demonstrate this. Instead, he states [p. 6], with reference to solar eclipses, that "in the absence of spherical geometry, very few successful predictions were possible ..."† Since the Han astronomers did not know the necessary mathematics, Sivin concludes, they could not have predicted solar eclipses.

However, we have just seen that eclipses can be predicted by using only the simple operations of arithmetic, so Sivin's argument does not apply.

If the Han astronomers did predict solar eclipses by a simple method like the one outlined above, we should like to know why they published accounts of individual solar eclipses but published only the method of prediction for lunar eclipses. There is a plausible explanation for this distinction between the two kinds of eclipse. The method used for predicting lunar eclipses fitted well into a grand scheme of astronomy and a description of the method formed a suitable part of a formal discourse on astronomy. The method used for solar eclipses, whether it was the one discussed above or not, almost surely depended upon making some ad hoc observations and making some simple calculations based upon them. Thus a method of predicting solar eclipses could not be made to fit into a grand mathematical scheme and would not have been published in the astronomical treatises.

We can now turn to specific records of eclipses. The eclipse of -197 August 27 was not used in AAO, but it would

---

†In writing "very few successful predictions", I think that Sivin is referring to the possibility that a prediction made by means of cycles would be successful a few times, but not often. This contrasts with the situation for lunar eclipses, for which cycles yield rather successful predictions.

have been except for some confusion about what the record
says. Dubs [1938, 1955] does not quote any remark which
indicates that this eclipse was large in either of his two
extensive publications about Former Han eclipses. I based
my omission in AAO upon this omission by Dubs. However, according to MS [p. 490] and M [p. 8.28], the record in the
astronomical treatise says that the eclipse was central.
Further, Gaubil [1732b, p. 256] says that the eclipse was
total according to the record, but he does not say whether
the record was in the treatise or the annals. Thus I conclude that Dubs accidentally omitted the relevant remark
and that the eclipse was described as central; this indicates
actual observation.

The eclipses of -187 July 17, -79 September 20, -27
June 19, and +120 January 18 were central according to one
of the Chinese sources and almost central or "like a hook"
according to the other. M and MS use only the statements
that occur in the astronomical treatise. I shall take all
of these to be eclipses that were recorded as being central
somewhere in heartland China. Thus the total collection of
eclipses during the Han dynasties that are recorded as central contains the eclipses of -197 August 7, -187 July 17,
-180 March 4, -79 September 20, -27 June 19, +2 November 23,
65 December 16, and 120 January 18. That is, I shall take
the standard deviation of the magnitude to be 0.071 for these
eclipses.

In the record for -88 September 29, we have two statements that indicate observation. There is a clear statement
that the eclipse was like a hook, and there is also a statement that the maximum and the end of the eclipse both came
in the late afternoon. Now "late afternoon" is not a vague
statement of time. The Chinese divided the day into 12 equal
parts, or dual-hours, which began and ended on the odd-num-
bered hours in our system of writing time. Dubs uses "late
afternoon" to denote the particular dual-hour that ran from
15 hours to 17 hours. An eclipse large enough to be like a
hook lasts for a considerable time, so this is a reasonably
precise statement that the midpoint between maximum and the
end of the eclipse came at 16 hours. In later work I shall
use this as a measure of the time of the eclipse. In this
work, however, I shall use only the fact that the eclipse
was like a hook and hence that it was partial, with the magnitude unspecified. That is, I shall take the magnitude to
be 0.81 ± 0.14.

In a record that Dubs [1938, p. 511] dates -34 November
1, the astronomical treatise says, in Dubs' translation: "It
was partial, like a hook; then it set." The date -34 November 1 was year 4 of Chien-chao, 9th month, last day, cycle
number 14. However the record says that the date was the
5th year of Chien-chao, 6th month, last day, cycle number 9,†
which is 9 months after the date that Dubs assigns. There

---

†Dubs says that the symbols for cycle number 14 "may easily
be read" as the symbols for cycle number 9. To my eye, the
symbols have almost no resemblance.

was no eclipse on the date recorded. Dubs calculates that the eclipse of -34 November 1 reached its largest magnitude of about 0.66 at sunset, so that it must be the eclipse referred to, in spite of the fact that the dates disagree in every important particular.

The date that Dubs assigns is not the only possibility. The eclipse of -48 August 9 was also a large eclipse at sunset, and it must be considered as a possibility. It comes in the right month, but the year is wrong and so is the cycle number; the cycle number on -48 August 9 was 37 rather than 9. Further, eclipses are recorded on -41 March 28 and -39 July 31. Thus if this is a record of -48 August 9, the record not only gives the wrong date but it has gotten out of order. It is possible for a compiler to rearrange the order of events by accident, but it is somewhat unlikely.

The date in the record is, I believe, -33 August 23, which had cycle number 9 as the record says. Further, this is the date of a new moon and it is only 1 lunar month away from the eclipse of -33 September 22.† Thus it could well be the result of an eclipse prediction. If this is the explanation, however, how does it happen that the record explicitly describes the magnitude and the setting, events that could not have been observed on -33 August 23?

The most likely explanation, it seems to me, is that the same thing happened here that happened with the records of 1039 August 22; these records were discussed in Section IV.5. That is, the sources originally recorded the eclipse of -34 November 1, with the remarks that have been quoted, and the "non-eclipse" of -33 August 23, with no accompanying remarks. The compiler then accidentally ran the records together, taking the date from -33 August 23 and the accompanying remarks from -34 November 1. Again this explanation replaces a complicated sequence of errors by a single error.

Thus I agree with Dubs that the eclipse of -34 November 1 was probably recorded, although I doubt his explanation of the errors involved. However, this conclusion cannot be regarded as one that is firmly established, and I shall not use this record.

MS originally took the place of observation for the eclipse of 65 December 16 to be Luoh-yang, which was the capital at the time. Their post-analysis study using their finally inferred parameters showed that the eclipse was not total there, and they thereupon studied the record again. The text refers to two specific Chinese states. By calculation with their final parameters, MS found that the eclipse was total in the capital of one of them. MS [p. 491] then say incorrectly that this is the only state referred to in the text. They also give a reading of the text which to my mind is rather forced but which, they say, clearly implies that this state capital was the point of observation.

---

†This was a penumbral eclipse that was probably not visible in China.

MS give the eclipse, as assigned to this state capital, a low weight and thus it does not affect their analysis seriously. M [pp. 8.34-8.35], however, takes the assignment as something that is well established and he uses it with a high weight. In fact, as we saw in Section IV.3, it is one of the critical records in his solution for the parameters. This action is close to circular reasoning.

I see no reason to assume that this eclipse was observed at any particular place, and I shall take the place to be anywhere in heartland China.†

The last eclipse in Table V.3 is the eclipse of +120 January 18. I used this eclipse in AAO and I shall use it again here. In Section IV.5 I discussed at length the way that MS and M use this record, along with the reasons why I shall not use it in the same way. It remains to describe the way in which I shall use it.

The record that MS and M use comes from the astronomical treatise. Gaubil [1732b, p. 277] quotes another record‡ of the same eclipse, which presumably comes from the annals. According to Gaubil, the eclipse was central, the stars appeared, and the eclipse gave fear to the people. I shall use this record of the eclipse. That is, I shall take this eclipse to be one that was recorded as central somewhere in heartland China.

Altogether the following records from the two Han dynasties state that the eclipses were large but not central: -146 November 10, -88 September 29, and -1 February 5. When I used these in AAO, I feared that we might not know the accelerations well enough to know which side of the eclipse path the observing region was on. Thus I ignored the statements that the eclipses were partial and treated these as ordinary records that give no clue to the magnitude. Now we know the accelerations well enough to treat these as partial, with the side of the path (the sign of η) known. Hence I shall assume here that these eclipses had magnitudes of 0.81, with $\sigma_\mu = 0.14$.

As I have already mentioned, the record for -15 November 1 says explicitly that the eclipse was observed at the capital, so I shall use Charng-an as the place of observation for this eclipse. The record for -14 March 29 says that the eclipse was not seen in the capital but that it was seen elsewhere. Hence I shall use heartland China as the place for this eclipse.

The records for these two eclipses do not say anything that implies a large eclipse, although the recorders seem to

---

†By calculation with the parameters that I finally infer, I find that the zone of totality went through the center of heartland China, and that the magnitude was almost exactly 1 at Luoh-yang.

‡Gaubil accidentally gives the year as 119, as I have already noted.

have been rather careful to note those which were fairly
large. Hence I shall assume that these eclipses were rather
small. Specifically, I shall assume that their magnitudes
were 0.5, with a standard deviation of 0.2. Thus the
eclipses will have a small weight, but there seems to be no
reason to ignore them.

All the eclipse records from the Han dynasties come
from compilations made long after the event. In the normal
way, I give a reliability of at most 0.5 to such records.
Here, however, we have two independent compilations (the
annals and the astronomical treatise), at least for almost
all the eclipses, so it is reasonable to give a full reliability of 1 to the Han records. Uncertainty about the place
of observation, which is one reason for lowering the reliability, will then instead be expressed by taking the place
of observation to be heartland China.

4. From the Han to the Tarng Dynasty

The Later Han dynasty fell in 220 and the Tarng dynasty
came to power in 618. The Tarng dynasty succeeded the Swei
dynasty, which came to power in 589, and which was the first
to rule a united China in several centuries. Between the
Later Han and the Swei, China was divided into northern and
southern kingdoms, and sometimes it was even more divided.

TABLE V.4

ECLIPSES BETWEEN THE HAN AND TARNG DYNASTIES

Designation	Standard deviation of the magnitude	Reliability
243 Jun 5 C	0.071	0.1
360 Aug 28 C	0.071	0.1
429 Dec 12 C	0.071	0.5
516 Apr 18 C	0.071	0.1
522 Jun 10 C	0.071	0.1

During this period, the Chinese annals record four
eclipses that were described as central (jih) and one eclipse
during which stars were seen. These eclipses are listed in
Table V.4, along with the values for the standard deviation
σ_μ of the magnitude and the reliability. I do not distinguish between jih eclipses and those in which stars were seen
in the matter of σ_μ for Chinese records.

The record for 429 December 12 is the one which says
that stars were seen. The others say nothing to indicate
actual observation except the use of the term jih. Previously
I have taken jih to be an indication of actual observations.
However, the prediction of eclipses even to the extent of
predicting whether they would be jih or not is by now a distinct possibility. Until the records from this period have

been studied in detail, I do not believe that one should give the other records as much reliability as the record for 429 December 12. Hence I shall use 0.5 for 429 December 12 (the maximum for a record from a late compilation) and 0.1 for the others.

In AAO, I did not use any Chinese records after the eclipse of 429 December 12.† The reason [AAO, p. 69] was that the eclipses could have been observed anywhere in heartland China, the uncertainty in position produces an uncertainty in the accelerations found, and this uncertainty would be so large for later eclipses that they would have little value. Even worse, if the center of the observing region shifted in later periods, as it might well have done, this shift could bias the results if I were unaware of it.

Now, after additional study, in particular of the Tarng period, I believe that the matter is not this bad. We shall see in the next section, which deals with the Tarng dynasty, that the main places of observation shifted little if at all between the Han and the Tarng, and there is little danger of a bias in using eclipses through the Tarng dynasty. Hence in this work I use the last two eclipses in Table V.4, which I did not use in AAO.

If we knew that a compiler of records in the period from the Han to the Swei compiled only records made in his own state, we could reduce the region of observation for a particular eclipse to the confines of the state for which the appropriate compiler worked. However, the assumption of purely local recording does not seem a safe one to me.‡ In particular, I cannot follow MS and M in assigning the eclipses of 516 April 18 and 522 June 10 to Nanking, just because it was one of the capitals at the time.

Gaubil [1732b] lists all the eclipses in Table V.4, as well as the remarks that indicate a large eclipse.

5.   Eclipses During and After the Tarng Dynasty

I shall use only one Chinese record of an eclipse after the fall of the Tarng dynasty in 907. This is the record designated as 1221 May 23a C in Table V.5. For convenience I shall discuss this record before I turn to the Tarng records.

---

†Except for the eclipse of 1221 May 23, which is a special case. See the next section.

‡Even if a particular set of original records had a purely local origin, it does not follow that these late compilations were drawn only from the local records of a particular region. Purely local compilation, if it existed, can be established only by an intensive study of the records, Pending such a study, the only conservative course is to assume that the records could have originated from a wide region.

TABLE V.5

TARNG OR LATER RECORDS THAT STATE
THE MAGNITUDE OR PLACE

Designation	Statement
702 Sep 26 C	Like a hook, nearly central. Seen in capital and provinces.
729 Oct 27 C	Not central, like a hook.
754 Jun 25 C	Like a hook, almost central.
756 Oct 28 C	Central.
761 Aug 5 C	Central; all the large stars were visible.
792 Nov 19 C[a]	The magnitude was only 3, although 8 was predicted.
808 Jul 27 C	Predicted, seen by the emperor.
879 Apr 25 C[b]	Central.
888 Apr 15 C	Central.
1221 May 23aC	The stars could be seen.

[a]This record will be used only in later work, for reasons described in the text.

[b]This was a predicted date, but there was no eclipse on this date. The compiler of the astronomical treatise accidentally transferred "central" from the date of 888 April 15 to this date in the imperial annals.

This record has been used by Curott [1966], by MS and M, and in AAO, and it was discussed at length by Wylie [1897]. In 1221 a Chinese expedition travelled from China across the region that we call Outer Mongolia to Samarkand, in what is now the Uzbek S.S.R. While the expedition was somewhere along the Kerulen River in Mongolia, the travellers saw a large eclipse of the sun. I have seen two translations of the record which differ in one important respect.

In AAO I used the following translation by Wylie [1897]: ". . at noon, an eclipse of the sun happened, while we were on the southern bank of the (Kerulen) river. It was so dark that the stars could be seen, but soon it brightened up again." Curott, MS, and M use a translation by Waley† which differs mainly by saying that the eclipse was total. Curott then adds the following remark: "He notes 'total eclipse' is translated from an archaic phrase!"‡ None of the people

---

†This reference is to A. Waley, The Travels of an Alchemist, page 66, Routledge, London, 1931. I have not consulted this reference.

‡"He" refers to the translator Waley. The exclamation point is Curott's.

who quote from Waley say what phrase was translated as "total". We saw in Section III.6 that jih, which did originally mean a complete or central eclipse, had probably lost this meaning by the Tarng dynasty if not sooner. Thus, even if Waley is correct and if Wylie accidentally omitted this phrase from his translation, we cannot safely assume that the eclipse was central. Because of the reference to the stars, I shall use 0.071 for the standard deviation of the magnitude.

The various users of this record have used various positions for the party at the time of the eclipse. Wylie concluded that it was probably at 48°.7N and 116°.25E. Curott concluded that it was some distance away, at 47°.5N and 112°E; I used the mean of these positions in AAO. MS give "a confident location as somewhere between the points 48° 10' N, 115° 51' E and 48° 12'N, 115° 57' E." Since they do not say how they reached this "confident" location, we cannot judge it, and since we cannot judge it we cannot accept it.

I shall take the place of observation to be somewhere along the line joining the places used by Wylie and by Curott. Luckily, the exact point does not matter much. The central line of the eclipse ran almost exactly parallel to the course of the river at any of the points mentioned. Hence all the choices give almost identical equations of condition.

The magnitude of this eclipse was measured at two places. One place cannot be identified, but the other was Samarkand. I shall use the Samarkand measurement in a later work.

M [p. 8.65] has a Chinese record of the annular eclipse of 1292 January 21, which he quotes as saying that "the body of the Sun looked like a golden ring." He does not use the record, however, because it is after his "cutoff date for inclusion of eastern records." As usual, he takes the place of observation to be the capital, which was Peking at this time. At this time in history, it may be correct to assume the capital, but we need to investigate the situation before we do so. Since I have not studied the records from this part of Chinese history, I shall not use this record.

We stated the general conclusions that we need about the Tarng records in Section III.6, in connection with deciding upon the region of observation to use with the Tarng records. The conclusions that we need here are:

a. The eclipse records could have originated from anywhere in heartland China, as this term is defined in Section V.1.

b. The term jih, which was once used to indicate a complete or central eclipse, had probably lost this meaning by the Tarng dynasty.

c. We may not assume that an eclipse was actually observed unless the record contains a remark that specifically implies observation.

Conclusions a and b are connected. If eclipse records routinely originated only from within heartland China, jih cannot denote centrality. If jih does continue to denote centrality, records (at least of large eclipses) must have routinely originated from points at an unlikely distance outside heartland China. Hence I shall use the assumptions stated, with the recognition that further research may require their modification.

The History of the Yuan Dynasty, which was completed in 1370,† states the times when the eclipse was a maximum for seven eclipses that occurred during the Tarng dynasty. These times are not found in any earlier source that is extant. Cohen and Newton show that these times were probably calculated by astronomers under the Yuan dynasty, about five centuries after the event. Thus there is no reason to conclude that these seven eclipses were actually observed.

Nine records from sources that are specifically related to the Tarng dynasty contain remarks that imply either the magnitude of the eclipse or the place where it was observed. These nine records, along with the record 1221 May 23a C, are listed in Table V.5. The table gives the designation of each eclipse record and the nature of the remark that caused its inclusion.

Most of the remarks in Table V.5 should be self-explanatory, but two remarks need special comment. The remark for 792 November 19 gives us a measurement of the magnitude. In this work, I am using only the places where eclipses were observed, and I shall defer the use of this measurement to a later work. The remark in the astronomical treatise that the "eclipse" of 879 April 25 was central was transferred accidentally from the record of 888 April 15 in the astronomical archives. Thus only seven records from the Tarng sources will be used in this work.

For most of the Tarng records in Table V.5, we have more than one independent source, and we have allowed an uncertainty in the place of observation. Hence I shall use a reliability of 1 for all the Tarng records, even though they come from compilations made centuries after the event. The record of 1221 May 23 comes from a contemporaneous account, and it too receives a reliability of 1. We shall discuss next the values of η that I shall assign, and the standard deviations of the magnitude.

The records 702 September 26 C, 729 October 27 C, and 754 June 25 C indicate a large eclipse while denying centrality. I shall treat these records in the same way as the analogous records in Section V.3. That is, I shall take their magnitude to be 0.81 ± 0.14. The records 756 October 28 C, 761 August 5 C, and 888 April 15 C all assert that the eclipse was central (jih), and the record 761 August 5 C adds that all the large stars were visible. I shall take σ_μ = 0.071 for all these records; this is the value of σ_μ for such records that was found in Section IV.1.

---

†See the discussion by Cohen and Newton [in preparation].

Finally we have the record 808 July 27 C, which says that an eclipse was predicted for that day. On the next day, the emperor said to the prime minister that the prediction had been made and that everything happened just as predicted. The annals indicate that the emperor was in Charng-an at the time, so we may take that as the place of observation. The record does not say anything about the magnitude; since the eclipse had been predicted, the observers were likely to see it even if it were small. I shall take the magnitude to be 0.5, with a standard deviation of 0.2.

We may take Charng-an as the place for the record 808 July 27 C, but we must use heartland China for all the other records from the Tarng dynasty that we shall use.

I must now comment on a choice that was first stated in Section IV.1 and that has been repeated several times. This is the choice to use the same value of σ_μ, namely 0.071, for eclipses described as jih (central) and eclipses that are not described as jih but that give specific indicators of a large eclipse such as the seeing of stars. The reason for this choice is discussed in detail by Cohen and Newton [in preparation], and I shall only summarize it here.

If we examine separately the eclipses described as jih and those which have specific indicators such as seeing stars but which are not described as jih, we find that the average magnitude is actually somewhat higher for the second group. It is possible that this reflects the meaning, or lack of meaning, of jih, but this conclusion does not sound plausible to me. It does not seem reasonable to me that the eclipses for which jih is not used should be larger than the eclipses for which it is used. Further, in view of the small samples involved, the difference is not very significant statistically. On the other hand, in view of the actual facts, it is not reasonable to take the jih eclipses as being larger. Pending further information, then, I decide to use the same σ_μ for all these eclipses.

It is possible that further study will cause a revision of this conclusion. Even if the conclusion is in error, however, it will not affect the results of this work appreciably. The error is small, it affects only a small fraction of the observations involved (about 6 out of more than 600), and it is not biased with respect to the inference of the accelerations. It is also unlikely that further study will cause a revision, because it is unlikely that there are enough data to let us reach a firm conclusion on the matter.

The places from which eclipses were reported, and the practices followed in recording eclipses, may well have changed after the Tarng dynasty. Thus, although there are many records after the Tarng period besides 1221 May 23a C, it is not safe to use them until a definitive study has clarified the necessary points. Nonetheless, some of the later records present interesting features that I shall discuss briefly.

In early times, a solar eclipse may have been regarded

as a serious omen or warning; I shall discuss this point in Section V.7. After the prediction of eclipses became customary, it is hard to believe that a predicted eclipse was considered seriously, and in fact the eclipse of 808 July 27, which the emperor himself saw, did not seem to cause the emperor concern. A possible corollary to this, as I shall suggest in Section V.7, is that a predicted eclipse that was not seen may have been considered as a good omen. Gaubil [1732b, pp. 333-346] gives confirmation of this suggestion. In the years 965, 1024, 1061, 1073, and 1106, eclipses were not seen as predicted,† and the emperor was congratulated in consequence. The record of 1061 is particularly interesting; it must refer to 1061 June 20. The prediction was that the magnitude would be 6 digits,‡ but the magnitude was only 4 digits when clouds prevented further observation. The emperor was thereupon congratulated, presumably because Heaven sent an eclipse that was smaller than the prediction.

The record of 1073 May 9 has another feature. An eclipse was predicted for that day. The emperor was warned of it and retired to ponder the misfortunes with which Heaven menaced him. However, clouds came and prevented observation of the eclipse, and the emperor was congratulated in consequence. Here we have almost a specific statement that the eclipse would be a warning from Heaven under certain circumstances, but, since the eclipse was not seen, the occasion was one for rejoicing.

We saw in Section V.2 that several eclipses before -480 caused the beating of drums and the sacrifice of a victim, presumably human. It is interesting to notice that at least some of this practice continued even after the Tarng dynasty. On 1054 May 18 an eclipse of 9 digits was observed, and on 1059 February 15 an eclipse of 3 digits was observed. On both occasions a sacrifice took place. There is no mention of drum beating in Gaubil's discussion of these records, nor do we know whether the victim was human. Since a sacrifice took place on these occasions, even though the eclipses were not very large, we may speculate that eclipses of the observed size had not been predicted, or at least that the court was not warned of them by the astronomers.

6.   Places of Observation for the Eclipses from Lu

We saw in Section V.2 that the records furnish a specific place where five eclipses from the Annals of Lu were observed. We must now consider what place or region to use for the other eclipses from the records of Lu.

---

† I have not tried to translate the dates into the Julian calendar. Since the "eclipses" may not have been real ones, we cannot infer the dates from a canon of eclipses, except for 1061 and 1073.

‡ I am not sure what the Chinese digit was, but I believe that it was 1/15 [AAO, p. 144]. The eclipse of 1061 June 20 was nearly total in heartland China.

Various writers have made various choices about the
place or region to use. According to the map which Legge
gives in the cited translation of Confucius [-479], Lu was
an oval-shaped region lying approximately between latitudes
$34\frac{1}{2}$ and $36\frac{1}{4}$ degrees north and between longitudes $115\frac{1}{2}$ and
118 degrees east. Curott [1966] assumed that the eclipses
could have been observed anywhere within the state of Lu, but
he used a range of longitude that differs slightly from the
range I gave above.[†]

Souciet [1729, pp. 18-20] specifically assigned the observations to Khaifeng which, according to Table V.1, is at
34°.78 north latitude and 114°.33 east longitude. This is
somewhat west of the confines of Lu and somewhat east of the
centroid of heartland China (Equation V.1). In writing AAO,
where I used only the three central eclipses because I had
studied only secondary sources, I followed Souciet in taking
Khaifeng as the center of the region. However, instead of
taking it as the specific spot, I used standard deviations of
2° in both latitude and longitude about it.

MS and M also use only the three eclipses that are stated
to be central. They assume that two of the eclipses were observed in Kufow, the capital of Lu, but that the other one,
the eclipse of -600 September 20, was observed at Ying, the
capital of the state of Ch'u.[‡] Ying is at latitude 30°.34,
longitude 112°.25, and it lies far to the southwest of Lu.
In spite of this fact, MS and M choose Ying as the point of
observation for a reason that needs explaining.

At a time earlier than that covered by the Annals of Lu,
China[‡] had been a country united under a single ruler whom I
shall call the emperor. However, by the time of these annals,
China was in a considerable, though not total, state of an-
archy. There was still an emperor, but he did not exert
central authority. In effect, China was divided into a number of states who nominally acknowledged the emperor but who
were in reality independent, warring, states.

From time to time, one state would acquire a condition
of supremacy. Sometimes, when this happened, the ruler of

[†]Curott used the three eclipses from the Annals of Lu which
were stated to be central. He then looked at the eclipse
maps in Oppolzer [1887], selected four other recorded eclipses
which were large near the capital according to the maps, and
calculated accelerations from these eclipses. He remarks
that the results from these eclipses are speculative; he
should have remarked that using these eclipses is reasoning
in a circle that tells us nothing. He uses names for all
seven eclipses in his table of results. He coined the
names by using "Lu" followed by the name of the ruler in
whose reign the eclipse occurred.

[‡]This is sometimes spelled Ts'oo in English.

[‡]Throughout this immediate discussion, China will mean only
the region that I have called heartland China, plus perhaps
some small areas adjacent to it. This is much smaller than
the area we call China today.

that state would proclaim himself [Legge, 1872, pp. 114ff] the "presiding chief"; if he were powerful enough, he could force the emperor to recognize him as a kind of viceroy or deputy. Legge [p. 115] names five people who claimed the position of presiding chief at one time or another. At the time of the eclipse of -600 September 20, the listed claimant was the ruler of Ch'u. Since one of the actions of the presiding chief was to convoke an assembly of the state rulers from time to time, MS [p. 490] conclude that an eclipse report from Ying (the capital of Ch'u) "might well have reached" the capital of Lu and been entered in its annals.

I cannot find an explicit statement in MS about how they chose Ying, but M [p. 8.18] writes: "The observation location originally assumed in M&S (Chu-fu)† contradicted the overall result of that study and was apparently an error." MS thereupon apparently reviewed the situation, came up with the point about the presiding chief, and chose Ying as the place in consequence. MS say only that the eclipse report might have come from Ch'u, and they give the report a low weight because, as they say explicitly, they are playing the identification game.

Before we go farther, we should comment upon the sources of the Annals of Lu, meaning the formal compilation attributed to Confucius. The sources clearly include the annals of Lu, meaning the collection of annalistic records kept in the court of Lu's ruler. Legge [1872, p. 9ff] concludes that the Annals of Lu are drawn only from records kept in Lu, and that the compiler did not use the records of other states. From a study of the Annals of Lu, I agree with this conclusion.

However, the various states, or at least those that maintained diplomatic relations, routinely exchanged notices about various events. The receiving state entered such notices in its own annals, "without regard", in Legge's words, "to whether they conveyed a correct account of the facts or not." Thus it is conceivable that an eclipse record from Ch'u would be sent to Lu, and this had nothing to do with the matter of who was presiding chief.

As I said, MS give the eclipse of -600 September 20 a low weight and thus they avoid any serious effect of the identification game. This is not so with M. M takes Ying as the place of observation and uses the eclipse with a high weight. Further, as we saw in Section IV.3, this eclipse is one of the critical eclipses in his solution for the accelerations. Then, after he has found the accelerations, M reviews all the records. Since his assumption about this eclipse dominates his solution, he necessarily finds that the "record" is consistent with his solution. Hence he concludes [M, p. 13.3] that the assignment of Ying as the location is "probably correct."

This seems to me like the purest example of reasoning in a circle that I have encountered in the study of ancient astronomical observations. M denies that it is reasoning in

---

† MS and M use this spelling for the place I spell Kufow.

a circle, saying [$\underline{M}$, p. 8.19]: ". . the above historical research could† have been done in the beginning."

I agree in principle with the idea expressed here. A research worker should revise his conclusions when additional research uncovers additional evidence which invalidates them. The trouble here is that I can find no evidence of historical research, either in the beginning or later. All that $\underline{MS}$ (or $\underline{M}$) have done is to state that the ruler of Ch'u was presiding chief in -600. On this basis alone, $\underline{M}$ assumes that the eclipse was observed in the capital of Ch'u, and he then "verifies" this assumption by reasoning in a circle. What is needed is to study the Annals of Lu to see whether entries from, let us say, state X customarily appear in the annals during a period when the ruler of X was presiding chief. I shall do this in a moment, but another point needs to be addressed first.

The statement that the ruler of Ch'u was the presiding chief is itself open to question. The arguments of $\underline{MS}$ and $\underline{M}$ start from a statement by Legge [1872, p. 115] that there were five presiding chiefs during the period covered by the Annals of Lu. Their terms of "office" were intermittent. There were times when there was no presiding chief, and there were times when two or even three persons competed for the title. The ruler of Ch'u in -600 is the fifth and last chief in Legge's list. Immediately after he lists the five chiefs, Legge writes: "The first two, however, are the best, and I think the only‡ representatives of the system." Thus, contrary to what $\underline{MS}$ and $\underline{M}$ state, Legge does not say that the ruler of Ch'u was the presiding chief in -600.

Legge then outlines the careers of the five "chiefs" named. The first two were forceful rulers and fully merit the title. The next two do not merit it. The fifth one, who is the one that concerns us, is intermediate. The emperor still retained enough independence that he did not recognize this person as chief. According to Legge, many states from fear of military punishment did recognize him, but many did not. He does not say which camp Lu belonged to.

I have studied the Annals of Lu in an attempt to answer this question. During the period when the first-named chief, the ruler of Ts'e, was the presiding chief, I find considerable evidence of contact between Lu and Ts'e. During the period around -600, however, when the ruler of Ch'u claimed to be the presiding chief, I find several instances when the ruler of Lu met with the rulers of various states, but I can find no occasion when he met with the ruler of Ch'u. Thus there are grounds for believing that Lu did not acknowledge the primacy of Ch'u. If this is correct, there is no reason for selecting the capital of Ch'u as the origin for the eclipse report of -600 September 20.

---

†The emphasis is $\underline{M}$'s.

‡The emphasis is mine.

-175-

Further, if this is correct, the record could have come
from Ch'u only if it were one of the annals that were routinely interchanged. In this regard, a look at Legge's map
[Legge, 1872, between pp. 112 and 113] is instructive. I
soon grew tired of counting, but I believe it is safe to say
that there were at least 50 independent states at the period
in question. So far as possibilities are concerned, the record "might well have reached" Lu from any of them.

I gather that MS and M could find no set of accelerations
which made all three eclipses† total in the same place. Unless
they made a numerical error, which is unlikely, we may accept
this as the situation. If two of the eclipses were central
in Lu, the eclipse of -600 September 20 was therefore central
only at points to the west of Lu.‡ It is somewhat unlikely
that Ch'u was the point of origin of the eclipse report, but
there is another and prime possibility.

The "royal domain" is one of the states on Legge's map.
It is about the size of Lu, but it lies well to the west,
with its capital at Luoh-yang (see Table V.1). While the
emperor had lost control of China as a whole, he apparently
was able to maintain rule over this small state. We find
from the annals that Lu professed submission to the emperor,
although we may question the sincerity of this submission in
practical matters.‡ Further, during the whole period, we
find a distinction between the ritual to be observed by the
emperor and that to be observed by the ruler of a state.*
Thus, if the report of the eclipse came from outside Lu, the
most likely place is the imperial capital Luoh-yang.

In summary, there is no historical reason for assigning
Ying, the capital of Ch'u, as the place where the eclipse of
-600 September 20 was observed. If we are forced to conclude
that Kufow was not the place, because of astronomical calculations, some place within the royal domain is more likely.

We have seen that Confucius probably based his Annals
of Lu only upon records found in the state annals and other
historical sources. We have also seen that five records
clearly imply that the corresponding eclipses were observed
in Lu, probably in its capital Kufow. Thus it is reasonable
to assume, from the purely historical point of view, that all

---

†Those of -708 July 17, -600 September 20, and -548 June 19.

‡I shall test this point in the analysis that will follow
the study of the eclipse records. See Section XIV.7.

‡For example, in -617, and on a few other occasions, the
emperor sent a special embassy to Lu to ask for financial
support. The implication of the text is, I believe, that
the request was not granted.

*See, for example, p. 109 of the cited translation of Con-
fucius [-479]. Both MS [p. 490] and M [p. 8.19] cite this
distinction in support of their contention that the record
of the eclipse of -600 September 20 came from Ch'u. For
this reason, I believe that they have confused the roles of
the emperor and the presiding chief.

the eclipses were observed in Kufow unless we suspect that
eclipse reports might have been sent from other states in
the general exchange of records. We must now ask whether
eclipses were among the events that one state would send to
another.

We have two possibilities. If an eclipse were a warning
from Heaven to the ruler of a state, it seems unlikely that
a ruler would tell his fellow rulers that he was, so to speak,
serving only under a probation imposed by Heaven. On the
other hand, if an eclipse was not viewed as a serious matter,
it is not likely that a report of one would be sent to other
states in the routine exchange of important records. Thus,
either way, it is unlikely that the record of -600 September
20 was sent to Lu from some other state as part of the systematic exchange.

This does not say that an origin from outside Lu is impossible. We must still admit the chance that an eclipse
record was formally sent from one state to another. There
is also a possibility that an emissary from Lu was in some
other state when the eclipse happened, and that he reported
it to the court historian as part of his formal report when
he finished his embassy. If so, the Annals of Lu make it
unlikely that Ch'u was the other state.

Thus I shall take Kufow as the place of observation for
all the eclipses from the annals of Lu, unless astronomical
calculations force the abandonment of this assumption. There
will be a final review of the question in Section XIV.7.

We should notice the consequences of an error in this
assumption. An error in the longitude is probably more important than an error in latitude, since it causes a bias in
the inferred accelerations. MS use both Kufow and Ying as
possible sites, and the average longitude of these places is
about $114°.6$. Souciet [1729] used a point with a longitude
of $114°.33$, as I have already noted. Curott [1966] used a
longitude which is essentially the same as that of Kufow,
namely $117°.02$.†

Thus the longitude of Kufow differs by $2°.7$ or less
from the longitude of any other point that has been used. We
saw in Section V.1 that an error of $1°$ in longitude gives an
error of 0.61 in the value of D'', for an eclipse with $T = -20$.
The eclipses in the Annals of Lu have an average T of about
-25, so the error in D'' is about 0.40 for a longitude error
of $1°$. An error of $2°.7$ thus gives an error of about 1.1 in
D''. This is the error that we would have if D'' were determined only by eclipses that have been wrongly assigned to
Kufow. Since D'' will also be affected by other eclipses, the
error in D'' is actually less than 1.1.

This is probably less than the statistical uncertainty
that we shall find in the value of D''. Therefore the error
in using Kufow, if any, is not statistically significant.

---

†The latitude of Kufow is $35°.67$ [Times Atlas, 1955].

The uncertainty in the position will be expressed by
using a reliability of 0.5 for the records, unless a record
gives a specific indication that the position was Kufow.

7.   The Significance of Eclipses in Ancient China

It is widely stated that an eclipse of the sun was a
warning sent to a ruler by Heaven. For example, MS [p. 477]
quote a statement written in China around +400: "Thus when
the Sun and Moon are veiled or eclipsed one can infer irreg-
ularities in the administration." However, it is hard to be
sure what the actual attitude toward eclipses was because
the sources themselves conflict; there may have been more
than one attitude.

For example, Sivin [1969, pp. 5-6] writes: "It is well
known, for instance, that in the early Standard Histories
observations of solar eclipses were recorded and interpreted
in the Imperial Annals or the Treatise on Five-Elements
Phenomena . . while for lunar eclipses, which could be predicted, it was sufficient to publish the method of their calculation ..." This contradicts the statement quoted in the
first paragraph, which attributes equal significance to solar
and lunar eclipses.

These quotations not only contradict each other but both
contradict a quotation given in connection with the eclipse
of -520 June 10 in Section V.2. That quotation was presumably written about -450,† and it in turn is presumably a quotation from a source written when the eclipse occurred. I
repeat the quotation here for convenience in contrasting it
with the others: "At the solstices and equinoxes, an eclipse
of the sun does not indicate calamity. The sun and the moon, in
their travelling, are, at the equinoxes, in the same path;
and at the solstices, they pass each other. On other months,
an eclipse indicates calamity."

I have pointed out in Section V.5 some strong reasons
for believing that the occurrence of a predicted eclipse was
not considered as a portent or a serious event of any sort
during the Tarng dynasty. The same attitude would be expected
in later times. Further, in Section V.3, we found strong but
not conclusive evidence that solar eclipses were being predicted under the Former Han dynasty, and that many of these
predictions found their way into the official annals. This
indicates that the general situation was the same in the Former
Han as in the Tarng dynasty; the main difference between the
dynasties in this regard seems to have been the accuracy of
the predictions.‡

† We have no knowledge of the text [Tso, ca. -450] earlier
than about -200. Thus we must admit the possibility that
the text may have been altered in some particulars, or had
interpolations made in it, between its composition and
-200. For this reason, I say the quotation was "presumably"
written about -450. See Legge [1872, p. 24].

‡ We expect an improvement in accuracy between the Han and
Tarng dynasties if the eclipses were predicted, and this
is just what we find.

The quotation from Tso that was just given looks like an early stage in the development of this attitude. Here we have an explicit statement that the normal courses of the sun and moon through the sky require the occurrence of solar eclipses at certain times. An eclipse that occurs at one of these times is not a portent, but an eclipse that occurs at some other time is a portent.

In Section V.2, we found eclipses in the Annals of Lu which were greeted by the performance of magic rituals. We also found one occasion (-520 June 10) on which the ruler watched an eclipse occurring near a solstice and was apparently satisfied that the eclipse did not portend anything. Thus we have confirmation that the attitude suggested above already existed by -450 and perhaps by -520. Some eclipses can be predicted and thus they are not portents. Eclipses that are not predicted are portents, however.

I suggest, then, that this was generally the attitude of Chinese rulers and historians toward solar eclipses from the time of the Lu annals onward. There may have been individuals who, for one reason or another, did not subscribe to this attitude. There was another aspect to this attitude for which we found evidence in Section III.6 and near the end of Section V.5: If a predicted eclipse did not occur, or if it was smaller than predicted, this was a sign of Heaven's approval of the ruler.

Thus we seem to have a sensible and symmetrical attitude toward solar eclipses. If an eclipse occurred more or less as predicted, it meant nothing. If it were smaller than predicted, including the case of not occurring at all, this was a favorable sign. If it were larger than predicted, including the case of an observed eclipse that was not predicted, this was an unfavorable sign.

If this conclusion is correct, it has an interesting consequence. When an eclipse was predicted, it was necessary to watch for it to see if the circumstances were worse or better than the prediction. At times when an eclipse was not predicted, it was still necessary to watch for one in case Heaven decided to send an "unnatural" one. Thus, even after the theory of eclipses became reasonably accurate, an intensive eclipse watch was just as necessary as when there was no theory of eclipses at all.

This conclusion raises a question about the authenticity of the eclipse records in the Annals of Lu. In later periods, when the prediction of eclipses for specific dates had become fairly common, we find that eclipses in the records are often predictions rather than observations. However, the theory outlined in the Lu period does not seem to predict eclipses for specific dates; it merely says that natural eclipses are confined to certain occasions. If this is so, the eclipse dates in the Annals of Lu come from observation rather than prediction, and they are therefore authentic.

However, we do not need to depend upon this conclusion. The Annals of Lu give records of 37 eclipses, as we have seen. In three records, there are scribal errors which keep

us from identifying the eclipses, but there are no reasons
to suspect that the erroneous records came from predictions;
we can give simple explanations of all the errors but one.
Further, the records specify that three of the eclipses were
central, and calculation indicates that all three were at
least very large. Thus the accuracy of the records is considerably better than the accuracy that we find a millenium
later under the Tarng dynasty. This does not seem possible
if many of the dates came from calculation.

Therefore we may conclude with high confidence that the
eclipse records in the Annals of Lu are the results of actual
observation. We may not conclude this for records from the
Han dynasties or later.

We end this section by stating another interesting consequence of the attitude toward eclipses that was suggested
a moment ago. I suggested that a predicted eclipse which
was not seen was a favorable sign for the emperor. Thus it
would be in the emperor's interests for the court astronomer
to "overpredict" solar eclipses, and Sivin [1969] points out
that Chinese eclipse theories had just this property. We
should also remember the conclusion of Bielenstein [1956]
that was discussed in Section III.6. According to this conclusion, the fraction of actual eclipses that was recorded
was proportional to the unpopularity of the current emperor.
We should also remember that one of Bielenstein's starting
points was the assumption that every recorded eclipse meant
a warning sent by Heaven to the emperor.

Since we have strong reasons to doubt this assumption,
we may also doubt Bielenstein's conclusion. Perhaps there
is a better measure of the emperor's popularity or unpopularity than the fraction of recorded eclipses. If a predicted
but not observed eclipse was a favorable sign, the court
astronomer could make an emperor look as if he had Heaven's
approval by recording far too many predicted eclipses. That
is, the number of unobservable eclipses in the records may
be a real measure of an emperor's popularity.

8.   The Work of Velikovsky

According to Velikovsky [1950], the sun-earth-moon system
has undergone two periods of extreme disturbance which have
changed the parameters of the system by large amounts within
the past few thousand years. In the first period, the earth
had two near-collisions with Venus, the first being in the
time of the Hebrew exodus from Egypt and the second being on
the day during Joshua's campaigns when the sun and moon stood
still; see the discussion of the "record" of the eclipse of
-1130 September 30 in Section II.2.

These near-collisions were about 50 years apart, and
they took place sometime around -1400, according to Velikovsky.
After the second one, the sun-earth-moon system was left with
a year of 360 days and a month of 30 days. Velikovsky [pp.
334ff] emphasizes that 30 days is not a mere approximation to
29.53 days, and that the month was almost exactly 30 days at
that time. Venus was left in a highly elliptic orbit.

Velikovsky also emphasizes that a calendar year as long as 365 days was unknown until sometime after -700, and that all known calendars in the period from -1400 to -700 had only 360 days. This, as I understand what he writes, is his basis for saying that the astronomical year was only 360 days during this period. On his page 340, in reference to the Egyptian calendar, he explicitly writes: "In the eighth or seventh century (before the common era)† the five epagomena days were added to the year under conditions which caused them to be regarded as unpropitious." By the five epagomena days (usually called the epagomenal days) he means the days that were added to a period containing 12 months of 30 days each (see APO, Section II.3), thus bringing the Egyptian calendar year to 365 days. Parker [1974] explicitly contradicts Velikovsky's statement. According to Parker, the Egyptians introduced a year of 365 days early in the third millenium before the common era. This is a time between -3000 and -2500. Yet according to Velikovsky [p. 333] the year at this time was even shorter than 360 days.

During the second period, the earth had two or more near-collisions with Mars, and Velikovsky [p. 218 and p. 241] gives exact dates for the first and last of them. The first was on -746 February 26 and the last was on -686 March 23.‡ Since -686 March 23, the parameters of the sun, moon, and planets have been undisturbed. For at least part of the time between -746 and -686, the month had 36 days while the year continued to have 360 days, according to Velikovsky; thus he explains the fact [Velikovsky, pp. 347ff] that the oldest known Roman calendar (see Appendix I) had only 10 named months.

At some time during the series of near-collisions with Mars, the earth's spin axis shifted with respect to the crust [Velikovsky, pp. 316ff] by 20° or so. The longitude of the older North Pole was about 90°W in terms of present maps. That is, the North Pole was once near the west end of Baffin Island. Also during this time, Venus and Mars had one or more near-collisions with each other. These changed the eccentricity of Venus's orbit from a large value to its present small value of only about 0.007. At the last of these near-collisions, Mars must have had an orbit that came close to the present orbit of Venus and that also extended as far out as the present distance of Mars from the sun. I did not find any place in which Velikovsky explained how the orbit of Mars assumed its present shape.

Data already presented in this chapter allow us to prove rigorously that Velikovsky's conclusions are wrong. Before I present this proof, however, I wish to discuss some aspects of Velikovsky's work and the nature of his arguments. Let us start with his "demonstration" that the last collision with

---

†I have added the parenthesis in order to clarify the meaning.

‡Velikovsky writes the years as -747 and -687, respectively. It is clear from the context that he uses the minus sign to denote a year before the common era instead of using it to denote a year in astronomical style. The years are -746 and -686 in the notation used in this work.

Venus occurred on the day that the sun and moon stood still so that Joshua could finish his battle.

As Velikovsky [p. 58] points out, the Biblical story of Joshua recounts a shower of large stones (Joshua 10:11) that occurred just before the day when the sun and moon stood still, and the stones killed more of Israel's enemies than Israel's soldiers killed in combat. Velikovsky also quotes from a non-scriptural poem which says that earthquakes and whirlwinds occurred at the same time, and he writes [p. 59]: "A torrent of large stones coming from the sky, an earthquake, a whirlwind, a disturbance in the movement of the earth - these four phenomena belong together.† It appears that a large comet must have passed very near to our planet and disrupted its movement; a part of the stones dispersed in the neck and tail of the comet smote the surface of our earth a shattering blow." He later develops the thesis that the "comet" was the planet Venus, and hence that comets are massive dense bodies and not light tenuous ones.

This shows vividly the danger in taking a passage out of context, or of considering only one passage without examining the implications of other similar passages. For example, as we saw in Section II.3, in a year near +30, there was darkness over the earth for three hours, the curtain of the temple (in Jerusalem) was rent in two, the earth shook, and rocks were split and tombs torn open. In AAO [p. 78], I discussed a passage from the Song of Roland about what happened at the death of Roland. There was immeasurable rain and hail.‡ There was an earthquake so severe that "there was not a house whose walls did not burst" over a large area of France, and there was also a great darkness.

These passages provide almost exact parallels to the passage from Joshua. In particular, in all three passages we have an event that is impossible according to orthodox astronomy. The main difference is that the event is the stopping of the sun and moon in Joshua while it is an impossible eclipse in the other passages.‡ Thus, if we accept Velikovsky's argument that there must have been a near-collision of the earth with a massive body in the time of Joshua, there must have been other near-collisions around +30 and around 780, when Roland died. However, we have ample evidence that no such thing happened in or near 30 and 780. Therefore there is no basis for assuming such an event in the time of Joshua.

---

† By a disturbance in the movement of the earth, Velikovsky means that the sun and moon would appear to stand still only if the earth's rotation stopped.

‡ The passage from Joshua says that the large stones were hailstones. Velikovsky ignores this point and describes them as meteorites. We are entitled to the same liberty.

‡ It is amusing to remember from Section II.2 that the stopping of the sun and moon in the Joshua passage is really an oblique description of an eclipse, according to Sawyer [1972a].

Velikovsky frequently misunderstands the nature of the evidence that he cites in support of his conclusions, and I shall give two examples. On pages 251ff of his work, he argues that the Trojan War took place sometime around -800 or -750, and part of his argument reads as follows: "The tradition about Aeneas who, saved when Troy was captured, went to Carthage (a city built in the ninth century)† and from there to Italy, where he founded Rome (a city first built in the middle of the eighth century),† implies that Troy was destroyed in the eighth or late in the ninth century."

According to ancient Roman tradition, Rome was founded in the middle of the eighth century before the common era, and -752 is the year most commonly given. We may join Velikovsky in assuming that this year is reasonably accurate. However, the founding of the city did not mean the first building there. On the contrary, according to Italian practice of the time, it meant the formal organization of a well-inhabited area into the legal entity known as a city. This implies, among other things, that there must have been a fairly large population already living there, and hence that Rome had probably first been built at an earlier period.

The second example concerns the date -746 February 26 that Velikovsky takes as the date of the first near-collision between the earth and Mars. Velikovsky [p. 218] introduces his discussion of this date by writing: "In -747‡ a new calendar was introduced in the Middle East, and that year is known as 'the beginning of the era of Nabonassar.' It is asserted that some astronomical event gave birth to this new calendar, ..." Velikovsky does not say who made this assertion.

The date in question is indeed the era of Nabonassar, which is discussed in connection with the Egyptian calendar in APO [Section II.3]. However, there is no evidence that it ever formed the basis of a new calendar, there is no evidence that it was ever used by anyone until it was introduced into astronomy about 900 years later, and there is no evidence known to me that it was chosen because of some singular event. Ptolemy needed an epoch from which he could count years and days, and there is no reason to believe that he chose the era of Nabonassar on any basis other than convenience. The most probable explanation is that the oldest astronomical records known to him were made in the reign of Nabonassar. If so, it would be natural to take the beginning of Nabonassar's first year as the basic epoch for astronomical calculations.

In sum, there is no reason to assume that the epoch of Nabonassar was chosen because of a catastrophic event. It was merely the first "New Year's Day" after the accession of Nabonassar.

---

†Both parentheses are Velikovsky's. By the ninth and eighth centuries, he means the ninth and eighth centuries before the common era.

‡We should remember that this is the year -746 in the conventions used in the present work.

Velikovsky also fails to take certain physical facts into account. For example, as we have already noted, Velikovsky [pp. 316ff] concludes that the North Pole was once near the point on the earth's crust that is now at latitude 70°N and longitude 90°W. As a result of a near-collision with Mars around the year -700, the axis suddenly shifted to its present position.

At the present time, the earth is almost exactly in hydrostatic equilibrium under the joint influence of its self-gravitation and its rotation at its present rate and about its present axis. If the earth has come to equilibrium in 2600 years, as Velikovsky's conclusions demand, we can safely assume that it was in equilibrium before -700 as well. Let us see what would happen to an earth in equilibrium if its axis suddenly shifted, ignoring the problem of finding a force system that could cause the shift.

One consequence is that Siberia shifted from a temperate part of the earth to a frigid one. In regard to this point, Velikovsky [p. 330] writes: "The sudden extermination of mammoths was caused by a catastrophe and probably resulted from asphyxiation or electrocution. The immediately subsequent movement of the Siberian continent (sic) into the polar region is probably responsible for the preservation of the corpses." Later on the same page he concludes that they died from a "lack of oxygen caused by fires raging high in the atmosphere." He does not explain why the lack of oxygen did not affect other areas. If it killed mammoths in Siberia, it should have devastated animal and human life over the entire earth, but Velikovsky implicitly assumes that this did not happen. Velikovsky does not explain what there is in the upper atmosphere that could support a raging fire.

If this shift of the axis occurred, I agree with Velikovsky that the mammoths would have died from asphyxiation, but not for the reason he gives. Siberia and China would have increased their latitude by 20° as a result of the shift, and this shift in latitude changes the equilibrium radius of the earth. The equilibrium radius at the new latitude is less than at the old latitude by about 7 kilometers for any point in Siberia or China. In other words, Siberia and China suddenly found themselves, presumably in the period of about a day, at an altitude of at least 7 kilometers.

The rigid massive crust and mantle could not adjust themselves to this change as rapidly as the light fluid atmosphere. Hence all the inhabitants of Siberia and China suddenly found themselves in the atmospheric conditions appropriate to an altitude of 7000 meters. At such an extreme altitude, it is doubtful that many of the Chinese people would have survived. However, there is no evidence in the records that there was any large discontinuity in the Chinese population or in its conditions of living.

Further, the ocean would also adjust more rapidly than the earth. Hence all the islands just east of mainland Asia would have become joined to the mainland by dry land. On the other hand, the temperate parts of North America would have found themselves at a depth of 7000 meters below sea level and hence they would have been immediately inundated by the oceans. Even the peaks of the Rocky Mountains would have been under water. We may safely say that no such thing happened.

From these and many other examples, we see that Velikovsky has not produced any important evidence to support his conclusions. From one point of view, this should be sufficient; speculation that is not supported by any significant evidence has no place in the scholarly literature. However, I prefer not to rely upon the negative proposition that Velikovsky has failed to support his conclusions. Instead, I wish to present positive proof that a major part of his conclusions is wrong. We can do this by using one of the sources that Velikovsky himself relies upon.

I mentioned above that Velikovsky takes −686 March 23[†] as the exact date of the last collision between the earth and Mars, and that the sun-earth-moon system has been undisturbed since then. He takes this date from the Annals of Lu [Confucius, −479], using the following passage [p. 241] to do so: "The year 687 B.C., in the summer, in the fourth moon, in the day *sin mao* (23rd of March) during the night, the fixed stars did not appear, though the night was clear [cloudless]. In the middle of the night stars fell like a rain." Velikovsky says that this translation is taken from a work by Edouard Biot that I have not consulted. The parenthesis, the brackets, and the Italics all appear in Velikovsky's text; I do not know whether they appear in Biot's text or not.

The term *sin mao* designates a day whose cycle number (Appendix II) is 28. The cycle number for −686 March 23 is indeed 28. If the year should be changed to −685, the day would have to be changed to March 18, whose cycle number in −685 was also 28.

Velikovsky concludes that this meteor shower was occasioned by the earth's near-collision with Mars, and hence he is able to assign an exact date to the event.

As it happens, the Annals of Lu contains many records of astronomical importance besides the record of the meteor shower. In particular, it contains the records of 34 solar eclipses that are listed in Table V.2; these are the records up to the one designated −480 April 19 C. Of these records, as I have noted in the table, the reading of the date that I have listed as −644 August 28 is questionable; this leaves 33 records. Calculation with provisional values of the accelerations shows that all the eclipses took place in accordance with the records with only one exception. The eclipse of −552 August 31 was not visible in China, so that the record must be the result of a scribal error of some sort, unless it represents an early attempt to predict an eclipse. This leaves us with 32 valid records.

Twenty-nine of these eclipses occurred after −686 March 23. Since Velikovsky concludes that there have been no

_____
[†]In a parenthesis on his page 240, Velikovsky says that the year is "less probably" −685.

−185−

significant perturbations since that date, these eclipses have no direct bearing upon his conclusions. However, the agreement of the calculations and the records for these 29 eclipses shows that the chronology of the <u>Annals of Lu</u> is correct for years after -686. Further, the internal chronology of the annals is certain. Thus, if the chronology is correct after -686, it must also be correct for years before -686.

Table V.2 shows us that 3 eclipses took place before -686, and the preceding paragraph shows us that the recorded dates of these eclipses are correct. Further, calculation with the provisional accelerations shows agreement with the records. The record of -708 July 17 is particularly valuable because the record asserts that the eclipse was central, and calculation shows that the eclipse must have been quite large. I estimate that the agreement for -708 July 17 could not occur if the calculated time of the eclipse were in error by more than about an hour.

Between -708 July 17 and -686 March 23 there are about 265 months. During this interval, the accumulated error in the ephemeris of the moon† cannot amount to more than about an hour, or 3600 seconds. Thus the length of the month cannot have changed by more than about 14 seconds in -686. Yet Velikovsky concludes that it changed by more than 6 days, from 36 days to $29\frac{1}{2}$ days. The change that he claims is not consistent with any of the records before -686.

Thus we have proved, by using a source that Velikovsky himself relies upon, that the "catastrophe" of -686 did not occur. Further, it proves that there could have been no such catastrophe as late as -719 February 22, the date of the earliest eclipse in Table V.2. Still further, Velikovsky's arguments connecting the "catastrophe" of -746 with that of -686, and in fact with the entire set of "near-collisions" with Mars, are tightly connected, and if we have proved that the near-collision of -686 did not occur, we have proved that the entire set did not occur.

I do not see any way to extend the argument based upon Table V.2 to the alleged near-collisions with Venus around -1400. However, since the same kinds of argument are used for both the Venus and Mars events, the case for the near-collisions with Venus is considerably weakened from its already frail condition.

---

†In the ensuing calculation I take the length of the year to be constant. Velikovsky claims that the length of the year, as well as the length of the month, changed in -686. Allowing for a change in the length of the year complicates the argument, but it does not change my basic conclusion.

CHAPTER VI

OTHER ECLIPSE RECORDS BEFORE THE YEAR +100

1. Discussions of Individual Records

Table VI.1 lists all the solar eclipse records before
the year +100, except the Chinese ones, that have reliabili-
ties different from 0. When I refer to eclipse records, I
am referring only to records which tell us the place where
an eclipse was observed, without giving us specific measurements of either the magnitudes or the times. No observations
in this time period contain measurements of the magnitudes,
so far as I know, but two contain careful measurements of
the times. These will be used in a later work.

TABLE VI.1

USEFUL ECLIPSE RECORDS BEFORE +100,
EXCEPT CHINESE ONES

Designation	Place	Standard deviation of the magnitude	Reliability
-430 Aug 3a M	Athens to Thasos	0.14^a	1
-430 Aug 3b M	Athens	0.18	0.1
-423 Mar 21 M	Athens to Thasos	0.14^a	1
-393 Aug 14 M	Chaeronea	0.14^a	1
-363 Jul 13 M	Thebes	0.18	0.1
-309 Aug 15a M	East coast of Sicily	0.054	0.1
-135 Apr 15 BA	Babylon	0.034	1
+ 59 Apr 30 E	Campania	0.18	1
+ 59 Apr 30 M	Armenia	0.18	1

[a]The eclipse is taken to be partial, with a magnitude of
0.81.
Note: The eclipse is taken to be central if no estimate
of the magnitude is given.

Seven of the nine records in Table VI.1 were used in
AAO. The records -423 March 21 and -135 April 15 were not
used there and will need new discussion. In addition,
many "records" not in the table will be discussed in order
to explain why they are not there.

I shall use the same system of designating these records
that I used in AAO; this system is described in Section I.6.
For records in later chapters, I shall use the system of MCRE,

with suitable extensions where they are needed. I shall introduce the discussion of each record in Table VI.1 by stating the designation of the record. I shall introduce the discussion of all absent "records" by stating an appropriate date, followed by a question mark to indicate that the "record" is not to be used.

-1374 May 3 ? I have already discussed this record in Section III.4. It is likely that the record does in fact describe an eclipse, but we have no information about its magnitude. The record gives two details that might help in determining the date, and the date -1374 May 3, which $\underline{M}$ and $\underline{MS}$ adopt, does not satisfy either; thus we should reject their date. Unless more documentary evidence is found, we can identify this eclipse only by means of the identification game, which is the way that $\underline{MS}$ "identify" it.

-1334 March 13 ? $\underline{M}$ [p. 8.12] mentions a text which refers to a solar portent, but he does not quote it. Present data are inadequate to date the event even if we assume that it is an eclipse. Since there seems to be no clue to the magnitude if it is an eclipse, it is clearly impossible to date the portent by astronomical means; unfortunately this will probably not prevent many attempts and subsequent claims of success.

-1130 September 30 ? This refers to the famous account of the sun and moon standing still while Joshua's forces conquered their enemies, which I considered in Section II.2. In my opinion, there is no reason to assume that this is an eclipse in disguise, and it can be dated only by means of the identification game if it is an eclipse. As we saw in Section IV.3, this "eclipse" plays a critical role in $\underline{M}$'s determination of the parameters. I shall ignore it.

-1062 July 31 ? This has often been called the eclipse of Babylon, and it has played a role in at least one determination of the accelerations of the sun and moon (see Section VI.3). $\underline{MS}$ do not mention it, but $\underline{M}$ [pp. 8.15-8.16] devotes about a page to it and $\underline{AAO}$ [pp. 58-60] devotes about two. $\underline{M}$ and I concur that there is no particular reason to believe that the event in question was an eclipse, and that there is no way to date it more closely than about a century if it were one. Let us hope that this particular ghost eclipse will cease to haunt us.

-762 June 15 ? Until the present, this date has been accepted for most of this century as a firmly established date, and I used it in $\underline{AAO}$. As we saw in Section II.5, the date must now be considered as one of the fatalities caused by Ptolemy's fraudulent work in astronomy. Unless the date can be established by means which are totally independent of Ptolemy, we have no way to date the eclipse, particularly since we have no clue to its magnitude.

Although I just wrote that this is one of the victims of Ptolemy's fraud, further study suggests that we could not date the eponym canon eclipse even if Ptolemy's king list were valid. I thank Mr. Philip Couture of Santee, California for pointing out to me that the accuracy of the eponym list is not beyond question. For example, the king Sargon II is believed from other evidence to have reigned only 17 years, but the number of limmu listed for his reign is 32, according to Mr. Couture† (private communication); I have not verified this number independently. Thus we must allow the possibility that there are gaps in the list. Further, many of the names occur more than once. Many multiple occurrences are probably genuine, but we must also allow the possibility that some of them are accidental repetitions and hence that there may be more names than years in some parts of the text.

Oppert [1898] tries to date the eponym canon eclipse by a method that is independent of Ptolemy. He says that there is an inscription on a stone erected by order of the king Assur-nasir-abal which was discovered and published by Layard; he does not cite the publication. This says that the sun god, in the year of the king's accession, made an eclipse that was propitious for him, and that he reigned mightily, we presume in consequence of the god's favor as shown by the eclipse. Oppert further says that there were 121 years between the accession of this king and the year of the eponym canon eclipse.

Oppert then points out that there was no eclipse visible in Assyria 121 years before -762, and hence this cannot be the correct year. He believes that the eponym canon eclipse occurred on -808 June 13, so that the earlier eclipse occurred on -929 June 2. His choice of date for the eponym canon eclipse is earlier than any date I listed in Table II.1. Horner [1898, p. 243], however, points out that we are justified in taking the interval to be 122 years rather than 121. If so, we can still accept -762 June 15 for the eponym canon eclipse, yielding -884 July 13 for the eclipse in the accession year of Assur-nasir-abal.

I have inspected all the dates in Table II.1 to see how many are compatible with an eclipse that occurred between 120 and 122 years earlier. I find that we can eliminate almost none of the dates. It is interesting that the eclipse of -762 June 15 was larger than any other possibility for the later eclipse and that its mate on -884 July 13 was larger than any other possibility for the earlier eclipse. However, this is not enough justification to accept these dates as established.

It is also hard to know how seriously we should take the "eclipse of Assur-nasir-abal". We seem to have a clear reference to an eclipse in his accession year, but the

---

†Mr. Couture, who is much better acquainted with the relevant literature than I am, has also spent much effort trying to determine how the eponym canon eclipse came to be accepted as the eclipse of -762 June 15. He has had no success.

eclipse seems to be used in a magical fashion. Kings tended to make claims about being the favored of the gods, and to record the signs of this favor in inscriptions. We may try to say that a king could select almost any event he wanted to and choose to use it as a sign of celestial favor; if so, he did not need to make up an event. Hence, we may say, Assur-nasir-abal truthfully recorded an eclipse in the year of his accession. However, he may have also chosen an eclipse as a sign on the basis of its impressiveness rather than on the basis of its truth.

Thus I do not think that it is safe to accept the "eclipse of Assur-nasir-abal" as an astronomical observation. If it ever becomes possible to date his accession by independent means, and if it turns out that there was an eclipse visible in Assyria in that year, we could then accept it.

In sum, the record of the eponym canon eclipse is probably genuine, but we cannot date it on the basis of present knowledge. The eclipse during the accession year of Assur-nasir-abal may or may not be genuine. If it is genuine, we likewise have no way to date it.

Neither $\underline{MS}$ nor $\underline{M}$ uses the eponym canon eclipse, but $\underline{M}$ [p. 13.3] comments on it in his post-analysis discussion. He says that the eclipse of -762 June 15 was total at Nineveh, and that this " .. adds further support to the presumption that eclipses were observed in the location where the record is found." This needs two comments. First, the record does not imply that the eclipse was total, so $\underline{M}$'s calculation does not create a correspondence with the record; thus there is no support for the presumption. Second, I calculate (Table II.1) that the magnitude on -762 June 15 was only 0.95 at Nineveh. It is clear that we do not know the accelerations well enough to justify $\underline{M}$'s statement.

-647 April 6 ? Fotheringham [1920] gives the following translation of a passage written by the early Greek poet Archilochus: "Nothing there is beyond hope, nothing that can be sworn impossible, nothing wonderful, since Zeus father of the Olympians made night from midday, hiding the light of the shining sun, and sore fear came upon men." Much of the literature on ancient eclipses has accepted this passage as the immediate and eye witness account of a total eclipse of the sun. Oppolzer [1882] tentatively identified the date as -647 April 6. Many later writers, including Fotheringham, ignored Oppolzer's cautions about the date and took it to be an established one. This is the archetype of a literary eclipse [AAO, Section III.2] that has been dated by means of the identification game.

I devoted about two pages to this "eclipse" in AAO [pp. 91-93]. Because so much of the standard literature on the subject accepted this as a record of an eclipse, I did not dare to reject this idea entirely, but I did give this eclipse and other literary eclipses a very low weight. I intended to make the weight so low that such eclipses would have a negligible effect upon the calculated accelerations. However, at the time, I did not realize how weakly determined the individual accelerations $\dot{n}_M$ and y are, and as a result (Section III.7 above) the literary eclipses did have an effect upon $\dot{n}_M$ and y individually. As I intended, though, their effect upon D'', which is the strongly determined parameter, is negligible.

In this work, I should have preferred to dismiss this "eclipse" in a few words, while giving it a weight of zero, but the discussions of this passage by MS and M make this course impossible.

In the discussion in AAO, I pointed out that nothing in the passage keeps it from referring to a "dark day".[†] In his discussion of this "eclipse", M [p. 8.17] refers to this by saying: ". . Newton's suggestion of a cloud or other darkening seems very weak." M does not seem to be aware of the phenomenon called a dark day, which is known well enough to have an independent entry in several dictionaries and encyclopedias.

The specific cause of the great darkening is known for some dark days and unknown for others. I have not been able to locate any statistics on dark days, but it is plausible that they are as common as total solar eclipses.[‡] If my experience is typical, the dark day is more common. I have seen a total solar eclipse because I purposely travelled in order to do so; if I had taken things as they come, I would never have seen a total eclipse yet and it is unlikely that I would see one in my lifetime. However, I have experienced a dark day whose cause is unknown; it was probably an unusual cloud phenomenon.

The day in question started out as a day that was cloudy in an ordinary fashion. There was no storm and no unusual amount of wind. Suddenly the day became extremely dark, much darker, I think, than during the average annular eclipse where it is central. It was possible to walk without artificial light but it was not possible to drive an automobile without using headlights. The little light that remained had an eerie quality that I am incompetent to describe. The great darkness came suddenly and left just as suddenly. I estimate that the total time involved was about 30 minutes, and the darkness ended shortly before noon.

Altogether it was an awe-producing experience. If I were a poet, this "dark day" would be just as likely to inspire me to write a poem as a solar eclipse would. Archilochus wrote: ". . Zeus father of the Olympians made night from midday, hiding the light of the shining sun, and sore fear came upon men." A Greek poet of the -7th century could write this about a "dark day" like the one I saw just as truthfully as he could write it about a total solar eclipse.

[†]Both here and in AAO I have used the quotation marks to indicate that I am talking about a specific phenomenon, and that I am not merely talking about a day that happens to be dark.

[‡]On the average, there is an eclipse that is total at a particular place once in about 300 years.

The general principle that should guide us is the
following: When a passage describes a great darkness, or
an unusual behaviour of the sun, if we can date the passage
exactly from historical evidence alone, if we can place the
event exactly from historical evidence alone, and if there
happened to be a total eclipse of the sun on that day and at
that place, it is reasonable to assume that the passage refers to the eclipse; it is unlikely that there would be two
independent events that could have led to the passage.
Otherwise we have no right to assume that the passage refers
to an eclipse. †

None of the "if's" in the preceding paragraph apply to
the "eclipse of Archilochus". We have no right to assume
that Archilochus was talking about an eclipse; we cannot even
say that the probability is particularly high that he was
talking about an eclipse.

M [p. 8.17] writes: ". . the poem seems to have the
ring of personal‡ impact to me." Of course it has, but this
is irrelevant. One of the functions of a poet is to use
words in a manner that produces a sense of personal impact.
The fact that Archilochus did so here does not provide
grounds for the slightest presumption that he had personally
seen a total eclipse of the sun. It merely shows that he
was a good poet.

In AAO [p. 45], I introduced the "eclipse of Mark
Twain". In A Connecticut Yankee in King Arthur's Court
[Clemens, 1889, Chapter VI], Mark Twain describes a total
eclipse of the sun in a manner that has the "ring of personal
impact". He describes the uncanny cold,‡ the stars that became visible, and the "silver rim" of the sun that first became visible as the total phase was ending. This is a late
example of the literary eclipse, but it is just as much an
example as the "eclipse of Archilochus" or the other literary eclipses that adorn the astronomical literature over the
past century. I suggest that journal editors and book editors adopt the following rule: No author shall be allowed
to advocate the reality of any literary eclipse unless he
has first "identified" the eclipse of Mark Twain by showing
us where and when Mark Twain saw the total eclipse which
inspired him to write A Connecticut Yankee in King Arthur's
Court.* Further, he must show this by using only fragments

---

†The reader should consult the discussion under the date
-556 May 19 below.

‡The emphasis is M's.

‡This is a valid item in an eye-witness account, but it is
one that few eye witnesses mention.

*Twain says that the eclipse happened on 528 June 21 in
England. Many people have thought that the relevant question is whether there was such an eclipse. The answer is
that there was no such eclipse, but the answer is also that
this is not the relevant question. The reasoning that
leads people to impute reality to the "eclipse of Archilochus" says that the account was inspired because the
author had recently experienced a total eclipse. Thus the

of Twain's writings; he may not use any biographical information beyond the most rudimentary.

We should keep the following in mind when judging any literary eclipse: Something suggested to the author that he use an eclipse or a great darkness in his writing. It is possible that the suggestion came from seeing an eclipse, and it is also possible that the suggestion came from hearing or reading about one; Twain himself suggests this origin for his eclipse near the end of Chapter V. Even if the author derived the idea from seeing an eclipse, the eclipse does not need to have been total or even large. Any eclipse could serve as the nucleus around which the full literary description crystallized. To assume that a poet could describe a total eclipse only if he had seen one is equivalent to assuming that he had no imagination, which is an odd accusation to make about a poet.

Neither M nor MS actually use the eclipse of Archilochus in deriving the accelerations, but both sources return to it in some post-analysis considerations. MS [p. 523] write: "Our solution implies Paros† as the place of observation on 6 April -647, and the uncertainties in the solution are not enough to admit of any alternative." M [p. 13.1] makes a similar remark, gives a highly dramatic account of how Archilochus must have behaved on the day in question, and concludes: "Historians are probably safe in placing Archilochus in Paros on this date."

A reader with a reasonable amount of caution must reject both of these statements. There is little reason to believe that Archilochus was inspired by any solar eclipse, and there is no reason to assume that he was inspired by a total eclipse. Further, we do not know that the eclipse of -647 April 6 was total on Paros. It may well be that it was if we use the accelerations that MS and M derive; I have not verified their calculation. It may also be true that no other eclipse was total on either Paros or Thasos‡ within the appropriate time period, for accelerations within the limits that MS and M find from their solutions. Since they do not tell us what possible dates they considered, we cannot judge this point. However, even if they are correct on this point, we cannot accept this as a meaningful conclusion, quite aside from the necessity to consider places other than Paros and Thasos. It is clear from earlier parts of this work that MS and M have been rather optimistic about the bounds to the accelerations. I have shown [Newton, 1974] that a reasonable set of accelerations makes the eclipse of

---

"eclipse of Mark Twain" "proves" that Twain himself saw an eclipse and was inspired by it to write his description, if we accept the usual reasoning.

†An island in the southern Aegean where Archilochus is believed to have been born. He also lived for some time on the island of Thasos in the northern Aegean.

‡Most writers on the subject assume that Archilochus had to see the eclipse on either Paros or Thasos, ignoring the fact that he lived elsewhere, such as in Sicily, for a considerable but unknown time. See AAO [pp.92-93].

-647 April 6 total on Paros and partial on Thasos. A small change, well within our firm knowledge of the accelerations, makes this eclipse partial on Paros but makes the eclipse of -656 April 15 total on Thasos. We have no basis, either astronomical or historical, for choosing between the two cases.

I have spent more space on the "eclipse of Archilochus" than it deserves on its own right. However, it is the archetype of the literary eclipse, and the way it is analyzed in many sources is typical of the way many literary and magical eclipses are analyzed in the literature. It is as valuable to remove an erroneous datum from the accepted body of data as it is to add a correct one. Thus, if this discussion has helped convince the reader to be extremely skeptical with regard to literary and magical eclipses, it is worth the space that it has taken. In my opinion, astronomers should not use personal subjective reactions to poems as if their reactions were authentic astronomical observations.

-584 May 28 ? This is called the "eclipse of Thales" in much earlier literature. In AAO [pp. 94-97], I attached a reliability of 0.04 to the idea that this account was inspired by an eclipse that occurred in Asia Minor in the appropriate historical period, while pointing out that there was no reason to believe any of the history connected with the account of the eclipse; I intended that the reliability would give negligible weight to this account in the formal analysis. I no longer see any reason to give the account any weight at all, and I join MS and M in dropping it from consideration.

-556 May 19 ? A passage by Xenophon in the Anabasis III, 4.8, can be translated in this way [AAO, p. 97]: "When the Persians obtained the empire from the Medes, the king of the Persians besieged this city, but could not in any way take it. But a cloud covered the sun and caused it to disappear completely ..." In spite of the clear statement about the cause of the darkness, at least one astronomer in the 19th century took this to be an unquestioned account of a total eclipse, but I have not seen any writer do so recently. I mention this passage only because the cloud is relevant to the "eclipse of Archilochus". Here we have a complete disappearing of the sun that explicitly results from a cloud and not an eclipse.

-477 February 17 ? Herodotus (see AAO [pp. 97-99]) claims that there was a total eclipse of the sun just as Xerxes began his campaign against Greece. We know the date of this event rather well and we know that no such eclipse was possible. I gave this account a trivial weight in AAO and I shall give it a weight of 0 here.

-462 April 30 ? The poet Pindar wrote a dramatic account of some sort of darkness, one that is reminiscent of the account written by Archilochus. M [pp. 8.22-8.23] quotes

extensively from an unpublished study of this account made
by Stephenson, who apparently concluded that Pindar was an
eye witness of a solar eclipse; we are not told what Stephenson concluded about the magnitude of the eclipse. Fotheringham [1920] writes that the "terms used seem to indicate
a total eclipse of the Sun." M [p. 13.4] says that Pindar
was apparently trying to tell us that the eclipse was partial;
this tells us something about the reliability of subjective
reactions to poetry.

All these writers seem to take it for granted that Pindar was giving an eye-witness account of an eclipse. Nothing
in the passage indicates this; we should remember the discussion of the eclipse of Archilochus. Both Fotheringham and M
take this to be an eclipse even though both quote a slightly
later passage which says that God can "shroud in a dark cloud
of gloom the pure light of day." Pindar does not say that
God did so shroud the light, but this is the closest thing
we have to an indication of what he had in mind. If this is
a valid indication, and if both passages refer to the same
event, he was not describing an eclipse, and nothing auth-
orizes us to assume that he was.

Those who take this to be an eclipse take Thebes to be
the place of observation. I wrote in AAO [p. 100] that this
is "based upon a tradition that Pindar was born near there.."
M [p. 8.22] says in reference to this statement that I was
apparently unaware of the fact that Pindar addressed his poem
to the Thebans. I was indeed aware of this fact, but I did
not mention it because it is not relevant to the question,
and it did not occur to me that anyone else would consider
it to be relevant. This point illustrates another characteristic feature of the traditional methods of analyzing literary eclipses, and it is worth the expenditure of a few lines.

A basic assumption that underlies these analyses, including those of Fotheringham, Muller, and apparently Stephenson, is that the author wrote his literary passage immediately upon coming indoors after watching the eclipse. As
I wrote elsewhere [Newton, 1974, p. 101]: "This is equivalent to the assumption that the writer had neither imagination nor memory."

Let us suppose for the sake of argument that Pindar did
base his poem upon having seen a solar eclipse, although I
see no particular basis for the supposition. He could have
seen the eclipse at any earlier time and kept it stored in
his memory to bring out when he found a use for it. Further,
he did not have to be in Thebes in order to write a passage
addressed to the Thebans. Addressing the poem to the Thebans
merely means that Pindar had some particular interest in
Thebes. He may have had this interest because he was born
there. Just as plausibly, he may have had this interest because he had just heard of some action by the Thebans which
particularly interested or irritated him.

I gave the "eclipse of Pindar" a weight of 0 in AAO
and I shall do the same here.

-430 August 3a M. Outside of China, this is the earliest solar eclipse† of which we have a definite and datable
record, and we actually have two independent records of it.
I used this record in AAO and I shall use it again here.

Thucydides [ca. -420, Chapter 2.28] says that the sun
was eclipsed after noon, and that after it had assumed the
form of a crescent, it returned to its normal shape. He
also says that some stars became visible during the eclipse.
When I wrote AAO, I did not feel that we knew the accelerations well enough to be sure of the direction from the observer to the eclipse path at the height of the eclipse.
Thus I used $\eta = 0$ as the best estimate that could be made
at that time, in spite of the clear statement that the sun
was a crescent at maximum eclipse. Here I shall use the
standard values for an eclipse that was clearly partial but
for which there is no suggestion about the magnitude.‡ That
is, I shall take the magnitude to be 0.81, with a standard
deviation of 0.14.

Thucydides is usually associated with Athens, but he
also owned important estates on the mainland opposite the
island of Thasos. Thus I assumed in AAO that the report
could have come from Athens, from Thasos, or from any point
in between. Specifically, I took the best estimate of the
place to be the midpoint between Athens and Thasos, and I
took the standard deviation of the error in this estimate to
be the distance from the midpoint to either Athens or Thasos.
I shall do the same here.

M [p. 8.24] writes that the weight which I assigned to
this record was "split between Athens and Thasos‡ — the
latter on a pure speculation engendered by the results of
computation." This is a serious and unwarranted misrepresentation of what I did, in three separate ways. First, I
have just described how I handled the place, Table IV.5 in
AAO makes the way explicit, and it was not the way that M
claims. If I had done what he said, this record would have
received the same total weight as a record with no uncertainty about the place of observation.* Instead, the large
standard deviation in position lowered the weight by a considerable amount. Second, the choice was not based upon pure
speculation; it was based upon the known facts of the life
of Thucydides, which necessarily leave an uncertainty in the
origin of his record. Third, the choice was not engendered
by the results of computation. On page xv of the preface to
AAO I explicitly stated that all choices such as this were
made well in advance of any computation. I stated that I

---

†We have a Babylonian record of the lunar eclipse of -522
July 16. Ptolemy gives accounts of still earlier lunar
eclipses, but they have been fabricated [Newton, 1977,
Chapter XIII].

‡Thucydides's statement that stars were visible is doubtful,
since the eclipse was annular. I shall return to this
matter in Section XIV.8.

‡The emphasis is M's.

*See Section III.7.

did this in order to eliminate any possibility of reasoning in a circle and I described the precautions I took to guarantee to the reader that I had done so.

-430 August 3b M. This account comes from the biography of Pericles by Plutarch [ca. 100], which I used in AAO [pp. 101-102]. The record says: ". . . and Pericles being gone aboard his own galley, it happened that the sun was eclipsed, and it grew dark on a sudden, to the affright of all, for this was looked upon as extremely ominous." The incident happened on a ship preparing to leave Athens for a raid on the Peloponnesus.

M also cites a work by Cicero in connection with this account that I have not consulted, because M's quotation indicates that Cicero's and Plutarch's accounts contain the same basic information. According to M [p. 8.24], Stephenson concluded in an unpublished work that ". . . it is doubtful whether there is any truth in the story." The only basis for this conclusion that I can find in M's discussion is the statement: "The Plutarch citation is clearly wrong as to the magnitude." Since Plutarch says nothing about the magnitude, I do not understand this statement.†

I see no reason to doubt the general truth of the account, although we should doubt that Pericles spoke the actual words that Plutarch imputes to him. Since we do not know Plutarch's (or Cicero's) sources, how they were transmitted, or how they were used, I give the record a reliability of only 0.1, as I did in AAO. It is possible that Plutarch, or Cicero before him, knowing of the eclipse from the account by Thucydides, decided that Pericles must have seen the eclipse also and hence decided to use it in connection with Pericles. The eclipse will be taken as central at Athens, with a standard deviation of 0.18 for the magnitude.

-423 March 21 M. Thucydides [ca. -420, Chapter IV.52] writes: "At the beginning of the following summer, about the new moon, there was a partial eclipse of the sun, and at the beginning of the same month an earthquake." This happened at the beginning of the eighth year of the Peloponnesian War and hence it must have been within a year of -423. The eclipse of -423 March 21 is the only possibility. This date seems incompatible with the beginning of summer until we realize how Thucydides divides the seasons. In his history, the only seasons he uses are summer and winter. He uses summer to mean the time within which military operations are possible, with winter meaning the rest of the year. He begins summer in what we would call early spring, and he continues it until about our October. Thus, as Thucydides uses the term, March 21 is at the beginning of summer. I use the same region of observation that I used for -430

---

†Since Plutarch says that "it grew dark on a sudden", M may have taken this to be a statement that the eclipse was total. I doubt that one should push the words this far. The reader should compare M's use of this record with his use of the record -393 August 14 M below.

August 3a M. Since it is clearly stated that the eclipse
was partial, with no other indication of the magnitude, I
shall treat the magnitude in the standard way.

        -399 June 21 ? I did not refer to this "eclipse" in
AAO. At the time I wrote AAO, I had not seen any reference
to it that had been written within this century, and I saw
no reason to revive this magical eclipse. However, M [p.
8.25] studies it and gives it a "secure" date, namely -399
June 21. Thus we must review the situation.

        The text in question [Cicero, ca. -53, Section I.16]
can be translated thus: ". . our Ennius . . writes thus
concerning the year which was about 350 after the founding†
of Rome -- on the nones of June the moon stood in front of
the sun, and night.‡ And in this thing is contained so
much method and cleverness that, from this date, which we
see vouched for in Ennius and in the Annales Maximi, earlier
eclipses of the sun were calculated as far back as that one
which was on the nones of July, when Romulus was reigning:
in which darkness, although nature carried Romulus to the
human end, virtue is said to have elevated him into heaven."

        Ennius was a Roman poet who died in about -170, and
one of his poems was a history of early Rome. Most of his
work is lost. Cicero [-54, Book II, Section 12.52] tells
us elsewhere about the Annales Maximi. He says that the
pontifex maximus, from the earliest Roman times down to
about -125, posted an annual notice of the main events of
the year, and the collection of these notices was called
the Annales Maximi. This collection is now lost but apparently it still existed in Cicero's time. Many, and I
believe most, students of Roman history strongly doubt the
authenticity of these annals, particularly for times as far
back as -400. It is likely that someone, perhaps around the
year -250, simply made up the earlier annals and then claimed
that he had just discovered this priceless set of records of
early Roman history. Thus it is likely that this record,
which Cicero and M think they are quoting, never really
existed.

        M ends his quotation of the record with the sentence‡
which directly describes the eclipse; it is: "Nonis Iunis
soli luna obstitit et nox." He then uses the following
translation: "In the month of June, the day was then the
fifth, the Moon and night obscured the shining Sun." The
most obvious explanation of this record, he says, is a large
eclipse at sunset. The only such eclipse near the year -400
was the eclipse of -399 June 21, which was probably large
but not total at Rome at sunset. Thus M takes the date to
be "secure". He does not use the record, however, because
even a small eclipse can be seen easily at sunset and thus

_____

†Thus the year is about -400.

‡This is neither a typographical error nor an error in
translation. I shall explain the meaning in a moment.

‡M says that he has taken his discussion from Stephenson's
unpublished dissertation.

we may not infer from the record that the eclipse was large.

Unfortunately this dating is based upon an error in translation. The verb obstitit is singular and it can have only one subject. The subject is clearly the moon (luna), and "night" (nox) clearly belongs to the clause that follows "and" (et); the poet has given us this clause only in ellipsis. An ellipsis at this point does not make sense if the missing predicate is independent of what has gone before Put another way, the ellipsis is allowable logically only if we can deduce the missing predicate; this means that it must be a consequence of the first clause in the sentence. Sunset is not a consequence of the first clause. Thus the reading that makes most sense goes something like this: On the nones of June the moon stood in front of the sun and night (was the result). I have supplied the words in parentheses.

However, this is a line of poetry taken out of context, and we cannot rely upon a logical reading being correct. All we can say is that the most probable meaning is this: the darkness, not necessarily complete, was the result of the moon's standing in front of the sun. This in turn means that the eclipse happened long enough before sunset (or long enough after sunrise) to make it clear that the darkness was not an ordinary night but the result of the eclipse. There was no such eclipse in June at any time within the necessary historical period. In fact, the only other large eclipse in Rome in the possible period is the eclipse of -401 January 18, which is in the wrong month.

Perhaps the main point to make is that any interpretation of this line of poetry is subjective, and as such it has no place in the literature of astronomy. We cannot assign the meaning with confidence, and we certainly cannot give the eclipse a secure date.

The text refers to two eclipses, and the one associated with Romulus is clearly magical. Altogether I know of three legends about the death of Romulus, who himself is probably historical in the sense that Arthur is, but whose feats are probably mythical in the whole. One legend says that Romulus was assassinated, one says that he died during a miraculous storm, and the other, which is the one in Cicero's text, says that he died during an eclipse and was transported into heaven. The eclipse that happened around -400 is not directly magical, but it is used to date a magical eclipse and thus it is indirectly magical. There is also a further argument which tells us that the eclipse is magical, with reasonable probability.

Appendix I to the present work is a review of the Roman calendars that were used from the times of the kings down to the final routine adoption of the Julian calendar under Augustus in the year +8. From it we learn that the Roman months in the time of Romulus were almost surely lunar months, and that the nones of a lunar month came at the first quarter of the moon. Thus it is impossible for there to be a solar eclipse on the nones of any month if that month is lunar. That is, we conclude that the date of the "eclipse of Romulus" is impossible, with high confidence.

We do not know when the months ceased to be lunar months, so that an eclipse could come on the nones of a month, but it was probably after -400. Thus, with less confidence, we can say that the date given for the "eclipse of Ennius" is also impossible.

In summary, the "record" of the "eclipse of Ennius" is probably a late forgery, although it was already an old forgery in the time of Cicero. The circumstances described in the record are impossible. The eclipse is probably magical. Even if it were real, the meaning of the passage is so uncertain that we cannot date the eclipse, and no eclipse fits the most likely meaning.

-393 August 14 M. I used this record in AAO [pp. 102-103], but I shall make two changes in the way I use it here. The record says that the sun took on the shape of a crescent while a certain army was invading the part of Greece known as Boeotia. Since the eclipse was definitely partial, I shall use the magnitude and standard deviation that are a-dopted as conventional for partial eclipses in this work. The other change concerns the place of observation. In AAO I took the place to be anywhere in Boeotia.† M [p. 8.25] says that a passage from Plutarch about the same event says that the army was near Chaeronea. I have not consulted the passage by Plutarch, but I see no particular reason to doubt it, since Chaeronea is in Boeotia. Thus I shall take Chaeronea as the place. Since Boeotia is a small region, this change has almost no effect.

It is interesting that M is willing to take Plutarch as a valid authority for this record but not for the record -430 August 3b M that was discussed above.

-363 July 13 M. The only change that I shall make from AAO [p. 103] in handling this record is to change σ_μ to 0.18.

-321 September 26 BA. The record of this eclipse gives a careful measurement of the time when the eclipse began, but it does not give any indication of the magnitude. In fact, the sun set while the eclipse was still rather small. Since the main information in the record is the measured time, I shall use this record in later work but not here. I mention it only so that the reader may know that it has not been forgotten.

-309 August 15a M.‡ This is the record of the eclipse seen on board the ship of the tyrant Agathocles [AAO,

---

†Not Thebes, as M [p. 8.26] erroneously says that I did.

‡The designation -309 August 15b M was used in AAO for one of the possible dates for the "eclipse of Hipparchus". This eclipse will be discussed under the heading -128 November 20 M.

pp. 103-104] as it was running a Carthaginian blockade.
Since the record says that stars were seen, I shall use
0.054 for σ_μ rather than 0.01 as I did in AAO. Otherwise
the characteristics are unchanged.

-189 March 14 ? In AAO [p. 70] I noted that the historical writings of Livy (Titus Livius) contain many references to eclipses, and I wrote about these references: "The
dating is usually vague. Sometimes the dating is detailed
enough to make us sure that there was no corresponding large
eclipse. More than half of the eclipses were accompanied by
a rain of stones (hailstorm?).† I presume that all these
reports are of magical eclipses."

I have read Livy in fair detail since I wrote this and
I see no reason to change my basic opinion. There has been
much debate about the worth of Livy as an historian and I
shall not venture an opinion on the general question. However, there can be no doubt that Livy was quite credulous
and that he made a specialty of collecting and transmitting
omens, marvels, and various prodigies. Some of his eclipses
may be true, but we know that some are not. There is no
basis in Livy's writing to distinguish the true from the
false. The only safe course in such a situation, it seems
to me, is to ignore all his eclipse reports.

M, basing his remarks upon Stephenson's unpublished
work, disagrees with this and says that there is no reason
to reject two records that will now be discussed. M dates
one record as -189 March 14 and the other as -187 July 17.

The record [Livy, ca. 0, Chapter XXXVII.4] which M
[pp. 8.28-8.29] dates as -189 March 14 can be translated
thus: "About that time, when the consul set out for the
war, during the Apollonian spectacles on the fifth day before the ides of July, in a clear sky in the daytime, the
light was darkened‡ because the moon passed under the sun."
The editor of the cited edition assigns the year as -189,
presumably on the basis of the consuls. I have not seen
any analysis of the accuracy with which the year can be
assigned in this way. I tend to doubt the accuracy because
we cannot date reliably by means of the consuls even several
centuries later. MCRE [pp. 456ff] gives one example of the
difficulty in trying to deduce the year from the names of
the consuls. The discussion of the record designated 71

---

† The parenthesis is in the original.

‡ The word used is obscurata. M translates this as "dimmed"
and he later says [p. 13.5] that this indicates a partial
eclipse. This translation is somewhat unusual; the word
is generally used with greater force and I see no reason
to doubt that it is being so used here. The word is often
used to indicate a total eclipse when it is used in an
astronomical context. See the record 1191 June 23f B,E in
Section VII.3 for an example.

March 20 E below supplies another.†

$\underline{M}$ assigns the date confidently as if it were obvious. However, the record tells us in two independent ways that the month is July,‡ and the nearest possible eclipse in July is -187 July 17. It is not likely that the annalist made two independent errors in writing the month and thus the date -189 March 14 is almost surely wrong.‡

The year -189 may be correct for the consuls named in spite of this fact. However, it is important to recognize one point. The eclipse may not be used to confirm that this is the correct year.

$\underline{M}$ implies that there are no magical aspects to this record. I cannot read it this way. In Livy's writing, it follows almost immediately after an extensive tabulation of miracles and prodigies that happened in the same year, whatever that year was.

$\underline{M}$ does not actually use this record because it does not give an indication of totality.

-187 July 17 ?   This is the second record from Livy that $\underline{M}$ [pp. 8.29-8.30] discusses, and he actually admits this one to his data sample. He gives the following translation of a passage from $\underline{Livy}$ [ca. 0, Chapter XXXVIII.36]: "Before the new magistrates departed for their provinces a three-day period of prayer was proclaimed in the name of the college of decemvirs at all the street-corner shrines because in the daytime darkness had covered everything (tenebrae obortae fuerant). Also a nine-day sacrifice was decreed because (so it was said)* there had been a shower of stones

---

†This discussion follows the discussion bearing the heading "71 March 20?", which refers to the so-called "eclipse of Plutarch".

‡The Apollonian spectacles (ludi Apollinares) were held annually a few days before the ides of July. I translate ludi as "spectacles" here because these particular ludi are believed to have been mostly theatrical shows.

‡Several writers have said that the discrepancy between July 11 and March 14 indicates the state of disorder in the Roman calendar in -189. It is true that the Roman calendar at this time, though probably intended to be solar rather than lunar, could slip or gain on the true solar year. However, even in -44, when Julius Caesar promulgated the Julian calendar, the calendar error was (Appendix I) only 80 days rather than 4 months, and this came from the accumulation of error over the years. An error of 4 months as early as -189 is quite unlikely.
    Incidentally, the months by -189 were probably conventional calendar months rather than lunar months, and an eclipse could come on any day of the month.

*$\underline{M}$ inserts the first parenthesis in order to give the original Latin phrase used by Livy. The second parenthesis is by Livy himself.

on the Aventine." I noted above that more than half of
Livy's eclipses are accompanied by a shower of stones, and
here we have an example. This is sufficient to counter M's
claim that there is no magical aspect to this record; I do
not believe that there is a natural correlation between
eclipses and showers of stones.

There are two errors in this translation. The phrase
tenebrae obortae fuerant should not be translated as "darkness had covered everything". "Obortae fuerant" is a compound tense of the deponent verb oborior, which simply means
to occur, to happen, or to arise. Thus tenebrae obortae
fuerant simply means that darkness happened.

The second error is one of omission. The record clearly
says that the darkness occurred between the 3rd and 4th hours
of the day. That is, it happened somewhat after mid-morning.
However, the circumstances given by Oppolzer [1887] indicate
that the eclipse of -187 July 17 came in the early morning
in Rome. I shall check the time by an independent calculation and report the result in Section XIV.8; in the meantime
I shall accept Oppolzer's circumstances as being reasonably
accurate.† To be sure, time keeping in Rome in -187 was not
very accurate by modern standards, but it could not have been
as bad as many have made out or the public business could not
have been transacted. I think there is little chance that
early morning could have been mistaken for a time after mid-
morning. Thus the date of -187 July 17 is not compatible
with the record.

M concludes that this passage from Livy is a record of
a total solar eclipse. He seems to base this conclusion
partly upon the error in translation that has been mentioned
and partly upon the following consideration: ". . the history of Rome is not littered with three-day decrees of prayer
from tenebrae obortae fuerant as we would expect if every er-
rant cloud was taken as an omen." This overlooks two important points.

One point is that this record simply says that darkness
occurred, while the preceding record says that the moon obscured the sun. Thus nothing in this record indicates a
solar eclipse at all, and the record is completely compatible
with a "dark day" (see the discussion under -647 April 6
above). This is particularly significant in view of the fact
that the preceding record gives an astronomically correct
description of an eclipse.‡ Since it was known by now that

---

†In a calculation made with provisional values of the accelerations, I find that maximum eclipse came at 1.35 hours of
the day and the end came at 2.25 hours. Thus the eclipse
was over before the time the record gives for the darkness.
While many eclipse records have errors in the hour, such
errors usually come when there is only a rather vague statement. Here the statement is fairly precise and the error
seems unlikely.

‡Note that the earlier eclipse did not elicit three days of
prayer.

eclipses occurred for perfectly natural reasons, we would
not expect one to be regarded as an omen. A dark day, on
the other hand, had no known origin and it could still be
portentous. Thus the decree of prayer suggests that the
event was not an eclipse. Of course, people are not consistent and this argument is not rigorous. Still it less-
ens significantly the chance that the record refers to an
eclipse.

The second point concerns the way in which such decrees
were made. The people who interpreted omens and issued such
decrees were people who owed their office to the political
process and who were likely to be strong political partisans.
Thus, particularly by this stage in history, it is likely
that decrees of prayer and similar actions were political in
origin. To be sure, such a decree probably had to have some
pretext, but the decision about which pretexts to follow and
which to ignore was a political decision. Thus the decree
tells us nothing about the seriousness of the "portent".

In both this record and the preceding one the event
occurred when new officers (who were chosen annually)† were
about to leave to assume their duties abroad. This suggests
that the time of year may have been the same for both events,
even though the specified officers are not the same in the
two accounts. Thus it is possible that Livy found two independent records of the same event, and that he put them in
different years without recognizing this. That is, the preceding record may refer to the eclipse of -187 July 17.

In summary, this record, unlike the preceding one, says
nothing to suggest that we are dealing with an eclipse, and
the official reaction of three days of prayer suggests that
we are not. We more likely have a "dark day". The only
eclipse that can match the year came at the wrong hour;
either we are not dealing with an eclipse or the circumstances are wrong and we cannot date it with useful confidence. Finally, Livy recounts many eclipses. Wherever we
can test him, he proves to be wrong. The most we can say
for this record and the preceding one is that we cannot test
him and thus we cannot prove him wrong. In view of his general credibility, however, this is not sufficient and the
only safe course is to ignore all of his possible eclipse
reports.

-135 April 15 BA. There are two records of this eclipse.
So far as I know, MS [p. 482] are the first people to publish
a translation of these records. One record has been published
in the form of a drawing of the cuneiform tablet; MS give the
citation. The other has not been published previously in any
form, and a translation was provided to MS in a private communication from Professor A. J. Sachs of Brown University.

---

†The American reader in particular should try to imagine the
fun that came from having annual elections for the highest
offices in the land.

One record reads: "On the 29th day there was a solar eclipse beginning on the south-west side. After 18 uš .. it became complete such that there was complete night at 24 uš after sunrise."† The month has already been identified as the intercalary (13th) month of the year, and MS say that the year is identified as -136/-135 by means of data for Venus and Mars. The calendar tables of Parker and Dubberstein [1956] give -135 March 19 as the first day of this month and hence -135 April 16 for the eclipse. Since the tables of Parker and Dubberstein are based upon theoretical calculations of the sunsets when the lunar crescent first became visible, the difference is not serious; presumably the first day of the month should be shifted to -135 March 18.‡

The second record gives the year explicitly, and it agrees with the date just given. Then it reads: "Daytime of the 29th, 24 uš after sunrise, a solar eclipse beginning on the south west side . . Venus, Mercury and the Normal Stars‡ were visible; Jupiter and Mars, which were in their period of disappearance were visible in that eclipse . . . moved from south west to north east.* 35 uš for obscuration and clearing up."

I find the statements of time somewhat confusing, but the following interpretation seems plausible: The eclipse became total at 24 uš (96 minutes) after sunrise. The beginning had been observed 18 uš earlier and hence at 6 uš 24 minutes) after sunrise. The entire duration was 35 uš, or 140 minutes. This seems short for the duration of a total eclipse, but a calculation with provisional accelerations gives 135 minutes, in excellent confirmation of the record and of this interpretation. I shall use the times in a later work; here I shall use only the fact that the eclipse was recorded as total.

We found in Section IV.1 that 0.034 is the standard deviation of the magnitude for records other than Chinese which state that an eclipse was total. Since this record was apparently made by trained astronomers who give a careful account of the visibility of the planets, we might think that we should use 0 for the standard deviation of this observation. Because of experience with the Chinese records, doing so might be optimistic. The Chinese records were also made by trained astronomers, and yet they often

---

†The uš was the "time degree" which equals 4 minutes. The dots indicate a break caused by damage to the tablet.

‡Since the Babylonian day began at sunset, this would mean starting the month at sunset on -135 March 17.

‡This denotes a set of stars which Babylonian astronomers used as references in stating the longitudes of the moon and planets. APO [Section III.10] gives an extensive discussion of them. Since many of them were of the 3rd magnitude or fainter, the statement that they were visible implies a quite dark eclipse if it is accurate.

*Presumably this refers to the eclipsed portion of the sun.

exaggerate the indications of totality. There is a risk
that the Babylonian astronomers did the same.† Hence I
shall continue to use 0.034 for the standard deviation.

The text does not say where the observation was made,
and three places have been used in the literature, namely
Babylon, Sippar, and Borsippa. Neugebauer [1955, volume 1,
pp. 5-6] reviews the evidence pertaining to a collection of
texts of which this one forms a part, and he concludes that
it is "at least very plausible" that they came from Babylon.
Luckily the possible places are close together and the uncertainty is not important for an eclipse this old. I shall
take the place to be Babylon, although there is a slight
uncertainty in the matter.

MS have checked the positions of the planets at the
time of the eclipse. They find that Mars and Jupiter were
indeed so close to the sun that they would be invisible except during the eclipse. They, along with Mercury and Venus,
were above the horizon, and hence they were visible during a
total eclipse. Saturn was below the horizon, in accordance
with its omission from the record. Thus the record is accurate in its statements about the planets. However, as we
have noted, Babylonian astronomers in -135 regularly calculated the planetary visibilities, and the accuracy of the
information is not a strong indication that the eclipse was
actually observed as total.

If we accept the place as Babylon, this is the earliest
record which tells us that an eclipse of known date was observed to be total at a known place, and it is the only such
record before +840. We should remember that there is actually a slight uncertainty about the place.

-128 November 20? This is the famous eclipse of Hipparchus about which so much has been written. Although the
eclipse was total at places near the Hellespont, the magnitude was only 4/5 at Alexandria. Until a few years agao,
most writers on ancient observations took it as a settled
matter that the date was -128 November 20. I studied the
evidence in AAO [pp. 104-110] and showed that the dates
-309 August 15, -281 August 6, -189 March 14, and -128
November 20 are all possible dates. In AAO, I split the
weight among these four possibilities. In this work, for
the reasons that were discussed in Section III.7, I shall
ignore an eclipse that cannot be dated unambiguously.

MS [p. 522] and M [pp. 8.32-8.33] both concur that the
eclipse cannot be dated, and van der Waerden [1961] reaches
the same conclusion. We may now hope that it will not again
be necessary to devote extensive discussions to the date of
this eclipse.

---

†It was rather simple to calculate the visibility conditions
of the planets during the eclipse, so the statements about
the planets do not prove actual observation.

Although M̲ reaches the correct conclusion, he makes a
serious error in reasoning in doing so. It is desirable to
discuss this point, because the argument in question is involved in other important matters.

Hipparchus used this eclipse in a study of the solar
and lunar parallaxes. His known career stretched from -161
September 27 to -127 March 23.† In AA̲O̲, I assumed that he
might have been active up to March of -126, a year after his
latest known observation; this is a reasonable assumption to
make for purposes of illustration, but it should not be taken
literally. Hipparchus had extensive compilations of old observations and he was not limited to using eclipses that came
in his own lifetime. Hence he could have done his study of
parallaxes at any time in his career with equal probability.
Thus he was far more likely to have done the work in question
in the long part of his career before -128 November 20 than
in the short part after that date.‡ This means that the odds
are high that the date was not -128 November 20. Put another
way, the odds are high that the eclipse came before Hipparchus was born.‡ Thus we can eliminate -128 November 20, but
the other dates listed must be given equal probabilities.

M̲ [p. 8.33] says that this reasoning is not valid, and
he "demonstrates" this by the following words: "There are
only two possibilities. Either Hipparchus had previous data,
or he didn't. We are concerned with the second possibility.
In this case, he could not have made his calculation until
the eclipse took place (obviously),* and no amount of argu-
ing about the length of his working life can affect this,
or be used to estimate a probability that the identification
of -128 is valid."

M̲ is wrong in saying that we are concerned with the
second possibility. We are concerned with the question of
which possibility is correct. We know from independent evidence that the first possibility is correct and that the
second possibility is wrong. Therefore M̲'s argument, which
is based upon the correctness of the second possibility, is
irrelevant. All that he has shown is that the date of -128
November 20 is correct if the eclipse happened in Hipparchus's
working lifetime, and we already knew that anyway.

Thus, if we had no other information, it would remain
true that a date before Hipparchus's lifetime is much more
probable than a date during it, simply because of the way
that the date -128 November 20 divides his lifetime. Actually we do have some other information that I did not know

---

†These are the dates of his earliest and latest known
observations.

‡-128 November 20 is the only possible date that came during
Hipparchus's lifetime.

‡ Speaking rigorously, we must admit the possibility that
Hipparchus was born by -189 March 14, but we may be sure
that this was before the beginning of his career.

*The parenthesis is M̲'s.

when I wrote AAO. We know from the commentaries of Pappus†
and others that Hipparchus made three studies of the solar
and lunar parallaxes, and he used the eclipse in question
in the first one of these. In fact, after he first used the
eclipse, he had time to do the necessary research and to
write two books on the solar and lunar parallaxes, and to do
other work besides. This much activity, which involved many
laborious calculations among other things, seems almost impossible if the eclipse he used were indeed the eclipse of
-128 November 20. Thus the odds against this date are even
longer than I thought in AAO, and the probability that -128
November 20 is correct is almost negligible.

The relation between the date of an astronomical observation and the birth and death dates of some particular
astronomer is involved in the interpretation of some other
ancient observations. This is the reason why I have spent
time in discussing this point.

-50 March 7 ? This is the so-called "eclipse of Caesar"
[AAO, p. 72] which has been used seriously by a few writers
on ancient astronomical observations. Let us hope that the
myth of this eclipse can now be forgotten.

29 November 24 ? This is a hardy weed that seems impossible to uproot. Its most recent growth is its identification as an eclipse observed in Antioch by S. Luke, and I
discussed this new development in Section II.3. The eclipse
is obviously magical and there is no reason to suppose that
a real eclipse lies behind it. Even if we assume that a
real eclipse does lie behind the account, we cannot date it,
contrary to the claims of Sawyer [1972b]; at least five
other dates are possible. Unfortunately, MS [p. 523] seem
to accept the "eclipse of Luke" as genuine, although they
do not use it for reasons that they do not state.

In its earlier form, this was called the "eclipse of
Phlegon". The patristic writer Eusebius [ca. 325] claims
that the 2nd-century historian Phlegon recorded an eclipse
and other marvelous events in a year that lies in our years
32 or 33, and he claims that Phlegon's account refers to the
magical darkness that accompanied the Crucifixion of Jesus.‡
No eclipse was possible at the Crucifixion because it came
within a day of a full moon. Further, there was no large
eclipse anywhere in the Roman world in either 32 or 33. Thus
from no viewpoint is there any reason to credit the "eclipse
of Phlegon".

M does not mention the "eclipse of Luke" so far as I
can discover. However, drawing in part from unpublished
work by Stephenson, he does refer [p. 8.33] to the "eclipse

---

†Pappus wrote around +300, and he has left valuable information about Hipparchus and his work. See Heath [1913, pp.
341ff] or Newton [1977, pp. 178ff].
‡See the discussion in AAO [pp. 110-113].

of Phlegon" in this way: "Stephenson (1972)† points out
that: 'Newton calls this a magical eclipse on account of
the association by Eusebius‡ of the account with the events
occurring at the Crucifixion. Phlegon makes no such association in the fragment quoted - 'by Eusebius. This does
not help, in the end however, (*sic*) since the place of observation could be anywhere in the ancient world.' "

From the last sentence, I deduce that M accepts the
"eclipse of Phlegon" as a genuine account of an eclipse,
although he does not use it because he cannot locate the
place of observation. Whether the deduction is correct or
not, the comment just quoted misses an important point. We
do not know what, if anything, Phlegon said about an eclipse.
All that we have comes through the pen of Eusebius, and it
does not matter which words he chooses to present as his own
and which as Phlegon's; Eusebius is the writer so far as we
are concerned. The eclipse is clearly magical and all the
information that is given about it is impossible.

Thus I ignored this "eclipse" in AAO and I shall ignore
it here.

45 August 1 E. The historian Dio Cassius refers to a
number of solar eclipses in his history of Rome, and two of
these are probably genuine. One of them is probably dated
either 5 March 28 or 5 September 22. The other [Dio Cassius,
ca. 230, Section LX.26] can be dated exactly: "Since there
was to be an eclipse of the sun on his birthday,‡ . . he
therefore issued a proclamation in which he stated not only
the fact that there was to be an eclipse and when, and for
how long, but also the reasons for which this was bound to
happen." Dio then proceeds to describe the causes of both
solar and lunar eclipses.

The birthday of Claudius was August 1, and only one
eclipse came on his birthday during the time he was emperor;
this was the eclipse of 45 August 1. Unfortunately Dio does
not tell us whether anyone observed the eclipse. It would
have been visible in many parts of the Roman Empire but
would not have been total anywhere within it. This account
is interesting mainly for what it tells us about the state
of astronomical knowledge, and the apparent desire of Claudius that the eclipse not be interpreted in a portentous
way. I shall not use this record because there is no statement about observing the eclipse.

59 April 30 E and 59 April 30 M. These record observations of the same eclipse made [AAO, pp. 73-74 and pp. 113-
114] in Campania and Armenia. I shall use the same characteristics for these records that I used in AAO except for the
standard deviation of the magnitude. I shall use the value

---

†This refers to Stephenson's unpublished work.

‡All the emphases in this passage occur in M. I do not know
 whether this one also occurs in Stephenson's writing or not.

‡This refers to the emperor Claudius.

0.18 in this work. We should note that there is a typographical error in AAO on page 114. The latitude range for the Armenia observation in my calculations was taken to be 40° to 42° rather than 30° to 42°.

71 March 20 ? This "eclipse" also seems to be immortal. It is based upon a passage from a book in which Plutarch [ca. 90] gives an exposition of astronomy aimed at the "intelligent layman". Fotheringham [1920], M [pp. 8.35-8.36], and AAO [p. 114] all give translations of the passage that agree on the essential points. The book is written in the form of a conversation involving about eight people. One of the people refers to a "recent" eclipse that began shortly after noon which chilled the air and let many stars shine out. After he mentions this eclipse, the speaker quotes descriptions of eclipses from several famous Greek works, going back to Archilochus (see the discussion under the date -647 April 6 above). Still later, the passage says that "a kind of light is visible about the rim which keeps the shadow (or darkness)† from being profound and absolute."

This literary eclipse differs from the other literary eclipses we have studied in an important feature. In the other works, an eclipse was not necessary; the writer could have used some other phenomenon or development just as well. Here, however, Plutarch is talking about astronomy in general, and in particular his plan is to present everything that was known about the moon in his time. Therefore an eclipse is an essential part of his book and its presence does not in any way suggest that Plutarch had seen an eclipse. He could just as well have taken everything about it from his reading.

In AAO, I gave marginal weights to some "eclipses" but I did not do even that for the "eclipse of Plutarch"; I gave it a weight of zero. I shall continue to do so, but it is necessary to discuss the "eclipse" some more because of the writing of MS and M on the subject.

M [p. 8.36] says that Stephenson, in his unpublished dissertation, rejected this eclipse "in no uncertain terms". However, Stephenson must have reconsidered because MS [pp. 522-523] say that the eclipse of 71 March 20 was total in "Plutarch's home city of Chaeronea. This could be coincidence, or accident, but the probability of this happening by chance is rather small. We believe, therefore, that Plutarch very probably witnessed this eclipse at home, . . . " This needs two comments.

The first comment is that the uncertainties in their accelerations are considerably greater than MS thought. Therefore we do not know yet whether the eclipse of 71 March 20 was total in Chaeronea or not.‡

---

†The parenthesis is one that I have added in explanation.

‡The calculations to be presented in Section XIV.8 indicate that it almost certainly was not.

The second comment concerns the location of Plutarch's
home. Almost every writer on the subject of this "eclipse"
asserts that Chaeronea was his home at the time, although it
probably was not. According to standard biographical sources,
Plutarch was born in Chaeronea in about 46. After he studied
philosophy in Athens, he went to Rome. He spent many years
in Rome, in other parts of Italy, or in travel. Finally he
returned to Chaeronea, where he apparently spent the rest of
his life. He was apparently acquainted with the emperors
Trajan and Hadrian,† who both bestowed various honors upon
him.

The usual age when a person left home for advanced education, then and now, was around 18 or 20. Thus Plutarch
probably left Chaeronea for Athens around the year 66. A
few years later, say around 68, he left Athens for Rome and
stayed in Rome, or elsewhere in Italy, for many years.
According to this chronology, he was in Italy on 71 March
20. If we assume that his home was still Chaeronea on 71
March 20, this means that he did not start his advanced
education until he was 25. While this is possible, it is
unlikely, and the odds are certainly against the idea that
he was living in Chaeronea on the day of the eclipse in
question.

M [pp. 8.35-8.38] actually admits the "eclipse of
Plutarch" to his data sample. He concludes that it was the
eclipse of 71 March 20, that it was observed to be total in
Chaeronea, and he assigns a reliability of 0.5 to this conclusion. He bases his conclusion, he says, upon an argument that he has not noticed in the earlier literature.

He first studies the passage from Plutarch on the assumption that the conversation was a real one. On this
basis, he finds that the only possibility was that the
eclipse had been seen only by the immediate speaker and not
by the other participants. The date was 75 January 5 and
the place was Carthage on this basis. However, M does not
believe that the conversation was real, and he rejects this
date and place.

From the conclusion that the conversation was not real
to the conclusion that the eclipse observation was real is
a short step in M's argument, but it is a step that I cannot
follow. So far as I can see, there are only two points in
this argument:‡ (a) "The entire account has, to my mind,‡

---

†There is a recurring story that Plutarch was entrusted with
the education of Hadrian, but there seems to be no source
for the story earlier than the Middle Ages. If the story
should be correct, Plutarch would still or again have been
in Rome for several years around 90.

‡Both points were in fact made by many writers before M.
Perhaps M meant to apply the novelty of his argument only
to the conclusion that the date was 75 January 5 if the
conversion was real.

‡That is, to Mueller's mind.

the definite flavor of personal experience and eye-witness description." (b) The "kind of light" which prevented total darkness is "a description of the corona, which he may† have taken from an earlier source, but our evidence today is that no such source is presently extant." Both points need comment.

(a) We need the same comment that was made about Archilochus in connection with the date -647 April 6 above. If this account has the "flavor of personal experience" to it, this suggests nothing about the observation of the eclipse. It merely tells us that Plutarch, like Archilochus and Mark Twain, was a skillful writer. I repeat what I said earlier: No one should advocate adopting a "literary eclipse" as a real astronomical observation until he has identified the "eclipse of Mark Twain". Further, he must do this without using a detailed biography of Mark Twain, in order to duplicate the situation with regard to Plutarch, Archilochus, and other ancient writers.

(b) If Plutarch did indeed base his account upon the eclipse of 71 March 20, the "kind of light" is probably not the corona. The eclipse of 71 March 20 was the relatively rare type called "annular-total". This type occurs when the angular size of the moon is closely matched to that of the sun. If an observer on the central line of the eclipse sees it near sunrise or sunset, he is as far from the moon as he can get and still see the eclipse as central; at these points the moon is too small to cover the sun. If an observer is on the central line where it is near noon, he is as close as he can get, and to him the moon is just large enough to cover the sun.

If Plutarch saw the corona during the eclipse of 71 March 20, he necessarily saw it in the highly restricted zone where the eclipse was total. We have one and (so far as I know) only one attested observation of an annular-total eclipse made with the naked eye at a place where it may have been total. The trained astronomer Clavius saw the annular-total eclipse of 1567 April 9 in Rome, where the eclipse, if central, was total.‡ However, although he was a trained astronomer, he reported that the eclipse was annular. I estimate that the magnitude of this eclipse did not exceed about 1.002 anywhere, and it was probably less than 1.001 at the latitude of Rome.‡

The chromosphere, which is bright but not as brilliant as the main solar disk, extends to a height of about 0.02 times the sun's radius. Thus the apparent area of the chromosphere is about 0.04 times the area of the disk. Although

---

†The emphasis is M's.

‡See Section X.3.

‡This calculation depends mostly upon the distances to the sun and moon, and thus it is almost completely unaffected by the accelerations.

the chromosphere is not as bright as the disk, it is likely
that the pale corona cannot be seen unless much of the chromosphere is covered. Tentatively, we say that the corona
cannot be seen unless the moon is large enough to make the
magnitude equal to 1.01 somewhere.† Whether this value is
correct or not, we may be virtually sure that Clavius did
not see the corona if he thought he was seeing the bright
rim of the sun's disk. The circumstances on 71 March 20 are
similar to the ones on 1567 April 9, and it does not seem
likely that any observer on that day saw the corona.

There is a consideration that I should have noticed
when I wrote AAO but failed to. Over a good portion of their
path lengths, including the portions that lie in Greece and
Italy, the eclipse paths for 71 March 20 and 75 January 5
are almost identical as drawn in Oppolzer's Canon. Thus, if
we assume that Plutarch wrote his passage because he had
seen an eclipse in Chaeronea, we have no way to choose between the two dates, and we cannot identify the eclipse until we know the accelerations better than we know them now.

Few people who have studied this "eclipse" have paid
proper attention to the time that Plutarch states.‡ He
clearly says that the eclipse began shortly after noon. Most
people dismiss this statement, saying that time keeping in
Plutarch's time was inaccurate. However, the eclipse of 71
March 20 began shortly after midmorning in Chaeronea, and
the maximum phase came well before noon. M dismisses this
as a discordance of no importance, but I cannot. I do not
believe that anyone who could see the sun clearly‡ could
mistake mid-morning for afternoon. I do not believe that
anyone could think that the eclipse began after noon when
the maximum in fact came before noon. I certainly do not
believe that anyone with Plutarch's knowledge of astronomy
could make such a mistake.

If the eclipse is real, the only possible dates I can
find are 71 March 20, 75 January 5, 80 March 10, and 83
December 27.* The only possible combination of date and

---

†I thank my colleague I. H. Schroader for suggesting a
different possibility. He agrees that an ordinary observer
would be unlikely to see the corona during an eclipse that
is but slightly more than total. He suggests, however, that
the reason is a matter of timing rather than of illumination
levels. If a person has been watching an eclipse in an
ordinary way, his eyes are not dark-adapted. Before they
become adapted well enough for him to see the corona, the
total phase is over and it is too late.

‡M [p. 8.38] says that S.H. Sandbach, in a work that I have
not consulted, emphasizes the importance of the time.

‡If the observer could not see the sun clearly, he could not
have seen the eclipse.

*The eclipse of 80 March 10 was annular, but I do not think
it can be eliminated for this reason. However, it is
eliminated for other reasons.

place which gives an eclipse beginning shortly after noon, so far as I can see, is the eclipse of 83 December 27 observed in Egypt. Since this eclipse was safely total, the corona could have been seen.

We can now summarize the situation.

(a)   There is no reason to assume that the "eclipse of Plutarch" is anything more than a literary invention. There is no reason to introduce it into the astronomical literature as an astronomical observation.

(b) $\underline{M}$ concludes that no eclipse meets the circumstances if the events described were real. Hence the eclipse was a real observation only if the other events were imaginary. I see no reason to take the eclipse as real if the rest was imaginary.

(c)   The light that keeps the darkness from being complete may be the corona and it may be something else. The sun's rim in an annular eclipse, the chromosphere in a barely total eclipse, and the corona form a continuous spectrum in brightness, and ancient observers may not have distinguished between them.

(d)   Plutarch lived in both Greece and Italy, and he may have travelled to other parts of the Roman empire. If we assume that the eclipse was real, and if we restrict ourselves to observations in Greece or Italy, the only possible dates are 71 March 20 and 75 January 5. If we allow other places, the dates of 80 March 10 and 83 December 27 are also possible.

(e)   Most people who take the eclipse as real assume that Plutarch was in Chaeronea. If so, only 71 March 20 and 75 January 5 were possibly total there during the possible period. However, on both these dates, the known facts of Plutarch's life make it almost certain that he was in Italy. If he saw either eclipse, it must have been somewhere other than Chaeronea.

(f)   If we do conclude that Plutarch saw the eclipse in Chaeronea, we cannot identify the eclipse because we cannot choose between the two possible dates if Oppolzer's map is accurate.

(g)   The preceding statements ignore the stated time of day. When we include the specifically stated time, the only possibility is the eclipse of 83 December 27 seen in Egypt. This choice is also consistent with seeing the corona; the usually accepted date of 71 March 20 probably is not.

(h)   Thus, if we assume that the eclipse observation was real, only the eclipse of 83 December 27 seen in Egypt is compatible with the stated circumstances.

I do not see any reason to believe that the observation was real, and therefore I shall not use it.

71 March 20 E. Pliny [77, Book II, para. 57] says that
it once happened that there was a lunar eclipse just as the
sun was rising and the moon was setting, and that both could
be seen even while the moon was eclipsed.† He also says
that it happened in his own time, when Vespasian was consul
for the third time and his son for the second time, that
there were two eclipses within 15 days.

Vespasian became emperor in 69, and emperors had already
developed the habit of repeatedly naming themselves as consuls. Thus the year is probably early in his reign. Further,
Pliny presumably wrote this in or before 77, and both he and
Vespasian died in 79. Within the possible time period, I
find only two possibilities for the solar eclipse, namely 71
March 20 and 75 January 5. If the latter were the solar
eclipse, the lunar eclipse could only be 74 December 22. This
was a small eclipse that began about 7.3 hours, Rome time,
according to the information in Oppolzer's Canon. It is
doubtful that this eclipse could be seen in Rome. If the
solar eclipse were on 71 March 20, the lunar eclipse was 71
March 4. This was also a small eclipse, with a magnitude of
about 0.4. Further, its maximum came at about 21 hours, Rome
time, and it was readily observable there. Hence, it is the
more likely of the two possibilities from the astronomical
viewpoint.

Let us now see what we can learn from the names of the
consuls. I have found three lists of consuls for the time
period in question. Two of them, Fasti Vindobonenses Priores
and Barbarus Scaligeri‡ are found in Mommsen [1892], and the
third is Paschale [ca. 628]. Each gives a year in which
Vespasian was consul for the third time. The first two say
that this was the year in which his son was also consul for
the third time while the other source says that his son was
consul for the first time in the same year. Thus our sources
disagree within themselves and all disagree with Pliny. This
shows that we cannot find the year reliably from the names of
the consuls, even though we are near the high tide of Roman
prosperity and interest in its own history. I see no reason
to assume that we could do this 2½ centuries earlier,‡ and
it is certain that we cannot do it more than 4 centuries
later, when the empire was reduced to its eastern half; see
MCRE [pp. 456ff]. All we can say from the consular lists is

---

†This phenomenon is called a horizontal eclipse. If it can
really happen, it is made possible by refraction, which
lifts both bodies above the horizon. Johnson [1889, p. 27]
says that this was the eclipse of 72 February 22. Even if
we assume that the eclipse happened in Pliny's time, I am
not sure that we can identify the eclipse from the circumstance that maximum eclipse was near sunrise. Further, I
do not think that Pliny implies that the eclipse was necessarily in his own time.

‡These names refer to medieval collections of manuscripts,
and have nothing to do with the compilers of the consular
lists.

‡See the discussions labelled "-189 March 14" and "-187 July
17" above.

that 71 is more likely than 75 for the year. Thus the historical and astronomical evidence both indicate that 71 is probably the correct year.

However, I shall not use this record, partly because there is some doubt about the date. A more important reason is that we do not know where the eclipse was observed (there is no reason to assume that Pliny observed it himself; after all, his work is an encyclopedia), although the region around Rome is more likely than distant parts of the empire.

This completes the discussion of eclipse reports before the year 100, with the exception of the Chinese records which were discussed in Chapter V. The discussion is rather discouraging at first inspection. When we look at the progress that has been made in the past decade, we see that a few myths about ancient eclipses may have been eliminated from the literature. An example of a myth that may now be gone is the claim that the "eclipse of Hipparchus" happened on -128 November 20, but it would be premature to conclude that it has truly vanished. However, other myths have sprung up in place of the ones that may be gone, and the total amount of mythology connected with ancient eclipses is probably larger than it was a decade ago.

Luckily, the amount of valid observation in the literature has also increased. Thus it is possible that the ratio of valid information to myth has at least held constant, and it may even have increased. Let us hope that this is the case.

2. The Principle of Averaging Solutions for the Accelerations

Many people who are interested in the subject of the astronomical accelerations do not want to work directly with the ancient or medieval data for one reason or another. What they do instead is to derive a new solution for the accelerations by combining values of the accelerations obtained by other people who have worked directly with the data. Sometimes they use a simple average and sometimes they use more complex combinatorial processes. For simplicity, I shall use the term "averaging" to denote any of these processes.

Most of the solutions that have been used for averaging purposes have been based, either in large measure or entirely, upon large solar eclipses observed before +100. Thus this chapter is a good place to consider the validity of averaging in this situation, to consider the validity of solutions that have been used in averaging, and hence to study the validity of solutions that have been obtained by averaging. In this section, we shall consider the circumstances that must be met if the averaging process is to be valid.

As an example, let us consider the measurement of the acceleration g due to gravity at a particular spot. We may measure g by measuring the time taken by a body in free fall to cover a certain distance, by measuring the period of a gravity pendulum, by measuring the deflection of a spring which supports a standard mass, and doubtless by other methods.

The types of error in the different methods are independent of each other. Further, if we build two pieces of apparatus to measure g using the same method, it is likely that we will find that the errors differ between the two pieces. Thus different measurements incorporate independent bodies of data, and they involve independent errors that have a tendency to cancel. Under these circumstances, a judicious average is expected to give us a better estimate than we can get from any single measurement.

This is merely an example of a well-known principle: If we introduce additional valid information, if the additional information is relevant and unbiased, and if we use it correctly, we improve our knowledge of the situation. This principle is so obvious that we expect people always to recognize it. Yet, as we have seen, $\underline{MS}$ and $\underline{M}$ claim that one of the errors in my work is that I have used certain kinds of valid and unbiased data which they prefer not to use.†

There are circumstances, however, in which a particular study should not be used in an averaging process. We can identify three, and further contemplation might yield others.

(a) We should not use the result of a study if it is based upon a theoretical error. To give an absurd example, suppose that an experimenter measures g by the free fall method, and that he reduces the data using the erroneous formula $s = 0.6gt^2$ instead of the correct formula $s = \frac{1}{2}gt^2$. We should obviously not average his value of g with other values. However, we could use his original data, deduce g therefrom by correct theory, and then use the new value of g in an averaging process.

(b) Similarly, we should not use the result of a study if it is based upon invalid data. Unlike a study based upon invalid theory but valid data, a study based upon invalid data cannot be rescued.

(c) We should not use studies that depend upon each other. For example, suppose A publishes a value of g obtained from measurements. Suppose that B makes some new measurements which he combines with those of A before he

---

†Another example is even more striking. A recent study, which I think it is kinder not to identify, analyzes the process of finding certain geodetic information from artificial satellites. Its authors, by means of computer simulation, reach an interesting conclusion. They find that the accuracy of the geodetic information is degraded by having accurate information about the satellite orbit. To get the best geodetic results, they claim, we must not know the orbit accurately. This is so ridiculous that I am surprised to find that such a study could ever be written.

deduces a value of g. We should not use the results of both A and B in finding an average g, because the results of A are automatically included if we use the results of B.

In the next section, I shall review some often-used determinations of the accelerations, in order to see which ones, if any, should be used in an averaging process.

3. Some Recent Studies of the Accelerations

It is useful to start with the study by Fotheringham [1920], which is probably still the most-quoted study of the accelerations of the sun and moon (or the earth and moon).

TABLE VI.2

SOLAR ECLIPSES CITED BY FOTHERINGHAM

Name	Date	Place Fotheringham assigned
Babylon	-1062 Jul 31	Babylon
Eponym Canon	- 762 Jun 15	Assyria
Archilochus	- 647 Apr 6	Paros or Thasos
Thales	- 584 May 28	Asia Minor
Pindar	- 462 Apr 30	Thebes
Thucydides	- 430 Aug 3	Athens
Agathocles	- 309 Aug 15	East coast of Sicily
Hipparchus	- 128 Nov 20	Hellespont
Phlegon	+ 29 Nov 24	Nicaea
Plutarch	71 Mar 20	Delphi or Chaeronea
Theon	364 Jun 16	Alexandria

Most people who cite this paper state that Fotheringham finds the accelerations by using ancient solar eclipses, thus showing that they have not read the paper. Fotheringham indeed starts by considering 11 solar eclipses. Following the custom of his time, he designates each eclipse by associating a name (except, ironically, for the eponym canon eclipse) with it. The terms associated with each of the 11 eclipses, their dates, and the places of observation that Fotheringham uses, are listed in Table VI.2. A few comments are necessary.

The astronomer Theon, in Alexandria, measured the times of the beginning, middle, and end of the eclipse of 364 June

-218-

16. At the time Fotheringham wrote, this was the earliest known measurement of the times of a solar eclipse. Fotheringham finds that the times are incompatible with the other data he uses, and MS [p. 519] do the same. This comes from the fact that these authors insist upon constant accelerations. When the acceleration parameter D'' is allowed to vary in the way that I have found from using both ancient and medieval data, the measurements of Theon seem to be rather accurate.

Fotheringham assumes that the eclipse of -430 August 3 was partial at Athens and that the eclipse of -309 August 15 was total somewhere off the east coast of Sicily. These assumptions are valid, and these observations, along with the measurements by Theon, are valid data. However, these are the only valid eclipse data in his study, and it is ironic that he rejects all of them. The other data, which he does use, are imaginary and should never have been used in a serious astronomical study.†

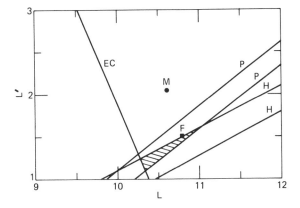

Figure VI.1. Fotheringham's solution for the accelerations. When he uses only solar eclipses, he finds that the point representing the solution must lie between the pair of lines marked with a P. It must simultaneously lie between the pair of lines marked with an H and be to the right of the line marked EC. Hence it must lie within the shaded triangle. From data other than solar eclipses, he finds that M is the correct point, but M does not lie within the triangle. He finally settles on the point marked with a square and labelled F.

---

†He had no reason to question the date of -762 June 15 assigned to the eponym canon eclipse, and he was conservative in taking the place of observation to be anywhere in Assyria. However, he had no basis for taking the eclipse to be total.

He analyzes the solar eclipses that he does use by the method of linear inequalities (Section IV.3), and the only eclipses he ends up using are those of the eponym canon, of Hipparchus, and of Plutarch, all of which he takes to be total. He works with parameters L and L' that I shall define in a moment. If we assume that the observer was on one edge of the zone of totality, we can find one linear relation between the parameters. Since each zone has two edges, each eclipse furnishes two linear relations. The resulting situation is shown in Figure VI.1.

The limiting relations furnished by the eclipse of Plutarch are the lines marked with a P; the solution must therefore lie between them, according to Fotheringham. The relations furnished by the eclipse of Hipparchus are marked with an H. One relation furnished by the eponym canon eclipse is the line marked EC; the other relation is far to the right off the scale of the figure. The only region that is compatible with all three eclipses is the cross-hatched triangle near the bottom of the figure, and the solution for the parameters must lie within this triangle.†

There is an interesting feature to this situation. The line marked EC cuts across the figure at a large angle and it does not play a really important role in the solution. In fact, in Fotheringham's final solution, as we shall see in a moment, it plays no role at all. Thus, as Fotheringham ends up, the only solar eclipses he uses are ones that have been dated by means of the identification game, and he rejects the only valid data in his set. The reason for this is probably that the "identification game" data, which come from computation, tend to have an artificially small scatter‡ while real data have a real scatter. Thus a person who selects his data by means of their consistency tends to end up by rejecting the real data and keeping only spurious ones.

Fotheringham now turns to data other than the solar eclipses. Altogether he uses [Fotheringham, 1920, p. 124] measured times of equinoxes, of lunar eclipses, and of lunar occultations. He also uses the measured magnitudes of partial lunar eclipses. From these data, he finds L = 10.61 and L' = 2.04. These values correspond to the point marked M in Figure VI.1, which is well above the triangle. He also says [p. 124] that it "would be difficult to shake the presumptions made concerning the eclipse of Hipparchus, . ." This makes him unwilling to let the solution lie above the

---

†Fotheringham finds that the eclipses of Babylon and of the eponym canon are not compatible. He thereupon rejects the eclipse of Babylon, concluding that the event described was not an eclipse or, if it were an eclipse, that it was not total at Babylon. Most people who cite his paper say erroneously that Fotheringham concluded that this was a total eclipse at Babylon.

‡If there were many possible eclipses that could be used in each playing of the identification game, the scatter would be very small. Sometimes it happens that only a few eclipses can take part in the game, and such a case may yield an "eclipse" that differs considerably from the others.

upper line marked H in the figure. He finally takes the
point marked F (L = 10.8, L′ = 1.5) as the best estimate
that can be made from all the data combined. He does not
say exactly how he chose this point; it is not the nearest
point on line H to point M. Perhaps he meant to use the
nearest point but did not think it worth the bother to find
the point exactly.

It is clear that his final solution F depends upon the
data other than eclipses, and upon the eclipse of Hipparchus,
but not upon any other solar eclipses. We know that there
is no basis for the date he assigned to the eclipse of Hipparchus, and we must now ask about the validity of the data
which lead to the point M.

These data comprise four main groups:

(a) The equinox measurements made by Hipparchus [AAO,
Section II.1], which come to us through Ptolemy's Syntaxis
[Ptolemy, ca. 142]. I concluded [Newton, 1977, Section
XIII.2] that these observations are probably genuine even
though they were used by Ptolemy. Fotheringham [1918] found
that the errors in the equinoxes that Ptolemy claims to have
measured himself were unreasonably large and he did not use
them, but he was puzzled by them. He makes no mention of
the demonstration by Delambre [1819, p. lxviii] that they
are fraudulent.

(b) The measured magnitudes of lunar eclipses found
in the Syntaxis. Six of these measurements were made in
Babylon and five were made by Greek astronomers. Fotheringham [1909] rejected the ones allegedly made in Babylon on
the grounds that the Babylonians had "reduced" the data before they prepared the records which Ptolemy used. He kept
only the five Greek measurements, which have dates ranging
from −173 April 30 to +136 March 6; Ptolemy claims to have
made the last one himself. All five are fraudulent [Newton,
1977, Sections XIII.1 and XIII.2] except possibly the one
dated 125 April 5, which may be genuine.†

(c) The measured times of lunar eclipses found in the
Syntaxis. Ten measurements, with dates ranging from −720
March 19 to −381 December 12, were allegedly observed in
Babylon. Fotheringham [1915b] rejected these for the reason
given above. We know now that the observations dated −501
November 19 and −490 April 25 may be genuine but that Ptolemy
fabricated the others. Nine measurements, with dates rang-
ing from −200 September 22 to +136 March 6, were allegedly

---

†Fotheringham also made a serious theoretical error in using
these magnitudes. He used a theory which says that the
magnitude of a lunar eclipse depends upon the orientation
of the earth about its axis but that it is independent of
the position of the moon in space. The correct situation
is obviously the opposite of this. I corrected this error
[AAO, Section IX.1] before I knew that most of the observations are fraudulent and hence that correction is useless.

made by Greek astronomers, and Fotheringham used these. The
observation dated 125 April 5 may be genuine but Ptolemy
fabricated the others. Thus three of the nineteen observations may be genuine, but it is not safe to use them since
they occur in such a suspicious context.

(d) The times of seven occultations of stars by the
moon found in the Syntaxis. Fotheringham [1915a] used these
to find the acceleration of the moon with respect to solar
time. Ptolemy fabricated all seven observations.

It is safest to ignore the few eclipse observations in
the Syntaxis which may be genuine. When we do so, we may
say that the only valid observations which appear in Fotheringham's final point F are the equinoxes. They yield the
value $L' = 1.93$, but they cannot give us an estimate of L.
Thus Fotheringham has no solution for the accelerations, and
we cannot use his results because he has none.

The quantity that Fotheringham calls L' is half of ν_S'.
The quantity that he calls L would be half of ν_M' except for
the fact that he includes both gravitational perturbations
and the non-gravitational effects in this parameter. We must
first subtract 6.1 from his parameter L [AAO, pp. 5-6]. If
we then double the result, we have the acceleration ν_M'. Thus
his values L = 10.8 and $L' = 1.5$ correspond to $\nu_S' = 3.0$ $"/cy^2$
and $\nu_M' = 9.4$ $"/cy^2$. These are reasonably accurate for the
average epoch of the "data" which he uses.

Since Fotheringham did not use any genuine data except
the equinox measurements, which have nothing to do with ν_M',
we must ask how it is possible for him to find a reasonably
accurate value of ν_M'. If a person has found "results"
without the use of data, we expect his results to have no
connection with reality. There are three parts to the answer.

(a) When he fabricated the data which Fotheringham and
others have used, Ptolemy had rather accurate values of the
mean motions of the sun, moon, and stars with respect to each
other, but he had poor values for their motions with respect
to the equinox, and he had rather large errors in the periodic terms in the moon's motion. Thus he had large errors
in his fabricated equinox times. On the other hand, the algebraic average error in his fabricated eclipses and occultations is small. The standard deviation in his fabricated
eclipse and occultation times, including the periodic errors,
is about 30 minutes. Thus, when we use his fabricated lunar
"data" we get answers that are in the right range even if
they are not right.

(b) Halley [1695] made the first studies of the astronomical accelerations, so far as I know. He based his studies upon Ptolemy's "observations" and upon various observations from the early Islamic period. Thus his results were
based upon a mixture of genuine and fabricated data, and

even the fabricated data did not lead him seriously astray, as we have just seen.† So far as I know, his general approach was followed by his successors in this work, down through Newcomb [1875]. Thus the work on the subject through 1875 was based upon sound principles, so far as I can find. Students of the subject were limited in accuracy mainly by their dependence upon Ptolemy for data, and they were not aware that his lunar data were fabricated.

(c) About the time of Newcomb's work, or perhaps sooner, we find the beginning of the movement which dominated the field for nearly a century. This was the use of ancient literary and magical eclipses, of which Table VI.2 gives us a typical sample, along with some valid data. The dates assigned to the literary and magical eclipses were obtained from the identification game. When the followers of this movement used these "eclipses" to find the accelerations, they naturally found the values that had originally been used in the dating. Hence the use of these eclipses did not lead to inaccurate values, because it was simply reasoning in a circle. Their use merely obscured the real points at issue.

For these reasons, we have the remarkable circumstance that Fotheringham's results are "in the right ball park" even though, in their essentials, they are based upon no data at all except Hipparchus's equinoxes. It is important that we ignore his results in all future work, except for his study of the equinoxes.

The next writer in chronological order whom I want to consider is de Sitter. For convenience, however, I shall first finish the survey of Fotheringham's work in this field. Fotheringham wrote several papers on the subject after his long paper of 1920, but only one has significant new results, so far as I know. In this paper [Fotheringham, 1935], he uses the measured times of the solar eclipse of -321 September 26 (Section VI.1) and of the lunar eclipse of -424 October 9. From these, he finds $L - L' = 9.5$. This agrees closely with the results from his early work (Figure VI.1 above), and he does not combine this result with earlier ones.

The record dated -321 September 26 is reliable and accurate, so far as we have any reason to believe. We must

---

†In the cited paper, Halley says about the accelerations only that he could establish the lunar acceleration accurately if he had good geodetic positions for the ancient places where observations had been made. His main work on the subject seems to have been given in an address to the Royal Society on 1692 October 19 [MacPike, 1932, p. 229]; this work was never written down, so far as I can discover. According to Martin [1969, p. 2], Dunthorne found the first accurate value for ν_M', namely 10 $''/cy^2$. Martin accidentally omits the citation. M lists a reference to Dunthorne, Philosophical Transactions of the Royal Society, 46, p. 162, 1749, which is probably the same reference. I have not consulted this reference.

conclude, however, that the record referring to -424 October 9 is a calculation and not an observation. Kugler [1914, pp. 233-242] pointed this out when he published the text, and I confirmed and solidified his conclusion in APO [pp. 127-130]. Fotheringham refers to this conclusion of Kugler's and denies that it is correct, but it seems to me that there is no room for question in the matter.

Successive lines in the text refer to the lunar eclipse of -424 October 9 and to the solar eclipse of -424 October 23. The text gives no basis for discriminating between the eclipses, and the text says nothing that implies observation. Hence we must conclude either that both eclipses were observed or that both were calculated. It is impossible to find a set of accelerations which allows both eclipses to be observed. If we adjust the accelerations so that the lunar eclipse occurred after moonrise, it turns out that the solar eclipse started after sunset. Similarly, if the solar eclipse was observable, the lunar eclipse was not. We may ultimately know the accelerations well enough to let us say that a particular one of these eclipses was observable while the other was not. This will not allow us to take the corresponding record as an observation, however. The incompatibility of the two eclipses, and many other features of the text, tell us that it is really an ephemeris text and not a set of observations.

de Sitter [1927] did not actually deal with the ancient observations himself. Instead, he noted that C. Schoch, in a work I have not consulted,† had already used the alleged eclipse observation of -424 October 9 just mentioned, and de Sitter takes Schoch's equation of condition for this alleged record. He also considers various categories of "observations" that were used by Fotheringham, and he derives a few equations of condition for each category.‡ Finally, he solves all these equations by the method of least squares and finds

$$\nu_M' = 10.44 \pm 0.60, \quad \nu_S' = 3.60 \pm 0.32.$$

Since de Sitter merely adds the "non-record" of -424 October 9 to the non-existent (except for the equinoxes) data used by Fotheringham, his results are meaningless and they should not be compared with or averaged in with other results. His work is of interest mainly because Spencer Jones [1939] used it in his famous study of the astronomical accelerations. If Spencer Jones had noticed that modern data require ν_M' to be negative, which makes them incompatible with de Sitter's value, the later history of this subject would probably have been quite different.

---

† de Sitter cites this work as Die Seculäre Acceleration des Mondes und der Sonne, Berlin, 1926. Fotheringham [1935] gives the same citation.

‡ For example, he derives a single equation for all the occultations of stars by the moon that Fotheringham had considered. He also replaces Fotheringham's method of linear inequalities for the solar eclipses by three equations of condition which represent all Fotheringham's eclipses.

Newcomb [1875] used, among other data, the measured
magnitudes and times of both solar and lunar eclipses that
Ibn Yunis [1008] collected. van der Waerden [1961] uses
these data and derives a single equation of condition to
represent them. He also uses three ancient "observations",
namely the lunar eclipse of -424 October 9, the lunar eclipse
of -382 December 23, and an occultation of α Virginis by the
moon on -282 November 9; the last two "observations" are
found in the writing of Ptolemy. He rightly does not use
any of the solar eclipses used by Fotheringham, and he specifically points out the nature of the error which Fotheringham and others make in using the eclipse of Hipparchus.

This paper is the first one I have seen after Newcomb
that is based upon sound principles of selecting data.† It
is van der Waerden's misfortune that his three ancient observations are all invalid. The reference to the eclipse
of -424 October 9, as we have seen, occurs in a calculated
ephemeris that has been mistaken for a set of observations,
while the observations dated -382 December 23 and -282
November 9 were fabricated by Ptolemy. van der Waerden had
no reason to suspect the observations used by Ptolemy. In
view of his knowledge of Babylonian astronomy, however, I
am mildly surprised that he accepted the eclipse of -424
October 9. Unfortunately, the fact that his work depends
upon non-existent observations means that it cannot be used
in studies of the accelerations, although his analytic methods are original and valuable.

Next we come to the work of Curott [1966]. Curott assumes that the value $\dot{n}_M$ = -22.44, which we derive from the
work of Spencer Jones [1939], is correct, and thus he infers
only values for the parameter y which relates to the earth's
spin acceleration. He uses only the places where solar
eclipses were observed. The data that he uses are about
equally balanced between valid Chinese observations‡ and the
literary or magical eclipses favored by Fotheringham and
others of that school.. Since he mixes valid and invalid
data, his overall results cannot be used; only his individual
calculations for the valid observations are useful. Since I
used all his valid observations in AAO, his correct results
involve only a subset of the data that I have used, and thus
his results may not be averaged with the results of AAO or
the results that we shall find in the present work.

The work of Dicke [1969] is based upon the calculations
of Curott. For each "eclipse" that he considered, Curott
gave a value of y and an accompanying standard deviation.
Dicke confines himself to the "eclipses" for which the standard deviation in the value of y that Curott calculates is
less than 1 (out of about 20).

---

†If I have overlooked any sound work in making this statement, I apologize to the authors. I ask the reader to send
me any relevant citations.
‡However, he makes an error in four of the Chinese records
that he uses, as we saw in Section V.6.

Dicke then argues that any given record that meets this condition has high precision but low validity, because the eclipses are but "vaguely described". However, the values from the valid descriptions should cluster closely together, because they are precise, but those from the invalid descriptions should scatter widely. Thus, if we mark the individual values of y on a line, and if we find a cluster of values, this cluster should mark the correct value of y.

He finds 15 values from Curott's calculations which have the required precision. Their values of y scatter over a range of about 7, but 5 of the 15 values lie within a range of slightly less than 1; this range is about 1/8 of the total range. The probability that this would happen by chance is only about 0.007, he says, so that the clustering has a high statistical significance. The mean value within the cluster is about -15.9, so he has found y = -15.9 ± 0.2.† Since he has assumed that $\dot{n}_M$ = -22.44, this corresponds to $\nu_M' = 5.2$. We must now review Dicke's argument.

Dicke starts his argument from the premise that the eclipse records studied by Curott are statements that the sun was totally eclipsed at a known place on a known date, so that the problem is to find which statements are true and which have been misunderstood. Actually, his premise is not correct. In not a single case considered by Curott does the record say that the sun was totally eclipsed at a known point on a known date. Thus Dicke's entire argument starts from an incorrect premise.

Unfortunately, his argument then ends in a theoretical error. He says that the cluster width is 1/8 of the total range of the distribution and he writes: "If a flat a priori distribution is assumed, the probability of 5 or more events out of a total of 15 falling by chance in a cell with a width of 1/8 of the total distribution is 0.007." If this were true, the cluster would be highly significant, in spite of the preceding paragraph.

However, 0.007 is not the probability in question. This is the probability that 6 (not 5) or more events out of 15 will fall in a cell of width 1/8 that is assigned in advance. The fact that Dicke accidentally calculated the probability for 6 rather than 5 events is not a matter of great moment. The main difficulty is that he calculated the probability for a cell that is assigned in advance.

To be more specific, suppose we take 7 black boxes and 1 red box, and let us distribute 15 objects among the boxes at random, with no attention to their color. We then look in the red box. There is a probability of 0.031 (rather than 0.007) that we will find 5 or more objects in it.

This is still a low probability, and thus the cluster would still be statistically significant if this were the correct calculation. However, it is not. What we do in

---

†Dicke does not explicitly state these values in any place that I noticed. I obtain them by scaling from his Figure 1.

looking for a cluster of values is to distribute the 15
objects at random among 8 boxes. We then look in every box
to see if we can find 5 or more objects in any box. What
we want is the probability that we will find 5 or more objects in at least 1 box, purely by chance.

This probability is rather high. To see why this should
be so, let us consider the probability that we will find less
than 5 objects in the first box. This probability is 1 -
0.031 = 0.969. We then look in the next box, and so on, and
we want the probability that we will fail with every box.
This probability is not high.

Feller [1957, p. 103] gives the answer to the following
problem: Suppose that r objects are distributed at random
among n cells. What is the probability that exactly m cells
are each occupied by exactly k things? In the present case,
we have r = 15 and n = 8. We get the answer we want by sum-
ming the probabilities, for these values of r and n, for m
equal to 1 or greater and for k equal to 5 or greater, omitting the impossible cases. † The resulting probability is
0.253.‡ Thus the cluster that Dicke found is not particularly significant if this probability is correct.

Actually, this probability is still far too low, for
the following reason: Let us number the cells from left to
right. Suppose we find 3 objects in the right half of cell
4 and 2 objects in the left half of cell 5. This gives 5
objects which cluster within the width of a cell (1/8 of the
total range in this instance), which is what we are looking
for, even though we do not find 5 objects in any one cell.

Thus we must avoid choosing a cell in advance, and we
must also avoid drawing the cell boundaries in advance. What
we do in using Dicke's method is the following: We distribute 15 points at random along a line. We then measure the
interval from the leftmost to the rightmost point, and we
construct a window whose width is 1/8 of this interval. We
then move the window along the line and count the number of
points that we can see for each position of the window. What
is the probability that there is some position in which we
can see 5 or more points? This probability is obviously much
higher than the value 0.253 that we calculated a short time
ago.

I have not been able to find a formal answer to this
problem, so I explored it empirically by using tables of
random numbers. I took 12 sets of random numbers, with 15
numbers in each set. I found a cluster meeting Dicke's
specifications in 6 of the sets, and I found two such clusters
in a 7th set. This sample is not large enough to give us a
precise answer to the problem, but it is clear that the desired probability is closer to 0.7 than to 0.007. I think

---

† Obviously we cannot have m = 4, for example, if there are
to be 5 objects in each cell and only 15 objects altogether.

‡ As a first approximation, the probability is $1 - (0.969)^8$,
which is 0.223.

it is safe to say that the probability is more than $\frac{1}{2}$.

In other words, we expect to find a cluster like the one Dicke found more than half the time if we are dealing with random numbers. Thus the fact that Dicke found such a cluster has no physical significance. Dicke's claim that he found a significant cluster therefore rests upon a theoretical error in applying the theory of probability.

Earlier, in the discussion of theoretical errors, I implied that we could correct a theoretical error if we had the original data. There I was thinking of errors in connecting the data with values of the parameters that we infer from them. Here we have a theoretical error of a different sort. This error deals with a method of selecting a set of valid data from a total body of numbers which contains both valid data and spurious data. When this theoretical error is corrected, we are left with a method of selection that does not work. Thus Dicke's method does not yield a set of valid data, therefore his inferred acceleration is not valid, and therefore the rest of his conclusions are not valid. Further, the difficulty with his method cannot be removed by correcting his theoretical error.

It follows that Dicke's results must not be combined with other results in an attempt to find the astronomical accelerations.

Next we come to AAO and MCRE, which have already been reviewed in earlier chapters. I believe that MCRE contains only minor errors in the data used. The data in AAO, however, need correcting in two ways. First, we need to remove the data that come to us only through Ptolemy without independent confirmation. In the present state of knowledge, this eliminates [Newton, 1977, Sections XIII.1 and XIII.2] all Greek lunar and planetary data, and leaves only some of the solar and stellar observations. Second, we should remove entirely the literary and magical eclipses that were given very low weights in AAO, as well as the eclipse of Hipparchus, which was given weights split between several dates. After the data are corrected, the remaining data must be analyzed with the use of the revised ephemerides that were discussed in Section I.3. Although the general nature of the results that were found in AAO and MCRE will remain valid, including a large time variation of D'', specific values will change. Thus the results of those works should not be averaged with other results.

Finally we come to the work of M and MS, which can be considered together in their essential features. We have reviewed these works extensively in this and earlier chapters. Both writers ignore the variation of D'' with time that is clearly established by the data, and therefore their quantitative results are incorrect. Further, as we have seen, they use a mixture of valid observations validly interpreted, valid observations that have been incorrectly interpreted (such as assigning an eclipse to a specific point of observation when we cannot assign it any more narrowly than to a large area), and literary or magical eclipses that have been dated by means of the identification game. Thus, for two

important reasons, their results must not be combined with
other results in an averaging process.

In summary, there is no earlier work whose results
should be combined with the results to be found here in
order to improve the accuracy with which we know the accelerations. In spite of its sound, this is not an expression of megalomania. Earlier works can be divided into two
classes. One class rests upon invalid theory, invalid selection of data, or both. Works in this class should naturally
not be used in finding the accelerations. With regard to
earlier works that are valid, I must choose between two possible approaches. I can avoid using data used in earlier
works, infer the accelerations only from data that have not
been used before, and combine the results with earlier results. Alternatively, I can take over the data from earlier
works, combine them with data that have not been used before,
and use the entire collection of data in a single inference
of the accelerations.

I have chosen the second approach because it poses a
simpler computational problem. Thus the results of earlier
works, even the valid ones that are useful in their own
right, should not be combined with the results of this work,
because the data used in valid earlier works are not independent of the data used here. If the reader does not want
to accept the results of this work, it will be necessary for
him to perform a new inference of the accelerations, using
his own selection of data and using his chosen method of
inference. I have tried to give all the information that
the reader will need if he does this.

Note added in proof: Peter Green has made a fresh study
of the "eclipse of Archilochus" in The Shadow of the Parthenon
(University of California Press, Berkeley, 1972). He starts
by estimating the date of the eclipse from textual and historical evidence alone, but using the assumption [p. 271] that
".. the only possible raison d'être for the mention of an
eclipse in this poem, given the context, is that it was local
and recent.." He concludes from his study that Archilochus
saw the eclipse on the island of Thasos at the age of about
25 between the years -700 and -685. I am not competent to
judge the validity of his conclusion, but it seems to be as
sound as estimates which date Archilochus 40 years later.

The annular eclipse of -688 January 11 was probably central in the Aegean (where the island of Thasos is) and thus
Green selects it as the "eclipse of Archilochus". Several
comments are needed.

(a) Green assumes that Archilochus could have had only
one possible reason for mentioning an eclipse. Now Archilochus is a poet whose works are mostly lost and who lived more
than 25 centuries ago in an intellectual environment that we
really know little about. I do not see how we can assume that
we understand him well enough to justify Green's assumption.
Green himself earlier warns us against assuming that we understand Archilochus well. He says [p. 160] that the line " 'O
that I might touch Neobule's hand' has become justly famous,

even though there is a distinct possibility that what it really means is 'Just let me get my hands on Neobule.' " (Neobule was a girl to whom Archilochus was once engaged, if this part of his poetry is indeed autobiographical.)

(b) In dating the eclipse, Green assumes that the fragments of Archilochus's poems are accurately autobiographical, although he himself earlier warns us [p. 156]: "It is not even certain how far we can trust Archilochus' own fragments as an autobiographical guide." Later on the same page he writes that ".. on several occasions Archilochus adopts a first-person persona not his own, without the reader or hearer being warned that this is so." How then may we assume that Archilochus personally saw the "eclipse" that he describes, or, if he saw it, that it occurred close in time to other events in the same context?

(c) As we saw on pages 190-194 above, we may not even conclude that Archilochus's description refers to an eclipse at all.

(d) In the years around -700 to -650, eclipses that were large or total in the Aegean came about 15 years apart. By the most optimistic reckoning, we cannot locate this "eclipse" more closely than 15 years from historical evidence alone. Thus we can find an eclipse to fit any reasonable dating, and the "eclipse of Archilochus" provides no support to any conclusions about his dates.

(e) Since possible eclipses come so close together, we cannot hope to date Archilochus by means of the "eclipse of Archilochus". In fact, I cannot at the moment think of any occasion on which an eclipse has made an important contribution to chronology. The most an eclipse has done is to provide precision to a date that we already knew within a few years.

(f) Implicit in most thinking about ancient eclipses is the idea that an ancient writer had no opportunity to learn about eclipses from reading. Thus, according to this style of thought, if Archilochus described an eclipse (and we must remember that nothing in Archilochus's passage implies an eclipse rather than some other type of darkness), it could only have been on the basis of personal experience. This overlooks the role of oral transmission. A dramatic event like a large eclipse could well live for a long time in oral tradition, and Archilochus could well have learned about eclipses this way. In particular, there were many large eclipses in the Aegean in his time, and the necessary information could easily have survived long enough in the form of casual family conversation.

(g) If we do assume that the "eclipse of Archilochus" was actually an eclipse, there is at least as much evidence for the eclipse of -688 January 11 seen on Thasos as for the eclipse of -647 April 6 seen on Paros. Thus M̲ is certainly wrong in his conclusion about the latter choice that I quoted on page 193 above: "Historians are probably safe in placing Archilochus in Paros on this date." We cannot place Archilochus at any place on any date.

CHAPTER VII

RECORDS OF SOLAR ECLIPSES FROM THE BRITISH ISLES

1. Records That Were Used in Earlier Work

The next six chapters will have the same pattern. Each chapter will deal with the eclipse records from a particular provenance or from a suitable collection of provenances. In the first section of each chapter I shall summarize the records from the appropriate provenances that were used in earlier work and that will be used again here without change in the assigned characteristics. Following this will be one or more sections which present the ancient or medieval sources that have not been used before. Finally there will be a section which presents the eclipse records found in the new sources, along with records that have been used before but that will be used here with changed characteristics.

It is desirable to explain in some detail what is meant by the characteristics of a record and by changes in the characteristics. In order to use a record of a solar eclipse, we must assign it a date, a place or region of observation, a reliability, an estimate of the eclipse magnitude that was observed, and a standard deviation of the error in this estimate. These quantities will be called the characteristics of a record.

I believe that there will be no cases in which the date assigned to a record will be changed, although there are a few cases in which I conclude that a once-accepted date has become questionable.† There will definitely be cases in which one or another of the other characteristics will be changed. In particular, I believe that the standard deviation σ_μ of the magnitude will be changed for almost every record. The reason for this was presented in Section IV.1: In earlier work, I based the values of σ_μ upon hypothetical considerations. In this work, I shall base them upon the distributions of magnitude that are shown by the actual records.

In Table IV.4 in Section IV.1 I give "old" and "new" values of σ_μ. By the old values I mean the values used in AAO and MCRE and by the new values I mean the ones to be used here. If the only change made in the characteristics of a record is to change σ_μ in accordance with Table IV.4, I shall say that the record is being used here without change.‡

---

†The so-called eponym canon eclipse, which was discussed in Section II.5, is an example.

‡One qualification must be made to Table IV.4. In AAO, I took σ_μ to be 0.06 for most ancient (that is, earlier than medieval) records, for reasons that I no longer accept. I shall change these values to 0.18, which means that ancient and medieval records will now be treated alike.

-231-

TABLE VII.1

ENGLISH ECLIPSE RECORDS THAT WERE USED BEFORE

Designation	Place	Standard deviation of the magnitude	Reliability
664 May 1 B,E	England	0.18	0.5
733 Aug 14a B,E	England	0.18	0.5
753 Jan 9 B,E	Jarrow	0.18	0.3
809 Jul 16 B,E	England	0.18	0.5
878 Oct 29 B,E	England	0.082	1
1023 Jan 24 B,E	Burton-on-Trent	0.18	1
1093 Sep 23a B,E	Worcester	0.18	0.1
1093 Sep 23b B,E	Durham	0.18	0.1
1124 Aug 11a B,E	London	0.14[a]	0.1
1124 Aug 11b B,E	S. Albans	0.14[a]	0.1
1124 Aug 11c B,E	Osney	0.14[a]	0.1
1133 Aug 2b B,E	Malmesbury	0.054	0.5
1133 Aug 2c B,E	England	0.082	1
1133 Aug 2d B,E	Coggeshall	0.054	0.5
1140 Mar 20a B,E	Peterborough or Canterbury	0.110	1
1140 Mar 20b B,E	Malmesbury	0.054	0.5
1140 Mar 20c B,E	Canterbury	0.054	0.5
1140 Mar 20d B,E	Worcester	0.18	1
1140 Mar 20e B,E	S. Albans	0.110	0.5
1140 Mar 20f B,E	Coggeshall	0.18	1
1140 Mar 20g B,E	Worcester	0.18	0.5
1178 Sep 13a B,E	London	0.18	1
1178 Sep 13c B,E	Worcester	0.18	0.5
1185 May 1a B,E	London	0.18	1
1185 May 1b B,E	Canterbury	0.14[a]	1

[a]The eclipse is taken to be partial, with a magnitude of 0.81.

TABLE VII.1 (Continued)

Designation	Place	Standard deviation of the magnitude	Reliability
1185 May 1c B,E	Coggeshall	0.18	1
1185 May 1d B,E	Farnham[b]	0.18	0.2
1186 Apr 21a B,E	London	0.18	1
1186 Apr 21b B,E	Canterbury	0.14[a]	1
1187 Sep 4 B,E	Canterbury	0.14[a]	1
1191 Jun 23a B,E	S. Albans	0.14[a]	0.5
1191 Jun 23b B,E	Coggeshall	0.14[a]	0.5
1191 Jun 23c B,E	Farnham[b]	0.18	0.2
1191 Jun 23d B,E	Bury S. Edmunds	0.054	0.5
1194 Apr 22 B,E	Canterbury	0.14[a]	1
1207 Feb 28a B,E	S. Albans	0.18	0.5
1207 Feb 28c B,E	London	0.18	0.5
1230 May 14 B,E	S. Albans	0.18	1
1239 Jun 3a B,E	S. Albans	0.18	1
1239 Jun 3b B,E	Tewkesbury	0.18	1
1241 Oct 6a B,E	S. Albans	0.18	1
1241 Oct 6b B,E	Tewkesbury	0.14[a]	1
1241 Oct 6c B,E	Worcester	0.18	1
1241 Oct 6d B,E	Osney	0.18	0.5
1261 Apr 1 B,E	Bury S. Edmunds	0.18	1
1263 Aug 5a B,E	Farnham[b]	0.18	1
1263 Aug 5b B,E	Osney	0.18	1
1288 Apr 2 B,E	Osney	0.18	1

[a]The eclipse is taken to be partial, with a magnitude of 0.81.

[b]In Surrey.

Note: The eclipse is taken to be central if no estimate of the magnitude is given.

TABLE VII.2

ECLIPSE RECORDS FROM IRELAND, SCOTLAND,
WALES, AND THE ISLE OF MAN THAT WERE USED BEFORE

Designation			Place	Standard deviation of the magnitude	Reliability
664 May	1a	B,I	Ireland	0.18	0.5
664 May	1b	B,I	Ireland	0.18	0.05
688 Jul	3	B,I	Ireland	0.14[a]	0.2
753 Jan	9	B,I	Ireland	0.18	0.2
764 Jun	4	B,I	Ireland	0.18	0.5
807 Feb	11	B,W	S. David's	0.18	0.2
865 Jan	1	B,I	Ireland	0.18	0.5
878 Oct	29	B,I	Ireland	0.18	0.5
878 Oct	29	B,W	S. David's	0.18	0.2
885 Jun	16	B,I	Ireland	0.054	0.3
885 Jun	16	B,SM	Brechin	0.18	0.5
1023 Jan	24	B,I	Ireland	0.18	0.5
1030 Aug	31	B,I	Ireland	0.18	0.5
1133 Aug	2	B,I	Ireland	0.18	0.5
1133 Aug	2	B,SM	Melrose	0.110	0.5
1140 Mar	20	B,SM	Melrose	0.18	0.5
1140 Mar	20	B,W	S. David's	0.18	0..5
1178 Sep	13a	B,W	Western Wales	0.18	0.2
1178 Sep	13b	B,W	Margan	0.18	0.2
1185 May	1a	B,SM	Melrose	0.054	1
1185 May	1b	B,SM	Isle of Man	0.18	0.5
1185 May	1a	B,W	S. David's	0.14[a]	0.2
1185 May	1b	B,W	Margan	0.14[a]	0.2
1191 Jun	23	B,SM	Melrose	0.18	1
1191 Jun	23	B,W	Margan	0.18	0.2
1288 Apr	2	B,W	Western Wales	0.14[a]	1

[a]The eclipse is taken to be partial, with a magnitude of 0.81.
Note: The eclipse is taken to be central if no estimate of the magnitude is given.

The records which say that an eclipse was partial but which give no specific indication of the magnitude also need particular mention. In MCRE, I adopted conventional values of the magnitude and standard deviation for such records that were based upon guesses about the eclipses people would notice, and why they would record that an eclipse was partial. In Table IV.3 of this work, and the accompanying discussion, I found estimates of the magnitude and its standard deviation by using tentative calculations of such eclipses, and I decided to take 0.81 ± 0.14 for the magnitude and its standard deviation. When I assign these values to such a record, I shall also say that the record is being used without change.

When we must assign a region rather than a specific spot as the place of observation, the exact extent of the region is somewhat arbitrary. Thus we are free to choose the exact limits of the region for convenience, provided that we preserve a reasonable agreement with geographical facts. Since the analytic methods to be used here differ in some details from those used in earlier works, changes in the ways used to describe regions are sometimes convenient. The limits to be used for various regions are defined exactly in Appendix III. If I change the exact limits from the limits used in earlier work to the limits in Appendix III, with no other change in the characteristics, I shall again say that the record is being used without change.

I shall discuss all other changes in the characteristics of a record in detail in the appropriate places. Two kinds of change occur most frequently. In one kind, I made an actual error in the coordinates of a place of observation. The other kind concerns records that give an estimate of the magnitude of an eclipse, but in a descriptive rather than a numerical way. Section IV.1 shows that records tend to exaggerate the magnitude of large eclipses. Accordingly, I shall revise most estimates of the magnitude in this work.

Table VII.1 lists the English records that have been used before and that will be used again without change, except for the routine changes that were described a moment ago. The table should be self-explanatory. Table VII.2 lists the records from Ireland, Scotland, Wales, and the Isle of Man that are being used again without change.

2.   Sources That Have Not Been Used Before

In this section, I shall describe the historical sources that have not been used in my earlier works. One source is Irish and the others are English. For simplicity, I shall discuss the Irish source in the same section as the English sources, instead of devoting a separate section to it.

In MCRE, the general rule governing the sources and eclipses was the following: The aim was to have a cutoff date of 1200. If a source contained records of eclipses before 1200, I also used records from that source with dates between 1200 and 1300, if any. I did not use records from any source unless it had some records earlier than 1200, and I did not use any records after 1300.

In searching for sources to use in this work, the emphasis was on sources that did not contain any eclipse records before 1200. In the search, I came across a few sources with earlier records which I had missed in preparing MCRE. Records from these sources will therefore be included in this work.

The value of a record in determining the accelerations obviously decreases steadily as we come forward in time. In fact, the value is proportional to the square of the age of the record before the present time. In MCRE, the loss in value of an individual record was countered by the tendency to have more records as we come toward the present. Thus, as Figure XVIII.1 of MCRE shows us, the uncertainty in the value of D'' inferred from the records of, say, a single half-century at least holds constant, and it may even improve slightly as we move from the early Middle Ages to the year 1300.

I expected the number of records to increase still more for records in the 14th and later centuries, so that we could continue to find accurate values for D'' in spite of the steadily decreasing sensitivity of a single record. In this respect, the present work has been a serious disappointment. I have found only 9 usable records from England for the 14th and 15th centuries combined. In contrast, MCRE had 38 usable records from England from the 12th century alone.

There are at least two reasons for this situation. One reason may be called the growing sophistication of historical writers. Sources in the earlier Middle Ages tended to be simple lists of events, with all the notable events of a single year grouped together, many of them under the wrong year. Many of the events were political in nature; in this term I include such matters as accessions and deaths of popes, bishops, and secular rulers, as well as wars, rebellions, and similar events. However, many of the events were those that interested the "common" man. That is, they were events that were likely to be noticed by the average person, or events that directly affected his well-being. These include weather events, such as storms or droughts, local fires, comets, and eclipses of the sun and moon.

The "sophisticated" medieval historian tended to ignore events unless they affected the fate of large numbers of people or the fate of "important" people. Thus he would ignore the weather except, perhaps, for a drought that severely affected the total food supply and hence the stability of the realm. He would also ignore comets, eclipses, and similar matters.† He focussed his attention upon struggles

---

†An historian who was sufficiently superstitious might believe that an astronomical happening was an omen of, for example, the death of a king. If so, he might record this happening because he believed that it was important politically.

between rulers for control over large areas, for example, or upon contests between kings and popes about the investiture of bishops. He also tried to explain the causes and effects of the events that he described. Thus he tended to scorn eclipses and the like as matters unworthy of his attention.

As we come toward the 14th century, historians tend to become more sophisticated in the sense I have just described. Thus they tend, in ever-increasing fashion, to omit eclipses from their histories.

At the same time, the production of histories and of historical sources seems to have decreased markedly in many parts of Europe. The introduction to the cited edition of S. Albani [ca. 1431] says that monastic annals reached their peak in the 13th century and then almost died out. Even the continuations that were continued spasmodically are almost gone by the beginning of the 15th century. The editor of John of Reading [ca. 1367] specifies the interval from 1356 to 1367 as the interval in which "the writing of history in England reached perhaps its lowest ebb before the fifteenth century."

Thus the number of sources decreased at the same time that they became less likely to record eclipses, both solar and lunar.† In consequence, English and other British records of eclipses decline in number at the same time that each individual record declines in value. Luckily, the situation is not so bad in some other parts of Europe, particularly in Italy.

When we have had time to study sources of all kinds in later work, it may well be that our ability to find the accelerations does not decline as we move from the 13th century toward the modern period. The reason is that astronomical activity seems to increase steadily in Europe from its beginning in 1092‡ up to the present time. However, astronomical sources from, say, 1100 to 1700 have not been published extensively, and a great deal of manuscript research may be needed before we get as much sensitivity in the accelerations for later centuries as we have for the 12th and 13th centuries. I shall study this situation in more detail in a later work.

---

†However, so far as I can judge without making a detailed statistical survey, the sources continued to record comets, haloes, and the like, as well as the vagaries of the weather, as frequently as ever. The importance of the weather is obvious, but I do not see why comets remained "memorable" while eclipses did not. It would be odd if comets had portentous value while eclipses, with all the emphasis on astrology that is found in this period, had lost theirs.

‡This is the earliest date I have found for an astronomical observation made in Europe with the aid of instruments constructed for the purpose. It was made by Walcher of Malvern and appears in his writing on the lunar motion [Walcher, ca. 1110].

We may now look at individual British sources. I shall discuss them in alphabetical order in this section and present the eclipse records in chronological order in the next section.

Angliae [ca. 1388]. The monastery of S. Albans, about 25 kilometers north-northwest of London, was an important center of historical writing during the medieval period. We used two chronicles in MCRE that originated there [Matthew Paris, ca. 1250 and Wendover, ca. 1235], and we shall use two more in this work. Angliae is one of them. It covers the period from 1328 to 1388, and it seems to be a contemporaneous record of events. There is no clue to the identity of the compiler or compilers.

Benedict [ca. 1192]. I mentioned this work in MCRE [p. 142] but I did not use it in order not to lengthen that work still further. Benedict was abbot of the monastery of Peterborough, about 100 kilometers north of London. Benedict seems to have been active in secular as well as monastic affairs and Richard I appointed him guardian of the seals in 1191. Because it was unlawful for a person in his position to accept this office, a letter was prepared excommunicating Benedict, but it is probable that the excommunication was never completed. When Richard was being held for ransom, he ordered Benedict to join him (1193). Benedict's history ends in the preceding year; apparently he interrupted writing it when he left to join Richard and never resumed it.

Bermondsey [ca. 1433]. I used this work in MCRE [Chapter VI], and it contained the records 1124 August 11a B,E and 1207 February 28c B,E. Although it runs for another 2 centuries, it does not contain another sure record of a solar eclipse; this illustrates one of the problems that were mentioned earlier in this section. The history describes two later occasions when the sun behaved in a peculiar manner, but I do not believe that either was an eclipse. These occasions will be described in the next section. Bermondsey was a monastery in the southeastern part of London.

Capgrave [ca. 1463]. I believe that this is the first history I have used which was written in English (I suppose one should say Middle English) prose.† Capgrave was born in Lynn,‡ Norfolk, on 1393 April 21 and he died there on 1464 August 12. His history does not contain any records of solar

---

†The Anglo-Saxon Chronicle was written in Anglo-Saxon. The history by Robert of Gloucester [ca. 1300] is in meter (I hesitate to say poetry) rather than prose. I found nothing in it that is useful for this work, and I shall not describe it.

‡This is the place name that the editor of Capgrave uses. I presume that it is the same as King's Lynn or Lynn Regis, but I have not investigated the matter.

eclipses. I mention it only because it has close relations
with Walsingham [ca. 1422], relations that are not mentioned
by the editors of either source.

These relations are shown by many parallel passages in
the two works. For example, both Walsingham and Capgrave
have accounts of a comet in March of 1368. Walsingham's
account is in Latin and Capgrave's account reads like a
translation of Walsingham's account into English. Since
Capgrave is almost surely about 40 years later than Walsingham, the latter is probably the original source so far as
Capgrave is concerned.

Cotton [ca. 1298]. As the editor says, all we know
about Cotton is that he "was a monk of Norwich, and that he
did not survive the year 1298, the last year of which the
annals are given, . ." This is a chronicle in the grand
manner, which begins at an early period and continues to
1298. Most of the work is simply copied from earlier sources,
often in the exact words. It seems to be original only for
the years 1264 to 1279 and from 1285 to its close, although
other parts do contain occasional events relating to Norwich
that are not found in the main sources. Cotton also repro-
duces a number of official documents, such as papal or royal
letters, so he had access to good sources of information for
the periods in which he is original.

Dorenses [ca. 1362]. These are from Dore, an abbey in
the county and diocese of Hereford, founded in 1147. The
Annales Dorenses follow a familiar pattern. They were compiled by a single person down to about 1280. Thereafter they
were continued by other writers on an intermittent basis down
to 1362. They seem to be local and original after 1280. The
abbey was at or near the place now called Abbeydore, which
can be found on a large-scale road map. By scaling from such
a map, I estimate that its latitude is 51°.97 and that its
longitude is 2°.89 west.

Flores Historiarum [ca. 1326]. This is one of the complex of historical writings associated with S. Albans, and
it is also only one of several writings to receive this
title. However, it is the first one for which I have not
found a more suitable means of citation. This is a continuation of the work that I cited as Matthew Paris [ca. 1250]
in MCRE, a work that is sometimes called Historia Minor. It
was begun at S. Albans and stayed there until about 1265.
At that point the manuscript, or a copy of it, was transferred to Westminster, where it was continued by different
people. The most likely thing, I believe, is that someone
associated with Westminster made a copy of it at S. Albans
and took the copy to Westminster where he and others continued it.

The continuations go to 1326, and were written by several monks of Westminster. The one who wrote the part associated with the reign of Edward II was named Robert of
Reading, but the editor believes that it is not correct to

attribute the entire Westminster continuation to him. Quite
a long time ago, someone noticed that the Flores Historiarum
was associated with someone named Matthew and that it was
also associated with Westminster. Consequently, the work
was attributed to "Matthew of Westminster" for a long time,
but it now seems safe to say that there was no such person.

The part of this work after the move to Westminster
seems to be original, and I have studied only this part.

John of Reading [ca. 1367]. This is the concluding
portion of what is essentially a continuous chronicle of
Westminster Abbey from 1299 to 1367. Robert of Reading, who
wrote part of Flores Historiarum [ca. 1326], wrote the part
from 1299 to 1325. John of Reading wrote the part from 1346
to 1367. Either this was one of the sources for Walsingham
[ca. 1422] or Walsingham used some of the sources that John
also used. Luckily the material that is shared by the two
writers does not include the eclipse records.

The editor of the cited edition does not think highly
of John's abilities, saying [p. 8]: "He falls decidedly below the level even of his bombastic predecessor Robert of
Reading. His chronicle possesses no literary and but moderate historical value." However, the editor further says,
John's is the fullest account known for the decade 1356-
1367, and then makes a remark that was already quoted: ". .
in which years the writing of history in England reached
perhaps its lowest ebb before the fifteenth century."

Loch Cè [ca. 1590]. We can put all the discussion of
Irish annals that we need under this heading. In MCRE, I
made extensive use of the Annals of Ulster [Ulster, ca.
1498], which has records of eclipses from 512 June 29 to
1023 January 24. Although the main part of Ulster runs to
1498, with continuations to 1588, there is no mention of a
solar eclipse that I can find later than 1023, although
there continue to be records of droughts, famines, and similar events.

The Annals of Loch Cè has a similar pattern. It seems
to be mostly the same as Ulster down to about 1225, and it
has the same account of the eclipse of 1023 January 24.
Although the main part of Loch Cè runs to 1590, with intermittent continuations to 1648, there is no further reference
to a solar eclipse. The editor of the cited edition thinks
that an entry under the year 1473 refers to an eclipse in or
near that year, but I think that it has some other meaning.
I shall discuss this entry in the next section.

Londonienses [ca. 1330]. Since London is a large place
and undoubtedly had many annals and chronicles, it is undesirable to use "London" in citing these annals. However,
there is no known author of these annals, and they are not
associated with any particular institution in London, so
there is no convenient alternative. These annals seem to be
a compilation and not a collection of running records, and

there are indications that the compiler changed about 1307.
The editor of the cited edition believes that the compiler
(speaking of only one for convenience) was connected with
the government of the city, but that he was certainly not a
skilled user of Latin. The material after about 1280 is
mostly independent of other known sources and is presumably
local to London.

The manuscript has an interesting history and provides
an amusing anecdote. The original formed part of the famous
collection of medieval materials made by Sir Robert Bruce
Cotton (1571-1631). After Cotton's death, his collection
was maintained as an independent library until 1753, when it
was transferred to the British Museum. In 1731 the collection suffered a serious fire which destroyed many important
manuscripts, including all but about 35 pages of Londonienses.
Luckily, someone had made a manuscript copy of it before the
fire. The parts of the copy that can be compared with the
original prove to be highly faithful, even, as the editor
says, to the errors in the original, and he gives an example.
The copy at one point exhibits the set of symbols "302 pavit".
Since the copy has Arabic numerals while the original almost
surely had Roman numerals, it is possible that we could guess
what happened, but we do not have to guess because this part
of the original has survived. It reads the nonsense "cccii.
pavit". We can now be sure that the compiler of the original
simply copied some source without attending to its meaning.
The original must have read "occupavit". The compiler misread "o" as "c" and "u" as "ii", thus creating an error
which the copyist faithfully followed.

Merton [ca. 1323]. In volume 3 of the cited edition
of Flores Historiarum [ca. 1326], the editor prints an appendix which consists of passages found in various copies
of Flores Historiarum that are not found in the original.
Many of these through the year 1323 are from a source which
the editor identifies as "MS. E. (Merton.)" It seems to me
that there is only one likely explanation of this situation.
Someone made a compilation which was drawn mostly from the
Flores Historiarum, but which also used independent sources.
Thus the additions constitute an additional source which I
take the liberty of naming Merton.

The editor does not say what "Merton" refers to, but
the most likely possibility is Merton College at Oxford; I
shall assume that this is true in the remaining discussion.
This does not mean that the source came from Oxford. In
fact, I noticed few references to Oxford and none whose
nature suggests an Oxford origin. On the other hand, I
noticed some statements which suggest a London origin. For
example, in 1259, King Alexander of Scotland "came to London."
From these usages, I deduce that Merton comes from London.
As it turns out, the astronomical computations are hardly
affected if we use Oxford instead.

Opus Chronicorum [ca. 1296]. I dislike using a general
term like this as a citation but, as with Londonienses,
there is no obvious alternative. Opus Chronicorum was

compiled at S. Albans, and if it seemed to be a running set of annals, I would call it S. Albans and distinguish it from other histories written there by the date. It was compiled by one person, apparently a monk. At one time, critics thought it was written by a monk named Rishanger, but the editor of the cited edition thinks that this is not correct.

The chronology is quite careless. At one point, to give an outstanding example, the writer lists events for three different years and calls them all 1267. None of the events that I checked in any of these years happened in 1267. I have given the year of compilation as "ca. 1296" because 1296 is the year listed for the last event. However, the work lists 1274 as the year in which a plague began among the sheep, and it says the plague lasted for 28 years, which would bring us to 1301 or 1302. This is one of the reasons why the editor thinks the work was compiled considerably later than 1296; if the writer were more accurate in his chronology, this would be a good argument. The editor is perhaps on firmer ground when he points out that the chronicler uses "all the days of his life" in writing about Edward I, from which the editor infers that this was written after Edward died in 1307. However, it seems to me that even this inference is not really safe, in view of the careless writing in Opus Chronicorum.

Paulini [ca. 1341]. This starts as a careful abridgement of some of the writing of Matthew Paris and his continuators down to 1306, with a continuation to 1341. However, it begins to insert some original material as early as 1280. The parts after 1306 are in a variety of hands, which suggests that they were being written contemporaneously with the events they describe. The original writer was closely connected with S. Paul's Cathedral in London, and we may safely take London as the place.

S. Albani [ca. 1431]. This collection of annals was made at S. Albans, and it is one of the few sets of annals which were made in the early 15th century. The editor of the cited edition believes that this was an informal journal which one of the monks kept for his own amusement. If so, it was not maintained continuously, because events are sometimes out of order. However, we may imagine that the writer in 1420, say, had his attention called to something that had happened in 1418, say, and decided to record it at that point. There is no reason to question that the records are essentially original.

S. Edmundi [ca. 1212]. A Benedictine abbey dedicated to S. Edmund was established in 1020 in the town of Bury, replacing an earlier monastery. The abbey was given jurisdiction over the town, a fact which caused much friction between town and abbey. This led to riots in 1327 in which the abbey was destroyed and the abbot was seized and carried off to the continent.

These annals begin with the Christian era and their

early parts are copied from known sources, which the editor
identifies. New entries taken from local records begin to
be inserted in 1044. The annals were apparently compiled
by a single person, presumably a monk, who used the known
sources and local records. The text breaks off abruptly in
1212; I do not know whether later parts have been lost or
whether the writer broke off abruptly for some reason and
never resumed his work.

Walsingham [ca. 1422]. Walsingham probably came from
somewhere in Norfolk. A passage referring to Oxford seems
to indicate that he was educated there, but the editor of
the cited edition thinks this passage may be copied from
some other source. Walsingham was at S. Albans for some
time, then he was prior of a monastery named Wymundham† from
1394 to 1400, and then he apparently returned to S. Albans.
The editor says that there is no certain mention of him after
1419. He presumably lived at least to 1422, since the last
event in his history occurred in that year.

Walsingham's history seems to be entirely derivative,
and the editor discusses the major sources that Walsingham
uses. They seem to originate from S. Albans, so we may assume that this is the location of the eclipse observations.

Walsingham is fond of summarizing the nature of the
year from the standpoint of agriculture. He gives such a
summary for almost every year from 1273 to 1341, and again
from 1383 to 1391. The reader who is interested in clima-
tology for this period should find his work helpful. I list
Walsingham's summaries in Appendix IV.

Wykes [ca. 1289]. I used this source in MCRE and I
mention it here only because there is a problem about the
place where the observations in it were made. The place is
Osney, in Oxfordshire, for which I have been unable to find
precise coordinates. Hence I use the coordinates of Bicester,
which I believe is close by.

Wyntershylle [ca. 1406]. This is either one of the
sources used by Walsingham or it is derived from sources
that he used. However, Walsingham drops this (or its
sources) in 1392 and later years. Thus it happens that
Walsingham does not record the eclipse of 1406 June 16,
which we find in Wyntershylle. This is another of the many
historical works written at S. Albans.

3.   Records from the British Isles

In this section, I shall discuss the records which have
been used before but whose characteristics will be changed
here, and I shall also discuss the records that were not

---

†This was presumably in the town of the same name in Norfolk.
There was a monastery there which was originally a depend-
ency of the monastery of S. Albans.

used in earlier work. Records from any part of the British Isles will be discussed together, in chronological order.

733 August 14b B,E. This record was used in MCRE [p. 152]. It says that "almost the entire disk of the sun appeared to be covered by a horrible black shield." This explicitly says that the eclipse was partial, but it also seems to indicate a large eclipse. I shall take the magnitude to be 0.90 with a standard deviation of 0.05. The other characteristics are unchanged from MCRE.

764 June 4 B,I. I shall give this record the same characteristics as in MCRE [p. 194], but I am discussing it here because I wish to comment on an argument that was used there. In deciding upon the date of this eclipse, which is put in the year 763 in the source, I used a record of a lunar eclipse which was listed in the first sentence of the entry for the year 762. From this fact, I deduced that the lunar eclipse occurred early in the year. I realize now that this argument is not valid. Annalists usually but by no means always kept events in the proper order so far as the year is concerned. However, many of them made no attempt to keep events in the correct order within a year, perhaps because they did not have the necessary information. Luckily, the dating did not depend in any important way upon my error, and there is no reason to question the date.

934 April 16 ? Reference: Capgrave [ca. 1463]. On page 117, the reference says that the sun appeared like blood in the time of Lothar II. If this is Lothar of Italy, which seems to be the case, the possible years are 931 to 950. It is possible that this is a reference to the eclipse of 934 April 16, but it could also mean something besides an eclipse. I shall not use this record.

1133 August 2a B,E. This record is from the Anglo-Saxon Chronicle, and I used it in MCRE [p. 160]. The record explicitly says that the eclipse was partial, but it also says that stars could be seen during it. The only other record I remember offhand that states this combination of circumstances is the record -430 August 3a M in Section VI.1 above, and the combination poses problems. In MCRE, I took the magnitude to be 1 in spite of the clear statement of partiality, with a standard deviation of 0.03. Here I shall take the magnitude to be 0.92, with a standard deviation of 0.05. This action is based upon my growing realization of the amount of exaggeration in eclipse records.

In discussing this record, M [p. 8.52] writes: "Among other notables, Henry II personally witnessed this eclipse (magnitude about .99) while crossing the channel to fight the French (Anglo-Saxon Chronicle). The producers of the film Becket missed their chance by failing to include this genuine† historical coincidence with an eclipse." Actually,

---

†The emphasis and the parentheses in this passage are M's.

Thomas Becket was appointed archbishop of Canterbury in 1162 and he was murdered in 1170. We are concerned with the year 1133 and King Henry I, not Henry II. Further, according to the record, Henry was asleep during the eclipse and did not see it.

1178 September 13b B,E. This record says that the eclipse was partial, but it also gives details which indicate a large eclipse [MCRE, p. 168]. In MCRE, I put high faith in the indications of a large eclipse and took the magnitude to be 0.95, with a standard deviation of 0.05. I now have less faith in the reliability of such indications, and I shall take this to be a partial eclipse with no indication of the magnitude. That is, I shall use 0.81 for the magnitude, with a standard deviation of 0.14.

1178 September 13 B,SM. This record [MCRE, p. 203] says that the sun "took on a pallor and was almost totally eclipsed . . ." This record is more explicit than the preceding one about the approach to totality, so I shall use a larger magnitude, perhaps wrongly. I shall use 0.90, with a standard deviation of 0.05. In MCRE, in contrast, I was afraid that we could not calculate the sign of η with certainty, so I took the magnitude to be 1, with a standard deviation of 0.05.

1185 May 1e B,E. Reference: Benedict [ca. 1192]. Under the year 1185, this source says: "Meanwhile, on the calends of May, on the very day of the apostles Philip and Jacob, about midday, an eclipse was seen through all England." All the calendrical details are correct. The eclipse probably began about midday, with maximum eclipse occurring somewhat later. Since Benedict was apparently at Peterborough at the time, I shall take Peterborough to be the place in spite of the reference to all England. Reliability: 1; standard deviation of the magnitude: 0.18.

1191 June 23e B,E. I used this record in both AAO [p. 56] and MCRE [p. 174]. The record says that the sun "shone with less than ordinary brightness." As I noted in both earlier works, this indicates a partial eclipse, but I took it to be central from fear that we could not calculate the sign of η with assurance. Now that we know the accelerations better, there should be no question about the sign of η. I shall take this to be a standard partial eclipse with the magnitude not specified. That is, the magnitude is taken as 0.81, with a standard deviation of 0.14.

1191 June 23f B,E. Reference: [Benedict, ca. 1192]. Under the year 1191, this source says: "In the same month of June, on the Lord's day, on the eve of the Nativity of S. John the Baptist, on the 9th calends of July, at about the hour of nones, the sun suffered an eclipse that lasted for three hours, and the sun was obscuratus, and there was darkness over the earth, and stars appeared in the sky; and

when the eclipse receded the sun returned to its usual
splendor." The date is correctly stated in two different
ways. Two words need comment.

The hour called nones originally meant the 9th hour of
the day, and it was the hour at which many medieval people
ate their main meal of the day. Later, many people changed
to eating their main meal at the middle of the day, but they
kept the word nones for the meal time, and from this we have
our word noon. There is no way I see to tell the meaning of
nones from the text, but Oppolzer's chart shows that the
eclipse should have begun near midday.

After saying that the sun was eclipsed, the record goes
on to say that the sun was obscuratus. This is the adjective
that was used in the discussion of the alleged eclipse record
of -189 March 14 in Section VI.1, which M translates as
"dimmed". As I noted there, the word usually has greater
force. Since the writer here has already told us that the
sun was eclipsed, I do not see why he would go on to say that
it was dimmed, and thus I conclude that the word is being
used with great force here. It would be reasonable to assume
that it means a total eclipse, but I shall not go this far.
I shall take the eclipse to be central, with 0.054 for the
standard deviation of the magnitude because of the stars. The
place is Peterborough and the reliability is 1.

1207 February 28b B,E and 1207 February 28d B,E. The
first record was used in MCRE [p. 174] and it comes from
Bury S. Edmund. The second record is from S. Edmundi [ca.
1212], under the year 1207. It is also geographically from
Bury S. Edmund. The records are really the same. In MCRE,
before I had found the record from S. Edmundi, I gave a reliability of 1 to the record 1207 February 28b B,E. I tend
to think that this record is indeed the original, but I see
no way to settle the question. Therefore I shall give both
records the same characteristics that I gave the first record in MCRE, except that I shall now assign a reliability of
0.5 to each. This makes no difference astronomically since
the two records taken together have the weight originally
assigned to one of them, but it reflects the historical
situation better.

1255 December 30 B,E. As I noted in Section II.1 above,
I used the correct characteristics of this record in MCRE
[p. 177], except that I used the wrong coordinates for the
place of observation, which is Oxnead.

The source Cotton [ca. 1298] has clearly copied this
record, which is from the source Oxenedes [ca. 1292]. However, the editor of the cited edition of Cotton says that he
copied the chronicler Matthew Paris. I am not sure which
work by Matthew Paris he means, but I did not see a record
of this eclipse in any work associated with him. I do not
know whether this is a slip on the part of the editor or
whether I have not consulted the work that he had in mind.
Cotton has also copied the record of the eclipse of 1261
April 1 [MCRE, p. 178] from the source that I designate as

Florence of Worcester [ca. 1118]. More precisely, it is
from the source that some writers designate as John of
Taxter's continuation of Florence of Worcester.

1263 August 5 ? On 1265 August 4, the popular leader
Simon de Montfort was killed in battle near the town of
Evesham, about 40 kilometers south of Birmingham. Within a
few years, we find chroniclers telling us that the event was
accompanied by tremendous thunder and lightning storms, by
earthquakes, and by an eclipse of the sun. All in all, there
was the most impressive turn-out of the elements for the
death of a hero that I have seen since the death of Roland in
778.† The sources that relate these events, not always con-
sistently, include Cotton [ca. 1298], Opus Chronicorum [ca.
1296], and Robert of Gloucester [ca. 1300]. It is possible
that these accounts were inspired by the eclipse of 1263
August 5, almost two years earlier to the day, but I doubt
it. In any case, there is no need to use these magical records, because we have two genuine English records of this
eclipse in MCRE [p. 179], and a third one follows.

1263 August 5c B,E. Reference: Merton [ca. 1323].
Under the year 1263, on page 251, the source says: "On the
nones of August, after the hour of nones, the sun suffered
an eclipse." The nones of August is the 5th, so the date is
exact. According to Oppolzer's map, maximum eclipse came
about mid-afternoon. Thus, when nones is used the second
time to denote the hour, it probably means the 9th hour of
the day rather than our noon. I shall take the place to be
London, although, as I said in discussing this source, it
would make little difference if the place were Oxford. The
reliability is 1. I shall take the eclipse to be central,
with 0.18 for the standard deviation of the magnitude.

1310 January 31a B,E. Reference: Flores Historiarum
[ca. 1326]. This is entered under the year 1309: "This
year there was an eclipse of the sun on the 31st day of the
month of January." This is one of the earliest records in
which the date is given in a modern style rather than the
Roman style. The year may not be an error; instead, the
writer may have followed a convention in which the year began in March. The reliability is 1 and the place is Westminster. I shall use 0.18 for the standard deviation of the
magnitude.

1310 January 31b B,E. Reference: Dorenses [ca. 1362].
This source also puts the eclipse under the year 1309: "In
the year 1309, there was an eclipse of the sun on the eve of
S. Brigid's day, about midday, on the Sabbath." S. Brigid's
day is February 1, so the dating is correct if the chronicler
began his year later than we do. The eclipse was on a Saturday, so the writer used "Sabbath" in the Jewish rather than
the usual Christian sense. The reliability is 1 and the
place is Dore. I take the eclipse to be central, with 0.18

---

†See AAO [pp. 78-80] or MCRE [p. 324].

for the standard deviation of the magnitude.

1330 July 16 B,E. Reference: <u>Paulini</u> [ca. 1341]. The
source reads, under the year 1330 (on page 350): "On the
following Monday, namely the 16th day of June, there was an
eclipse of the sun at the 4th hour <u>post meridiem</u>."[†] The hour
looks reasonable. The place is London, the reliability is
1, the value of η is 0, and the standard deviation of the
magnitude is 0.18.

1339 July 7 B,E. Reference: <u>Angliae</u> [ca. 1388]. This
source reads on page 9, under the year 1339: "This year, on
the day of the Translation of S. Thomas, archbishop of Canterbury, there was an eclipse of the sun immediately after midday, and it lasted for two hours. It was eclipsed unto its
fourth part." <u>Londonienses</u> [ca. 1330] tells us under the
year 1220 that the translation[‡] of S. Thomas occurred that
year on July 7, by authority of Pope Honorius. Thus the date
is stated correctly, and the hour looks reasonable. The main
problem lies in deciding how to handle the magnitude and its
standard deviation.

I think the record means that the sun was eclipsed until
only a fourth of it was shining, which would mean a magnitude
of 0.75. However, from the text alone we cannot rule out the
possibility that it was eclipsed until a fourth of it had
vanished, meaning a magnitude of 0.25. Luckily the possibilities are so far apart that we can settle between them by
calculation, even with provisional values of the accelerations. Calculation shows clearly that a magnitude of 0.75
was meant,[‡] so I shall use this value. I shall use 0.075
for the standard deviation. The reliability is 1 and the
place is S. Albans.

1360 April 14 ? Under the year 1360, <u>Bermondsey</u> [ca.
1433] has a passage which reads: "This year, on the fourteenth day of April, thus on the day after Easter, when King
Edward III was living with his (<u>sic</u>) outside the city of
Paris, was a day of extreme cloudiness and intolerable cold,
which was called Black Monday." This is undoubtedly a dark
day rather than an eclipse, but the dating is not possible.
In the years 1358 through 1362, April 14 was never on a
Monday nor was it ever in the week of Easter. If the year
was 1360 and if the day was the day after Easter, the correct

---

[†]Note the increasing frequency of modern usage in stating
the day of the week, the day of the month, and the hour.
Literally, the week day is given as "day of the moon". The
printed edition gives the month as July, but a footnote
says that the manuscript gives it as June, an obvious slip
of no importance.

[‡]The translation of a saint usually refers to a transfer of
his remains from one place to another, probably from his
first place of burial to a special shrine.

[‡]Actually the magnitude was close to 7/8.

date was April 6. However, the day is written out in words, and this error of writing is unlikely. If we are to determine the date of this Black Monday, it will have to be from other sources.

1361 May 5 B,E. Reference: John of Reading [ca. 1367]. The reference reads: ". . . on the 6th day of May, on the eve of the Ascension of the Lord, the day being bright till then, there was an eclipse of the sun." Actually, Ascension Day was on May 6, so the writer was probably thinking of that day when he wrote the date; this slip does not affect the identification. The reliability is 1, the place is Westminster, and I shall take the eclipse to be central, with 0.18 for the standard deviation of the magnitude.

Angliae [ca. 1388] and Walsingham [ca. 1422], both presumably writing in S. Albans, have identical short records of this eclipse. This common record could be original, but its wording suggests that it could have been abridged from John of Reading. For safety, I shall not use it.

1384 August 17a B,E. Reference: Angliae [ca. 1388]. The record occurs under the correct year: "In this year, on the 17th day of the month of August, there was an eclipse of the sun in the afternoon." The reliability is 1, the place is S. Albans, and I shall use $\eta = 0$, with 0.18 for the standard deviation of the magnitude.

1384 August 17b B,E. Reference: Walsingham [ca. 1422]. We seem to have two independent records of this eclipse from S. Albans. Walsingham was probably living at S. Albans by now, and this may be his own observation, although most of his history is derivative. It is also possible that he copied it from some source at S. Albans other than Angliae.

Under the year 1384, in a section headed "About an Eclipse of the Sun", we find: "In this year, late on the 16th day of the month of August and through the following night, horrible thunder was heard in England and terrible lightning was seen; which terrified mortals greatly. On the next day, in the afternoon, for about an hour there was an eclipse of the sun and a conjunction simultaneously, which were watched by many." The wording suggests that the eclipse and the conjunction were unconnected phenomena in the mind of the writer, and that they just happened to come at the same time. If so, however, how did it happen that he knew about the conjunction? It is not something that would usually occur to an untrained person, it seems to me.

These considerations do not interfere with the fact that we have an independent record of the eclipse. This record had the same characteristics as the preceding one.

On page 202 of volume 2 of the cited edition of Walsingham, we find a statement that the sun on 1391 July 9 appeared

red through thick clouds, that it was dark from nones† to
sunset, and that the sun was mostly obscured for the next
six weeks. I do not take this to be a disguised record of
an eclipse.

1406 June 16 B,E. Reference: Wyntershylle [ca. 1406].
This record is under the year 1406 and it is almost the last
event in Wyntershylle's annals: "This year, on the 16th
day of the month of June, in the early morning, during 6
o'clock,‡ that is, about 6 hours after midnight, there was
an eclipse of the sun." All the details are accurate. This
has the same characteristics as the records of 1384, including the place.

1424 June 26 B,E. Reference: S. Albani [ca. 1431].
This record, found on page 7 of the cited edition, reads:
"In the 4th year of the reign of King Henry Sixth, on the
day of S. John and Paul, was an eclipse of the sun." This
is the 4th year of the life of Henry VI rather than of his
reign.‡ Otherwise the dating is accurate. This record
continues to have the same characteristics, including the
place.

1473 April 27? Reference: Loch Cè [ca. 1590]. On
page 173 of the cited edition, under the year 1473, the
printed translation reads: "Ruaidhri, the son of Art
O'Neill, died this year; and the harvest of the black day."
A footnote then reads: "An annular eclipse of the sun is
recorded to have occurred on the 27th of April in this year;
but there is no mention of an eclipse of the sun in harvest,
although such an event must be intended by the 'black day.'"
I believe that this note is wrong in most of its major
aspects.

There was an annular eclipse on 1473 April 27, but most
of its path lay in Siberia and I doubt that it was visible
at all in Ireland. There was no eclipse in the harvest season this year, but the editor neglects to mention an important fact. There was no eclipse large enough to produce
significant darkening in Ireland nearer in time than 1491
May 8, and there was none in the harvest season nearer than
1502 October 1. Finally, I see no reason to assume that an
eclipse was intended by the writer. He probably means what
I called a "dark day" in Section VI.1, under the date -647
April 6, but it is possible that he means that the harvest

---

†I see no way to decide whether this means midday or the
9th hour of the day.

‡The actual phrase is inter sex de campana (during 6 of the
bell), which, it seems to me, is appropriately and almost
literally translated as 6 o'clock. I believe that this is
the earliest usage of this form that I have seen. I presume that he means during the 6th hour, and not during the
time that the bells were striking six times.

‡He became king when he was not quite 10 months old.

-250-

TABLE VII.3

ECLIPSE RECORDS FROM THE BRITISH ISLES
THAT ARE NEW OR THAT HAVE BEEN RE-INTERPRETED

Designation	Place	Standard deviation of the magnitude	Reliability
733 Aug 14b B,E	Jarrow	0.05^a	0.5
1133 Aug 2a B,E	Peterborough or Canterbury	0.05^b	0.5
1178 Sep 13b B,E	Canterbury	0.14^c	1
1178 Sep 13 B,SM	Melrose	0.05^a	1
1185 May 1e B,E	Peterborough	0.18	1
1191 Jun 23e B,E	Winchester	0.14^c	1
1191 Jun 23f B,E	Peterborough	0.054	1
1207 Feb 28b B,E	Bury S. Edmund	0.18	0.5
1207 Feb 28d B,E	Bury S. Edmund	0.18	0.5
1255 Dec 30 B,E	Oxnead	0.18	1
1263 Aug 5c B,E	London	0.18	1
1310 Jan 31a B,E	Westminster	0.18	1
1310 Jan 31b B,E	Dore	0.18	1
1330 Jul 16 B,E	London	0.18	1
1339 Jul 7 B,E	S. Albans	0.075^d	1
1361 May 5 B,E	Westminster	0.18	1
1384 Aug 17a B,E	S. Albans	0.18	1
1384 Aug 17b B,E	S. Albans	0.18	1
1406 Jun 16 B,E	S. Albans	0.18	1
1424 Jun 26 B,E	S. Albans	0.18	1

[a]The eclipse is taken to be partial, with a magnitude of 0.90.
[b]The magnitude is taken to be 0.92.
[c]The magnitude is taken to be 0.81.
[d]The magnitude is taken to be 0.75.
Note: The eclipse is taken to be central if no estimate of
the magnitude is given.

was extremely bad, so bad that it was a black day when it was brought in. This record clearly does not refer to an eclipse, at least not to one that can be dated.

The records that have been discussed in this section are summarized in Table VII.3, which has the same format as Table VI.1.

CHAPTER VIII

FRENCH RECORDS OF SOLAR ECLIPSES

1. Records That Were Used in Earlier Work

The records from France that will be used again without
change are listed in Table VIII.1, which has the same format
as Table VII.1. This includes all but six of the records
that were used in MCRE. These six will be used again, but
with changes that relate to the magnitudes of partial eclipses.

The situation with regard to records after 1300 is even
worse for France than for the British Isles. I have found
only 4 usable records after 1300, of which 3 are for the
eclipse of 1310 January 31 and 1 is for the eclipse of 1433
June 17. However, in the course of the search for these
records, I found many additional records for eclipses before
1300.

2. Sources That Were Not Used Before

In this section I shall describe the French sources
that were not used in MCRE.

Albericus [ca. 1241]. This source is a chronicle which
presents events from both eastern and western Europe, but
with the main attention given to affairs of France. The editors say that Albericus is valuable because his sources
differ from those used by Guillelmus de Nangiaco [ca. 1368],
whose history will be discussed below. Albericus is unusually precise in giving his dates, but the reader should remember that he begins his year with Christmas.

The editors describe Albericus as a monk of Trois-
Fontaines. I can find only two places called by this name.
One is a group of three small islands just off the coast of
northwestern Sicily, near the town of Mazara del Vallo. This
is not a likely place for a chronicler who emphasized French
history. The other place is a small town between Nancy and
Strasbourg, and I assume that it is the correct one.

Alpertus [ca. 1020]. This source is found in the series
often cited as Bouquet. By the time we reach the volume that
contains Alpertus, the series was being edited by some monks
of S. Maur, with no individuals being named. As often happens with this series, only parts of the source are presented,
often spread over several volumes, and I do not know when
Alpertus really ends. The last event I noted was the ʿ. ᷄
pearance of a comet in 1020. All we know of Alpertus is
that he was a monk of the monastery of S. Symphorianus in
Metz. The editors say that he was attentive to celestial
phenomena, to which he accorded the "prejudices of a vile
astrology."

TABLE VIII.1

FRENCH ECLIPSE RECORDS THAT WERE USED BEFORE

Designation			Place	Standard deviation of the magnitude	Reliability
563 Oct	3	E,F	Tours	0.1[a]	0.5
590 Oct	4	E,F	Tours	0.1[a]	1
592 Mar	19	E,F	Burgundy	0.17[b]	0.25
603 Aug	12	E,F	Burgundy	0.18	0.5
764 Jun	4	E,F	Flavigny	0.18	0.5
807 Feb	11	E,F	S. Omer	0.18	0.5
840 May	5a	E,F	Sens	0.18	0.5
840 May	5b	E,F	Lyon	0.054	1
840 May	5c	E,F	Wissembourg	0.18	1
840 May	5d	E,F	Angoulême	0.18	1
841 Oct	18	E,F	Orléans to Tours	0.18	1
878 Oct	29a	E,F	Anjou	0.18	0.2
878 Oct	29b	E,F	Arras	0.18	1
878 Oct	29c	E,F	Beze	0.054	1
878 Oct	29d	E,F	S. Benoit-sur-Loire	0.18	1
878 Oct	29e	E,F	S. Amand-les-Eaux	0.054	0.5
878 Oct	29f	E,F	S. Omer	0.18	0.5
891 Aug	8	E,F	S. Omer	0.18	0.5
968 Dec	22	E,F	S. Benoit-sur-Loire	0.054	1
1023 Jan	24	E,F	Limoges	0.18	0.5
1033 Jun	29b	E,F	Wissembourg	0.18	1
1033 Jun	29c	E,F	Dijon	0.054	0.2
1033 Jun	29e	E,F	Nevers	0.18	1
1039 Aug	22a	E,F	Pont-à-Mousson	0.18	1
1039 Aug	22b	E,F	Cluny	0.18	1

[a] The magnitude is taken to be 0.8.
[b] The magnitude is taken to be 0.67.

TABLE VIII.1 (Continued)

Designation	Place	Standard deviation of the magnitude	Reliability
1044 Nov 22a E,F	Nevers	0.18	1
1044 Nov 22b E,F	S. Benoit-sur-Loire	0.18	0.5
1044 Nov 22c E,F	Cluny	0.18	1
1133 Aug 2a E,F	Bourbourg	0.054	1
1133 Aug 2b E,F	Laon	0.18	1
1133 Aug 2c E,F	Cambrai	0.054	1
1133 Aug 2d E,F	Autun	0.034	1
1133 Aug 2e E,F	Rouen	0.18	1
1133 Aug 2f E,F	Rouen	0.18	0.2
1140 Mar 20 E,F	Bourbourg	0.18	0.5
1147 Oct 26a E,F	Beauvais	0.18	1
1147 Oct 26b E,F	Laon	0.18	1
1147 Oct 26c E,F	Nevers	0.18	1
1153 Jan 26a E,F	Douai[c]	0.18	1
1153 Jan 26b E,F	Mont S. Michel	0.18	1
1178 Sep 13a E,F	Saumur	0.18	0.5
1178 Sep 13b E,F	Douai[c]	0.18	1
1178 Sep 13d E,F	Dijon	0.18	1
1185 May 1a E,F	Saumur	0.18	0.5
1191 Jun 23a E,F	Douai[c]	0.05[d]	1
1191 Jun 23b E,F	Angers	0.18	1
1191 Jun 23c E,F	Vendôme	0.18	1
1191 Jun 23d E,F	S. Amand-les-Eaux	0.124	1
1191 Jun 23e E,F	Metz	0.18	1
1207 Feb 28 E,F	Angers	0.18	0.5

[c]The correct place is Anchin, but I use the nearby place Douai because I cannot find the coordinates of Anchin.

[d]The magnitude is taken to be 0.94.

Note: The eclipse is taken to be central if no estimate of the magnitude is given.

Autissiodorense [ca. 1174]. This is a small collection
of annals that runs from 1005 to 1174 in most copies, with a
continuation to 1190 in one copy. The entries are often
quite intermittent; I counted 13 entries for the years 1033
to 1076, inclusive, for example. Collections of this sort
tend to be original, in my experience. This collection was
made in Auxerre.

Colmarienses [ca. 1300]. These are described as minor
annals from Colmar; if there are major ones, either I have
not seen them or I made no record of them because they contained no eclipse records. They are bound with other documents pertaining to Colmar and Basel in a codex now in
Stuttgart. The annals as published run from 1211 to 1300
and are written in three hands. However, the editor says,
many entries down to about 1240 are derivative, and a few
entries after that date are. The year 1300 appears, but no
event is recorded for that year.

Girardus de Fracheto [ca. 1328]. Note the interesting
mixture of languages shown in this chronicler's name. The
source we want is not really by Girardus, but there is no
convenient way to cite it except by using his name. Girardus
compiled a chronicle for which I forgot to record the final
date, and it was then continued by some unknown person to
1328. The part we want is in the continuation. The editors
say that the chronicle of Girardus is hardly worth publishing, but the continuation is valuable because it often cor-
rects the chronicle of Guillelmus de Nangiaco [ca. 1368],
which will be discussed below. The continuation was apparently made at the monastery of S. Denis, near Paris.

Godel [ca. 1320]. This source presents the same problem in citation as the preceding one. Godel was a chronicler of Limoges, whose chronicle had an anonymous continuation, and the information we want is in the continuation.
The continuation consists of a few sparse entries ranging
from 1276 to 1320.

Guidonis [ca. 1327]. As I understand what the editors
say, Guidonis wrote two works, one called Flores Historiarum
and the other called Chronicon Regum Francorum. What they
have published is not either source, but a mixture of excerpts from both. The principal goal of Guidonis seems to.
have been the establishment of an accurate chronology. If
he encountered a date in his sources that seemed doubtful,
he undertook an extensive investigation and discussion.
However, much of his material seems to be independent.

If the editors of Guidonis say where Guidonis did his
writing, I failed to note it. The nature of his work suggests Paris, and I shall use this as the place.

<u>Guillelmus Armoricus</u> [ca. 1224]. Guillelmus tells us
explicitly in the first paragraph of his work that he is
continuing the life of King Philip† at the point where
<u>Rigordus</u> [ca. 1208] left off; Rigordus will be discussed
below. Guillelmus also provides a short summary of earlier
events which he has taken from Rigordus. I found no explicit statement by the editors about the location of
Guillelmus. However, Rigordus was at the monastery of S.
Denis, so Guillelmus presumably was there also.

    <u>Guillelmus de Nangiaco</u> [ca. 1368]. I have already
mentioned this writer in connection with <u>Girardus de Fracheto</u>
[ca. 1328] and <u>Albericus</u> [ca. 1241]. Nangiaco seems to refer
to Nangis, which may be a family name but is probably the
place where Guillelmus was born. Guillelmus was a monk of
S. Denis who lived under Philip IV (king 1285-1314). Guillelmus wrote several historical works, including lives of
S. Louis (Louis IX, king 1226-1270) and Philip III (king
1270-1285). This work is a chronicle that runs from 1113 to
1300, with anonymous continuations to 1368 which I include
under the same citation. The chronicle is in Latin, but
some authorities suspect that Guillelmus wrote a separate
French version, and that the French version is incorporated
into the source called <u>S. Denis</u> [ca. 1328] below.

    This chronicle starts in 1113 because the purpose of
Guillelmus is to write a continuation of the famous chronicle
of <u>Sigebertus</u> [ca. 1111], which was discussed in <u>MCRE</u> [p.
223]. Both Sigebertus and Guillelmus seem to be rather
careless in their chronicling, at least in my experience of
them, and I have mentioned that the main merit of <u>Girardus</u>
<u>de Fracheto</u> [ca. 1328] resides in its corrections of Guil-
lelmus. The early parts of Guillelmus's chronicle are necessarily derivative, but it seems to be an independent authority from about 1225 on. The continuations after 1300
seem to be independent and located at S. Denis.

    <u>Guillelmus de Podio-Laurentii</u> [ca. 1272]. This Guil-
lelmus is presumably from the town of Puylaurens, which is
about 50 kilometers south of Albi and about 50 kilometers
east of Toulouse. About all we know about Guillelmus, except for the dates covered in his history, is that he was
chaplain to Count Raimond VII of Toulouse, who was count
from 1222 to 1249. This work is a history of the Albigeois
region. I do not know the exact boundaries of the region,
but Toulouse is in it or close to it. It is probable that
Guillelmus used local records in preparing this history,
and I shall take Toulouse as the location of his observations.

    <u>Iohannes Longus</u> [ca. 1365]. This is a history of the
monastery dedicated to S. Bertini, which is located on an
island called Sithiu at the town of S. Omer in northern

---

†This is King Philip II of France, sometimes called Philip
Augustus.

France. We used several sources from this monastery in MCRE [Section X.2]. Iohannes was born in the town which the editor calls Ipra (which may be the modern Ypres or Ieper) around the beginning of the 14th century. He became a monk at the monastery probably around 1325, was sent to Paris to study for awhile around 1340, and then returned. He became abbot on 1365 March 24. I have guessed 1365 as the year when he finished his history, with the idea that he may have quit writing when he became abbot, but this may well be wrong. Since the history only goes to 1294, it is not strictly original, but it seems to be based upon local and independent sources.

Iterius [1225]. Iterius is one of many chroniclers from Limoges, all from the monastery of S. Martial there. The monastery was founded in the early 9th century. We learn from his chronicle that he was born in 1163, that he held a position called armarius in the monastery, and that he died on 1225 January 27. The chronicle is continued anonymously to 1297, and it is naturally from the continuation that we learn the date of his death.

Johannis à S. Victore [ca. 1322]. This Johannis was at S. Victor's in Paris, and we know nothing else about him. He began his chronicle by copying Guillelmus de Nangiaco [ca. 1368] above, who was at the nearby monastery of S. Denis. From about 1300 on, he is an independent authority. Unfortunately, say the editors, he put his dates in the margins. Thus there is some doubt about whether they have correctly associated events with the dates that he intended.

Lambertus Waterlos [ca. 1170]. Lambertus compiled the annals of the town of Cambrai in northern France, and he gives us his own biography under the year 1108, the year of his birth. If I have identified the place names correctly, he was born in the town of Wattrelos near Tournai. Both places are near the modern French-Belgian border, with Wattrelos in France and Tournai in Belgium, and they are 40 to 50 kilometers north of Cambrai. According to Lambertus, he was the scion of a noble family with extensive church connections. This claim seems to be borne out by his career. He became a canon in 1119, at the age of 11, and other promotions which I shall not enumerate followed rapidly. The earlier part of his annals is derivative, but he seems to be an independent authority from about 1150 on. The annals are fairly extensive; each year after 1150 takes about two printed pages.

Lemovicense [ca. 1342]. This is described as a major chronicle of Limoges, which is published in fragments in three different volumes of the Bouquet series. It was written by several monks in the monastery of S. Martial there, and the editors identify one of them as Petrus Coral.

Lemovicensia [ca. 1658]. As I understand the matter, this is not a single source except to the extent that the editor has made it one. As printed, it consists of fragments of various chronicles maintained in Limoges for years from 848 to 1658. The editor says that the provenance of these fragments is certain, and that it is nearly certain that they were written contemporaneously.

Mettensis [ca. 1447]. The title of this source, when translated, is A Universal Chronicle of Metz. The editor of the cited edition says that he has already discussed the source at length in a work which he identifies as "N. Archiv III, p. 67 sqq."; I have not tried to locate this reference. In spite of the title, the work is mostly a list of popes, Roman emperors, and French kings through the year 1250. Then there are some annals from 1251 to 1274. Following this is a continuation which has annals from 1153 to 1447. This is the only section of the work with information relative to our purposes, so I have used the date 1447 in the citation. There is another continuation with annals from 1451 to 1509.

Ordericus Vitalis [1141]. Ordericus is a famous historian whom we know a fair amount about. He was born in Shrewsbury in England in 1075, the son of a French priest[†] and an English woman; thus he seems to represent an early stage in the reconciliation of the Norman conquerors with the conquered Anglo-Saxons. He entered the Norman monastery of S. Evroul-en-Ouche at the age of 10 and lived there the rest of his life. However, he did spend a moderate amount of time in travel in both England and France. He wrote this history allegedly upon the instructions of his superiors, but we may guess that the instructions were welcome. It was supposed to be a history of the monastery, but it ended up as an ecclesiastical history of England and Normandy. It is considered to be an independent authority from about 1070 on.

I have had some difficulty in finding the coordinates of S. Evroul. One encyclopedia says that it is on the river Carentan about 10 kilometers from Lisieux. In contradiction, Muirhead's Guide to Northwestern France for 1926 says that it is almost to S. Gauburge - S. Colombe on the road from Bernay. The description in this guidebook is detailed, it seems to be accurate, and I accept it. As well as I can scale from a map, the coordinates are latitude 48°.8N, 0°.5E.

Remense [ca. 1190]. I know nothing about this source except that it was apparently written in Reims. Excerpts

---

[†]Celibacy was not a requirement of Catholic clergy until about the middle of the 12th century. At the moment of writing, I cannot lay my hands upon the exact date when the requirement was imposed.

from it are printed in both volumes IX and X of the Bouquet
series. All the information that we want appears on page
271 of volume X.

Rigordus [ca. 1208]. This is the history of the reign
of Philip II (Philip Augustus) that I mentioned earlier in
connection with Guillelmus Armoricus [ca. 1224]. Rigordus
was from Languedoc, and he was originally a physician by
profession. Because of his poverty (!) he became a monk in
the monastery of S. Denis just north of Paris. He dedicated
his history to Prince Louis (later Louis VIII) in 1200, although it goes through 1208. As we have seen, Guillelmus
Armoricus continued the history from the point where Rigordus
left off. The editor of the cited edition says that Rigordus
did not have access to the court and its records. Thus he
merely related events that he heard of, mixing them with his
own accounts of dreams, visions, and prodigies of various
kinds. His records of solar eclipses suggest to me that he
had an elementary knowledge of astronomy, more than most
people of the time probably had.

Robertus de Monte [ca. 1186]. I used this source in
MCRE [Section X.2], but by some accident I overlooked 3 of
his records of solar eclipses. For 2 of these 3 eclipses,
he in fact has the only records that I have found. Here I
cite a different edition from the one I used in MCRE in
order to emphasize that this accident is not a matter of the
edition used. Robertus was the abbot of Mont S. Michel from
1154 until his death in about 1186.

S. Denis [ca. 1328]. It is doubtful that this should
be described as a single source, but there is no convenient
alternative. As well as I can gather, the editors found
many independent chronicles at S. Denis that were written
in French rather than Latin. They have then published excerpts from various ones of these all mixed together, with
no attempt to publish all of any one source. Many parts of
these French chronicles seem to be translations from Guillelmus de Nangiaco [ca. 1368], which was discussed earlier.
However, many parts are independent of other known sources.

S. Nicasii [ca. 1309]. This is one of several sets of
annals written in Reims that appear in the same volume of
the Monumenta Germaniae Historica series. It is the only
one I noticed with a record of a solar eclipse. It runs,
with many omitted years, from 1197 to 1309.

S. Victoris [ca. 1542]. This is a set of annals kept
in the monastery of S. Victoris in Marseilles. After some
short early notes, which deal mostly with the deaths of
successive abbots, there are fairly frequent entries from
1178 to 1265. Following this, there are only a few entries
from 1334 to 1542. The last two entries, for 1524 and 1542,
are in French rather than Latin, and this may be the reason
that the editor printed them in Italics. The annals seem

to be independent and original.

Tolosanum [ca. 1271]. This is a brief chronicle that runs from 1096 to 1271. It was apparently compiled by one person near the end of the 13th century. In spite of this, it has records of 2 eclipses that are independent of any other records I have seen. It is printed in 2 different volumes of the Bouquet series.

3. Records from France

This section includes French records of solar eclipses that have not been used before, as well as some that were used before but whose characteristics will be changed in this work. The changes all relate to the method of handling eclipses which were stated to be partial.

1009 March 29. Under the year 1009, the printed edition of Remense [ca. 1190] has the following passage: "There was an eclipse of the sun at the 2nd hour of the day [on the 15th calends of April]." The editor says in the margin that the year is 1010 rather than 1009. Now there was an eclipse that was visible in France on 1010 March 18, which is the 15th calends of April. Since it came in the late afternoon, almost at sunset, it is not the eclipse being reported. However, I think that the editors supplied "on the 15th calends of April" and that they used the brackets to indicate this fact. The record surely refers to the eclipse of 1009 March 29, which did happen at about the 2nd hour of the day. Thus the record is accurate as it stands, and there is no need to supply material in emendation. Unfortunately, this record is almost identical with the record 1009 March 29a E,BN in MCRE [p. 238], which is from Liège. Since Liège is fairly close to Reims, there is a good chance that the Reims record found in Remense is a copy, and I shall not use it.

1018 April 18 E,F. A passage on page 139 of the cited edition of Alpertus [ca. 1020] reads: "In the year before the Council at Noviomagus† was proclaimed, the moon was eclipsed after the middle of the night in winter, and when the king the following year was staying in the same place in Easter week there was an eclipse of the sun." No pair of eclipses fits this passage exactly. The lunar eclipse "in winter" on 1017 November 6 came well before midnight. The lunar eclipse that fits best is the one of 1016 November 17, which was a large eclipse with a maximum at about 5 hours, local time. The only solar eclipse that fits is the one of 1018 April 18, except for the fact that Easter was on April 6 that year. However, I see no way to decide whether the observation was made at Metz, where the writer presumably

---

†The present Lisieux. I do not know whether the Council was held at Lisieux or whether the proclamation was made there.

was, or whether it was reported from Lisieux, where the king was staying. Thus I shall not use the record.

Remense [ca. 1190] has a short record of this eclipse that is identical with the record of Sigebertus [ca. 1111], except for a minor change in word order. Thus I shall not use this record either.

1033 June 29a E,F. This is one of the records in MCRE [pp. 334-336] in which the magnitude is estimated by a comparison with a crescent moon. In this case, the sun was like the moon on its 4th night, which corresponds to a magnitude that I shall round to 0.88. In MCRE, I said that I would use this value for the magnitude provided we could tell the sign of η unambiguously, and that I would use η = 0 if we could not tell. However, I did not record the final choice I made when writing MCRE, and I have not preserved it in my notes. I think that there is no problem about telling the sign of η with the knowledge we now have, and I shall take the magnitude to be 0.88, with a standard deviation of 0.06.

This is one of the records that I discussed in Section IV.1, which I shall include in a special study in Section XIV.6 after I have inferred the accelerations. M [p. 8.50] says that the magnitude was 0.994 ± 0.001 and that the low estimate of the magnitude was the result of irradiation. He also says that the eclipse was annular, and he makes a further interesting remark.

There are two records of this eclipse [MCRE, pp. 334-336] which say that the sun became a sapphire color. For this reason, I pointed out the possibility that the eclipse was actually total and that this was an attempt to describe the corona. However, it now seems sure that the eclipse was barely annular and that this interpretation is impossible.

M says that the reference to sapphire strongly suggests that the observer saw the solar chromosphere. I have never heard of the chromosphere appearing blue, and this suggestion seems impossible to me. It may be that the sky became a deep blue around the sun, and that the observer was trying to describe this situation.†

There is still another difficulty with this record. When M calculates that the magnitude was 0.994 ± 0.001, he uses C̄luny as the place, and he cites MCRE as the authority for this conclusion. I explicitly did not reach this conclusion. As I pointed out [MCRE, pp. 334-335], this record is found both in Cluny and Beze, and we cannot tell which is the correct place of observation. If we assume that the magnitude is indeed about 0.88, it does not matter much which place we use, and in MCRE I used Dijon, which is about midway between Beze and Cluny. However, if there is a large difference between the circumstances at the two places, this is not a legitimate procedure. In this work, I shall

---

†The author of the record 1239 June 3e E,F below says that the sky became deep blue during the eclipse.

perform the necessary calculations for both places and in-
spect the results. If the circumstances differ seriously
at the two places, I shall not use this record.

The record 1033 June 29d E,F [MCRE, p. 337] says that
the sun first looked green and then yellow during this
eclipse. Thus we have seen the sun called three colors,
namely blue, green, and yellow. Since colors during an
eclipse are usually not mentioned at all, there may have
been some unusual phenomenon connected with this eclipse.

1033 June 29d E,F. As I have just mentioned, this record [MCRE, p. 337] says that the sun first looked green and
then yellow during this eclipse. The record in fact gives
such a vigorous description of the eclipse that I concluded
that the eclipse had been seen as total. For example, it
says that the sun during its recovery from the eclipse at
first looked like the crescent new moon on its first night.
It is hard to see how this could really have happened unless
the eclipse was very close to total. Now I tend to doubt
the importance of such descriptions. I shall, however, take
the standard deviation of the magnitude to be 0.054, which
is the value used when stars are seen, but I do so with some
misgivings.

1033 June 29f E,F. The source Autissiodorense [ca.
1174] has the following entry: "In the year 1033, the sun
was obscured (obscuratus) on the "natale" of S. Peter, on
the 3rd calends of July. There was a peace Council in this
city." The city is Auxerre. The 3rd calends of July is
June 29, which is more often called the day of S. Peter and
S. Paul than the natale of S. Peter. Although this report
is rather short, it seems to be independent of any others I
have found, and the nature of the source suggests that it
consists of entries made contemporaneously with the events
they describe. Nonetheless, for safety, I shall use a reliability of only 0.5. I shall take the eclipse to be central,
with 0.18 for the standard deviation of the magnitude.

Remense [ca. 1190] has a record of this eclipse which,
like the record in the same source for 1018 April 18, seems
to be copied from Sigebertus [ca. 1111]. I shall not use
this record.

1037 April 18a E,F. This record [MCRE, p. 338] says
that the sun at one time appeared like the moon on its 2nd
day, that later it looked like the moon on its 5th day, and
still later like the moon on its 8th day. The writer is
clearly describing the recovery from the eclipse, with the
maximum magnitude corresponding to a crescent moon on its
2nd day. For this reason, I took the magnitude to be 0.98
in MCRE, with a standard deviation of 0.01. M [p. 5.25]
believes that eclipse records of this sort always underestimate the magnitude,† but it is my suspicion that the

────────────────
†See also the discussion of the record 1033 June 29a E,F
above.

-263-

estimates are more likely to exaggerate the amount of eclipsing. I shall revise the magnitude estimate to 0.94, with a standard deviation of 0.05.

1037 April 18b E,F. This record [MCRE, pp. 338-339] indicates the recovery from the eclipse by giving a sequence of drawings of the sun. The drawing that indicates maximum eclipse suggests a very large eclipse, so I used 0.95 for the magnitude, with a standard deviation of 0.03. Here I revise these figures to 0.90 for the magnitude, with a standard deviation of 0.05.

1098 December 25 E,F. On page 665 of the cited edition, Ordericus Vitalis [1141] writes, under the year 1098: ". . on the Sabbath, on the Nativity of the Lord, the sun was thrown into blackness." Christmas was on Saturday that year, so Ordericus used Sabbath in its Jewish sense. The place is S. Evroul-en-Ouche, the reliability is 1, the value of η will be taken as 0, and the standard deviation of the magnitude is 0.18.

1153 January 26c E,F. I used two French records of this eclipse in MCRE. This record is from Lambertus Waterlos [ca. 1170]. The next to the last sentence under 1152 reads: "On the 4th calends of January there was an eclipse of the sun about the middle hour of the day." Then the first entry under 1153 reads: "In the year 1153, a very great splendor appeared in the night, which was a terrible sign for mortals, on the 4th calends of January. After a moderate interval, there was a remarkable thunder. On the 7th calends of February a solar eclipse took place." Note that the first solar eclipse took place, according to the record, on the same day as the "splendor" in the night, which was probably an aurora. However, the annalist correctly put this date in 1152 on one occasion and put it incorrectly in 1153 in the other.

The editor says that the two solar eclipses were those of 1152 February 7 and 1152 August 2, but he is surely mistaken. Neither eclipse was visible in France. In identifying the eclipses, we should start from the second eclipse. The date given for it is 1153 January 26, which is an eclipse that was visible in France and which took place about the middle hour of the day. Thus both statements about a solar eclipse seem to refer to the same eclipse. We may make the following speculation about the origin of this composite and garbled record: The writer started to put down the record of the eclipse of 1153 January 26 but accidentally copied the date of the aurora. Realizing this, he then recorded the aurora and the later thunderstorm. Then, forgetting to erase the first statement about the eclipse, he again entered the eclipse under the correct date.

The record is contemporaneous and receives a reliability of 1. The place is Cambrai, the value of η is 0, and the standard deviation of the magnitude is 0.18.

1178 September 13c E,F and 1178 September 13e E,F.
These records [MCRE, pp. 348 and 349] both give an estimate
of the magnitude by comparing the sun with the moon. I
shall revise my earlier optimistic estimates to 0.90 for
the magnitude, with 0.05 for its standard deviation, for
both records.

1178 September 13f E,F. Before he records this eclipse,
Robertus de Monte [ca. 1186] has a brief notice of the
eclipse of 1147 October 26. This notice could have been
copied from any one of several sources, or it could be original. For safety, I shall not use it.

On page 325 of the cited edition, Robertus gives a long
paragraph in which he records a number of eclipses, not in
chronological order. Apparently he had collected a number
of notes about eclipses and decided to put them all down at
once without bothering to order them. The eclipse of 1147
October 26 is not in this group; it is on page 291. For the
year 1168 (MCLXVIII), he says that "we saw" an eclipse of the
moon on the night after the 3rd of the nones of March, in the
first hour of the night, another lunar eclipse on the 4th
calends of September, and an eclipse of the sun on the ides
of September. There are two scribal errors in this record.
He left an X out of the year, which should be 1178 rather
than 1168, and he wrote the 4th calends of September when he
meant the 3rd. When we make these changes, the dates become
1178 March 5 and 1178 August 30 for the lunar eclipses and
1178 September 13 for the solar eclipse.

He also says that the solar eclipse took place about
midday, which is reasonably accurate. He then adds that the
sun was almost totally eclipsed. I shall take the magnitude
to be 0.90, with a standard deviation of 0.05. The record
is contemporaneous and receives a reliability of 1, and the
place is Mont S. Michel.

1178 September 13g E,F. On page 199 (volume XIX) of
the cited edition of Guillelmus de Podio-Laurentii [ca. 1272],
the writer has been talking about the year 1196. Then he
writes: "Some time before, in the year of the Lord 1188
(MCLXXXVIII), on the ides of September, on the 4th day of
the week, in the 6th hour, there was an eclipse of the sun
which was exceedingly terrible and dark." The record does
not say explicitly that the eclipse was total, nor does it
deny totality as the preceding record did. I think it is
reasonable to take the eclipse as central, with 0.068 for
the standard deviation of the magnitude. Since the record
is not contemporaneous, it has a reliability of only 0.5.
The place is Toulouse.

Tolosanum [ca. 1271], which is also from Toulouse, has
a brief notice of this eclipse which could have been copied
from many places. I shall not use it for safety. Guillelmus
de Nangiaco [ca. 1368] also has a brief notice that is almost
surely derivative.

1178 September 13h E,F. One of the entries in <u>Iterius</u>
[ca. 1225] says: "In the year of grace 1178, an eclipse of
the sun on the exaltation of the Holy Cross." Actually the
day of the exaltation is September 14. Iterius was only 15
at the time, and this does not seem like a likely mistake if
the record were contemporaneous. Hence I give it a reliability of 0.5. The place is Limoges, and we may use $\eta = 0$
with a standard deviation of 0.18 for the magnitude.

1178 September 13i E,F. <u>Lemovicensia</u> [ca. 1658] has a
long passage about this eclipse: "In the year 1178 from the
Incarnation of the Lord, in September, on the ides thereof,
on the eve of the exaltation of the Holy Cross, with us and
the whole people watching, the sun was made as black as
coarse cloth,† except that it kept a narrow line like the
moon on its second or third. It was such an eclipse that
everything shimmered like water, and those standing in one
part of the cloister could hardly recognize someone in another part of the cloister. The eclipse lasted from the
4th hour until the full 6th hour,‡ that is, from the beginning of the major mass until the brothers entered the re-
fectory. Before 15 days were accomplished, there was also
an eclipse . . ." The last part seems to indicate a lunar
eclipse at the full moon following 1178 September 13, but
there was no such eclipse. Perhaps the writer meant to say
that the lunar eclipse was 15 days before.

In any case, this is a clear description of an eclipse
that was large but not total, and it is apparently a contem-
porary account of an eye witness. The place is Limoges and
the reliability is 1. I shall use 0.94 for the magnitude,
with 0.06 for its standard deviation.

1180 January 28 E,F. <u>Robertus de Monte</u> [ca. 1186]
records this eclipse in the same long paragraph as the
eclipse of 1178 September 13. This record says simply: "In
the year 1130‡ was an eclipse in the sun, on the 5th calends
of February, on the third day of the week." Actually it was
on the second day of the week, but this error does not affect
the reliability of the identification, and the record is
contemporaneous. The reliability is 1, the place is Mont S.
Michel, the value of η is 0, and the standard deviation of
the magnitude is 0.18.

---

†The word used is <u>cilicium</u>. This is a form of the adjective
derived from Cilicia, and it also means a coarse cloth made
from goats' hair; I can find no other meanings. Perhaps it
is a variant spelling of some other word that I have not
recognized.

‡ I presume this means until the end of the 6th hour. This
would be the time we call noon, that is, 12 hours.

‡ I have MCXXX in my notes. Either Robertus or I has accidentally left out L.

1181 July 13 E,F. This is the last solar eclipse in the long paragraph just mentioned. Robertus writes about this one: "In the year 1181 of the Lord, on the 3rd of the ides, at the hour of nones, a solar eclipse; and almost a third part of the sun was obscured; and from the beginning to the end of the eclipse was the space of 1 equal hour† and 24 minutes. Nones apparently means the 9th hour, or mid-afternoon. This record is remarkable for its use of the equal hour, for the precise statement of the duration, and for the small magnitude of the eclipse. These things suggest either that Robertus had just been studying astronomy or that someone versed in astronomy had just joined the monastery.

Calculation with provisional accelerations shows that the duration was about 1 hour and 48 minutes, so the precise statement of the duration is not particularly accurate. Nonetheless, this is a contemporaneous record with a reliability of 1, the place is Mont S. Michel, the magnitude is 0.3 (since almost a third was eclipsed), and I shall take 0.18 for the standard deviation of the magnitude.‡ Given its time, place, and author, this is a remarkable record.

1185 May 1c E,F. Under the year 1185, Rigordus [ca. 1208, p. 15] writes: "In this year there was a partial eclipse of the sun on the first day of May, at nones, with the sun being in Taurus." The eclipse was maximum after midday but before midafternoon, so far as I can judge from Oppolzer's map, and thus I cannot tell the meaning of nones in this context. This is a record that seems to be contemporaneous and original, and it receives a reliability of 1; the place is S. Denis. When a record says that an eclipse is partial without giving an estimate of the magnitude, I conventionally take the magnitude to be 0.81, with a standard deviation of 0.14.

1187 September 4 E,F. Several pages after the preceding record, Rigordus [ca. 1208, p. 24] writes: "In the year of the Lord's incarnation 1187, on the 4th day of September, at the 3rd hour, was a partial eclipse of the sun in the 18th degree of Virgo, and it lasted for two hours." The stated longitude of the sun is 168°, which is correct when the value is rounded to the nearest degree. The record receives the same characteristics as the preceding one.

---

†This means the 24th part of a day, not an hour of the day. I presume that Robertus is using the term in its technically correct sense even though he makes a grammatical error. He has spatium unius horae aequaliter, but the last word should be aequalis. That is, he has accidentally used the adverb where he should have used the genitive singular of the adjective, as I interpret the remark.

‡I intended to use the round value 0.2, but by inadvertence I used 0.18 in the calculations. It is not worth the trouble to change to 0.2.

1191 June 23f E,F. Rigordus stayed busy recording
eclipses. On page 34, under the year 1191, he has: "This
year, on the 23rd of June, on the eve of S. John the Baptist,
while the king was at the siege of Acre,† there was an
eclipse of the sun in the 7th degree of Cancer, the moon
being in the 6th degree of the same sign, and the tail of
the Dragon in the 12th, and it lasted for four hours."

The "tail of the Dragon" means the descending node of
the moon's orbit in the astronomical terminology of the time.
This is one of the few European records of this time which
gives the position of the node at the time of an eclipse,
and the position given is accurate when rounded to the nearest degree. It would be interesting to know where Rigordus
got his astronomical data.

In his two preceding eclipse records, and in his only
remaining one (1207 February 28a E,F below), Rigordus specifically says that the eclipses were partial. He makes no
reference to the magnitude in this record, but I do not
think that we can safely infer a central eclipse from this
omission. The omission may be accidental rather than intended.

Since Rigordus does not say specifically that this
eclipse was partial, I shall use $\eta = 0$ for it, with a standard deviation of 0.18 for the magnitude. The place is S.
Denis and the reliability is unity.

1191 June 23g E,F. Under the year 1187, Iohannes
Longus [ca. 1365] says, with the aid of some unusual spelling: "Also the sun was eclipsed in an extremely large
eclipse on the 9th calends of July (June 23), from the 3rd
hour to the 9th." The same author says that the eclipse of
1239 June 3 (see the record 1239 June 3c E,F below) was also
extremely large, but it could not have been. Thus we must
ignore his description. Although this record is short, it
seems to be independent of other records of this eclipse,
and I shall give it a reliability of 0.5. The place is S.
Omer in northern France. I shall use $\eta = 0$, with 0.18 for
the standard deviation of the magnitude.

1207 February 28a E,F. This illustrates an inadequacy
of the method of designation that I have adopted. I decided
to use a small letter after the date to distinguish different records of the same eclipse from the same provenance,
but I used no letter in MCRE if I had found only one such
record. There is a record designated 1207 February 28 E,F
in MCRE [p. 351]. In order to have a consistent notation, I
should either change it to 1207 February 28a E,F or I should
drop the letter a from the designation for the first record

---

†The name used for the place in question is Accii in the
genitive case. Since this is the right time for the re-
conquest of Acre by the Crusaders, I assume that the place
is Acre, although I have not previously seen this spelling.

for all eclipses, and change all other letters accordingly. The probability of doing this without error is quite small and further, if I did it, I should then have different designations for the same eclipse record in different writings. I believe that it is better to continue without making any changes. At least each record will then have a unique designation.

This eclipse is the last one that Rigordus [ca. 1208] records, saying: "In the year 1206, on the day before the calends of March, there was a partial eclipse of the sun, at the 6th hour of the day, in the 16th degree of Pisces." This means that the longitude of the sun was 346°, which is correct. Rigordus has made an error of a year in the date, but this does not affect the reliability; it is possible that he did not begin the year in January. The reliability is 1, the place is S. Denis, and the magnitude will be taken as 0.81 with a standard deviation of 0.14.

1207 February 28b E,F. This source [Tolosanum, ca. 1271] gives the same year as the preceding one, but it is nonetheless independent. It says: "1206, on the day before the calends of March, on the 5th day of the week, there was an eclipse of the sun from the 3rd hour until nones, but not as terrible as the first one." Here nones seems to mean our noon. The week day was Wednesday rather than Thursday, but there is no question about the identification.

The statement that this eclipse was "not as terrible as the first one" poses an interesting problem. The editor says that the writer was remembering the eclipse of 1178 September 13. However, the record of 1178 September 13 in Tolosanum merely says that an eclipse occurred, with no hint that it was terrible. I think the writer must have read the record 1178 September 13g E,F above, which is also from Toulouse, and which does describe that eclipse as "exceedingly terrible and dark."

All this means, I believe, that the eclipse of 1207 February 28 was not central at Toulouse. Hence I shall take the magnitude to be 0.81, with a standard deviation of 0.14. The reliability is 1.

1207 February 28c E,F. Mettensis [ca. 1447] has under the year 1207: "An eclipse of the sun on the day before the calends of March, namely the 4th day of the week, about the 6th hour, with the moon then at 27." All this information is correct. The last remark means that it was the 27th day of the month in the ecclesiastical lunar calendar, as opposed to the Julian solar calendar. The record seems to be contemporaneous. The reliability is 1 and the place is Metz. I shall take the eclipse to be central, with 0.18 for the standard deviation of the magnitude.

1209 July 3 ? The entry under the year 1210 in S. Nicasii [ca. 1309] reads: "This year the church of Reims

was burned on the day of S. John before the Latin gate. On
the same day there was an eclipse of the sun about midday."
The day of S. John before the Latin gate is May 5. Even if
we assume that the writer meant "the same year" rather than
"the same day" the account is completely garbled. I can
find no eclipse visible in Reims near midday in any year
near 1210. The editor says that this refers to the eclipse
of 1209 July 3, but I believe that this is wrong. It is
doubtful that this eclipse was visible at all in Reims, but
if it were, it would have been near sunset.

1236 August 3 E,F. Albericus [ca. 1241] records this
eclipse on page 617 of the cited edition: "On the 3rd nones
of August, there was an eclipse of the sun about the 6th
hour of the day, on the Lord's day." He has already identified the year as 1236. The day stated is August 3, which
was a Sunday, and the eclipse should have been a maximum in
Trois-Fontaines at about the 6th hour of the day, which is
our noon. The reliability is 1, the place is Trois-Fontaines,
and the eclipse will be taken as central, with 0.18 for the
standard deviation of the magnitude.

1239 June 3a E,F. On page 766 of the cited edition,
Guillelmus de Podio-Laurentii [ca. 1272] writes: "There
was an eclipse of the sun in the year of the Lord 1239, on
the 3rd nones of June (June 3), on the 6th day of the week,
at the 6th hour. Also in the same year was a solar eclipse
on the day of S. Jacob and the sun was obscured beyond pale-
ness;† but not as much as in the one before; then it was
darkened so much that stars were seen." This refers to two
eclipses, and the first one is 1239 June 3, which was on a
Friday. I cannot identify the second one. The only day of
S. Jacob I find in the calendar is July 25, and there was
no eclipse on 1239 July 25. Even if we assume that July 25
is wrong, there was no later eclipse in 1239. After 1239
June 3, the next eclipse I can find that was visible in Toulouse was that of 1241 October 6, which should have been
moderately large. The next eclipse after that was 1245 July
25, which is indeed on the right day of the year. Calculation with provisional accelerations gives this eclipse a
magnitude of about 0.4 in Toulouse, which hardly seems to
fit the description. In spite of the agreement about the
day of the year with the 1245 eclipse, I conclude that the
second eclipse cannot be identified.

The eclipse of 1239 June 3 occurred during the adult
life of the author, at a time when he was in Toulouse. If
it were not for the mix-up about the second eclipse, we
would probably conclude that the records were of personal
observations, and that stars were seen at Toulouse during
the eclipse of 1239 June 3. However, the garbling of the
second record seems possible only if the writer were working
from sources which he misunderstood. For this reason, I

---

†The phrase is supra pallorem. I do not like this translation, but I think of no better one. I have supplied June
3 where it appears in parentheses.

shall assume that the record gives no clue to the magnitude on 1239 June 3, and I shall give the record a reliability of 0.5. I shall take the eclipse to be central, with 0.18 for the standard deviation of the magnitude.

1239 June 3b E,F. This record is also from Toulouse. Tolosanum [ca. 1271] has the following passage: "1229, on the 3rd nones of June, on the day of Venus (Friday), between the 6th and 9th (hours), was a great marvelous sign in the sky, namely an eclipse from the 6th to the 9th hour: but it was not such as had been predicted; because all the rotundity of the sun became as pale as mortal flesh, and in that rotundity was contained the sign of the lunar crescent red as if fire, and nearby was a very bright star, and afterward the moon and star before the eyes of those watching slowly receded from the sun and returned to their usual place, and then the sun covered itself with its own light and shone forth naturally as before."†

I suppose it is possible that the red crescent was a prominence, but it is more likely to be a small crescent of the sun in an almost total eclipse. If so, then the prediction was probably for a total eclipse which almost but not quite happened. The most likely interpretation of this record is, I believe, that the eclipse was very large but not total. I shall take the magnitude to be 0.995 with a standard deviation of 0.034. Although Tolosanum in its present form was not compiled until about 50 years later, this record seems completely reliable and it will receive a reliability of 1.

The compiler accidentally omitted an X when he wrote the date, making the year 1229. On this basis, the editor concluded that the correct date of the eclipse was 1228 December 28, thus ignoring all aspects of the record.

In speaking of this eclipse in the record 1239 June 3a E,F above, Guillelmus de Podio-Laurentii says that stars could be seen during it, but his own record of the eclipse does not say this. His source for the statement may be this record from Tolosanum. If so, he has exaggerated, because Tolosanum specifically refers to a single star.

1239 June 3c E,F. Iohannes Longus [ca. 1365] says of this eclipse on page 841: "This year on the 3rd nones of June (June 3) about midday there was an extremely large‡ eclipse and largest in parts of the kingdom of Navarre and around Pamplona, so much so that the day, which first shone brightly, in the hour of this eclipse became so dark that obscure night was seen." He then goes on to describe a miracle: Some pilgrims were at a river that was so full

---

†I have supplied the words in parentheses.

‡The term is permaxima, which is the term that the same writer used in the record 1191 June 23g E,F above. I have supplied the date in the form June 3.

they could not cross, but at the hour of the eclipse the
river ran backward and the pilgrims crossed dry-shod. However, Iohannes Longus then goes on to explain the astronomical cause of eclipses, so he is not completely superstitious.

It is probably safe to say that the writer is describing an eclipse at Pamplona that was total according to his
information. However, I do not think that it is safe to
use this aspect of the record, because Pamplona is far from
northern France and we do not know how the information reached
the writer; some exaggeration beyond the story of the pilgrims
might have been encountered along the way. We know that the
eclipse was not total in northern France, so that "extremely
large" from the pen of this writer does not mean much. Since
the eclipse is said to be greater in Pamplona than at S.
Omer, where the writer was, we may safely assume that he is
in fact describing a partial eclipse at S. Omer. Hence I
shall use 0.81 for the magnitude, with 0.14 for its standard
deviation. Since Iohannes Longus compiled his chronicle in
the next century, this record has a reliability of 0.5.

1239 June 3d E,F. S. Victoris [ca. 1542, p. 5] reads:
"1239 . . . nones of June, on the 6th weekday, on the 28th
of the moon, was an eclipse of the sun." The number of days
before the nones is missing, but the eclipse was on a Friday
and it was on the 28th day of the lunar month, according to
the ecclesiastical lunar calendar. Hence the date is accurately specified even though the day of the calendar month
is missing. The place is Marseilles, the reliability is 1,
$\eta = 0$, and the standard deviation of the magnitude is 0.18.

1239 June 3e E,F. This record is from Lemovicense [ca.
1342, volume XXI, p. 764]: "1239, on the 3rd of the nones
of June, on the day of Venus (Friday) after Pentecost, there
was an eclipse of the sun about the 6th, on the 26th of the
moon. And the sun was livid, making of itself a circle as
the moon does on its first or second night, and the sky was
deep blue† and the earth was deep blue, which blinded human
vision." I have supplied "Friday". "About the 6th" means
about the 6th hour of the day, I believe, which is about our
noon. Actually the day was the 28th rather than the 26th
day of the ecclesiastical moon. We have an attempt to estimate the magnitude of a partial eclipse by comparing the sun
with a new moon. I think it is reasonable from the description to take 0.94 for the magnitude with 0.05 for the standard deviation. The place is Limoges. Since this source was
written by several people, the record may be contemporaneous,
and I shall use 1 for the reliability.

1241 October 6a E,F. Colmarienses [ca. 1300] has the
following entry under 1241: "An eclipse of the sun happened
on the nones of October." The nones of October comes on

---

†This reminds us of the use of sapphire in the record 1033
June 29a E,F above.

October 7, so the date is wrong by one day. This is a brief notice of an eclipse that is recorded in many places and it could well be copied from somewhere else. However, the wording, though brief, differs from that of any other record I have found, and there are no other records at nearby places, so this record has a good chance of being original. I shall use it with a reliability of 0.5. The place is Colmar, the value of η is 0, and the standard deviation of the magnitude is 0.18.

1241 October 6b E,F. Albericus [ca. 1241] records this eclipse on page 630 of the cited edition, under the year 1241: "There was an eclipse of the sun on the Lord's day after the day of S. Remigius, and on the 5th day of the week following, it is said, the flight of a dragon was seen." S. Remigius is one of several saints whose day is October 1, and October 6 was the following Sunday in 1241. I wonder if the reference to a dragon was inspired by an aurora. This is almost the last entry in Albericus's chronicle, so it was certainly recorded in a timely fashion. The reliability is 1, the place is Trois-Fontaines, the value of η is 0, and the standard deviation of the magnitude is 0.18.

1290 September 5a E,F. There are two independent French records of this eclipse, both from Limoges. Godel [ca. 1320] has the following entry under the year 1290: "Also this year, on the day of Mars,† before the Nativity of the Blessed Mary, in the first hour, with the moon at the 27th, there was an eclipse of the sun, with the sun itself being in Virgo, in the time of Pope Nicholas IIII." All the calendrical information is correct. The Nativity of Mary is celebrated on September 8. The reliability is 1 and the eclipse will be taken as central, with 0.18 for the standard deviation of the magnitude.

1290 September 5b E,F. This is the other Limoges record, from the source Lemovicensia [ca. 1658], under the year 1290: "Also this year, others running as‡ upon the day of Mars before the Nativity of the Blessed Mary, before the first,‡ there was a partial eclipse of the sun." Although there are strong resemblances between the records, I believe that they result from similar thought patterns on the part of the ecclesiastical writers and not from copying. In fact, each record has information not found in the other, although the extra information found in Godel could have been calculated. I believe that both records are independent and contemporaneous, and both receive a reliability of 1. Since

---

†Tuesday.

‡The preceding words are aliis currentibus ut. I do not understand what they mean in this context.

‡The words used here are ante primam. They cannot mean before the first hour of the day, because that would mean before sunrise. They may mean before the end of the first hour, or, more likely, before the office of Prime Song.

this record states that the eclipse was partial, with no clues to the magnitude, I shall use 0.81 for the magnitude, with 0.14 for its standard deviation.

1310 January 31a E,F. <u>Guillelmus de Nangiaco</u> [ca. 1368, p. 600] has this record under the year 1309:† "On the last day of January, at 1 hour and 24 minutes <u>post meridiem</u>, was seen an eclipse of the sun located in its center, doubtless because the center of the moon was next to the center of the sun, and then was a conjunction of the sun and moon in the 20th degree of Aquarius;‡ from beginning to end the eclipse lasted more than two natural hours, and the air in the hour of the eclipse appeared a yellow or red color." He goes on to say that the colors are connected with the dominance of Jupiter at the time of the eclipse; I do not understand astrology well enough to know what he means.

This may be an attempt to describe an annular eclipse that was central as the writer saw it, and this eclipse was indeed annular. However, such an interpretation may require more precision than he intended with some of his words. Hence I shall interpret the record in a conventional manner, using η = 0 with 0.18 for the standard deviation of the magnitude. This record actually occurs in a contemporaneous continuation of Guillelmus's chronicle and the reliability is 1. The place is S. Denis, just north of Paris.

I shall not use the stated hour of the eclipse, but I shall test it for accuracy in Section XIV.7.

Two other records of this eclipse written at S. Denis are adapted from this one. The record in <u>Girardus de Fracheto</u> [ca. 1328] is clearly an abridgement of it. The record in <u>S. Denis</u> [ca. 1328] reads like a poor translation of it from Latin to French. I shall not use either of these two records.

1310 January 31b E,F. <u>Johannis à S. Victore</u> [ca. 1322], writing in Paris, says the following under the year 1309: "On the Sabbath day, on the day before the calends of February, about the meridian hour, in the 18th degree of Aquarius,‡ there was a partial eclipse of the sun, found out about and predicted many days before by certain Parisian clerks who are expert in the facility of astronomy; . ." The eclipse was on a Saturday. This is contemporaneous and receives a reliability of 1. I shall take the magnitude to be 0.81, with 0.14 for the standard deviation of the magnitude.

---

†I think that giving the year as 1309 is a matter of the convention about when the year began rather than an error.

‡This means a longitude of 320°. The longitude was 319°.508 according to Oppolzer's tables.

‡This is a longitude of 318°. Johannis's tables are not as accurate as those we have seen in other records.

TABLE VIII.2

FRENCH ECLIPSE RECORDS THAT ARE NEW OR THAT HAVE BEEN RE-INTERPRETED

Designation	Place	Standard deviation of the magnitude	Reliability
1033 Jun 29a E,F	Beze or Cluny	0.06[a]	1
1033 Jun 29d E,F	S. Amand-les-Eaux	0.054	1
1033 Jun 29f E,F	Auxerre	0.18	0.5
1037 Apr 18a E,F	S. Benoit-sur-Loire	0.05[b]	0.5
1037 Apr 18b E,F	Limoges	0.05[c]	0.5
1098 Dec 25 E,F	S. Evroul-en-Ouche	0.18	1
1153 Jan 26c E,F	Cambrai	0.18	1
1178 Sep 13c E,F	Angers	0.05[c]	1
1178 Sep 13e E,F	Vigeois	0.05[c]	1
1178 Sep 13f E,F	Mont S. Michel	0.05[c]	1
1178 Sep 13g E,F	Toulouse	0.068	0.5
1178 Sep 13h E,F	Limoges	0.18	0.5
1178 Sep 13i E,F	Limoges	0.06[b]	1
1180 Jan 28 E,F	Mont S. Michel	0.18	1
1181 Jul 13 E,F	Mont S. Michel	0.18[d]	1
1185 May 1c E,F	S. Denis	0.14[e]	1
1187 Sep 4 E,F	S. Denis	0.14[e]	1
1191 Jun 23f E,F	S. Denis	0.18	1
1191 Jun 23g E,F	S. Omer	0.18	0.5
1207 Feb 28a E,F	S. Denis	0.14[e]	1
1207 Feb 28b E,F	Toulouse	0.14[e]	1
1207 Feb 28c E,F	Metz	0.18	1
1236 Aug 3 E,F	Trois-Fontaines	0.18	1
1239 Jun 3a E,F	Toulouse	0.18	0.5
1239 Jun 3b E,F	Toulouse	0.034[f]	1

[a] The eclipse will be taken as partial, with a magnitude of 0.88.
[b] The magnitude will be taken as 0.94.
[c] The magnitude will be taken as 0.90.
[d] The magnitude will be taken as 0.30.
[e] The magnitude will be taken as 0.81.
[f] The magnitude will be taken as 0.995.

TABLE VIII.2 (Continued)

Designation	Place	Standard deviation of the magnitude	Reliability
1239 Jun 3c E,F	S. Omer	0.14^e	0.5
1239 Jun 3d E,F	Marseilles	0.18	1
1239 Jun 3e E,F	Limoges	0.05^b	1
1241 Oct 6a E,F	Colmar	0.18	0.5
1241 Oct 6b E,F	Trois-Fontaines	0.18	1
1290 Sep 5a E,F	Limoges	0.18	1
1290 Sep 5b E,F	Limoges	0.14^e	1
1310 Jan 31a E,F	S. Denis	0.18	1
1310 Jan 31b E,F	Paris	0.14^e	1
1310 Jan 31c E,F	Paris	0.14^e	1
1433 Jun 17 E,F	Marseilles	0.2^g	1

^{b}The magnitude will be taken as 0.94.

^{e}The magnitude will be taken as 0.81.

^{g}The magnitude will be taken as 0.5.

Note: The eclipse is taken to be central if no estimate of the magnitude is given.

1310 January 31c E,F. Guidonis [ca. 1327, p. 719] also puts this eclipse under 1309: "This year, on the last day of January, about midday was an eclipse of the sun, not enough however that the day was extraordinarily darkened, which had earlier been predicted by many." I do not know whether he means simply that the eclipse had been predicted, or whether he means that it had been predicted to be an extraordinarily dark one and that it wasn't. I think that this description is compatible with a magnitude of 0.9 or perhaps even greater. However, I shall take the magnitude to be 0.81, with a standard deviation of 0.14, in the conventional fashion. The reliability is 1 and I assume that the place is Paris.

1433 June 17 E,F. S. Victoris [ca. 1542, p. 7] has this record: "In the year from the Nativity of the Lord 1433, on the 17th day of the month of June, was a small eclipse of the sun in the city of Marseilles. According to some astrologers the said eclipse should have been a large one, but this was not so." I shall treat the magnitude for this record in the same way as for the Chinese records for -15 November 1 and -14 March 29 (Section V.3), in which the magnitude seems to be smaller than usual. That is, I shall use 0.5 for the magnitude, with 0.2 for its standard deviation. This record will contribute little to finding the accelerations, but this is no reason to neglect it totally. The reliability is

1 and the place, as the record says, is Marseilles.

The records that have been discussed in this section are summarized in Table VIII.2, which has the same format as Table VI.1.

GERMAN RECORDS OF SOLAR ECLIPSES

1. Records That Were Used in Earlier Work

Table IX.1 lists the German records of solar eclipses
that were used in MCRE or AAO and that will be used again
here with only the routine changes enumerated in Section
VII.1. A few records that were used before will be used
here with small but not routine changes in the way that the
magnitude is handled. Table IX.1 has the same format as the
earlier tables with the same purpose.

With regard to eclipses after 1300, the situation is
slightly better in Germany than it was in France or in the
British Isles. I have found 10 usable records from Germany,
compared with 9 from the British Isles and only 4 from
France. In finding these 10 records, I have found many new
records of eclipses from the 13th century.

2. Sources That Have Not Been Used Before

Altahenses [ca. 1073]. These annals are from a monastery whose medieval name was something like Altaha; I have
not found an exact spelling that I can carry over into English. It was at the place that is now called Nieder-Alteich,†
on the Danube just below its junction with the Isar. Nieder-
Alteich does not appear in the Times Atlas, but it does appear on a large scale road map that I have, one that fortu-
nately gives latitude and longitude lines. I find that its
latitude is 48°.76 and that its longitude is 13°.04. M
[p. A2.2] gives 48°.77 and 13°.03, in close agreement.

The first part of the annals runs from 708 to 1032, and
it is attributed to Wolfherius of Hildesheim. This part is
obviously derivative and I did not use it. The second part,
which is contemporaneous with events or nearly so, is written
by an unknown monk of Nieder-Alteich.

Altahenses [ca. 1585]. Properly speaking, this is not
a medieval source. It is rather a collection of notes that
Philip Jaffé, around 1860, extracted from various medieval
documents which originated in the monastery at Nieder-Alteich.,
I gather that the notes are found in scattered sources, none
of them particularly extensive, that contain contemporaneous
records of various events. We are fortunate that Jaffé considered eclipses to be within the proper interest of an

---

†The last half of this name is sometimes spelled Altaich,
but Alteich is the spelling on the map I have. There was
also a monastery at Ober-Alteich, but it is much younger
and it is not the one that concerns us at the moment. See
the discussion of Windbergenses [ca. 1407] later in this
section.

TABLE IX.1

GERMAN ECLIPSE RECORDS THAT WERE USED BEFORE

Designation	Place	Standard deviation of the magnitude	Reliability
764 Jun 4 E,G	Lorsch	0.18	0.5
787 Sep 16a E,G	Lorsch	0.18	1
787 Sep 16b E,G	Quedlinburg	0.18	0.5
807 Feb 11 E,G	Lorsch	0.18	1
810 Nov 30 E,G	Lorsch	0.18	1
812 May 14a E,G	Lorsch	0.18	0.5
812 May 14b E,G	Fulda	0.18	0.5
818 Jul 7a E,G	Lorsch	0.18	0.5
818 Jul 7b E,G	Fulda	0.18	0.5
840 May 5a E,G	Reichenau	0.18	0.2
840 May 5c E,G	Fulda	0.054	1
840 May 5e E,G	Ingolstadt	0.034	0.5
878 Oct 29a E,G	Reichenau	0.18	0.5
878 Oct 29c E,G	Fulda	0.034	1
878 Oct 29d E,G	Prum	0.18	1
939 Jul 19 E,G	Corvei	0.14[a]	1
961 May 17a E,G	Trier	0.18	1
961 May 17b E,G	Trier	0.18	1
968 Dec 22 E,G	Reichenau	0.18	0.5
990 Oct 21a E,G	Quedlinburg	0.18	1
990 Oct 21b E,G	Hildesheim	0.18	1
1033 Jun 29a E,G	Hildesheim	0.18	1
1033 Jun 29b E,G	Reichenau	0.18	1
1033 Jun 29c E,G	Stuttgart	0.18	1
1039 Aug 22a E,G	Stuttgart	0.18	0.2

[a]The eclipse will be taken as partial, with a magnitude of 0.81.

TABLE IX.1 (Continued)

Designation	Place	Standard deviation of the magnitude	Reliability
1039 Aug 22b E,G	Fulda	0.18	1
1044 Nov 22a E,G	Fulda	0.18	1
1044 Nov 22b E,G	Regensburg	0.18	1
1093 Sep 23a E,G	Hildesheim	0.18	1
1093 Sep 23b E,G	Augsburg	0.18	1
1093 Sep 23c E,G	Zwiefalten	0.18	0.5
1093 Sep 23d E,G	Zwiefalten	0.18	1
1093 Sep 23e E,G	Bamberg	0.18	1
1098 Dec 25a E,G	Augsburg	0.18	1
1124 Aug 11a E,G	Paderborn	0.18	1
1124 Aug 11b E,G	Harsefeld	0.18	1
1133 Aug 2b E,G	Paderborn	0.054	1
1133 Aug 2c E,G	Eichstätt	0.18	0.5
1133 Aug 2d E,G	Zwiefalten	0.18	0.5
1133 Aug 2e E,G	Herzogenrath	0.034	1
1133 Aug 2f E,G	Bamberg	0.18	0.5
1133 Aug 2g E,G	Corvei	0.054	1
1133 Aug 2h E,G	Disibodenberg	0.054	0.5
1133 Aug 2i E,G	Heilsbronn	0.18	0.5
1133 Aug 2j E,G	Heilsbronn	0.034	1
1133 Aug 2k E,G	Würzburg	0.034	1
1133 Aug 2ℓ E,G	Pegau	0.18	0.5
1133 Aug 2m E,G	Brauweiler	0.054	0.5
1140 Mar 20 E,G	S. Blasien	0.18	1
1147 Oct 26b E,G	Magdeburg	0.05[b]	1
1147 Oct 26c E,G	Aachen	0.18	1
1147 Oct 26d E,G	Brauweiler	0.18	1
1147 Oct 26e E,G	Cologne	0.18	1
1147 Oct 26f E,G	S. Blasien	0.18	0.5
1153 Jan 26a E,G	Zwiefalten	0.18	1

[b]The magnitude will be taken as 0.9.

TABLE IX.1 (Continued)

Designation	Place	Standard deviation of the magnitude	Reliability
1153 Jan 26b E,G	Freising	0.18	1
1153 Jan 26d E,G	Cologne	0.18	1
1187 Sep 4a E,G	Augsburg	0.18	1
1187 Sep 4b E,G	Munich	0.18	1
1187 Sep 4c E,G	Pohlde	0.18	1
1187 Sep 4d E,G	Pegau	0.18	1
1187 Sep 4e E,G	Cologne	0.18	1
1187 Sep 4f E,G	Scheftlarn	0.18	1
1191 Jun 23a E,G	Zwiefalten	0.18	0.5
1191 Jun 23b E,G	Munich	0.11	1
1191 Jun 23c E,G	Freising	0.18	0.5
1191 Jun 23d E,G	Cologne	0.18	1
1191 Jun 23e E,G	S. Trudperti	0.14^c	0.5
1207 Feb 28a E,G	Zwiefalten	0.18	1
1207 Feb 28b E,G	Freising	0.18	1
1207 Feb 28c E,G	Cologne	0.18	1
1207 Feb 28d E,G	Scheftlarn	0.18	1
1230 May 14 E,G	Cologne	0.14^c	1
1239 Jun 3a E,G	Neresheim	0.14^c	1
1239 Jun 3b E,G	Freising	0.18	1
1241 Oct 6a E,G	Neresheim	0.068	1
1241 Oct 6b E,G	Augsburg	0.18	1
1241 Oct 6c E,G	Corvei	0.034	1
1241 Oct 6d E,G	Freising	0.054	1
1241 Oct 6e E,G	Heilsbronn	0.18	1
1241 Oct 6f E,G	Stade	0.034	1
1241 Oct 6g E,G	Weltenburg	0.034	1
1241 Oct 6h E,G	Scheftlarn	0.034	0.5
1241 Oct 6i E,G	Scheftlarn	0.034	1
1241 Oct 6ℓ E,G	Pohlde	0.18	1

[c]The eclipse will be taken as partial, with a magnitude of 0.81.

TABLE IX.1 (Continued)

Designation	Place	Standard deviation of the magnitude	Reliability
1245 Jul 25 E,G	Stade	0.18	1
1261 Apr 1 E,G	Freising	0.18	0.5
1263 Aug 5 E,G	Freising	0.18	1
1267 May 25 E,G	Scheftlarn	0.18	1

Note: The eclipse is taken to be central if no estimate of the magnitude is given.

historian; he included records of four eclipses in his collection.

Burchardus and Cuonradus [ca. 1229]. Burchardus was the abbot of the monastery of Ursperg, if that is its correct modern spelling, from 1215 to 1226, and Cuonradus was his successor. I forgot to note the year of Cuonradus's demise. Burchardus wrote the first part of this chronicle which, after a brief discussion of ancient times, really begins in 1126. He brought it up to 1222 and Cuonradus carried it to its close.

The editors of the cited edition say of the location of Ursperg only that it is about midway between Ulm and Augsburg. Luckily, Ginzel [1884] gives its position, using the spelling Ursberg. Its latitude is 48° 16' and its longitude is 10° 32'. This is about midway between Ulm and Augsburg.

Chounradus [ca. 1226]. Chounradus was the abbot of the monastery of Scheirn from 1206 to 1225. He wrote several books which he probably did not name himself. This one is named Annales in the cited edition, but I think this name was given by the editor rather than by Chounradus. Chounradus actually records the election of his successor, which is somewhat unusual; the date was 1225 February 14. I gather that he resigned because of old age, and his annals were probably terminated by his death.

Scheirn is described as a monastery in the diocese of Freising, but I cannot find it on a large map. Thus, if I were going to use any eclipse records from this source, I would use the coordinates of Freising. However, for reasons that will appear in the next section, I shall not use any of Chounradus's records.

Coloniensis [ca. 1220]. The editor of this source says that it is a continuation of the Annales Colonienses Maximi, which is the source that I designated as Colonienses [ca. 1238] in MCRE [p. 356]. I do not understand his reason for saying this, since Coloniensis stops in 1220 while Colonienses goes on to 1238. Both sources are from Cologne.

Diessenses [ca. 1432]. This source is the type of
document called a necrology kept in the monastery of Diessen.
A necrology is a document that lists the dates of the deaths
of personages who had either a general interest or a particular interest to the people where the necrology was prepared.
A necrology often has notes of other events as well, and
this one records two eclipses of the sun. The events listed
run from 1122 to 1432, and the notes seem to be contemporaneous with the events.

Diessen is at the south end of the Ammersee near Munich.
The monastery there was founded about 1100.

Elwacense [ca. 1477]. This source is closely connected
with the source Neresheimenses [ca. 1296]† that was used in
MCRE [p. 370]. There is a set of annals called Annales
Elwangenses which was prepared in the monastery at Ellwangen.
I referred to the source in passing in MCRE [p. 371], but I
do not cite it in full either here or there because it has no
information that is useful to us. Neresheimenses in its
earlier parts relies heavily on the Annales Elwangenses. Thus
it is fitting that a continuation of Neresheimenses was in
turn kept at Ellwangen. Elwacense is this continuation. It
is interesting that the two sources from Ellwangen use quite
different spellings even though both were edited by the same
person. I do not know whether the sources themselves had
different spellings or whether the editor adopted two spellings for some reason.

Elwacense seems to become independent of Neresheimenses
in the second half of the 13th century. However, it cannot
be an original source until near its end, at least in its
present form, because it is written in a single hand of the
15th century.

Ensdorfenses [ca. 1368]. A monk named Heimo in the
monastery of S. Michael at Bamberg wrote a chronicle of the
world [Heimo, 1135; see MCRE, p. 361] from the Creation until
his own time. Heimo's work was fairly popular, and it was
copied and continued in several places. A copy made in Bamberg says that Heimo died on 1139 July 31, but this copy
contains no records of eclipses. The continuation that
interests us here was made at Ensdorf, which is fairly close
to Bamberg. It continues for more than two centuries after
the death of Heimo.

---

†This source illustrates the difficulty that I mentioned in
Section VII.2. Together with its direct continuation, it
runs to 1572, making it one of the longest medieval sources
that we have with regard to the time covered. However, it
records no solar eclipses after 1241 October 6.

Hamburgenses [ca. 1265]. This set of annals is found
in a manuscript in the library at Hamburg. It contains an
early portion that is derived from other sources. Its original part is a set of annals written in a contemporaneous
hand that runs only from 1251 to 1265.

Herbipolenses [ca. 1400]. The editor calls these annals
the Annales Herbipolenses Minores in order to distinguish
them from a more extensive set that I used in MCRE [p. 362].
I distinguish them instead by using a different date in the
citation. It is hard to make out the reason for the choice
of material found in this source. It contains a few entries,
which are mostly derivative, for a few years between 688 and
1245. Then it contains entries that seem to be original for
the isolated years 1250, 1266, 1241, and 1400. These annals
were written in Würzburg.

Hermannus [1273]. Hermannus was the abbot of the monastery at Nieder-Alteich (see Altahenses [ca. 1073] above)
for more than thirty years until he resigned in 1273 because
of old age. He wrote these annals while he was in office.
They run from 1137 to 1273, and they seem to be mostly original from about 1239 on.

Lubicenses [ca. 1324]. This is a set of annals by one
person which cover the period from 1264 to 1324 rather extensively. Thus they are probably not original except in
their later portions. They are from the Hanseatic seaport
of Lübeck in Germany.

Moguntini [ca. 1309]. This is a set of annals compiled
by one person that runs altogether from 1083 to 1309. Until
about 1285 the entries are sparse and short and we may assume that they were copied. Even from 1285 to the close in
1309 there is not an entry for every year. Since the ancient
and medieval name of Mainz was Mogontiacum or Maguntiacum, I
assume that Mainz is the place for these annals.

Ottenburani [ca. 1298]. The first monastery of Ottenburanus was founded in 764 in honor of S. Alexander. A new
one was founded in 1121, and this seemed to furnish the occasion for the beginning of these annals. The editor
divides them into three sections. The first part runs from
1121 to 1168. The second part contains only four entries
for years between 1145 and 1157; apparently someone inserted
items that had been omitted which he thought worthy of mention. Then someone else decided to start again at 1145.
This third part runs regularly from 1145 to 1298, after
which it has only isolated entries for 1353, 1371, and 1416.
Unfortunately these are not eclipses, and in fact the only
eclipse recorded is that of 1241 October 6.

The monastery is described as being at the 10th mile-
stone east of Memmingen in Bavaria. We do not know exactly
what mile was used, but we can make at most a trivial error

if we use the English mile.

Ratisbonenses [ca. 1201]. These are from the city of
Regensburg. I referred to these annals on page 370 of MCRE
and said that they contained nothing useful in the study of
eclipses; I do not know how I came to make this error. These
annals are found in two codices which are, or at least were
in 1861, found in Vienna and Munich. These codices are not
mere copies since they often differ considerably. The annals
begin with the birth of Jesus and go to 1201, at least in
their present form. From about 900 on, they seem to be an
independent authority.

S. Blasii [ca. 1173]. There were at least two religious institutions in Germany that were dedicated to S.
Blasius or Blasien. One was a monastery in the Black Forest
and I used some annals from it in MCRE [p. 373]. This one
is described as coming from S. Blasius in Brunswick, but
this does not tell us whether Brunswick refers to the town
or to the entire duchy. Since the cathedral in the town is
dedicated to S. Blasius, we may presume that the annals in
fact come from the town. However, the annals are not connected directly with the cathedral, because they stop in
1173 when the construction of the cathedral began. Apparently they are records that were kept in the forerunner of
the present cathedral. Even so, we may not assume that they
are local to Brunswick.

S. Blasii [ca. 1314]. This is another set of annals
from the same place. It is a short set containing about 40
entries dating from 905 to 1314. It seems to have been compiled by one person, using criteria of selection that are
not obvious, and we should be prepared to find that the
entries are not independent of other known sources.

S. Blasii [ca. 1482]. These are also from the same
place, but the editor describes them as notes rather than
annals. This set of notes starts with a list of relics
contained in the cathedral, then it records events in six
different years from 1312 to 1482, and it ends with a list
of dukes of Brunswick.

S. Nicolai [ca. 1287]. These annals are from the church
or monastery of S. Nicola in a suburb of Passau. They are
highly intermittent, with about 20 entries between 1067 and
1287 made by unknown people. A deacon named Wolfelmus added
a few personal notes about his own career in the years 1265,
1266, 1270, and 1275.

S. Petri Erfordensis [ca. 1353]. The town of Erfurt
is an old one that, according to tradition, was founded in
the 6th century. It produced many historical writings in
the medieval period, some of which were used in MCRE [Section
XI.1]. This is a major work which starts with the birth of

Christ. At the year 1070 someone in the 14th century wrote: "Note that the modern chronicle begins from this point." Except for a few excerpts, the editor of the cited edition has printed only the "modern" part because the earlier part is derived from known sources. From about 1070 on, so far as I can judge, this is an independent authority.

The main part of the work runs to 1335 and exists in many copies. One of them was kept at the Minorite convent of S. Elizabeth at Eisenach and contains information that is not in the others. It is reasonable to conclude that this information constitutes a continuation that was made at Eisenach.

Saxonici [ca. 1271]. At one point, I planned not to use the Annales Saxonici because they cannot be localized more closely than the region of Saxony. I finally decided to use them because they record the eclipse of 1267 May 25, something that few other sources do. I shall take Saxony to be the rectangle that extends from $50°.5$ to $51°.25$ in latitude and from $12°.5$ to $14°.5$ in longitude.

The main part of this source consists of lists of patriarchs, emperors, and other eminent persons; these lists are not published in the cited edition. Along with these lists are intermittent annals running from 1186 to 1271. Annals of this sort are usually made by various people contemporaneously with the events. These annals, however, are all written in a single hand and thus they cannot be original in their entirety. However, most of the eclipse records seem to be independent of other known sources.

Saxonicum [ca. 1139]. Parts of this are published in volumes VI, VII, VIII, X, and XIII of the series often designated as Bouquet; the exact designation is given in the references. The only part that concerns us appears in volume X, but the discussion of the source is in the preface to volume VI. Like Saxonici [ca. 1271], this source cannot be localized more closely than Saxony, with exceptions that will be mentioned in a moment.

The preface to volume VI of the Bouquet series, which was published in 1744, gives the following information: The original of the chronicle is in the library of S. Germain-des-Prés in Paris; this accounts for the fact that it appears in a French collection. Around 1700 a copy was made and sent, to Leibnitz after he had begun his work as an historian, who promised that he would not publish it.† After Leibnitz died, George Eccard did not feel bound by this promise and he published it in a collection called Corpus Historicum, published in Leipzig in 1723. Apparently Bouquet felt that Eccard's publication did not give him true priority because of the

---

†He was asked to make this promise because a French scholar was preparing an edition of it. I do not know whether this was Bouquet or whether some earlier French scholar was working with it. The date sounds too early for Bouquet.

earlier promise, and thus he began a new edition.

The chronicle runs from 741 to 1139 and has several authors. When this happens, we usually find that the earlier parts are derivative introductions to original later parts. Here we have almost the opposite situation. The part of the work that deals with the latter part of the 10th century is sometimes original, but both earlier and later parts are copied, usually from known sources.

Sifridus [ca. 1307]. The title of this work, in translation, is A Universal History and Compendium of History, and it is written by Sifridus, who describes himself as a priest of Balnhusin. This is the place now called Gross-Ballhausen, which is close to Weimar; I shall use the coordinates of Weimar in the calculations. Most of the work is necessarily second-hand, but some of the material after about 1230 seems to be independent of other known sources.

Thuringici [ca. 1291]. Thuringia denotes a region of Germany that lies just to the northwest of Bohemia. We can approximate it reasonably well by the rectangle that stretches from longitude $10°.5E$ to $12°.5E$, between latitudes $50°.0$ and $50°.5$. These annals from Thuringia are highly intermittent. They contain an entry for 1099, and then they have 19 entries for years between 1187 and 1291. They are printed on a single page of the cited edition.

Treverorum [ca. 1242]. Some annals that were published in volume VIII of the Monumenta Germaniae Historica, Scriptores series were apparently quite popular and had many continuations.† Several continuations were made at the monastery in Trier, and the editor presents them in the same place. The one that interests us is the one that he identifies as Continuation IV, which runs only from 1212 to 1242.

Veterocellenses [ca. 1484]. These are from a Cistercian monastery at the place that is sometimes called Cella S. Mariae and sometimes called Cella Veteris. It is at Freiberg, about 30 kilometers from Dresden. The monastery, as the annals tell us, was founded in 1175 and the annals, although they contain earlier copied portions, form an independent authority from 1175 on. The annals continue to 1484 with an entry for about every third year on the average.

Windbergenses [ca. 1407]. This set of annals is from a monastery in the diocese of Regensburg, and northeast of Straubing. Ober-Alteich‡ fits this description and I find

---

†These annals contain nothing relevant to this work and consequently I do not cite them.

‡See the discussion of Altahenses [ca. 1073] at the beginning of this section.

no other ecclesiastical establishment that does. One copy
of the annals is found in the library of Munich, written in
several hands of the 14th century. It seems likely to me
that this is the original. The editor says that several
entries have been copied from Hermannus [1273], which comes
from Nieder-Alteich.

Zwetlenses 285 [ca. 1281]. The library of the monastery
at Zwettl, Austria contained a great number of medieval documents which the editor distinguished by using the codex
number; I followed his practice in MCRE and shall do the same
here. The source involved here is a minor one which the
editor entitles Very Brief Annals from Zwettl. It contains
only six entries in Latin for years from 1239 to 1281, with
an isolated entry in mixed Latin and German for 1461; I have
ignored this entry in assigning a date. The last entry says
that the king came to Regensburg on 1281 June 7, so this
entry must have been written in that city. It is unlikely
that such a minor source would have been started in Zwettl
and taken to Regensburg for continuation. It is more likely
that the entire set of entries was made in Regensburg and
that the document was taken to Zwettl for some reason, probably in company with other documents. I saw nothing in any
entry which contradicts this idea, and Regensburg is close
enough to Zwettl to make such an action plausible. Hence I
shall use Regensburg as the place.

3.   Records from Germany

     As in the preceding chapters, this section will contain
records that have not been used before, or records that have
been used before but whose use here will be based upon a
different interpretation. Most changes involve the magnitude
or its standard deviation.

     810 July 5 E,G. I quoted a record of this eclipse in
MCRE [pp. 395-396] but did not use it because the eclipse
was not visible [MCRE, pp. 595ff] at the assigned place,
which was Lorsch. It seems to me that I should use such a
record in spite of this fact, because invisibility in one
sense merely means an unusually large error in η. However,
it is not a simple programming matter to make my computing
programs yield a value for η when an eclipse is not visible,
so I shall not use this record here either.

     840 May 5b E,G. I do not actually change the interpre-
tation that I used in MCRE [p. 398] but, as I said in Section
II.1 above, I used wrong coordinates for Weingarten there.

     840 May 5c E,G. This record is from Fulda and I used
it in MCRE [p. 398]. The record says that stars could be
seen during the eclipse and that the color of things was
changed. I speculated in MCRE that the eclipse was not
total, because colors cannot be seen during a total eclipse,
I believe. However, it is not safe to rely upon the interpretation to this extent, because the writer could have used

the word in a figurative sense. Hence I shall continue to
use η = 0, with 0.054 for the standard deviation of the
magnitude.

840 May 5d E,G. This record is from the town of Xanten.
It says [MCRE, p. 399] that stars were seen clearly in the
sky just as at night. Because of the reference to night, I
took the standard deviation of the magnitude to be 0, but
this is probably not safe. Hence I shall change the standard
deviation to 0.054. We know now that this eclipse was actually far from total in Xanten and hence that this record may
have been brought from somewhere else. If I were dealing
with a subject that depended but little upon the subjective
interpretation of the data, I would feel justified in omitting this record from the analysis. Here, however, I am too
afraid of being trapped in circular reasoning, and I shall
retain the record. I prefer to retain such records and to
rely upon the force of the statistics of unbiased errors to
reduce the errors to a harmless level. See Section XIV.7
for further discussion.

878 October 29b E,G. This is another record from Weingarten for which I used wrong coordinates in MCRE [p. 400].

990 October 21c E,G. This is the earliest record from
Germany that has not been used before. Saxonicum [ca. 1139]
has two records of this eclipse, both on page 228, volume X
of the cited edition. The first record appears under the
year 989 and it has been copied, with trivial changes in
wording, from the record of the same eclipse in Hildesheimenses [ca. 1137].†

The other record appears under the correct year: "There
was a solar eclipse at the 5th hour (of the day).‡ Duke
Carol died and his son Otto succeeded him." The hour of the
day is accurate. It is possible that this record is adapted
from some other source, but it comes from a part of the
source that has some original material and it does not seem
to be derived from any other known source. I shall use it,
but with a reliability of only 0.5. The place is Saxony, the
value of η is 0, and the standard deviation of the magnitude
is 0.18.

---

†This is the record 990 October 21b E,G on page 403 of MCRE.
There are two reasons for concluding that Hildesheimenses
is the original and that Saxonicum is the copy. One reason
comes from our general knowledge of these sources based
upon evidence other than the eclipse records; this evidence
suggests that the relevant part of Hildesheimenses is original
but that much of Saxonicum is copied. The other is that
Hildesheimenses has the year right while Saxonicum has it
wrong. It is easy for a copier to put an event under the
wrong year but it is unlikely for him to correct an error
in his original.
‡My parenthesis.

As I remarked in the preceding section, Saxonicum has the odd property that some of its middle part is original but that all or at least almost all of its early and late parts are copied. This property is illustrated by its eclipse records. In addition to the eclipse of 990 October 21, it records the eclipses of 1009 March 29, 1093 September 23, 1124 August 11, and 1133 August 2, but all these records are copied from sources that were used in MCRE.

1018 April 18, E,G. This record [MCRE, p. 404] says that the sun appeared to be halved. Thus in MCRE I took the magnitude to be 0.5, with a standard deviation of 0.1. I feel that this standard deviation is too small for such a statement and shall change it to 0.2.

1023 January 24 ? Altahenses [ca. 1073] writes the following under the year 1023: "Three eclipses of the moon in one year, an eclipse of the sun after the Nativity of the Lord at the 10th hour of the day." If the writer began his year on January 1, 1023 was one of the rare years with three lunar eclipses, on January 9, July 5, and December 29. However, the solar eclipse cannot be identified reliably. There was a solar eclipse on 1023 December 15, and we might think the scribe accidentally wrote "after" when he meant "before" Christmas except for the fact that the eclipse was visible only in the southern hemisphere. It seems unlikely to me that a writer would describe 1023 January 24 as being after Christmas if he began his year on January 1, as the lunar eclipses indicate that he did. Further, this eclipse should have been greatest at about noon, rather than after the middle of the afternoon. Thus the solar eclipse cannot be identified safely.

1033 June 29. Ratisbonenses [ca. 1201] has a brief notice of the eclipse of 1033 June 29 which could be original but which also could have been abbreviated from many other sources. It is not safe to use this record.

1039 August 22c E,G. Altahenses [ca. 1073] has a brief note under the year 1039 which reads: "An eclipse of the sun occurred on the 11th calends of September." The date equals August 22. This record, though brief, does not read like any other known record of this eclipse, and this part of the source is believed to be an original authority. Hence I accept this record, which seems to be contemporaneous. The reliability is 1, the place is Nieder-Alteich, the value of η is 0, and the standard deviation of the magnitude is 0.18.

1044 November 22a E,G. This record was used in MCRE [p. 407]. I do not intend to change its interpretation, but I wish to comment on a point that I overlooked there. The record, in the cited edition, says that the eclipse lasted "from the first hour to 9," so that it lasted 8 hours of the day. I noted that 8 hours of the day on November 22 amounted to only about 6 ordinary hours, but that this still seemed unreasonably long. I should have realized that the original almost surely had <u>nones</u> rather than the numeral 9, and that <u>nones</u> here probably meant our noon. Hence the eclipse, according to the record, lasted only 5 hours of the day, which is somewhat less than 4 ordinary hours. This is more reasonable, especially when we consider that neither hour was intended to be a precise measurement.

1044 November 22c E,G. This is a new record of this eclipse, from <u>Altahenses</u> [ca. 1073]. The source says, under the year 1044: "An eclipse of the moon on the 4th nones of November and of the sun on the 10th calends of December." The first date equals November 2, but the lunar eclipse was on November 8; the writer should have said the 6th ides of November. The second date equals November 22, which is correct. The record is contemporaneous or nearly so and receives a reliability of 1. The place is Nieder-Alteich, and I shall take the eclipse to be central, with 0.18 for the standard deviation of the magnitude.

1133 August 2m E,G. This record is from <u>Brunwilarenses</u> [ca. 1179] and it was used in <u>MCRE</u> [p. 417]. It says that stars appeared during the eclipse. By accident I used 0.1 for the standard deviation when I meant to use 0.01 because of the reference to the stars. Here I shall change 0.01 to the value 0.054 from Table IV.4.

1133 August 2n E,G. We have already noted that <u>Ratisbonenses</u> [ca. 1201] has a record of the eclipse of 1033 June 29 that could have been abbreviated from any one of numerous sources. It also has a record of the eclipse of 1093 September 23 that is certainly copied, with trivial changes, from <u>Frutolf</u> [1103]. It is thus a pleasure to find that its last eclipse record is original. Under the year 1133 the recorder has written: "This year on the 4th nones August† a solar eclipse took place which was a prodigious miracle. For the moon on its 27th in the sign of Cancer was found under the sun in the sign of Leo, when according to calculation the moon should not have reached it until the 30th." That is, according to the ecclesiastical tables, the moon on August 2 should have been at only its 27th day after the new moon and it should still have been in the sign of Cancer. However, on this day, it passed under the sun which was in the sign of Leo (its longitude was in the 15th degree of that sign), although according to calculation this should not have happened until the 30th day of the moon, on August 5. It is unfortunate that the writer, in his excitement that the moon was not following its ecclesiastical schedule, did not give us any details about the eclipse itself. The reliability is 1, the place is Regensburg, the value of η is 0, and the standard deviation of the magnitude is 0.18.

---

†This is August 2.

Chounradus [ca. 1226] has copied this record with the
omission of an "X" from the year, thus converting it into
1123. Burchardus and Cuonradus [ca. 1229] and S. Blasii
[ca. 1173] both have short notices of the eclipse that cannot be used safely, even though we cannot identify the specific sources from which they took the information.

1147 October 26g E,G. S. Petri Erfordensis [ca. 1353]
records this eclipse under the year 1147 on page 366 of the
cited edition: "An eclipse of the sun happened on the 7th
calends of November at the _____ hour of the day, and in the
same year on the same day . . ." The text then goes on to
describe an event that happened during the Crusade. The date
given is October 26. Where I have left a blank, the editor
has supplied (6th) in parentheses on the basis of other annals. He does not say whether there was an actual blank in
the original or whether it reads "at the hour of the day".

The main trouble in understanding records of this
eclipse is that it was also observed in Asia Minor where it
played a role in the history of the Second Crusade.† Thus
we must know whether an observation appearing in a European
source represents a local observation or whether it appears
there as part of the story of the Crusade. Since the writer
at S. Petrus Erfordensis refers to the Crusade in the same
sentence, this strengthens the suspicion that the observation
comes from Asia rather than from Europe. If we knew the hour
of the observation, we might be able to tell the location
from it, but we do not. For safety, I shall not use this
record.

1153 January 26c E,G. This record is also from Erfurt
and I used it in MCRE [p. 422]. The record includes a
sketch of the eclipse which clearly shows a narrow crescent
of the sun. One should not take the size of the sun that
appears in the sketch as a scale drawing, of course, but it
nonetheless suggests a magnitude of about 0.9. Hence I used
this estimate of the magnitude, with a standard deviation of
0.05, in MCRE.

M [p. 8.55], however, says that the eclipse, which was
an annular one, was central at Erfurt. In order to account
for the fact that all the drawings‡ show a partial eclipse,
M suggests an overcast which attenuated the light and made
the solar disk easily visible until a short time before
maximum eclipse. Then the overcast changed and prevented a
clear sight of the central phase of the eclipse.

I do not know how M calculated the circumstances of the
eclipse. He says [p. 10.16] that only the record of the

---

†See MCRE [p. 556].

‡There is in fact only one drawing. M has apparently been
misled by the fact that several other sources have copied
the record from the original source, including the drawing
in question. See the discussion in MCRE.

eclipse of 1133 August 2 made at Klosterrath† fails to fit
his solution for the accelerations, out of the records he
used. I have tested this statement only for the important
record 1567 April 9 E,I to be given in Section X.3 below.
It is a record on which M̲ puts considerable stress, and it
has already been discussed some in Section III.7 above. It
fails to fit his solution by a large amount, although it fits
my solution excellently. Thus I shall ignore M̲'s statement
about the eclipse of 1153 January 26 at Erfurt. However, I
shall not use the magnitude characteristics that I used in
MCRE. Instead, I shall take this to be a record which says
that an eclipse was partial without indicating the magnitude.
That is, I shall take the magnitude to be 0.81, with a standard deviation of 0.14. I shall study the magnitude of this
eclipse further in Section XIV.8.

1178 September 13 E,G. Burchardus and Cuonradus [ca.
1229] record this eclipse under the year 1180: "On the ides
of September at the hour of nones the sun suffered an eclipse."
The only record I have found which gives the hour this way is
the record 1178 September 13d E,F [MCRE, p. 349], which comes
from Dijon. It is not likely that Burchardus, who wrote this
portion of the chronicle, copied a source from Dijon, so I
shall take this record to be independent of any others known.
However, it cannot be original because Burchardus wrote almost 50 years after the event. Hence I use a reliability of
0.5. The place is Ursperg, the value of η is 0, and the
standard deviation of the magnitude is 0.18.

The editor says that the date is 1180 September 5, but
this is neither the date stated in the record, which is September 13, nor is it the date of an eclipse. Perhaps he
misread the date as the nones of September, which is September 5. However, nones clearly refers to the hour, and in
this case it must mean the hour that we call noon.

1187 September 4g E,G. S. Petri Erfordensis [ca. 1353]
records an eclipse under the year 1187 in this way: "The
sun was obscured, in fact in the way that the moon is on its
24th, with a clear sky, on the 2nd nones of September at the
sixth hour of the day." On the day of the eclipse, the ecclesiastical moon was at its 28th day, so the writer is not
giving the calculated position of the moon. The editor says
in a footnote that this means a quarter moon. If he means
the quarter moon in the sense of being midway between full
and new, this is not correct. However, about a fourth of
the moon is visible on the 24th day of a lunar month,‡ and
I shall assume that this is what the writer means. Hence I
shall take the magnitude to be 0.75 with a standard deviation
of 0.125. We believe that events in this part of the chronicle were being recorded contemporaneously, so the reliability is 1. The place is Erfurt.

---

†This is the record 1133 August 2e E,G on pages 412-413 of
MCRE.
‡That is, on the 24th day of a lunar month which begins with
a visible crescent.

Chounradus [ca. 1226] has a brief record of this eclipse that gives no details and that could have been abridged from any one of many sources. Hence I shall not use it. The same remarks apply to the record in Burchardus and Cuonradus [ca. 1229], except that the latter source does state the hour.

1191 June 23f E,G. This record is from Wormatienses [ca. 1295] and I used it in MCRE [p. 427]. The record describes the eclipse as total (generalis), and I used 0 for the standard deviation of the magnitude in MCRE. Here, for reasons that will be discussed in connection with the record 1321 June 26c E,G later in this section, I shall use 0.054.

1191 June 23g E,G. S. Petri Erfordensis [ca. 1353] writes: "An eclipse of the sun occurred on the eve of S. John the Baptist at the sixth hour." The eve of the feast of S. John is indeed June 23, so the date is correct, and the hour is accurate if it applies to the middle of the eclipse. This record differs from the one in Burchardus and Cuonradus that was just mentioned only in the fact that the latter omit "eve" and thus put the eclipse on June 24. Aside from the matter of the date, there is another reason for using the record from S. Petri Erfordensis but not the one from Burchardus and Cuonradus. We know that the latter was not compiled until several centuries after the eclipse but we believe that the former consists of records that were made at the time of the events described.

Hence I shall use this record with a reliability of 1. The place is Erfurt, the value of η is 0, and the standard deviation of the magnitude is 0.18.

Thuringici [ca. 1291] records this eclipse in the following way: "There was an eclipse of the sun on the eve of John the Baptist." This record is so much like other records that it is not safe to use.

1207 February 28c E,G. The source that I used for this record in MCRE [p. 428] was Colonienses [ca. 1238]. Since then, I have discovered Coloniensis [ca. 1220], which contains the same record under the year 1207. Since Coloniensis is a slightly earlier source, it is probably the original and Colonienses is the copy. The question does not concern us here, because the characteristics of the record, including the place, are the same in either case.

1207 February 28e E,G. Burchardus and Cuonradus [ca. 1229] report this eclipse as follows: "The year of our Lord 1208. There was an eclipse of the sun in the month of February, at the hour of nones." The circumstances of the eclipse suggest that nones means our noon. Since we are now in Burchardus's own time, this record receives a reliability of 1. The place is Ursperg, the value of η is 0, and the standard deviation of the magnitude is 0.18.

1232 October 15, E,G. This record is also from Colonienses [ca. 1238] and it appears in MCRE [p. 429]. In this
case, the record specifically says that the eclipse was not
very notable. Hence I shall use the characteristics that I
introduced in Section V.3 for records that imply a small
eclipse. That is, I take the magnitude to be 0.5 with a
standard deviation of 0.2.

1241 October 6k E,G. I used this record in MCRE [p. 434]
and I shall use it here with a slight change. However, the
comments that M [p. 8.64] makes about it require extensive
discussion. The record is from Wormatienses [ca. 1295] and
the translation in MCRE reads: "1241, there was a total
eclipse of the sun." The word that I have translated as
"total" is generalis, and I leave the word untranslated in
the discussion that immediately follows.

M writes the following about the translation of generalis:
"This latin (sic) word is not used to describe a known total
eclipse in any record of our† acquaintance, and is probably
better translated as 'general' meaning 'over a wide area'."

I do not know why M did not refer to the record 1191
June 23f E,G on page 427 of MCRE. This record is also from
Wormatienses and it also says that the eclipse was generalis.‡
Further, I discussed there the possibility that generalis has
the meaning which M proposes and concluded that it does not.
The matter seemed so obvious that it did not seem necessary
to give many details, but M's comment makes it clear that
some more details are needed.

The adjectives generalis, universalis, and particularis
are used in dozens of medieval Latin records of solar eclipses.
It is obvious from the usage that generalis and universalis
are synonyms and that they are antonyms to particularis. For
example, in the record 1232 October 15 E,G that was just mentioned, Colonienses [ca. 1238] says that the eclipse was
particularis . . non tamen multum notabilis. I see no plausible translation for this except "partial . . and not very
notable, however." Gervase [ca. 1199], writing in Kent, uses
either universalis or particularis about each of the 5
eclipses that he records during his lifetime, and he uses
both of them in describing [p. 277] the eclipse of 1178 September 13. In the first part of his lengthy description, he
says that the eclipse in Kent was not universaliter but
particulariter.‡ This establishes that the words are used
with opposite meanings. Toward the end of his description,
Gervase then says: "Elsewhere, on the contrary, the eclipse
was universalis, so that there was dark night in the middle

---

†I think he is using "our" to refer to Stephenson and himself. He says that he has had a private communication from
Stephenson about this matter.

‡However MS [p. 481] do point out this fact.

‡ These are the adverbial forms of the obvious adjectives.
The sentence structure that Gervase uses here requires the
adverbs, but this does not affect the basic meaning.

of the day, and a person could not see his neighbor who was standing next to him."† This clearly refers to a total eclipse.

These usages establish the meanings of universalis and particularis, and it only remains to establish the meaning of generalis. This is perhaps easiest done by using the records 1241 October 6m E,G and 1290 September 5 E,G from underline{S}. Petri Erfordensis [ca. 1353], which will appear below. This source uses either particularis or generalis in describing many of the eclipses that it records, and it is clear that the words are used as contrasting adjectives. Since particularis means partial, generalis means total.‡ Q. E. D.

However, I shall show in connection with the record 1321 June 26c E,G below that annalists did not necessarily use the adjectives universalis and generalis with the rigor that a trained astronomer would exhibit. Hence I shall use 0.054 rather than 0.034 for the standard deviation of the magnitude when one of these adjectives is the only indicator of the magnitude.

1241 October 6m E,G. I mentioned this record in the preceding discussion. On page 394 of the cited edition, underline{S}. Petri Erfordensis [ca. 1353] says: "MCCXLI. on the day before the nones of October there was a total (generalis) eclipse of the sun at the 11th hour." Since October 7 is the nones, the date given is October 6. The 11th hour looks somewhat late unless it is intended to apply to the time when the eclipse was over. The reliability is 1, the place is Erfurt, and the value of η is 0, with 0.054 for the standard deviation of the magnitude.

1241 October 6n E,G. Ensdorfenses [ca. 1368] describes this eclipse in the following way under the year 1241: "There was an eclipse of the sun on the 2nd nones of October‡ about the hour of nones, and the sun was darkened, and there was darkness over the whole earth, and the stars appeared clearly." I believe that this is intended to be a description of a total eclipse, and hence I shall use 0.034 for the standard deviation of the magnitude, in accordance with Table IV.2. The place is Ensdorf and the reliability is 1.

1241 October 6o E,G. Windbergenses [ca. 1407] records this eclipse as follows: "In the year 1241 from the Incarnation of the Lord, the sun died at midday and stars were seen on the day before the nones of October." The picturesque statement about the sun sounds equivalent to a statement of

---

† Gervase's description forms the record 1178 September 13b B,E on page 168 of MCRE.

± We should also allow the possibility that generalis can mean a central but annular eclipse.

‡ That is, on October 6.

totality, and I shall take it to be so. Hence I take the
standard deviation of the magnitude to be 0.034. The place
is Ober-Alteich. In order to decide upon the reliability,
we go back to the editor's description given in the preceding
section. One copy of the annals is written in various hands
of the 14th century. It does not seem likely that a text
with this property would be a copy, so I assume that it is
the original. If so, the record is not contemporaneous, and
it receives a reliability of only 0.5.

1241 October 6p E,G. This record is from Altahenses
[ca. 1585]: "1241. On the second nones of October, on the
octave of S. Michael,† on the Lord's day, in the middle of
the day there was an eclipse of the sun and there was darkness, with stars appearing in the sky; the sun however was
very clear before and after the darkness." This is not quite
strong enough to let us assume that the eclipse was total,
but neither does it deny totality. Hence I shall use 0 for
the value of η, with 0.054 for the standard deviation of the
magnitude because of the reference to stars. The place is
Nieder-Alteich and the reliability is 1.

Hermannus [1273] reports this eclipse in these words:
"There was an eclipse of the sun in the middle of the day on
the octave of S. Michael, and stars were seen." Hermannus
was the abbot of Nieder-Alteich and he probably did not begin
to compile his annals until some years after 1241. In other
words, his report must have been taken from some older source
at Nieder-Alteich, and he may well have written an abbreviated
version of the record just given. Hence I shall not use the
report of Hermannus.

Elwacense [ca. 1477] reports this eclipse this way:
"There was a very great eclipse of the sun on the day before
the nones of October, and stars were seen, and very great
darkness at the time of midday." We remember that this source
is a continuation of Neresheimenses [ca. 1296] and the editor
of Elwacense says in a note that the record in Elwacense is
copied from Neresheimenses. This is almost but not quite correct, and the difference between the records poses an interesting problem. The records are indeed the same except for
the fact that Elwacense, the presumed copy, has information
that Neresheimenses‡ does not have. This information is contained in the clause "and stars were seen" which does not
appear in Neresheimenses.

Usually we would conclude that the record with additional
information is the original and that the abbreviated text is
the copy. Here, however, important textual evidence tells us
that Neresheimenses is the original in spite of this consideration. I have thought of two possible explanations of this

---

†This means the 8th day after the day dedicated to him, in
inclusive counting. October 6 is indeed the octave of S.
Michael. It was also a Sunday.

‡The record in Neresheimenses is the record 1241 October 6a
E,G on page 430 of MCRE.

situation, and the correct explanation may well be neither of these.

The first, and more likely, is that the compiler of El-wacense copied his report from two sources rather than one. In doing so, he took most of the record from Neresheimenses but took the remark about the stars from some unknown source.

The second explanation is based upon the fact that the manuscript of Neresheimenses has vanished,† so that the cited edition is based only upon an earlier printed edition that I have not had the opportunity to see. Thus it is possible that the clause in question actually appeared in the manu-script of Neresheimenses and that it was omitted by accident from the first printed edition.

1241 October 6q E,G. This record is from Diessenses [ca. 1432]: "In the year 1241 of the Incarnation of the Lord, on the 2nd nones of October, a solar eclipse was seen at the hour of nones." This could have been taken from any of several sources, so far as we can judge from the wording, but our knowledge of the text indicates that this is an original entry. The reliability is 1, the place is Diessen, the value of η is 0, and the standard deviation of the magni-tude is 0.18.

1241 October 6r E,G. Ottenburani [ca. 1298] records this eclipse in the following way: "On the 2nd nones of October there was a solar eclipse on the octave of Blessed Michael in the year 1241 of the Incarnation of the Lord in the period between midday and vespers."‡ This is in a con-tinuation of the original annals, which ran only to 1168, and this usually indicates an original source. The date is accurate and the time of day is reasonable. The record re-ceives a reliability of 1, the place is 10 miles east of Memmingen, the value of η is 0, and the standard deviation of the magnitude is 0.18.

1241 October 6s E,G. In Moguntini [ca. 1309], under the year 1241, we find: "In the same year, Sergus and Bacchus, the sun suffered an eclipse." I assume that Sergus and Bac-chus are saints whose day was October 6, but they are not listed on my church calendar. Since this is a late compila-tion, the record receives a reliability of 0.5. The place is Mainz, the value of η is 0, and the standard deviation of the magnitude is 0.18.

Saxonici [ca. 1271] has a brief notice that there was a solar eclipse on the octave of S. Michael (hence on October 6) in the year 1240. Since this is a late compilation, and since the record could have been extracted from any one of

---

†See MCRE [pp. 370-371].

‡The phrase used is in meditullio circa meridiem et vesperam. My translation is as free as the original.

many sources, it is not safe to use this record. S. Blasii [ca. 1314] has a record that is almost identical with the one in Saxonici, and it should not be used either. Neither should the record from Thuringici [ca. 1291], which reads simply: "Item 1241. There was an eclipse of the sun on the octave of the day of Michael."

1241 October 6t E,G. Veterocellenses [ca. 1484] has the following passage under the year 1241: "On the day before the nones of October there was an eclipse of the sun after midday, and stars appeared." The date is accurate and this has been an original source since about 1175. The reliability is 1, the place is Freiberg, to the southwest of Dresden, the value of η is 0, and the standard deviation of the magnitude is 0.054 because of the reference to the stars.

1241 October 6u E,G. Herbipolenses [ca. 1400] has this passage: "In the year of the Lord 1241, on the day before the nones of October, on the Lord's day, at midday, there was such an eclipse of the sun that stars appeared." Although all the information in this account appears in other places, it does not seem to be a copy of any of them. Since the source has some original material by this time, I shall give this record a reliability of 1. The place is Würzburg, the value of η is 0, and the standard deviation of the magnitude is 0.054.

1241 October 6v E,G. We find this in Treverorum [ca. 1242] under the year 1241: "In this year there was a great and dark eclipse of the sun in the octave of S. Michael, after midday on the Lord's day, signifying great trials for the wonderful church of God." Although the eclipse is described as great and dark, there are no specific indicators of a large or total eclipse. Hence I shall use η = 0 with 0.10 for the standard deviation of the magnitude. The reliability is 1 and the place is Trier.

1241 October 6w E,G. Sifridus [ca. 1307] has a record of this eclipse under the year 1241 that reads: "This year there was a total (generalis) eclipse of the sun on the octave of S. Michael." Since Sifridus did not write this until 50 years or more after the eclipse, this cannot be original, so the question is whether it is independent of other known sources. The report, though short, does not read exactly like any other report I have noticed, but Sifridus could have written an abbreviated paraphrase of other sources. In fact, the record 1241 October 6m E,G above used the word generalis, and that record is from Erfurt, only a few kilometers from Weimar where Sifridus wrote. On the other hand, Sifridus gives the date in a quite different fashion, and, I believe, in a fashion that is not obvious to one who is simply copying material. This creates a reasonable presumption that Sifridus was copying an independent record that had been written at the time. I shall compromise by using a reliability of 0.5. The place is Weimar, and I shall take η to be 0, with 0.054 for the standard deviation of the magnitude.

1241 October 6x E,G. On page 61 of the cited edition, S. Nicolai [ca. 1287] has an original passage which reads: "Year of Grace 1241, a little after midday, when the sun was in its brightness, it was suddenly covered by a great darkness such that no part of it could be seen and stars were seen just as at night for almost four hours." I think that the time interval is meant to apply to the total duration of the eclipse and not just to the duration of totality. This passage seems to be an explicit statement of a total eclipse, so I shall use 0.034 for the standard deviation of the magnitude. The reliability is 1 and the place is Passau.

1241 October 6y E,G. This is the last record of this eclipse from Germany that I shall use, at least in this work. It is possible to have one more record before resorting to double letters in order to designate individual records. This record is from Zwetlenses 285 [ca. 1281] but, for the reasons that I gave in the discussion of this source in the preceding section, I shall take the place to be Regensburg rather than Zwettl. The record reads: "In the year of the Lord 1239, on the octave of S. Michael, there was an eclipse of the sun at midday such that stars appeared." In spite of the error in the year, the identification is certain. The reliability is 1, the value of η is 0, and the standard deviation of the magnitude is 0.054 because of the reference to the stars.

I mentioned in MCRE [p. 435] that Neresheimenses [ca. 1296, p. 24] records an eclipse on 1249 August 16. Elwacense [ca. 1477] has copied this record. It is possible that this refers to the eclipse of 1254 August 14, but this is unlikely since that eclipse was probably not visible in Germany.

1263 August 5a E,G. There is also a record of this eclipse designated as 1263 August 5 E,G on page 435 of MCRE. Saxonici [ca. 1271] records the eclipse under the year 1262: "An eclipse of the sun on the day of the blessed Dominic, which was then on the Lord's day, and it lasted until vespers." The writer has the day of the month wrong as well as the year, for S. Dominic's day is August 4. However, the day of the eclipse was indeed on a Sunday, so he at least has the correct day of the week. Since this record is almost at the end of the annals, it should be contemporaneous and thus it receives a reliability of 1. The place is Saxony, as it was defined in the discussion of this source in the preceding section. The value of η is 0 and the standard deviation of the magnitude is 0.18.

1263 August 5b E,G. Veterocellenses [ca. 1484] says under the year 1263 that there was a solar eclipse on the nones of August (August 5) after midday. Although this is a brief account, it seems to be original and receives a reliability of 1. The place is Freiberg, the value of η is 0, and the standard deviation of the magnitude is 0.18.

1263 August 5c E,G. Hamburgenses [ca. 1265] writes
under the year 1263: "There was an eclipse of the sun on
the Lord's day and on the day of King Oswald." August 5 is
indeed the day of the king S. Oswald, and it was on Sunday
in 1263. The reliability is 1 since this is in the original
part of the annals, the place is Hamburg, the value of η is
0, and the standard deviation of the magnitude is 0.18.

   Thuringici [ca. 1291] writes: "In the year 1263 there
was a solar eclipse." This brief record could have been
taken from any of several places and it is not safe to use.

   1267 May 25a E,G. There is a record of this eclipse
designated 1267 May 25 E,G on page 436 of MCRE. This record
is from Saxonici [ca. 1271] under the year 1268, so that the
annalist has the year wrong in every one of his records of
solar eclipses: "An eclipse of the sun in the 8th degree of
Gemini on the day of Urban." S. Urban's day is May 25, so
the day of the year is correct. The sun was actually in the
11th degree of Gemini, but it was in the 8th degree according to the ecclesiastical tables of the sun. Hence the solar
position given is merely another way of stating the date. The
reliability is 1, the place is Saxony, the value of η is 0,
and the standard deviation of the magnitude is 0.18.

   1270 March 23 E,G. Lubicenses [ca. 1324] has a brief
note about this eclipse under the year 1270: "This year
there was an eclipse of the sun in dominica Laetare." I
think the underlined phrase refers to the ceremony that is
sometimes called Laetare Sunday, which comes on the fourth
Sunday in Lent and which was on March 23 in 1270. If so,
this source uses the same method of designating the day as
the record 1270 March 23 E,Sc on page 490 of MCRE, which
comes from Rye, Germany, Further, Rye and Lübeck are rather
close together, so that copying is a geographical possibility.
However, I know of no other evidence that Lubicenses used the
Rye annals as a source, and the exact words are in fact rather
different. Hence I shall take this as an independent record,
but not as an original one because it occurs in a late compilation. The reliability is 0.5, the place is Lübeck, the
value of η is 0, and the standard deviation of the magnitude
is 0.18.

   1290 September 5, E,G. S. Petri Erfordensis [ca. 1353]
devotes $2\frac{1}{2}$ pages in the printed edition to the year 1290, including an interesting record of the eclipse of 1290 September 5 on page 423: "Also in the same year on the 5th day of
the month of September about the third hour there was a partial eclipse of the sun, which a certain astrologer who came
to the king with the Marquis of Brandenburg had predicted a
long time ago, and he wrote in a certain storeroom† that it

---

†The word is apoteca. The only possibility I have thought
of is that this is a variant spelling of apotheca, which
means a storeroom, but this obviously does not make sense.

would take place in such a month on such a day and at such an hour; and it all happened as he had predicted." It is odd that this eclipse is so poorly reported (only four records of it are known), since its path in Europe is almost identical with that of 1133 August 2; perhaps the weather was mostly cloudy. The reliability is 1 and the place is Erfurt. I shall give this record the conventional characteristics for a partial eclipse. That is, I shall take 0.81 for the magnitude, with a standard deviation of 0.14.

1310 January 31a E,G. S. Petri Erfordensis [ca. 1353] has only a prosaic report of this eclipse, under the correct year: "In the same year there was a partial† eclipse of the sun on the eve of S. Brigid." Since S. Brigid's day is February 1, the dating is correct. The reliability is 1, the place is Erfurt, the magnitude is 0.81, and the standard deviation is 0.14.

1310 January 31b E,G. I mentioned in MCRE [p. 436] that three out of the last five entries in Augustani Minores [ca. 1321] record solar eclipses; this is the first of those three: "1310. A mighty (valida) eclipse of the sun on the last day of January which is the eve of S. Brigid, about the hour of vespers, on the Sabbath day." Sabbath is used here in its Hebrew sense. The reliability is 1 and the place is Augsburg. Since the size of the eclipse is emphasized, without either affirming or denying totality, I shall use η = 0, with 0.10 for the standard deviation of the magnitude.

1312 July 5 E,G. This is the second of the last three eclipses in Augustani Minores: "1312. A partial eclipse of the sun on the 6th day of July, which is the octave of the apostles Peter and Paul, before the lunch hour, on the 5th day of the week." The annalist was wrong but consistent about the date. The octave of S. Peter and Paul is July 6, which was on a Thursday that year. The lunch (prandium) hour was probably at our noon, but Oppolzer's map suggests that the hour of maximum eclipse was about 9:00 AM. The reliability is 1, the place is Augsburg, the magnitude is 0.81 since the record says specifically that the eclipse was partial, and the standard deviation is 0.14.

1321 June 26a E,G. This is the last eclipse record in Augustani Minores: "1321. A total (universalis) eclipse of the sun and darkness over the face of the earth through all Germany in June on the day of John and Paul, at the time of high mass, on the 6th day of the week." The week day is correct, and my calendar lists June 26 for the day of John and Paul. Since the annalist specifically says that the eclipse was seen over a wide area (that is, all Germany), he

---

†The word used both here and in the preceding record is particularis.

is probably not using <u>universalis</u> to repeat the fact. Instead, he is almost surely using it to mean that the eclipse
was total, or at least central.† The problem is whether we
can assume that the annalist used the term accurately. As
we shall see, three German records of this eclipse describe
it as being either <u>universalis</u> or <u>generalis</u>, there are good
reasons for believing that at least two of them are independent, and it is not possible for the eclipse to have been
central at all three places. Either the annalists did not
use the words accurately or they used information brought
from somewhere else. I shall assign the characteristics of
this record after the discussion of the next one.

1321 June 26b E,G. <u>Elwacense</u> [ca. 1477] records this
eclipse in the following words: "1321. This year on the
day of the holy martyrs John and Paul there was a total
(<u>universalis</u>) eclipse of the sun through all Germany, lasting
from the first hour to the third." Since this record has
information that is not contained in the preceding one, I
should have taken it to be independent except for the problem
of the word <u>universalis</u>, which alerted my attention to the
question of independence. Since Ellwangen and Augsburg are
only about 50 kilometers apart, most eclipses could have been
central in both places. However, the eclipse of 1321 June 26
was annular-total, meaning that it was annular to some observers and total to others. The central zone of such eclipses
is always very narrow and it is doubtful if it could have been
central in both places, especially since the line joining them
is almost perpendicular to the eclipse path.

<u>Augustani Minores</u> has the characteristics that we expect
of a set of annals that were noted as the events happen. <u>El-
wacense</u>, on the other hand, is written in a single hand and
it was written about a century after this eclipse. The most
plausible explanation of the situation seems to be that the
annalist at Ellwangen used two sources. One said that the
eclipse was <u>universalis</u> through all Germany and the other
said that it lasted from the first to the third hour. Therefore I shall give a reliability of 0 to the record from <u>El-
wacense</u> and a reliability of 1 to the record from Augsburg.
I shall use $\eta = 0$ for this record, with 0.054 for the standard deviation of the magnitude. The reason for this choice
of the standard deviation will appear in connection with the
next record.

1321 June 26c E,G. On page 826 of the cited edition,
<u>S. Blasii</u> [ca. 1482] writes: "In the year of our Lord 1322,
on the day of S. John and Paul at the first hour, there was
a total (<u>generalis</u>) eclipse of the sun, and the grain per-
ished that year and a great dearth followed the next year,
such that a <u>chorus</u>‡ of wheat sold for a mark‡ of pure silver,

---

†See the record 1241 October 6k E,G above.

‡I do not know what units are implied by these terms. The
mark has changed its value many times since this was written.

and the weather† was very rainy for the entire year." We
remember from the discussion of this source in the preceding
section that it comes from the cathedral of S. Blasius in
Brunswick rather than from the monastery of S. Blasius in
the Black Forest. Brunswick is so far from Augsburg that
there is no possibility that the eclipse was central in both
places.

There seem to be only two possible explanations of this
situation: (a) The records from S. Blasii and Augustani
Minores were based upon information obtained from elsewhere
or (b) The words generalis and universalis were not used with
enough rigor to qualify them as technical terms. Both sources
appear to be collections of notes and records that were written as the events happened, as I have already remarked about
Augustani Minores, and they do not seem to be based upon the
copying of records that originated in other places. If so,
the information is local unless it was brought by a traveler.
Thus we may imagine that the recorder at Augsburg, for example, first noted the occurrence of the eclipse and then
modified it to include the word universalis because a visitor
told him that he had seen it as a total eclipse.

If this is what happened, we may still say that a medieval annalist did not necessarily use universalis or generalis
to mean strictly total (or possibly central, for an annular
eclipse) at the place where he was writing. If we say this,
we automatically include the second possibility mentioned
above. Hence I shall assume that these words, when they are
used by a person who is not known to have had astronomical
training, mean a very large eclipse but not necessarily one
that is central. I shall use 0.054 for the standard deviation
of the magnitude‡ in such cases, with 0 for the value of η,
if there are no indications of totality except the use of
generalis or universalis. For the record from S. Blasii,
the reliability is 1 and the place is Brunswick. I shall return to these records in Section XIV.8.

1321 June 26d E,G. This record, from S. Petri Erfordensis
[ca. 1353]. is the only German record of this eclipse I have
found that does not describe the eclipse as total. It is interesting that it comes from a place midway between Brunswick
and Augsburg. After identifying the year as 1321, the record
says: "This year there was an eclipse of the sun at the first
hour on the day of S. John and Paul." The main body of these
annals comes from Erfurt. However, according to the editor,
this record occurs only in a copy that was kept at the neigh-
boring convent of S. Elizabeth outside the walls of Eisenach,
so it is reasonable to assume that this record in fact originated at Eisenach. The reliability is 1, the value of η is
0, and the standard deviation of the magnitude is 0.18.

---

†Since the original uses the word aura, I should probably
use a term that is more figurative than weather.

‡This is the value used for records which say that stars
became visible.

1406 June 16 E,G. Note that it has been 85 years since the preceding record. Altahenses [ca. 1585] records this eclipse in the following way: "1406. There was a solar eclipse on the day of S. Vitus martyr at the 3rd hour." June 15 is the day of S. Vitus, so the annalist has made an error of a day, but the hour is reasonable. The reliability is 1, the place is Nieder-Alteich, the value of η is 0, and the standard deviation of the magnitude is 0.18.

1415 June 7a E,G. Altahenses [ca. 1585] records this eclipse in the following picturesque manner: "1415. There was a dark eclipse of the sun, such that the sun completely lost its light for about two 'Miserere-mei', at the 6th hour of the day, on the 7th day of June." The eclipse occurred in the early morning at Nieder-Alteich. Thus it seems that the time was measured in the same kind of hours that we now use, measured from midnight. This is reasonable, since mechanical clocks were fairly common by this time, which is many decades later than the first pendulum clock with an escapement.[†] "Miserere mei" are the first two words of the 51st Psalm. M [p. 8.66] suggests that the frightened annalist managed to get through a recitation of the 51st Psalm "twice before totality ended!" (The exclamation point is M's.) He also says that the time required to do this is close to the computed duration of totality, which is 220 seconds. However, I had no trouble in reciting the psalm twice in 140 seconds, so the duration does not seem to be particularly accurate, and there is no reason why we should expect it to be.

Since the annalist seems to be stating that the eclipse was total, I shall use 0.034 for the standard deviation of the magnitude. The reliability is 1 and the place is Nieder-Alteich.

1415 June 7b E,G. Diessenses [ca. 1432] provides the only other German record of this eclipse: "In the year 1415 of the Incarnation of the Lord, on the 7th ides ..., an eclipse of the sun was seen at the 7th hour of the day." The month is missing, as the dots indicate; I do not know whether there is a blank space, an illegible word, or whether the rest of the record follows immediately after the word ides. In Diessen as well as in Nieder-Alteich, it seems that the hours were now measured from midnight. The reliability is 1, the place is Diessen, the value of η is 0, and the standard deviation of the magnitude is 0.18.

1544 January 24 E,G. This is the last medieval eclipse record I have found from Germany, and it is also from Altahenses [ca. 1585], as two of the preceding three records have

---

[†]This kind of clock, often called a foliot clock, had a torsion pendulum rather than a gravity pendulum. The moment of inertia of the pendulum was supplied by a transverse bar with two weights hanging from it. As the pendulum oscillated, the weights appeared to dance madly. The word foliot appears to be related to the Old French fol (mad).

TABLE IX.2

GERMAN ECLIPSE RECORDS THAT ARE NEW
OR THAT HAVE BEEN RE-INTERPRETED

Designation	Place	Standard deviation of the magnitude	Reliability
840 May 5b E,G	Weingarten	0.18	0.2
840 May 5d E,G	Xanten	0.054	1
878 Oct 29b E,G	Weingarten	0.18	0.5
990 Oct 21c E,G	Saxony	0.18	0.5
1018 Apr 18 E,G	Merseburg	0.2^c	1
1039 Aug 22c E,G	Nieder-Alteich	0.18	1
1044 Nov 22c E,G	Nieder-Alteich	0.18	1
1133 Aug 2n E,G	Regensburg	0.18	1
1153 Jan 26c E,G	Erfurt	0.14^a	1
1178 Sep 13 E,G	Ursperg	0.18	0.5
1187 Sep 4g E,G	Erfurt	0.125^b	1
1191 Jun 23f E,G	Worms	0.054	0.5
1191 Jun 23g E,G	Erfurt	0.18	1
1207 Feb 28e E,G	Ursperg	0.18	1
1232 Oct 15 E,G	Cologne	0.2^c	1
1241 Oct 6k E,G	Worms	0.054	0.5
1241 Oct 6m E,G	Erfurt	0.054	1
1241 Oct 6n E,G	Ensdorf	0.034	1
1241 Oct 6o E,G	Ober-Alteich	0.034	0.5
1241 Oct 6p E,G	Nieder-Alteich	0.054	1
1241 Oct 6q E,G	Diessen	0.18	1
1241 Oct 6r E,G	Memmingen^d	0.18	1
1241 Oct 6s E,G	Mainz	0.18	0.5
1241 Oct 6t E,G	Freiberg	0.054	1
1241 Oct 6u E,G	Würzburg	0.054	1

[a]The eclipse will be taken as partial, with a magnitude of 0.81.

[b]The magnitude will be taken as 0.75.

[c]The magnitude will be taken as 0.5.

[d]The place is actually 10 miles (16 kilometers) east of Memmingen.

TABLE IX.2 (Continued)

Designation	Place	Standard deviation of the magnitude	Reliability
1241 Oct 6v E,G	Trier	0.10	1
1241 Oct 6w E,G	Weimar	0.054	0.5
1241 Oct 6x E,G	Passau	0.034	1
1241 Oct 6y E,G	Regensburg	0.054	1
1263 Aug 5a E,G	Saxony	0.18	1
1263 Aug 5b E,G	Freiberg	0.18	1
1263 Aug 5c E,G	Hamburg	0.18	1
1267 May 25a E,G	Saxony	0.18	1
1270 Mar 23 E,G	Lübeck	0.18	0.5
1290 Sep 5 E,G	Erfurt	0.14[a]	1
1310 Jan 31a E,G	Erfurt	0.14[a]	1
1310 Jan 31b E,G	Augsburg	0.10	1
1312 Jul 5 E,G	Augsburg	0.14[a]	1
1321 Jun 26a E,G	Augsburg	0.054	1
1321 Jun 26c E,G	Brunswick	0.054	1
1321 Jun 26d E,G	Eisenach	0.18	1
1406 Jun 16 E,G	Nieder-Alteich	0.18	1
1415 Jun 7a E,G	Nieder-Alteich	0.034	1
1415 Jun 7b E,G	Diessen	0.18	1
1544 Jan 24 E,G	Nieder-Alteich	0.034	1

[a]The eclipse will be taken as partial, with a magnitude of 0.81.
Note: The eclipse is taken to be central if no estimate of the magnitude is given.

been: "It must be known that in the year of the Lord 1544, on the festival of S. Timothy Apostle, there was an eclipse before lunch at the hour of nones, and there was darkness over almost all the earth just if it were completely night, such that one person could not rightly see another. There has not been such an eclipse in the memory of men. May God be favorable toward us." Nones in this case seems to mean our noon; however the word prandium that I have translated as lunch† may by now have meant the main meal. Some sources give January 24 for the day of S. Timothy while others give

---

†This word is also used sometimes to mean a late breakfast.

April 7, so we may take it that the dating of the eclipse is accurate.

It is interesting that $\underline{M}$ uses the record of 1415 June 7 from Altahenses but that he does not mention this record, even though both give clear descriptions of total eclipses. The reliability is 1, the place is Nieder-Alteich, and the standard deviation of the magnitude is 0.034, since the annalist seems to imply that the eclipse was total.

The German records that have not been used before, or that are used here with the characteristics changed from earlier uses, are summarized in Table IX.2.

CHAPTER X

RECORDS OF SOLAR ECLIPSES FROM ITALY

1. Records That Were Used in Earlier Work

Table X.1 lists the Italian records of solar eclipses
that were used in MCRE or AAO and that will be used here
without change. In addition, there is one record of a partial eclipse for which the magnitude will be handled in a
different way. Table X.1 has the same format as the corresponding tables in earlier chapters.

The term "Italy" in Table X.1 means the region that I
called "most of Italy" in MCRE, but I shall use slightly different boundaries for it here for convenience. In MCRE it
was convenient to take it to be a circle. Here I shall take
it to be the quadrilateral whose corners have the latitudes
and longitudes 41°.0N and 14°.0E, 42°.0N and 15°.0E, 45°.0N
and 9°.0E, and 46°.0N and 10°.0E. Roughly, this corresponds
to the part of Italy north of a line running from Naples
across the peninsula and perpendicular to it.

With regard to eclipses after 1300, the situation is
far better in Italy than it has been anywhere else in Europe
that we have studied so far. I used only 24 records from
Italy in MCRE; of these, 23 appear in Table X.1 and 1 will
appear later. In contrast, I have found 33 useful Italian
records after the year 1300, considerably more than in all
the preceding chapters put together. I have also found many
records before the year 1300 that I have not used before.

2. Sources That Have Not Been Used Before

Benevenutus [ca. 1341]. Foligno is a town of moderate
size about 100 kilometers north of Rome that was founded in
the 8th century. Benevenutus held the office called notarius
in Foligno in the first part of the 14th century, and he prepared a set of annals relating to Foligno that runs from 1198
to 1341. For many years, particularly the early ones, the
only item that appears for a given year is the name of the
podesta. The autograph copy of the annals was found in a
library in Foligno in 1886.

Bergomates [ca. 1241]. This is a small set of annals
found in a library in Bergamo that appears in two printed
editions. The earlier edition has two events listed in 1167,
one in 1176, and then fairly regular entries from 1209 to
1241, including a record of the eclipse of 1239 June 3. The
later edition has somewhat more material, but the eclipse
record is the same. I do not know why the earlier edition
omitted some of the material. The pattern suggests that the
annals were kept by one person from 1209 to 1241, but that
he happened to be interested in the events in 1167 and 1176
for some reason.

-311-

TABLE X.1

ECLIPSE RECORDS FROM ITALY THAT WERE USED BEFORE

Designation	Place	Standard deviation of the magnitude	Reliability
118 Sep 3 E,I	Italy	0.18	0.2
393 Nov 20 E,I	Italy	0.18	0.2
418 Jul 19a E,I	Italy	0.18	0.5
512 Jun 29 E,I	Naples	0.18	0.5
534 Apr 29 E,I	Italy	0.18	0.2
538 Feb 15 E,I	Italy	0.18	0.05
540 Jun 20 E,I	Italy	0.054	0.05
807 Feb 11 E,I	Farfa	0.18	0.5
840 May 5a E,I	Venice	0.18	0.5
840 May 5b E,I	Ravenna	0.068	1
840 May 5c E,I	Bergamo	0.034	0.5
939 Jul 19a E,I	Monte Cassino	0.14[a]	1
968 Dec 22a E,I	Benevento	0.18	0.5
968 Dec 22b E,I	La Cava	0.18	0.5
1033 Jun 29a E,I	Benevento	0.068	0.5
1033 Jun 29c E,I	Milan	0.18	0.5
1044 Nov 22 E,I	Rome	0.18	1
1178 Sep 13a E,I	Salerno	0.18	1
1178 Sep 13b E,I	La Cava	0.14[a]	1
1191 Jun 23a E,I	Venice	0.18	0.5
1191 Jun 23b E,I	Bologna	0.18	0.5
1239 Jun 3a E,I	Venice	0.18	0.5
1239 Jun 3b E,I	Bologna	0.034	0.5

[a]The eclipse will be taken as partial with a magnitude of 0.81.
Note: The eclipse is taken to be central if no estimate of the magnitude is given.

Cantinellus [ca. 1306]. This chronicle has two distinct parts. The first part deals mostly with the affairs of Bologna. The beginning is missing, but the surviving pages of this part run from 1228 to 1278. The second part deals mainly with Faenza (near Ravenna) and runs from 1270 to 1306. Authorities disagree

on the explanation for this situation. Since there was a
Bolognese family named Cantinelli, one plausible explanation
is that the writer lived in Bologna until about 1278 and
that he then moved to Faenza, taking his uncompleted chronicle with him.

Casinenses [ca. 1098]. This is not actually a medieval
source. Instead, it is a collection of annals excerpted by
a modern editor from a wide variety of annals and other
sources preserved at Monte Cassino. All the material seems
to be local to the monastery there.

Casinenses [ca. 1212]. This is a complex set of annals
that is also from Monte Cassino. In fact, it is not clear
that it should be looked on as a single source. There are
five "copies" which differ considerably from each other both
in the material included and in the period of time covered.
As with the preceding source, all the material seems to be
local.

Ceccanenses [ca. 1217]. This set of annals concludes
by saying "End of the Chronicle of Fossa Nova". In spite of
this, the editor calls this source the Ceccanian Annals. The
reason for this is that the annals have been attributed to
Iohannis, count of Ceccano, and the editor uses the traditional name even though he thinks the count did not compile
the annals. Ceccano is about 30 kilometers from Terracina,
while Fossa Nova is described as being a short distance from
Terracina on the road to the south (the Appian Way). I shall
use the coordinates of Terracina, since I cannot find the
exact location of Fossa Nova.

Clavius [1593]. Christopher Clavius was a trained
astronomer and a member of the Society of Jesus, who came
from Bamberg and who lived for some time in Portugal and for
some time in Rome. One of his works is a long chronicle of
the world from the Creation up to his own time. Another
work, which is the one that concerns us here, is a commentary
on the astronomical writing of Ioannis de Sacro Bosco, as
Clavius spells his name. Clavius had the good luck to see
two remarkable eclipses of the sun, apparently without plan-
ning to do so, and his records of the eclipses are used in
this work.

Cremonenses [ca. 1232]. The editor of the cited edition
says that the Modena library owns an 18th century copy of
these annals, by Phillip Garbellus. This leaves it unclear
whether the annals or the copy is by Garbellus. The editor
further identifies Garbellus as an abbot of Pons Vicus, which
is slightly north of Cremona. If Garbellus were merely a
copier, it is unlikely that the editor would identify him in
so much detail, so he was probably the compiler. I shall
take the place for these annals to be Cremona, since Pons
Vicus is described as being close to it.

degli Unti [1440]. This is a short set of annals about
Foligno† from 1424 to 1440, when the writer died. He was a
member of a merchant family of Foligno and his home was still
standing when the cited edition was prepared; so far as I
know it is standing yet. An inscription on it reads: Ber-
nardinus de Untis de Fulgineo Arti et Medic Doc Astrologus.

dello Mastro [ca. 1484]. dello Mastro kept this chronicle
of affairs pertaining to the papacy from 1422 to 1484. The
place is Rome. I do not know anything about the author.

Ferrarese [ca. 1502]. This is a highly detailed history
of Ferrara from 1409 to 1502, and the editor believes that it
was written by one person. If so, it cannot be original in
its earlier parts. It begins to become quite detailed about
1467, so we may guess that this is about the time when it becomes original. More than half of the work deals with the
years from 1495 to 1502.

Forolivienses [ca. 1473]. These are some annals from
Forli, which is a short distance to the southwest of Ravenna.
They run from Roman times to 1473, with a few short additions
running to 1616 that contain nothing of interest to us. The
annals are rather sparse until about 1275, so this is probably the date when they become original. The existing copy
is entirely in a hand of the late 15th century, but various
annals show different literary styles and different interests. Hence the original was probably compiled by a number
of different people. Three of the five records of solar
eclipses show the same interesting phenomenon. In each case,
a space was originally left for the year but no event was
recorded. Then someone in the 16th century inserted the
record of the eclipse.

Ghirardacci [ca. 1509]. The editor describes Ghirardacci as a member of the Augustinian hermits; I believe we
would say friars in English. The cited edition has a picture
of him, and I believe he is the first chronicler of whom I
have seen a portrait. The work is an enormously detailed
history of Bologna from 1426 to 1509.

Guerrierus [ca. 1472]. Guerrierus seems to have been a
fairly prominent citizen of the town now called Gubbio, in
Umbria. His work is a history of the town from 1350 to 1472.
Thus it can be original only in its later parts, but its
eclipse records seem to be independent of any others known.

Hieronymus [ca. 1433]. The author is identified more
fully as Frater Hieronymus Foroliviensis Ordinis Praedicatorum,
sacrae Theologiae Magister. This is a chronicle of the town
of Forli from 1397 to 1433. So far as we can judge, the work

---

†See the discussion of Benevenutus [ca. 1341] above.

is entirely original.

Hyvanus [ca. 1478]. Hyvanus was a leading citizen of
Volterra in the second half of the 15th century, when Volterra
was under the domination of Florence, and he wrote several
books about the history of the town. This work is something
of an oddity in that it deals with the history of the single
year 1478.

Malaterra [ca. 1090]. All we can say of this work is
that Malaterra wrote it toward the end of the 11th century,
and that it deals mostly with Sicily under the Normans.

Mantuana [ca. 1250]. This is called a chronicle of the
popes and emperors, compiled in Mantua. The main part runs
to 1250, but there is a brief continuation that runs to 1309.
Since the early parts are obviously derivative, the editor
does not print any part of it before 1156. The implication
is that later parts are at least independent of other known
sources.

Mantuani [ca. 1299]. The editor describes these as a
set of annals collected from public records and acts, and
says that they were written contemporaneously from about 1268
on. I think he means that the collecting was done by one or
more Mantuans in the medieval period, not that he did the
collecting himself.

Maragone [1182]. Bernardo Maragone was a prominent
citizen of Pisa who was born around 1110 and died around 1190.
He held several offices in the city, notably ambassador to
Rome around 1150 and provisor in 1158; I do not know what
this term means. He tells us that he finished compiling
these annals of Pisa in 1182. His son Salem wrote a short
extension that goes to 1192. Unless there has been damage
to the manuscript, Salem stopped writing in the middle of a
sentence. Most of the writing is in Latin, but some is in
Italian.

Matthaeus [ca. 1472]. This writer's name is given in
full as Matthaeus de Griffonibus. He continues the tradition
of the Bolognese chroniclers (Bolognetti [ca. 1420], Rampona
[ca. 1425], Varignana [ca. 1425], and Villola [ca. 1376])
that I used extensively in MCRE and that I shall continue ,to
use to some extent here. Even though Matthaeus overlaps the
other chroniclers for much of his work, he did not copy from
them in any place that I noticed. He was a member of an old
Bolognese family, and the editors give his family tree extending over a period of three centuries.

Mediolanenses [ca. 1251]. This is actually a set of
notes that a modern editor has extracted from the necrology
and other minor sources kept at S. George's in Milan.

Miliolus [ca. 1286]. Albertus Miliolus was a member of
a family that, according to town records, had lived in Reggio
nell'Emilia for a long time. Miliolus wrote this chronicle
of town affairs that runs from the year 1 to 1286. Reggio
nell'Emilia is on the main road between Milan and Bologna,
and it should not be confused with Reggio di Calabria, which
is a seaport on the toe of Italy facing Sicily.

Palmerius [ca. 1474]. Mattheus Palmerius was a Florentine historian who was born in 1405 and who died in April of
1475; his death is noted in several sources of the time. He
wrote two historical works. One, called Liber de Temporibus,
is a history of the world from the year 1 to 1448. In the
part of his history after 1200, which is the only part I studied, I found five records of solar eclipses, all brief and
none safe to use. Hence I shall not use that work. The work
cited here is a set of annals pertaining to Florence from
1429 to 1474, the year before Palmerius died. Both the annals
and the Liber de Temporibus have records of the eclipse of
1448 August 29, but the two records differ considerably. It
seems safe to use the record found in the annals.

Parmenses [ca. 1335]. This set of annals from Parma
exists in two modern editions that I shall designate as MGH
and RIS† for the purposes of this discussion. The MGH edition runs from 1165 to 1335 and is in Latin. The RIS edition runs from 1038 to 1338 and parts of it are in Italian
rather than Latin. In particular, the part after 1335 is
almost entirely in Italian. In other parts, sentences or
parts of sentences that are in Latin in MGH are often in
Italian in RIS. In other words, the language in RIS changes
frequently, often within a single sentence.

Further, we have the circumstance that the copy edited
in MGH is written in a single hand of the 14th century, but
there is good textual evidence that the original was written
by many different people. The explanation of these facts
may be something like the following: There was an original,
probably in Latin, that is now lost. Then there was a Latin
copy which was the basis of the MGH edition, and which was
probably shortened to start in 1165 rather than 1038 for a
reason that we do not know. Finally, someone, working with
the original and not the copy, made a revision that is partly
in Italian, and that was extended to 1338 rather than stopping in 1335. It is possible, of course, that this reviser
is also responsible for the part between 1038 and 1165.

Petrus [ca. 1417]. The author's name appears as Antonius
Petrus in one place and as Antonio di Pietro dello Schiavo in
another. He was what is called a Beneficiatus in the Vatican,
and he wrote this detailed account of Roman affairs from 1404
October 19 to 1417 September 25. Most of the language is
Latin, with bits in Italian, but the Latin seems to me to be

---

†These are the initials of the series titles Monumenta
Germaniae Historica and Rerum Italicarum Scriptores.

strongly affected by Italian. Many passages end in "etcetera", which suggests that he is copying and does not want to be bothered with copying all of a long passage.

Pisanum [ca. 1136]. This gives a collection of events running from 688 to 1136, but they are so intermittent that it seems better to call them notes rather than annals. The notes seem to have been made by one person working entirely from Pisan sources. Except for the deaths of the emperors, the events seem to be local.

Placentini [ca. 1284]. This set of annals is called the Ghibelline Annals of Piacenza. Another set, printed in the same volume, is called the Guelph Annals of Piacenza; the latter contain no records of solar eclipses that I noticed. The Ghibelline annals were apparently compiled by a single person working in the late 13th century. The identity of the person is unknown, but he apparently worked only from local materials. The annals are quite long. They cover 131 years, from 1154 to 1284, but occupy 124 pages in the printed edition. In contrast, most annals take only a few lines for each year.

Rolandinus [1262]. In the prologue to this work, the author Rolandinus tells us that he was born in 1200. The last paragraph says that the work was inspected and approved by certain persons whom the author names, and that this action took place in Padua on 1262 April 13. This is a lengthy history of Padua during the lifetime of Rolandinus.

Ryccardus [ca. 1243]. Ryccardus is described more fully as Ryccardus de Sancto Germano. This refers to the town that was anciently known as Casinum. During the Middle Ages it was known as S. Germano, but it returned to its ancient name in 1871, using the Italian spelling Cassino. Ryccardus was a native of S. Germano and he held the office of notarius there. This work is a history of Italian affairs from 1189 to 1243, but we may take Cassino as the place of observation for the single solar eclipse that is recorded in it.

S. Georgii [ca. 1295]. This is a set of notes attached to a calendar that was originally kept in the church of S. George in Milan, and that is now in the Ambrosian library there. The editor says that he has put the notes in chronological order. I imagine that the events were originally noted on a calendar on the day of the year on which they occurred, with, we hope, a record of the year. The editor has then ordered them by year rather than by day of the year.

S. Georgii [ca. 1415]. This is a set of annals from the Benedictine monastery of Mons S. Georgii† in the Italian

---

†This is identified as S. Giorgio d.E. on my map of the region. I do not know what "E." stands for.

Tyrol; it is just east of the road from Bolzano to Bressanone.
With the aid of a large-scale map, I estimate that its coordinates are 46°.68N in latitude and 11°.63E in longitude. The
annals contain an isolated entry for the year 880 about the
death of Carloman, king of Bavaria and Carinthia, and another
isolated entry for 1197 about the death of Barbatus, whom I
have not identified. The main part of the annals runs from
1208 to 1260, with an isolated entry for 1415. I would have
cited the annals by using the date 1260 except for the fact
that the isolated entry records the eclipse of 1415 June 7.

S. Iustinae [ca. 1270]. This is a history of Padua
from 1207 to 1270, which is almost exactly the period that
Rolandinus [1262] covers. However, the works are entirely
independent, at least with respect to the events that I noted.
The editor says that the history was written by at least two
people.

Senenses [ca. 1479]. According to this source, Rainerius,
who was bishop of Siena from 1129 to 1170, ordered the preparation of a calendar that would serve as a necrologium for the
diocese in future years. The calendar that was prepared in
response to his orders was used for the recording of memorable
events for nearly four centuries. The first entry records the
translation of S. Ampsanus to Siena in 1107. Use of the calendar slackened noticeably in the 15th century, for which there
are only four entries; the last entry is for 1479.

Stefani [ca. 1386]. Marchionne Stefani, whose first name
is also spelled Melchionne, was a member of a Florentine family that held some civic offices. His chronicle has some
early parts going back to Genesis, but it becomes a chronicle
of Florentine affairs about 1250. It is highly detailed for
much of the 14th century, probably during the period of the
author's life. It really stops in 1384; after 1384 there are
only two short paragraphs that record a conjunction of Jupiter
and Saturn, a solar eclipse, and a lunar eclipse, all in 1386.
This chronicle was apparently quite popular, since many copies
of it still exist.

Urbevetani [ca. 1260]. It is convenient to discuss this
source and the next one at the same time.

Urbevetani [ca. 1276]. In the edition cited for both
this source and the preceding one, the editor begins with a
list of the podesta's of Orvieto from 1194 to 1224. He follows this, as part of the same work, by three sets of annals
which, it seems to me, are three distinct works. The first
is a set that runs from 1161 to 1276, which is the set that
I am designating as Urbevetani [ca. 1276]. It was written
by several people, and thus there is a good chance that it is
an original and contemporaneous source. The second set runs
from 1233 to 1260. It seems to be by only one person but,
since it covers only a short period of time, it is probably
an original source also. The third set of annals runs from

1255 to 1322, but I do not cite it because it has no records
of solar eclipses.

Urbevetani [ca. 1313]. This source is from Orvieto,
but otherwise it has no connection with the two preceding
sources. The editor calls it An Ancient Chronicle of Orvieto.
It has been preserved only because someone named Tommaso di
Silvestro kept a diary from 1482 to 1514 and, for some reason, copied this old set of annals as a prologue to his diary.
The editor says that the chronicle is contemporaneous with
the events it describes from about 1257 on.

Vincentina [ca. 1241]. This chronicle from Vicenza runs
from the birth of Christ through 1241, although the last
entry is erroneously dated 1242. The early part is copied
from various sources that the editor lists. Except for a
few introductory remarks, the cited edition begins in 1143,
when the chronicle apparently became an independent source.
The part that is printed consists mostly of a list of popes,
along with the records of two solar eclipses, one lunar
eclipse, and a comet.

3. Records from Italy

This section contains eclipse records that have not been
used in earlier work, one record that has been used but whose
characteristics will be changed, and a few records that will
be used unchanged but that need further discussion.

840 May 5c E,I. This record by Andreas Bergomatis [ca.
877, p. 226], which I used in MCRE [p. 465], says that stars
were seen, that the eclipse caused great tribulation, and
that such an eclipse had not come in this century. Because
of these remarks, I took the standard deviation of the magnitude to be 0 in MCRE, but this value will be altered to
0.034 in accordance with the standard practice of this work.

Afterward Andreas says that the sun seemed to tremble
as it began to recover its light. M [pp. 8.46-8.47] believes
that this is a reference to Baily's beads which, he says, do
seem to flash on and off when an eclipse is barely total.
This possibility cannot be denied and, if it were accepted,
I should have to use 0 rather than 0.034 for the standard
deviation of the magnitude. However, there are a number of
examples of records that use picturesque language which seems
to describe some astronomical phenomenon but which turns out
not to do so. For example, the record 1147 October 26a E,BN
[MCRE, p. 245] says that the sun at the height of the eclipse
was surrounded by a misty cover. This sounds like a good
description of the corona but it cannot be because the eclipse
was annular by a considerable margin. What the observer saw
instead was the dark circle of the moon surrounded by the
bright rim of the sun. Thus I think it is not safe to read
a description of Baily's beads into Andreas's remark, and I
think that we should use 0.034 for the standard deviation of
the magnitude.

939 July 19c E,I. Pisanum [ca. 1136] writes:
"DCCCCXXXIX. On the 14th calends of August, on the 6th day
of the week, at the 6th hour of the day, the sun was eclipsed
even to the farthest parts of the earth." The calendrical
data are correct, but the eclipse should have had its maximum
well before the 6th hour, which is our noon. The remark
about the farthest parts of the earth creates a temptation to
say that the eclipse must have been quite large. However,
the annalist may have realized that even an ordinary eclipse
can be seen over a wide area. Further, the hour that he
gives, though wrong, is the same as the hour given for the
great darkness that occurred at the Crucifixion of Jesus.†
Thus I strongly suspect that the annalist was being influenced
by the Biblical account when he wrote this record. Accordingly, I shall use the conventional value of 0.18 for the
standard deviation of the magnitude, with 0 for the value of
η. The place is Pisa, but, since the compilation was made
some time after the event, I shall use a reliability of only
0.5.

1010 March 18 ? Casinenses [ca. 1098] has the following
on page 1411 of the cited edition: "MXI. ind. VIIII. The
sun was eclipsed, and there was a great famine." I don't
believe the annalist intends to imply a causal connection
between the eclipse and the famine. "ind. VIIII" refers to
the cycle of the Indiction,‡ and 9 is correct for the Indiction if the year is 1011. Casinenses [ca. 1212] says the
same thing except that it omits the statement of the Indiction.

There was no eclipse in 1011 that was visible in Italy.
The only possibilities I can find are 1009 March 29 and 1010
March 18. Since the record gives no details that help in
the identification, we cannot identify this eclipse.

1033 June 29b E,I. I shall make two changes from the
characteristics that I used for this record in MCRE [p. 468].
The first concerns the source and hence the place. In MCRE,
I attributed the record to a source from the town of La
Cava, a few kilometers from Salerno. However, the identical
record occurs in Casinenses [ca. 1098], which is from Monte
Cassino. Since the material from Monte Cassino is considerably older than that from La Cava, I shall take the place
to be Monte Cassino. Since the two places are only about
100 kilometers apart, the change has little effect.

The other change concerns the treatment of the magnitude.
The record makes it clear that the eclipse was reasonably

---

†See the discussion of the eclipse of 29 November 24 in
Section II.3.

‡The Romans in late imperial times introduced a 15-year
cycle called the Indiction. It seems to have been used for
purposes connected with taxation, but its exact use is not
clear [MCRE, p. 516, or Explanatory Supplement, 1961,
p. 431].

large but nonetheless partial. Hence I shall use the conventional values adopted here for partial eclipses, rather
than trying to estimate the magnitude as I did in MCRE.

1067 February 16 E,I. On page 469 of MCRE I gave a record of this eclipse from the town of Benevento, near Naples.
It turned out [MCRE, p. 596] that the eclipse was not visible
there, although it failed to be visible by only a small margin. There is no simple way to use this record, and I
shall accordingly omit it.

1084 October 2 ? Malaterra [ca. 1090] gives a detailed
description of what sounds like a total eclipse, and he dates
it 1084 February 6, between the 6th and 9th hours. There
were eclipses on February 6 in both 1068 and 1087, but neither
was visible in Sicily or anywhere near there; the one in 1068
might have been visible in Egypt at sunrise. There were two
eclipses near this date that should have been large but not
total in Sicily, so far as we can judge from Oppolzer's charts,
and both would have occurred at about the hours that Malaterra
states. One was on 1086 February 16, and we can see plausible
errors in writing that would convert this into 1084 February
6. Since Malaterra gave the year in Roman numerals, an accidental change of VI into IV at the end of the numeral would
explain the year. He wrote the day of the month in words as
sexto, but he could have easily omitted decimo from sexto
decimo. The eclipse of 1084 October 2 looks as if it would
be larger than that of 1086 February 16, but it could not
have been total because it was annular. Further, I see no
plausible way to convert secundo die Octobris (October 2)
into sexto die Februarii.

As a matter of interest, I shall calculate the circumstances of the eclipses of 1086 February 16 and 1084 October
2 for a point in the center of Sicily, and I shall discuss
the results in Section XIV.7. I shall not use the record,
whatever the results may be.

1093 September 23 E,I. Casinenses [ca. 1212] writes,
under the year 1094: "The greatest part of the sun was obscured on the 8th calends of October." The day specified
is actually September 24, so the annalist has missed both
the day and the year by 1. Nonetheless, I think we may
safely accept the identification. The place is Monte Cassino and the reliability is 0.5 since the record is not contemporaneous in its existing form. Since the eclipse is
partial, but there is no estimate of the magnitude, I shall
use 0.81 for the magnitude with 0.14 for the standard deviation.

1133 August 2 E,I. The same source says, under the
year 1133: "The sun was almost totally eclipsed from the
6th hour to the 9th on the 4th nones of August." The day
specified is August 2, and the hours look reasonable. It is
remarkable that this is the only Italian record I have found
of this famous eclipse. This record is found in the copy of

Casinenses that the editor numbers 2, and the records in this copy now seem to be contemporaneous.† Hence I give the record a reliability of 1. The place is Monte Cassino, the magnitude is 0.81, and the standard deviation is 0.14.

Casinenses [ca. 1212] says under 1147 that the sun was eclipsed on October 29, and it says under 1153 that the sun was eclipsed, without giving the day. These are probably records of the eclipses of 1147 October 26 and 1153 January 26. However, there are many other records of these eclipses, and it is quite possible that these brief notes are copied from somewhere else. Hence I shall not use them.

1163 July 3 E,I. This is the last eclipse record in Casinenses [ca. 1212], and it is the only record of this eclipse I have found. Since it is thus necessarily independent of any other records known, this creates a reasonable chance that the records of 1147 and 1153 are also independent. Nonetheless, I shall adopt a conservative approach and not use those records. This record says: "In July on the 4th day of the week the sun was eclipsed." The day of the week is correct for 1163 July 3. The reliability is 1, the place is Monte Cassino, the value of η is 0, and the standard deviation of the magnitude is 0.18.

1178 September 13c E,I. Vincentina [ca. 1241], on page 150 of the cited edition, has a long paragraph that starts with the election of Alexander III as pope. The year of this event was 1159, but the chronicle erroneously gives it as 1160. It then records the pertinent facts about three antipopes, the first of whom was elected on almost the same day as Alexander III and the third of whom submitted to Alexander on 1178 August 29. Then it says: "This year on the 14th calends of September the moon was eclipsed, and the same year on the ides of September the sun underwent an eclipse." The first day is August 19 and the second is September 13. The annalist has become confused because he has dealt with so many years in the same passage. The lunar eclipse must be that of 1160 August 18, which he thinks is in the first year of these events, and the solar eclipse must be that of 1178 September 13, the last year dealt with in the entry. The reliability is 0.5, the place is Vicenza, the value of η is 0, and the standard deviation of the magnitude is 0.18.

1178 September 13d E,I. Ceccanenses [ca. 1217] has a short note about this eclipse: "On the ides of September the sun was eclipsed." Although this is a note that could have been derived from many sources, Ceccanenses is now at a stage where it seems to be independent. The reliability is 0.5, the place is Fossa Nova, but I shall use Terracina since I canot locate Fossa Nova exactly. The value of η is 0 and the standard deviation of the magnitude is 0.18.

---

†The record also appears in the copy numbered 3.

1178 September 13e E,I. In the middle of a paragraph about the year 1179, Maragone [1182, p. 67] writes: "In the year 1179, indiction 11, the sun was darkened from the hour of tierce until the hour of nones,† and no more of it could be seen than one sees of the horned moon after its return, and that was on the 13th day of September, and the moon was then in conjunction with the sun." That is, Maragone is comparing the sun to the crescent new moon. I think it is reasonable to take this as a magnitude of 0.90, with a standard deviation of 0.05. The place is Pisa and the reliability is 1.

1187 September 4a E,I. Cremonenses [ca. 1232, p. 802] records this eclipse in the following way: "And at that time Saladin captured Jerusalem, and the sun was darkened from the third hour to nones." The eclipse had its maximum in Italy at about our noon, so the hours given are symmetrical about the maximum if nones means the 9th hour, although the duration is much too long. We seem to have the same exaggeration as in the preceding record.

Since this eclipse is connected with an important event in the Crusades,‡ and since European historians often commented on this connection, our first task with any European record of this eclipse is to decide whether the historian is reporting a local observation or an observation made in the Holy Land. The hours given in this record are not possible for an observation made in the Holy Land, for either meaning of nones, so the observation must have been made in Cremona. The reliability is 0.5 since the record may not be contemporaneous. The value of η is 0 and the standard deviation of the magnitude is 0.18.

1187 September 4b E,I. S. Georgii [ca. 1295, p. 387] has a brief notice of this eclipse: "On the 2nd nones of September, 1187, the sun was eclipsed." Although this notice has no features that indicate local observation, we may conclude that it does represent an observation made in Milan. There is no Italian record that I have found from which it is likely to have been copied, and the annalist does not connect it with the capture of Jerusalem. I shall give the record a reliability of 0.5. The value of η is 0 and the standard deviation of the magnitude is 0.18.

1239 June 3c E,I. Although this eclipse is widely reported in Italian sources, I used only two Italian records of it in MCRE because most of the sources that record it have no eclipses before 1200. Vincentina [ca. 1241, p. 150]

---

†I think Maragone is referring to the ecclesiastical offices performed at the third and ninth hours of the day. If so, the duration given for the eclipse is much too long. Maximum eclipse was at about noon, and thus nones may mean our noon.

‡See MCRE [p. 562].

first notes the election of Pope Gregory IX in 1227. He
then says: "In the 13th year of his papacy, he excommunicated Emperor Frederick in March, and in June of the same
year the sun was totally eclipsed, and the following year
in February a comet appeared like a streak." This is near
the end of this chronicle, so the report receives a reliability of 1. The place is Vicenza. Since the record says
that the eclipse was total, I shall use 0.034 for the standard deviation of the magnitude, in accordance with the
practice of this work.

1239 June 3d E,I. Senenses [ca. 1479, p. 229] has this
account: "On the 3rd nones of June in the year of the Lord
1239, on Friday at the 6th hour the sun gradually began to
be darkened and covered over in a clear sky.† And in the
hour of nones it was totally obscured and so dark that no
light came back. And it became like dark night, so that the
sky appeared starry just as on a clear night. And people
kindled lights in their homes and stores. And after a while
it began little by little to be uncovered and to shine again
on the earth, so that at the hour of vespers it was restored
in all its splendor." It is a little hard to make sense of
the hours. If nones means midday, this is the same as the
6th hour. I believe that we should read the times this way:
The eclipse was first noticed in, not at, the 6th hour, and
this could be before 11:00 AM as we measure time on a day
this close to the summer solstice. The eclipse should have
been a maximum shortly after our noon, so we can take the
time when the eclipse was stated to be total to be in the
hour following noon.

The records in this source seem to have been made at
the time of the events, so the reliability is 1, and the
place is Siena. Since the record says that the eclipse was
total, I use 0.034 for the standard deviation of the magnitude.

1239 June 3e E,I. S. Iustinae [ca. 1270, p. 157] says
in reference to the year 1239: "This year at the beginning
of June was a solar eclipse." The reliability is 1, the
place is Padua, the value of η is 0, and the standard deviation of the magnitude is 0.18.

1239 June 3f E,I. Rolandinus [1262, p. 73] says of
this eclipse: "On the 3rd intrante‡ of June immediately

---

†The writer uses the word aer, but I believe that sky is
implied by the context.

‡In Italy, the days of the month in the medieval period were
often numbered intrante and stante. The Nth day intrante
was the Nth day from the first, in inclusive counting. That
is, it was the Nth day of the month in ordinary usage. The
Nth day stante was the Nth day before the last day of the
month, again in inclusive counting. Thus the 3rd stante of
June would be June 28.

after nones in the aforesaid year of the Lord 1239 the sun was seen to be darkened, with everybody looking on, and the eclipse lasted for about two hours." The writer goes on to say that the eclipse had been sent to warn the emperor, in the opinion of some people. We remember that the emperor had just been excommunicated; see the record 1239 June 3c above.

Since everybody was looking on, this eclipse may have been larger than the usual recorded eclipse. Hence I shall take the standard deviation of the magnitude to be 0.11 rather than 0.18. The reliability is 1 and the place is Padua.

1239 June 3g E,I. The entire entry for 1239 in Mantuani [ca. 1299] reads: "1239. Guido de Coregia was the podesta of Mantua, and in his time there was an eclipse of the sun on a Friday in June about the middle of the day." The eclipse was indeed on a Friday. The reliability is 1, the place is Mantua, the value of η is 0, and the standard deviation of the magnitude is 0.18.

1239 June 3h E,I. This record from Mantua [Mantuana, ca. 1250] reads quite differently: "And indeed in 1239, on Friday the 3rd intrante of June, the sun was darkened at about the hour of nones, and the day was so darkened that many stars were seen in the sky, and this lasted for the time that someone could ride a horse for 4 miles." I wonder how the annalist made this time estimate; I doubt that he jumped on a horse and measured how far he could ride during some interval connected with the eclipse. The fact that he measured a fairly short time interval, whether he measured it accurately or not, poses the interesting problem of what he measured. It is not likely that he measured an interval other than the interval of totality, because that is the interval that is most sharply defined for an observer who lacks extensive instrumentation. However, I believe that it is not safe to assume totality on the basis of this consideration. Hence I shall take the standard deviation of the magnitude to be 0.054, because of the reference to the stars. The reliability is 1.

1239 June 3i E,I. Bergomates [ca. 1241] has the following record of this eclipse: "In the year 1239, on the 3rd intrante of June at the hour of nones, the sun was obscured although no cloud shadowed it, and it stayed dark for a long time." The reliability is 1, the place is Bergamo, the value of η is 0, and the standard deviation of the magnitude is 0.18.

1239 June 3j E,I. Placentini [ca. 1248] records, under the year 1239: "And then on Friday,† the 3rd day of June,

---

†In all these records where I have written Friday or a similar day, the record gives the day of the week by using the name of a pagan god rather than giving the number of the week day. In Latin, Friday is called the day of Venus (dies Veneris).

at the hour of <u>nones</u> the sun in every respect† suffered an
eclipse, that is, a disappearing, and there was darkness."
This record poses a problem in judging the magnitude. The
use of <u>undique</u> implies totality by one meaning but not by
another. Aside from this, there are three different statements about the eclipsing of the sun, and this, combined
with a likely meaning of <u>undique</u>, suggests a rather large
eclipse. I shall compromise by using 0.14 for the standard
deviation of the magnitude. The reliability is 1 and the
place is Piacenza.

1239 June 3k E,I. <u>Mediolanenses</u> [ca. 1251, p. 402]
records the eclipse in this way: "1239,‡ on Friday, in the
middle of the day, when the sky was clear, the sun was
eclipsed almost totally." I shall guess that almost totally
means a magnitude of 0.94, with a standard deviation of 0.05.
The reliability is 1 and the place is Milan.

1239 June 3ℓ E,I. This eclipse is also recorded in
<u>Parmenses</u> [ca. 1335, p. 669] under the year 1239: "Also
that year, on Friday the 3rd of June about <u>nones</u>, the sun
appeared so darkened that stars were seen just as at night."
The editor believes that the entries in these annals were
made by many different people, which implies that the records
have a good chance of being contemporaneous. Hence the reliability is 1; the place is Parma. I shall take the standard deviation of the magnitude to be 0.054 because of the
reference to the stars, although it is tempting to use 0.034
because of the statement that they were seen just as at
night.

1239 June 3m E,I. This record comes from Reggio nell'
Emilia, which is about 25 kilometers southeast of Parma.
Chapter CCXLI of <u>Miliolus</u> [ca. 1286, p. 513], which deals
with the year 1239, contains the following: "And in the
hour of <u>nones</u> the sun was obscured and stars appeared in
the month of June on the 3rd <u>intrante</u>, and it looked just
as if it were dark night." In both this record and the preceding one, <u>nones</u> is probably the same as our noon. I believe we may safely take this as a statement that the eclipse
was total. Hence we use 0.034 for the standard deviation of
the magnitude. The reliability is 1 and the place is Reggio
nell'Emilia, as I just said.

---

†The word used here is <u>undique</u>, which can also mean everywhere.

‡I follow the practice of giving numerals in the same form
as the printed edition that I use. However, I imagine that
the original manuscripts used words or Roman numerals for
the year, but not Arabic numerals.

1239 June 3n E,I. Ryccardus [ca. 1243, p. 200] has a record of this eclipse that is not very useful, but it costs little effort to use it anyway: "In the month of June of the 3rd intrante, which was a Friday, about the hour of nones the sun was eclipsed." Since Monte Cassino is well to the south of the other places from which we have records, it is unfortunate that Ryccardus did not tell us whether the eclipse was total or not. The reliability is 1, the value of η is 0, and the standard deviation of the magnitude is 0.18.

1239 June 3o E,I. We have two records of this eclipse from Orvieto, of which the first, from Urbevetani [ca. 1276], is rather short: "MCCXXXVIIIJ.† - Calends of January. The podesta was Petrus Anibaldi of Rome,‡ in whose time the sun was obscured." I don't believe that the writer intends January 1 to be the date of the eclipse. I think he means that Petrus Anibaldi became podesta on January 1, and the eclipse happened during his term of office, which was often a year. The reliability is 1, the value of η is 0, and the standard deviation of the magnitude is 0.18.

1239 June 3p E,I. The other record from Orvieto [Urbe-vetani, ca. 1260] is more unusual and more informative: "MCCXXXVIIIJ. - Lord Petrus Anibaldi of Rome was podesta. In his time, the sun was eclipsed, whence the verse:

In the year thirty twice hundred thousand nine,
June entered, whose light three times stood,
The shadowed sun through its whole was darkened.
On the sixth feria was this miracle done.‡

Then stars were seen in the sky by day." I think we may take this to be a description of a total eclipse, so we use 0.034 for the standard deviation of the magnitude. The reliability is 1.

1239 June 3q E,I. Under the year 1239, Matthaeus [ca. 1472] records the eclipse this way: "In the same year, on Friday the 4th of June. - The sun was obscured at the hour of nones and the sky became starry." Note that he has the

---

†J was often used as the last element in a string of 1's in Roman numerals in medieval times.

‡Italian cities that had a podesta often brought one in from outside, with the idea that an outsider would be impartial in settling local disputes.

‡In the original, each line has an internal rhyme, but there is no rhyming between lines. I have tried to give only a rather literal translation. In particular, I have preserved accurately the way in which the year is stated. The word feria in this context means day of the week.

correct day of the week but the wrong day of the month. Here
nones must mean noon. Since this is a late compilation, the
record receives a reliability of only 0.5. The place is
Bologna, the value of η is 0, and the standard deviation of
the magnitude is 0.054.

1239 June 3r E,I. Cantinellus [ca. 1306] writes, with
regard to the year 1239: "This year the sun was obscured on
the 3rd day of June, . . ." The rest of the sentence deals
with a different subject. This is also in a late compilation and has a reliability of 0.5. The place is Bologna, the
value of η is 0, and the standard deviation of the magnitude
is 0.18.

1239 June 3s E,I. Benevenutus [ca. 1341] gives the
following record of this eclipse: "MCCXXXVIIII. Lord Thomas
Odoriscii. The sun was darkened per totum orbem in the month
of June on Friday the third about the hour of nones, and it
stayed darkened for an hour, and that was at the renewing of
the moon." I suppose that the name given is that of some new
official, possibly a new podesta. Interpreting per totum
orbem is like interpreting universalis,† except that here we
have no large body of usage to help us. The phrase means
through or for the entire disk, and the question is whether
the disk is that of the sun or of the earth. In other words,
was the eclipse total or was it merely seen "everywhere"?

Strictly speaking, orbis means a circle or disk, and it
was specifically used to mean the disk of the sun or moon.
It was not strictly used to mean the earth unless it was
used together with the word for earth, but this rule was
often violated. Further, in every place I have found where
the writer certainly means to say "in all the earth" or the
equivalent, he says in omnem terram or something like this.
Thus per totum orbem is more likely to mean "over the whole
disk of the sun" than "over the entire earth", and I shall
take this to be the meaning. Since the record occurs in a
late compilation, the reliability is 0.5, and the place is
Foligno. The value of η is 0. We saw in connection with
the record 1321 June 26c E,G in Section IX.3 that the word
generalis cannot be taken literally even though it means
"total" when applied to an eclipse, if the description is
given by someone who is not trained in astronomy. We should
use the same caution here and take the standard deviation of
the magnitude to be 0.054 rather than 0.034.

M [pp. 8.61-8.62] cites records of the eclipse of 1239
June 3 from the Italian towns of Cesena and Arezzo, and I
cited the same records in AAO [pp. 87-88]. Both M and I took
these records from the secondary source Celoria [1877a and
1877b]. In MCRE I decided not to use eclipse records from
secondary sources for several reasons, but chiefly for the
reason described in Section III.3 above. I shall continue
the same policy here. I am sure it would be possible to
locate the primary sources that Celoria used, but I have not
taken the trouble.

---

†See the record 1241 October 6k E,G in Section IX.3.

I noted above that I have found only one Italian record of the famous eclipse of 1133 August 2, and I have also found only one of the famous eclipse of 1241 October 6. Neither eclipse should have been total anywhere in Italy, but both should have been fairly large in many parts of Italy and we should expect many records of them. Perhaps the weather was cloudy over much of Italy on both days although it was clear over much of Europe. The record of 1241 October 6 follows.

1241 October 6 E,I. <u>S. Georgii</u> [ca. 1415] notes on page 722: "Year of the Lord 1241, an eclipse of the sun occurred about the feast of Michael." The annals seem to be original, so this record has a reliability of 1. The place is S. Giorgio d. E. The value of η is 0 and the standard deviation of the magnitude is 0.18.

1263 August 5a E,I. <u>S. Iustinae</u> [ca. 1270] has a fairly long note about this eclipse under the year 1263: "In the circuit of this year, on the 5th <u>intrante</u> of August, when the moon was at its 27th, there was an eclipse of the sun a little before vespers." This puts the eclipse in late afternoon, which is correct. According to the ecclesiastical tables of the lunar motion, the moon was in its 27th day since the new moon became visible, the day of the visible crescent being counted as 1. The reliability is 1, the place is Padua, the value of η is 0, and the standard deviation of the magnitude is 0.18.

1263 August 5b E,I. <u>Urbevetani</u> [ca. 1313] has the only other Italian record of this eclipse I have found: "1263. In the month of August the sun was obscured." This has the same characteristics as the preceding record, except that the place is Orvieto.

1290 September 5 E,I. <u>Cantinellus</u> [ca. 1306, p. 61] gives us this record under the year 1290: "Also this year, on Tuesday the 5th <u>intrante</u> of September, in the middle of the 3rd hour, the sun suffered a diminishing in an eclipse, for a good two parts of the sun were obscured, and it looked like a moon when it is seen to have two horns, and almost everybody was seen <u>extra calli</u>,† and that lasted for a great hour." <u>Cantinellus</u> does not say two parts out of how many, but I believe the magnitude had to be more than 0.5 if the sun seemed to have horns. Hence I shall assume that the eclipsing was two parts out of three. That is, I take the magnitude to be 0.67, with a standard deviation of 0.2. The reliability is 1, since we are in the author's active writing career. He has by now moved to Faenza.

1310 January 31a E,I. <u>Rampona</u> [ca. 1425, p. 313, v. 2] is the only one of the Bolognese chroniclers who mentions

---

†There seems to be a grammatical error here, and I do not know what the writer meant.

this eclipse. Under the year 1310, he records: "The same
year there was an eclipse of the sun in the month of April
at the 10th hour of the day." Since the hour is reasonable,
I believe that this refers to 1310 January 31 in spite of
the error in the month. I do not think of a logical explanation of this error; it is probably just one of those odd
things that happens occasionally to every writer. I shall
assign a reliability of 0.25 after we consider the next record. The value of η is 0 and the standard deviation of the
magnitude is 0.18.

1310 January 31b E,I. This record is from the nearby
town of Forli [Forolivienses, ca. 1473], and it reads: "Year
of the Lord 1310. There was an eclipse of the sun in the
month of April at the 10th hour of the day." In view of the
error about the month, it does not seem possible that this
record and the preceding one are independent. In assessing
the situation, we should note that both sources record four
solar eclipses between 1300 and 1425, when Rampona stops,
and that this is the only one that appears in both. Thus it
does not seem possible that either used the other as a reference. Instead, both writers must have used another source
that is now unknown to us, but we have no way to tell whether
the source came from Bologna or Forli. Since the record is
from a late compilation, it would receive a reliability of
0.5 if we knew its origin. Since there are two possibilities
with equal likelihood, I shall assign a reliability of 0.25
to each record. The value of η is 0 and the standard deviation of the magnitude is 0.18.

1330 July 16a E,I. Benevenutus [ca. 1341] records this
eclipse under the year 1330: "On the 16th day of July near
the hour of vespers the sun was eclipsed through its center
from the northern side, and it stayed eclipsed for about an
hour, and that was at the new moon." This eclipse came close
to the end of this chronicle and thus Benevenutus should have
seen it himself. The reliability is 1, the place is Foligno,
the magnitude is 0.5, and I shall assign a standard deviation
of 0.2 to this estimate.

1330 July 16b E,I. Rampona [ca. 1425] writes with regard to the year 1330: "The same year, on the 16th of July,
the sun was eclipsed after nones." Here nones probably means
the 9th hour of the day, or midafternoon, rather than our
noon. We are still a long way from the compilation of this
source, so the reliability is 0.5, and the place is Bologna.
The value of η is 0 and the standard deviation of the magnitude is 0.18.

Varignana [ca. 1425] and Bolognetti [ca. 1420] report
this eclipse in almost the same words as Rampona. Since all
three sources are from Bologna, there is little doubt that
they have all used the same original source that is now lost.
I am actually treating the set of three records as a single
record which happens to exist in three copies.

1330 July 16c E,I. Stefani [ca. 1386, p. 166] reports
this eclipse from Florence under the year 1330: "This year
in the month of July on the 16th day, about the 20th hour,
the sun was eclipsed by half of its body; . . ." The rest
of the sentence deals with the question of the meaning of
the eclipse as a sign. The only interpretation I can give
to the 20th hour is that the hours were counted continuously
from the preceding sunset. We shall see other examples of
this same method of numbering the hours. It is interesting
that two of the three Italian records of this eclipse give
an estimate of the magnitude, and that both estimate it as
0.5. As with the record 1330 July 16a E,I, I shall take 0.2
for the standard deviation of this estimate. The reliability
is 1, since we are in or at least close to Stefani's own time.

1333 May 14 E,I. After a reference to Sunday, 1333 May
16, Parmenses [ca. 1335, p. 786] has: "That same Friday
after nones appeared an eclipse in a certain part of the sun,
and it looked like the moon when it is in a circle and not
full." This sentence is in Latin in the MGH edition and
Italian in the RIS edition. The next sentence says that the
eclipse had been calculated by learned persons from both
Parma and Bologna. This sentence is entirely in Latin in MGH
and in a mixture of Latin and Italian in RIS. It seems odd
that the annalist describes the moon as circular when it is
not full; perhaps he accidentally used a word he did not mean.
The eclipse was on a Friday, as the record says. This record
is contemporaneous and receives a reliability of 1. The
place is Parma. I shall take the magnitude to be 0.81 with
a standard deviation of 0.14, since the eclipse was partial
and there is no specific estimate of the magnitude.

1339 July 7 E,I. Senenses [ca. 1479] has recorded this
eclipse: "In the year 1339 from the Incarnation of the Lord,
on Wednesday, the 7th day of July, the sun underwent an
eclipse between nones and vespers in this measure." The week
day is correct and nones is reasonable for the hour if it
means the 9th hour of the day. I suppose that the writer
said "in this measure" when he really meant "in some measure";
hence I assume that the eclipse was partial. The reliability
is 1 since the entries in this source were made contemporaneously, and the place is Siena. I shall use the conventional
values for a partial eclipse, namely 0.81 for the magnitude
with a standard deviation of 0.14.

1354 September 17 E,I. Rampona [ca. 1425, p. 41, v. 3]
gives us a record of this eclipse from Bologna under the year
1354: "In the said millesimo,† on the 17th of September, a
Wednesday, after the 3rd hour, the sun was strongly obscured,
and this lasted for an hour or more." Varignana [ca. 1425,

---

†This means a thousandth, although the context clearly calls
for a year. I believe that I have seen millesimo used
elsewhere with the meaning of a year, although my dictionary
does not list this usage and I cannot find an example at the
time of writing.

p. 46, v. 3] also has a short record that could easily be abbreviated from this one, and which I shall not use. The day was indeed a Wednesday. I believe the implication is that the eclipse was partial, so I shall use 0.81 for the magnitude, with 0.14 for the standard deviation. The reliability is still only 0.5.

1361 May 5 E,I. On page 233 of the cited edition, Senenses [ca. 1479] records: "On the 3rd nones of May in the year of the Lord 1361 on Wednesday at the 6th hour the sun began little by little to be obscured and arrive at an eclipse, because the moon opposed itself to the sun, and darkness stood for half an hour." The wording suggests a large eclipse, but I shall nonetheless use the conventional values, which are 0 for η and 0.18 for the standard deviation of the magnitude. The place is Siena and the reliability is 1.

1384 August 17, E,I. This record, from Stefani [ca. 1386, p. 434] has an interesting feature. After giving other events in 1384, the author writes: "In this year in the month of August, on the 17th, after the middle of the day, the sun was obscured by 17 parts of 23; that was good weather, and the sun seemed to shine hardly at all." The hour looks reasonable. The interesting feature is the statement of the magnitude. It seems unlikely that anyone measured the magnitude in units of twenty-thirds. However, it was common to measure it in units of twelfths, the so-called digit, and it would be reasonable for a person striving for precision to halve this. Thus I shall assume that the writer meant 17 parts out of 24 and that either he or a later copyist or editor accidentally changed 24 to 23.† Thus I take the magnitude to be 0.71, with a standard deviation of 0.1. The reliability is 1 and the place is Florence.

1386 January 1a E,I. In the year 1386, Matthaeus [ca. 1472, p. 80] first tells us about the death of a prominent citizen, and then he writes: ". . . and then the sun was eclipsed to such a degree that it was necessary to light candles for breakfast through all Bologna; and all said it was for love of . . ." the person who had died. The word prandium, which I have translated as breakfast, should probably be translated as a late breakfast in this passage.‡ Since the eclipse was a maximum in Bologna at the middle of the morning, this makes for a reasonable timing. The need to use candles does not imply a total eclipse. An eclipse less than total would make it dark enough indoors, where people ate breakfast, to make candles necessary. I have no

---

†The original probably used Roman numerals, and it is easy to write XXIII instead of XXIIII.

‡I translated prandium as lunch in connection with the record 1544 January 24 E,G in Section IX.3. I believe that prandium, like dinner and supper, eventually became a movable feast. One must guess at the time it occurred in a given passage from the context and circumstances.

idea what degree of totality is needed, but I shall use a
standard deviation of 0.11 for the magnitude, with 0 for the
value of η. The reliability is 1.

1386 January 1b E,I. This record is from Stefani [ca.
1386], but it appears in what looks like a short addition to
the main work, and it may have been made by someone else.
This possibility is enhanced by the fact that the record
does not give the year, whereas Stefani was careful to give
the year in all the main part of the chronicle that I noticed.
The record reads: "The sun was obscured the first day of
January between the sixth and the third,† and it was not
cloudy, such that almost all the body of the sun was hidden,
for about three fourths of it was hidden; afterward, when all
was cloudy, it grew very dark." I tend to think that the
eclipse never became total, because the darkness would have
been like night if there had been a total eclipse behind
cloud cover. I do not think there is a rigorous way to assign the magnitude, but it seems reasonable to use 0.9, with
0.05 for the standard deviation. The reliability is 1 and
the place is Florence.

1399 October 29 E,I. In the original text of Forolivienses [ca. 1473], there was a line for the year 1399 but no
event was entered. In the space that was left, someone in
the 16th century added this: "There was an eclipse of the
sun on the 2nd calends of October; also stars in the manner
of fire falling from heaven were seen in many parts of Italy."
I imagine that the stars were a meteor shower. The correct
date of the eclipse is the 4th calends of November rather
than the second calends of October. However, as we shall
see, the same person recorded two other eclipses, and he made
the same error with regard to the month in all three records.
In other words, it looks as if he did not understand how to
use the Roman calendar and that, if we grant him his method
of misusing it, he made a mistake only in the day of the
month. Since the record was made at least a century after
the event, the reliability is only 0.5, and the place is
Forli. The value of η is 0 and the standard deviation of
the magnitude is 0.18.

1406 June 16 E,I. Under the year 1406, on page 517 of
the cited edition, Rampona [ca. 1425] has: "And on the 16th
of June was an eclipse of the sun, and it obscured about
three parts of the sun." Unfortunately the writer does not
tell us what kind of parts these were. I think the most
plausible meaning is that the parts were quarters, and I
shall thus use a magnitude of 0.75, but I shall take the
standard deviation of this estimate to be 0.15 because of
the uncertainty. The reliability is 1 because we are now
close to the time when the annals were compiled, and the
place is Bologna.

---

†This is the way the text appears. I presume the writer
meant between the third and sixth hours of the day.

1408 October 19a E,I. Again the year (1408) was left blank in the original of <u>Forolivienses</u> [ca. 1473], and someone in the 16th century added a record of a solar eclipse: "An eclipse of the sun on the 14th calends of October." He should have said the 14th calends of November, so he made the same error in the month that he made in the record 1399 October 29 E,I. This time, however, he gave the day of the month correctly. The characteristics of this record are the same as those of the earlier one by the same person.

1408 October 19b E,I. <u>Hieronymus</u> [ca. 1433] gives four eclipse records. The first two have a style that is quite different from the last two, and I believe that he copied them from some source unknown to us. The first one appears on page 11 under the year 1408 and reads: "The same year on the 19th day of October there was an eclipse of the sun about an hour before the middle of the day, and it lasted for an hour and fifty minutes." Both this record and the next one give the duration to an unlikely precision, while Hieronymus's last two records do not give the duration at all. I believe that the records were originally made by someone who calculated the durations. For this reason, I shall use a reliability of only 0.5 for this record. The place is Forli, but this record seems to have nothing in common with the preceding one. If we could rely upon the duration, we could estimate the magnitude. Even though we cannot do this, I believe we may safely take the eclipse to be partial since it was so short. I shall use 0.81 for the magnitude and 0.14 for the standard deviation; these are conventional values for a partial eclipse.

1409 April 15a E,I. As with the eclipse of 1408 October 19, we have two records of this eclipse, and both are from Forli. Again the year in <u>Forolivienses</u> [ca. 1473] was originally left blank and the same person from the 16th century added a record of an eclipse: "An eclipse of the sun on the 17th calends of April." Again the recorder has the wrong month but the right day; he should have said the 17th calends of May. The characteristics are the same as for the record 1399 October 29 E,I.

1409 April 15b E,I. On page 13 of <u>Hieronymus</u> [ca. 1433] we find: "Year of the Lord 1409, day the 15th of April, there was an eclipse of the sun, and it lasted 2 hours, 8 minutes, and 24 seconds; and the eclipse was in Taurus." The sun was in the zodiacal sign of Taurus at the time of the eclipse. I suppose it is possible that some unknown person in Forli in 1409 was using a clock to measure the duration of an eclipse to a precision of a second, but it is highly unlikely. Bernard Walther† is supposed to be the first person who used a clock as an adjunct to astronomical observations. Thus I think that the duration was calculated. Again

---

†Walther observed in Nuremberg from 1475 to 1504, and he first used a clock in 1484. See <u>Beaver</u> [1970].

I think we may safely assume that the eclipse was partial because the duration was so short, even though I do not think we can safely use the duration in a quantitative way.†
This record receives the same characteristics as 1408 October 19b E,I.

1415 June 7a E,I. This is the first record in Hierony-
mus [ca. 1433] that we can safely attribute to Hieronymus himself, I believe. It is on page 25 of the cited edition, under the year 1415: "The same year was a small eclipse of the sun, on the 8th day of June, on Friday, in the revolution of the moon; and the sun was then in the sign of Gemini, in the 23rd degree, and it was at two hours of the day." The hour looks reasonable and the eclipse was on a Friday. Thus Hieronymus made a simple slip when he said that the date was June 8, and he did not take the week day from a calendar while using the wrong date. It is also interesting that the sun was in fact in the 23rd degree of Gemini (longitude 83°) at the time of the eclipse. This shows that Hieronymus had some knowledge of astronomy and that he did not use the ec-clesiastical tables of the sun which often appear on church calendars.‡ When he says that the eclipse was in the "revo-lution" of the moon, I imagine he meant the renovation (the new moon). However, it may be that revolution was used with this peculiar technical meaning at this time. Note that Hieronymus does not give the duration of the eclipse in this record.

Perhaps the most interesting feature is that Hieronymus says that the eclipse was small. We can be sure that the eclipse was in fact larger in Forli than the eclipses of 1408 October 19 and 1409 April 15. This conclusion is based upon calculation, but the calculation of the relative magni-tudes does not depend upon the accelerations and we can use the conclusion without reasoning in a circle. Thus we should take small to mean merely that the eclipse was less than total, and I shall use the conventional values for a partial eclipse. That is, I shall use a magnitude of 0.81 with a standard deviation of 0.14. The place is again Forli, but I think we can use a reliability of 1 for this record.

1415 June 7b E,I. Petrus [ca. 1417, p. 98] records the eclipse as one of the events of 1415: "Also on Friday the 7th of June at the 3rd hour of the morning or about,

---

†According to calculations with provisional accelerations, the duration was 2h 34m, and the magnitude was 0.73. Thus the stated duration is not highly accurate.

‡A church calendar often gave the date on which the sun entered each sign of the zodiac. These dates were calcu-lated from astronomical tables prepared in the 4th century, on the assumption that the year contained exactly $365\frac{1}{4}$ days. The error in this assumption amounted to about 6° in the position of the sun by 1415.

there was an eclipse,† and the sun obscured, but not totally, etcetera." The record actually ends with the word etcetera. As we noted in the preceding record, the eclipse was on a Friday. The reliability is 1, the place is Rome, the magnitude is 0.81 since the eclipse was not total, and the standard deviation is 0.14.

1415 June 7c E,I. On page 723 of the cited edition of S. Georgii [ca. 1415] we find: "In the year of the Lord 1415 there was an eclipse of the sun on the 3rd day after Boniface, and the dominical letter that year was F, and that eclipse was on the 6th day of the week, after the octave of Corpus Christi." The day dedicated to S. Boniface is June 5, so June 7 was the 3rd day after, in inclusive counting. Corpus Christi is the Thursday following the 8th Sunday after Easter, counting Easter itself as 0 rather than 1. The octave of Corpus Christi is the Thursday after that, so it came on June 6 in 1415. The annalist says correctly that June 7 was the Friday (6th day of the week) after June 6.

Finally, with regard to calendrical matters, the dominical letter corresponds to the day of January that comes on the first Sunday (dies Dominica) of January, in a year that is not a leap year. In 1415, January 6 was the first Sunday, so the dominical letter was F. The definition of the dominical letter for a leap year is more complicated, and the interested reader should consult pages 421-423 of the Explanatory Supplement [1961].

The reliability of this record is 1, and the place is S. Giorgio d. E. The value of η is 0 and the standard deviation of the magnitude is 0.18.

1431 February 12a E,I. The last eclipse record in Forolivienses [ca. 1473] is in the original and not in an addition made in the 16th century: "In the year of the Lord 1431, the 12th day of February, papa Martin V the holy pontifex of happy memory went to Christ; and the same day there was an eclipse of the sun after his death, on the 21st day." The writer is quite confused about the chronology of these events, but we can easily see what happened. According to Mercati [1947], Martin V died on 1431 February 20, eight days after the eclipse. There is a good chance that the news reached Forli on the next day, which is one of the dates that appears in the record. The chronicler then accidentally interchanged the dates and concluded that the eclipse came after Martin's death. Finally he wrote "the same day" when he meant "the same month". These errors prove that the record was not made at the time of the events, so the reliability is only 0.5; the place is Forli. The value of η is 0

†The set of letters "glisis" actually appears in the text, at least in the printed edition. I imagine that an initial "e" has been omitted, so that eglisis was intended. The word for eclipse was spelled in many ways in medieval writing, and I have not usually noted the variant spellings. The spelling eglisis is no more unusual than others that were used.

and the standard deviation of the magnitude is 0.18.

1431 February 12b E,I. Hieronymus [ca. 1433, p. 53]
also gives a record of this eclipse that comes from Forli:
". . and in the same month, namely February, on the 12th
day, on Monday† was an eclipse of the sun at the 21st hour,
whence followed an eclipse seen by day." The eclipse was on
a Monday, and the 21st hour is reasonable if the hours are
counted from the preceding sunset. This record receives a
reliability of 1. Otherwise it has the same characteristics
as the preceding record.

1431 February 12c E,I. The chronology is also confused
in this record from Guerrierus [ca. 1472, p. 48]: "During
the year 1431, on the 18th of February, Pope Martin V died.
On the same day was a very large eclipse." Since the eclipse
was very large, I shall take the magnitude to be 0.90, with
a standard deviation of 0.05. The reliability is 0.5 and the
place is Gubbio.

1431 February 12d E,I. dello Mastro [ca. 1484, p. 85]
writes: "Paolo recorded that in the year 1431 on the 11th
day of February in the 20th hour the sun was eclipsed, which
indicated the death of Pope Martin V." The Paolo in question
seems to be the author himself. It is interesting that the
chroniclers had so much trouble with the date of this eclipse.
I believe that the chronicler in this case at least means to
put the eclipse earlier than the death of Pope Martin, although the wording leaves a question on this point. The hour
seems to be counted from the preceding sunset. The record is
from Rome and it receives a reliability of 1 since it seems
to have been made by the chronicler. The value of η is 0
and the standard deviation of the magnitude is 0.18.

1431 February 12e E,I. degli Unti [1440] records this
eclipse in this way: "MCCCCXXXI. Note, on the 12th of February at 21 hours, which was the Monday of Carnival, the sun
was so eclipsed that it appeared like dark night, and the
face of the sun became black as coal. That was at the turning of the moon." The use of volto here, which I have translated as turning, may be parallel to the use of revolution
(revolutio) in some earlier records. The day of the eclipse
was indeed the Monday before Mardi Gras, so it was Carnival
Monday. The imaginative comparison of the sun to a coal
seems equivalent to a statement that the eclipse was total,
so the standard deviation of the magnitude is 0.034. The
reliability is 1 and the place is Foligno.

---

†At this point in the printed edition, the word quando,
meaning when, follows in parentheses. The occurrence of
parentheses usually indicates something the editor has
supplied because he thinks it is necessary to complete the
sense. Perhaps the word occurs in the original in this
case, and the editor has put it into parentheses in order
to indicate that it must be omitted in order for the passage to make sense.

1448 August 29a E,I. Palmerius [ca. 1474, p. 157] has
recorded this eclipse: "On the 29th day of August at the 6th
hour the sun was obscured by about two thirds." He has already identified the year as 1448. This eclipse was within
the lifetime of the author and we may give it a reliability
of 1; the place is Florence. We may accept his estimate of
the magnitude, and I shall use a standard deviation of 0.1
for his estimate.

1448 August 29b E,I. Ghirardacci [ca. 1509, p. 126]
also records this eclipse under the year 1448: "On the 29th
of August, on Thursday, at 15 hours the sun was obscured for
the space of three quarters of an hour." The eclipse was on
a Thursday and the hour is reasonable if it is measured from
the preceding sunset. The duration is not reasonable; it is
far too short. However, we can at least take the stated duration as an indication that the eclipse was partial. The
reliability is 0.5 since the record occurs in a late compilation, and the place is Bologna. I shall take the magnitude
to be 0.81 with a standard deviation of 0.14.

1460 July 18 E,I. Ghirardacci [ca. 1509, p. 174] also
records this eclipse, under the year 1460: "On the 18th
day of July, a Friday, at 11 hours and a half, was an eclipse
of the sun in which it was thus obscured for a half." The
eclipse should have been a maximum at about sunrise, so the
time is reasonable if the hour is measured from sunset. The
reading is ambiguous, but I believe that "obscured for a half"
means that the eclipse lasted for half an hour, not that the
magnitude was 0.5. Here we have a duration that is much too
short, just as we did in the preceding record by the same
author. One wonders how he defined the beginning and end
of an eclipse. I shall use the same characteristics that I
used for the preceding record.

1465 September 20a E,I. This record is from Guerrierus
[ca. 1472, p. 80]: "On the 20th of September 1465, was an
eclipse at the 23rd hour and it lasted until sunset; three
quarters of the sun was eclipsed." We are close to the end
of the work and the record should be original. Hence the
reliability is 1; the place is Gubbio. The estimated magnitude is 0.75, and I shall use a standard deviation of 0.1.

1465 September 20b E,I. Palmerius [ca. 1474, p. 183]
gives the only other record of this eclipse. Under 1465 he
has: "On the 24th of September in the evening at about sunset the sun was eclipsed, and afterward on the 4th of October
the moon was eclipsed as it was coming out over the mountain."
He has the right date for the lunar eclipse but a wrong one
for the solar eclipse. Apparently moonrise occurred over a
mountain and when the moon came over the mountain on October
4 it was already eclipsed. The reliability is 1 and the
place is Florence. Since there is no indication about the
magnitude of the eclipse, I shall take η to be 0, with 0.18
for the standard deviation of the magnitude.

1478 July 29 E,I.   As we noted in the preceding section, Hyvanus [ca. 1478] deals only with the year 1478.   On page 30, we have this record of an eclipse: "A solar eclipse, on the 3rd calends of August, occurring about two hours of the day in the sign of Leo, struck terror into the people."   The date is wrong by 1 day, but the sun was in the sign of Leo, as stated.   However, the eclipse was about two hours after noon, not at about two hours of the day (after sunrise).   I don't know whether this represents a shift of meaning or whether it is a mistake of the writer.   The reliability is 1, the place is Volterra, the value of η is 0, and the standard deviation of the magnitude is 0.18.

1485 March 16 E,I.   Ferrarese [ca. 1502, p. 119] has a record of this eclipse under the year 1485:   "On the 16th day of March, appeared a great eclipse of the sun at the 21st hour and lasted to the end of the 23rd hour, and it foretold the death of many people from unknown evil."   The eclipse was late in the day, so we seem to have again the practice of measuring time from sunset.   The reliability is 1, and the place is Ferrara.   Since the eclipse was great, I shall use 0 for η, with 0.10 for the standard deviation of the magnitude.

1567 April 9 E,I.   This is a valuable record because the eclipse involved had an extremely narrow zone of totality and the observer Clavius was close to, and perhaps in, that zone. Clavius in fact observed two total eclipses, apparently without arranging to do so.   One was the eclipse of 1560 August 21.   Since he observed it in Portugal, the record of that eclipse will be discussed in Section XI.7.   His account of the eclipse of 1567 April 9 [Clavius, pp. 508-509] reads: "I observed the other one in Rome in the year 1567, also about noon,† in which the moon was again placed between the sight and the sun, but it did not however obscure the entire sun as in the first eclipse, but (which perhaps never happened at another time) there remained on the sun a thin circle surrounding the entire moon on all sides."

Ginzel does not mention this record in any of his works that I know, but Johnson [1889, pp. 60-61] discusses it at some length.   M and MS also use it.   One interesting feature of the eclipse concerns whether it was total or annular.   The trained astronomer Clavius thought it was annular.   Johnson concluded from calculation that it was annular, but he notes that "the augmentation of the moon's semi-diameter would almost produce totality."   That is, the eclipse barely remained annular for all choices of parameters that Johnson used.   However, Oppolzer [1887], M, MS, and I all agree that the eclipse was annular-total and that the point where the eclipse was total at local noon was near or perhaps at Rome. The presumption is that Clavius saw the chromosphere and

---

†The word used here is meridiem, which is our noon.   Clavius says "also" because the eclipse of 1560 August 21 had been total at about noon where he saw it.

-339-

mistook it for part of the sun's disk.†

The reliability of the record is safely 1, and we may
take 0 for the value of η. The place and the standard deviation of the magnitude both need discussion.

The observation was certainly made in Rome, but M [p.
8.67] assumes a more precise location. Since Clavius was
teaching in the Collegio Romano at the time, M assumes that
the observation was made there, and he gives the coordinates
of the Collegio Romano to the thousandth of a degree.‡ I do
not know what the arrangements at the Collegio were at the
time, but it is likely that Clavius in fact lived there. However, even if he did, he did not have to be on its premises
at the time of the eclipse, and it seems to me that he could
readily have been taking a walk and hence have been a kilometer or so from the Collegio. Nonetheless I shall follow
M in using the Collegio as the place, since the region that
we should use is certainly centered on the Collegio, and the
effect of a kilometer's uncertainty in the place is negligible compared with the effect of the uncertainty in the magnitude.

Here we are dealing with an unusual eclipse and an unusual observer, and we should not follow the conventional
rules in assigning the standard deviation of the magnitude.
When an eclipse is total with a reasonable amount to spare,
there is little ambiguity about whether an observer sees the
eclipse as total or partial. Here the situation is quite
different, as we see from the fact that Clavius mistook a
total eclipse for an annular one. This presumably means that
he saw the chromosphere on all sides of the moon even at
maximum eclipse.

However it is not necessary for the eclipse to be total
in the strict astronomical sense in order for this to happen,
and I cannot follow M (and MS) in assuming that the eclipse
was strictly total. Although Clavius refers to a thin circle
of light surrounding the moon, I do not think we can interpret this as necessarily meaning a symmetrical situation; we
cannot assume that he chose his words this carefully.‡ We
can only assume that some of the chromosphere remained visible
on all sides of the moon. I have not been able to find information on the limits that permit this to happen. However,
the height of the chromosphere is about 0.02 times the sun's
radius. Hence, in an eclipse in which the sun's and moon's
apparent sizes are so nearly matched, I believe that the
eclipse could easily be 0.01 from strict totality and still
allow the chromosphere to be bright on all sides of the moon.

---

†See the discussion of the eclipse of 71 March 20 in Section
VI.1.

‡He gives a latitude of 41°.896 and a longitude of 12°.482
[p. A2.2].

‡The word used is circulus. I think a writer might well
apply it to an annulus whose width was not uniform. The
record 1333 May 14 E,I above in fact uses this word to
mean a crescent.

TABLE X.2

ITALIAN ECLIPSE RECORDS THAT ARE NEW
OR THAT HAVE BEEN RE-INTERPRETED

Designation	Place	Standard deviation of the magnitude	Reliability
939 Jul 19c E, I	Pisa	0.18	0.5
1033 Jun 29b E, I	Monte Cassino	0.14[a]	0.5
1093 Sep 23 E, I	Monte Cassino	0.14[a]	0.5
1133 Aug 2 E, I	Monte Cassino	0.14[a]	1
1163 Jul 3 E, I	Monte Cassino	0.18	1
1178 Sep 13c E, I	Vicenza	0.18	0.5
1178 Sep 13d E, I	Terracina	0.18	0.5
1178 Sep 13e E, I	Pisa	0.05[b]	1
1187 Sep 4a E, I	Cremona	0.18	0.5
1187 Sep 4b E, I	Milan	0.18	0.5
1239 Jun 3c E, I	Vicenza	0.034	1
1239 Jun 3d E, I	Siena	0.034	1
1239 Jun 3e E, I	Padua	0.18	1
1239 Jun 3f E, I	Padua	0.11	1
1239 Jun 3g E, I	Mantua	0.18	1
1239 Jun 3h E, I	Mantua	0.054	1
1239 Jun 3i E, I	Bergamo	0.18	1
1239 Jun 3j E, I	Piacenza	0.11	1
1239 Jun 3k E, I	Milan	0.05[c]	1
1239 Jun 3𝑙 E, I	Parma	0.054	1
1239 Jun 3m E, I	Reggio nell'Emilia	0.034	1
1239 Jun 3n E, I	Monte Cassino	0.18	1
1239 Jun 3o E, I	Orvieto	0.18	1
1239 Jun 3p E, I	Orvieto	0.034	1
1239 Jun 3q E, I	Bologna	0.054	0.5

[a] The eclipse will be taken as partial, with a magnitude of 0.81.

[b] The magnitude will be taken as 0.90.

[c] The magnitude will be taken as 0.94.

TABLE X.2 (Continued)

Designation	Place	Standard deviation of the magnitude	Reliability
1239 Jun 3r E, I	Bologna	0.18	0.5
1239 Jun 3s E, I	Foligno	0.054	0.5
1241 Oct 6 E, I	S. Giorgi d.E.	0.18	1
1263 Aug 5a E, I	Padua	0.18	1
1263 Aug 5b E, I	Orvieto	0.18	1
1290 Sep 5 E, I	Faenza	0.2^d	1
1310 Jan 31a E, I	Bologna	0.18	0.25
1310 Jan 31b E, I	Forli	0.18	0.25
1330 Jul 16a E, I	Foligno	0.2^e	1
1330 Jul 16b E, I	Bologna	0.18	0.5
1330 Jul 16c E, I	Florence	0.2^e	1
1333 May 14 E, I	Parma	0.14^a	1
1339 Jul 7 E, I	Siena	0.14^a	1
1354 Sep 17 E, I	Bologna	0.14^a	0.5
1361 May 5 E, I	Siena	0.18	1
1384 Aug 17 E, I	Florence	0.1^f	1
1386 Jan 1a E, I	Bologna	0.11	1
1386 Jan 1b E, I	Florence	0.05^b	1
1399 Oct 29 E, I	Forli	0.18	0.5
1406 Jun 16 E, I	Bologna	0.15^g	1
1408 Oct 19a E, I	Forli	0.18	0.5
1408 Oct 19b E, I	Forli	0.14^a	0.5
1409 Apr 15a E, I	Forli	0.18	0.5
1409 Apr 15b E, I	Forli	0.14^a	0.5
1415 Jun 7a E, I	Forli	0.14^a	1

[a]The eclipse will be taken as partial, with a magnitude of 0.81.

[b]The magnitude will be taken as 0.90.

[d]The magnitude will be taken as 0.67.

[e]The magnitude will be taken as 0.5.

[f]The magnitude will be taken as 0.71.

[g]The magnitude will be taken as 0.75.

TABLE X.2 (Continued)

Designation	Place	Standard deviation of the magnitude	Reliability
1415 Jun 7b E,I	Rome	0.14^a	1
1415 Jun 7c E,I	S. Giorgio d.E.	0.18	1
1431 Feb 12a E,I	Forli	0.18	0.5
1431 Feb 12b E,I	Forli	0.18	1
1431 Feb 12c E,I	Gubbio	0.05^b	0.5
1431 Feb 12d E,I	Rome	0.18	1
1431 Feb 12e E,I	Foligno	0.034	1
1448 Aug 29a E,I	Florence	0.10^d	1
1448 Aug 29b E,I	Bologna	0.14^a	0.5
1460 Jul 18 E,I	Bologna	0.14^a	0.5
1465 Sep 20a E,I	Gubbio	0.1^g	1
1465 Sep 20b E,I	Florence	0.18	1
1478 Jul 29 E,I	Volterra	0.18	1
1485 Mar 16 E,I	Ferrara	0.10	1
1567 Apr 9 E,I	Collegio Romano	0.01	1

[a]The eclipse will be taken as partial, with a magnitude of 0.81.
[b]The magnitude will be taken as 0.90.
[d]The magnitude will be taken as 0.67.
[g]The magnitude will be taken as 0.75.
Note: The eclipse is taken to be central if no estimate of the magnitude is given.

Thus I shall use 0.01 for the standard deviation of the magnitude. By the method of Section XIII.1, I find that 0.0063 is the corresponding value of η.

The Italian records that have not been used before, and those that are being used again but with changed characteristics, are listed in Table X.2.

CHAPTER XI

ECLIPSE RECORDS FROM OTHER PARTS
OF EUROPE

1. Preliminary Remarks

In this chapter I shall present the eclipse records
from parts of Europe that have not been considered in earlier
chapters, for years later than 100. In doing so, however, I
exclude records from the Byzantine Empire, even though some
of them come from places that are normally considered European.
I do this because much of the Byzantine Empire lay outside of
Europe, and it is more convenient to take up all Byzantine
records together in the next chapter.

The parts of Europe included in this chapter are (a)
Belgium and the Netherlands, designated by the letters BN,
(b) Central Europe, designated by the letters CE, including
Austria, Czechoslovakia, and Switzerland, (c) Poland and
Russia, designated by the letters PR, (d) the Scandinavian
countries, including Iceland, designated by Sc, and (e)
Spain and Portugal, designated by SP.

In MCRE I had records from Spain, which I designated by
Sp, but I did not have any from Portugal. Here, for convenience, I shall include Spain and Portugal in a single provenance that I shall designate by SP, as I have already mentioned. If the reader will consider that "Sp" has been
changed into "SP" in all records that have been used before,
he can keep all the eclipse designations used in MCRE.

The other parts of Europe are fairly useful for eclipse
records after 1300. We shall find 18 such records from the
other parts of Europe, compared with 9 from the British Isles,
4 from France, 10 from Germany, and 33 from Italy.

Table XI.1 lists the eclipse records from the other
parts of Europe that will be used here without change. It
has the usual format for such tables.

2. Sources from Belgium and the Netherlands That Have Not Been Used Before

Balduinus [ca. 1294]. All we know about Balduinus is
what he says about himself in his chronicle. He was educated
at the Praemonstratensian monastery in Ninove, which is a
small place in Flanders, and he later became a deacon and
canon there. The monastery was founded in 1137, and I ignored all of the chronicle before that date, since it could
not be original. The chronicle as written by Balduinus runs
to 1294, and all the eclipse records in it come before then.
Some items that do not concern us, such as a roster of the
abbots of the monastery, were added by others.

Menkonis [ca. 1273]. This chronicle was written in a

TABLE XI.1

ECLIPSE RECORDS FROM OTHER PARTS OF EUROPE
THAT HAVE BEEN USED BEFORE

Designation			Place	Standard deviation of the magnitude	Reliability
402 Nov 11	E,SP		Galicia	0.18	0.5
418 Jul 19	E,SP		Galicia	0.18	0.2
447 Dec 23	E,SP		Galicia	0.18	1
458 May 28	E,SP		Galicia	0.1^a	1
655 Apr 12	E,SP		Spain	0.054	0.5
760 Aug 15	E,CE		Gottweih	0.18	0.1
764 Jun 4	E,CE		Gottweih	0.18	0.1
787 Sep 16	E,CE		Melk	0.18	0.1
807 Feb 11a	E,CE		Salzburg	0.18	0.5
807 Feb 11b	E,CE		Melk	0.18	0.1
810 Nov 30a	E,CE		Salzburg	0.18	0.5
810 Nov 30b	E,CE		Gottweih	0.18	0.1
891 Aug 8	E,CE		S. Gall	0.18	0.5
939 Jul 19	E,CE		S. Gall	0.18	1
968 Dec 22	E,CE		S. Gall	0.18	1
968 Dec 22a	E,BN		Liège	0.034	0.05
968 Dec 22b	E,BN		Liège	0.18	0.05
1009 Mar 29a	E,BN		Liège	0.18	1
1018 Apr 18	E,BN		Gembloux	0.18	0.1
1087 Aug 1	E,CE		Zwettl	0.18	0.5
1093 Sep 23a	E,CE		Admont	0.18	0.2
1093 Sep 23b	E,CE		Prague	0.18	1
1093 Sep 23c	E,CE		Schaffhausen	0.054	1
1109 May 31	E,BN		Vormezeele	0.18	1
1118 May 22	E,BN		Vormezeele	0.18	1

---

[a]The eclipse will be taken as partial, with a magnitude of
0.8.

TABLE XI.1 (Continued)

Designation	Place	Standard deviation of the magnitude	Reliability
1124 Aug 11 E,BN	Belgium	0.14[b]	1
1124 Aug 11 E,CE	Prague	0.18	1
1133 Aug 2a E,BN	Gembloux	0.054	1
1133 Aug 2b E,BN	Fosse	0.054	1
1133 Aug 2d E,BN	Liège	0.054	1
1133 Aug 2e E,BN	Egmond aan Zee	0.034	1
1133 Aug 2a E,CE	Melk	0.082	1
1133 Aug 2b E,CE	Salzburg	0.034	1
1133 Aug 2c E,CE	Admont	0.034	1
1133 Aug 2f E,CE	Reichersberg	0.11	0.5
1133 Aug 2g E,CE	Sazava	0.18	0.5
1133 Aug 2i E,CE	Hradisch	0.18	1
1140 Mar 20a E,BN	Gembloux	0.034	1
1140 Mar 20a E,Sc	Lund	0.18	1
1140 Mar 20b E,Sc	Denmark	0.068	0.5
1140 Mar 20c E,Sc	Denmark	0.068	0.5
1147 Oct 26b E,BN	Liège	0.05	1
1147 Oct 26c E,BN	Egmond aan Zee	0.18	1
1147 Oct 26 E,CE	Engelberg	0.18	0.5
1153 Jan 26 E,CE	Admont	0.18	0.5
1178 Sep 13 E,Sc	Denmark	0.18	0.5
1185 May 1 E,Sc	Oslo to Frankfurt	0.18	0.5
1187 Sep 4a E,CE	Melk	0.18	1
1187 Sep 4b E,CE	Salzburg	0.18	1
1187 Sep 4c E,CE	Kremsmünster	0.054	0.5
1187 Sep 4a E,Sc	Denmark	0.18	0.5
1191 Jun 23a E,BN	Fosse	0.18	1
1191 Jun 23b E,BN	Liège	0.18	1
1191 Jun 23a E,CE	Melk	0.18	1
1191 Jun 23b E,CE	Salzburg	0.18	1

[b]The magnitude will be taken as 0.81.

TABLE XI.1 (Continued)

Designation	Place	Standard deviation of the magnitude	Reliability
1191 Jun 23c E,CE	Kremsmünster	0.18	0.5
1191 Jun 23d E,CE	Reichersberg	0.18	1
1230 May 14 E,Sc	Denmark	0.18	0.5
1236 Aug 3 E,Sc	Iceland	0.18	1
1239 Jun 3a E,CE	Salzburg	0.18	1
1239 Jun 3b E,CE	Lambach	0.18	1
1241 Oct 6 E,BN	Fosse	0.18	1
1241 Oct 6a E,CE	Salzburg	0.054	1
1241 Oct 6b E,CE	Admont	0.054	1
1241 Oct 6c E,CE	Lambach	0.034	1
1241 Oct 6d E,CE	Reichersberg	0.034	1
1241 Oct 6e E,CE	Prague	0.068	1
1241 Oct 6 E,Sc	Denmark	0.18	0.5
1255 Dec 30 E,CE	Prague	0.18	1
1263 Aug 5 E,CE	Salzburg	0.14[b]	1
1263 Aug 5a E,Sc	Bergen	0.18	1
1263 Aug 5b E,Sc	Rye	0.18	1
1263 Aug 5c E,Sc	Essenbek	0.18	0.5
1267 May 25 E,CE	Salzburg	0.18	1
1270 Mar 23 E,Sc	Rye	0.18	1

[b]The magnitude will be taken as 0.81.
Note: The eclipse is taken to be central if no estimate of
  the magnitude is given.

monastery in Frisia whose name is Floridi Orti in the genitive case. I did not see an occurrence of the name in the
nominative case, but I imagine that it is Floridus Ortus.
It was located in a place whose name has the root Werum. I
believe it is the place now known as Werm, in Belgium, at
latitude 50° 50'N, longitude 5° 28'E. The chronicle was begun by Emo, the first abbot, who carried it to his death in
1237. Menkonis succeeded him as abbot, and a note added to
the chronicle says that he continued it to 1273. An anonymous continuation carries it to 1296, but the continuation
does not concern us.

S. Iacobi [ca. 1393]. The Annales S. Iacobi come from

the monastery of S. Jacob, located on an island in Liège
[MCRE, p. 223]. The part of the annals from 1 to 1055 was
compiled by a single person and contains nothing useful for
our purposes. From 1056 to 1174, the annals were compiled
by many different people; this is the pattern we expect for
annals that are being kept as the events occur. I designated
this part of the annals as S. Iacobi [ca. 1174] and used it
in MCRE. After 1174 the annals for some reason revert to
1164, and the third part consists of intermittent records
running from 1164 to 1393; this is the part being cited as
S. Iacobi [ca. 1393]. This part also seems to be original
and contemporaneous.

3. **Sources from Central Europe That Have Not Been Used
    Before**

     Basileenses [ca. 1277]. I did not notice any statement
by the editor about the origin of these annals. However, they
make frequent references to the Praedicatorian monastery and
to Basel, so that is probably the place. The annals run only
from 1266 to 1277, but they are rather extensive to run for
such a short time. There is a continuation that was made in
Colmar from 1278 to 1305, but it contains nothing that is
useful for this work.

     Claustroneoburgensis [ca. 1383]. I discussed the com-
plicated set of annals called Mellicenses [ca. 1564] in MCRE
[pp. 252-254]. They were kept in a monastery in Melk, which
is about 70 kilometers from Vienna on the way to Salzburg.
The annals that were first prepared there run through 1123.
At this time, many copies of the annals were made and taken
to other places where they were continued for various periods
of time. One copy was taken to Klosterneuburg, which is only
a few kilometers northwest of Vienna on the Danube. There
they were continued until well into the 15th century, and the
editor identifies seven parts that differ distinctly from each
other. The date 1383 is the close of what seemed to me the
most significant part, and I use that date for purposes of
citation, even though there are some entries with later dates.

     Florianensis [ca. 1310]. This is another continuation
of Mellicenses [ca. 1564]; this one was made in the monastery
of S. Florian, a few kilometers southeast of Linz. A town
called Market S. Florian can be found on large-scale maps.
According to Ginzel [1884], its coordinates are 48° 13′N by
14° 29′E. Some entries as late as 1289 are taken from other
sources such as Claustroneoburgensis, even though there is
much original material before that date.

     Frisacenses [ca. 1300]. These are described as being
from the Praedicatorian monastery of Frisacus in Carinthia
(Kärnter). I believe that this is the place now called
Friesach. The annals are rather intermittent, having only
about 30 entries from 1217 to 1300.

Juvavenses [ca. 956]. The editor calls these annals
the Annales Juvavenses Maximi, and he prints them in parallel
with Annales Juvavenses Maiores and two other sets of annals.
I used the Annales Juvavenses Maiores in MCRE [p. 251], where
I designated them as Juvavenses [ca. 975]. However, there
are some differences between this edition and the one used
for MCRE. This edition is from volume XXX, part II of the
Monumenta Germaniae Historica, Scriptores series while the
edition used before is from volume I. In particular, the
last entry in volume I is dated 975 but the last entry in
the volume XXX edition is dated 976. The annals being introduced here [Juvavenses, ca. 956] run from 725 to 829, with a
continuation to 956; they do not occur in volume I.

Novimontensis [ca. 1396]. This is still another continuation of Mellicenses [ca. 1564]. It was written in a monastery at Neuberg, and the first entry in the continuation
tells us that the monastery was founded by Otto, duke of
Austria. Oddly the entry does not tell us the year, but we
find from a later entry that the first abbot died in 1333.
Ginzel [1884] calls the place Neuberg and gives its coordinates as 47° 4'N in latitude and 15° 39'E in longitude. This
is close to Graz, but I cannot find a place with this name
on a map. The continuation exists in two codices which frequently differ considerably from each other.

Praedicatorum Vindobonensium [ca. 1283]. This is also
a continuation of Mellicenses [ca. 1564], made at the Praedicatorian monastery in Vienna. In fact, we might call it a
continuation twice removed. To a large extent, it is copied
from other continuations. However, it does have some original material.

S. Rudberti [ca. 1327]. These are from the abbey now
called S. Peter's in Salzburg, which was known for a long
time by the name of S. Rudbertus who founded it about 700.
I used two sets of annals from this abbey in MCRE [p. 256],
and this set is a continuation of one of them. It runs from
1286 to 1327 and obviously has at least two authors. The
editor in fact prints this continuation in two parts, which
I have condensed into one source because they seem to form a
single entity.

Sancrucensis [ca. 1310]. Leopold, duke of Austria,
called the Pious, founded the monasteries of Klosterneuburg
and Heiligenkreuz near Vienna about 1114. Klosterneuburg
(see the discussion of Claustroneoburgensis [ca. 1383]) lies
a few kilometers north of Vienna while Heiligenkreuz (Holy
Cross) lies about the same distance to the southwest. Sancrucensis is yet another continuation of Mellicenses [ca.
1564], this one being made at Heiligenkreuz. It comes in
four separate parts which the editor distinguishes in the
printed edition. He designates them, in order, as Part I,
Part II, Historia Annorum 1264-1279, and Part III. I do not
know why he chose this particular scheme.

The first three parts are annals kept by various people
in a contemporaneous manner. The fourth part, the part that
the editor calls Part III, has an odd property, however. It
runs from 1279 to 1310, but the entries from 1279 to 1301 are
not original. Instead, they have been excerpted from Vindobonensis [ca. 1327], which will be described below. After
this, we find original entries from 1302 to the end of the
annals in 1310.

According to my large-scale map of Austria, Heiligenkreuz is at latitude 48° 03' and longitude 16° 08'. Ginzel
[1884] gives latitude 48° 04' and longitude 16° 14', in reasonable agreement. Times Atlas [1955], however, gives the
latitude as 46° 31', which I feel sure is seriously in error.

Stirensis [ca. 1346]. This is a continuation of the
French source Honorius [ca. 1137], which was used in MCRE
[p. 302]. From the nature of the contents, we can tell that
this continuation was written in Styria and probably in Graz;
I shall assume that Graz is the place. The manuscript is
now in a library in Vienna. The source contains about 30
entries for years from 1220 to 1346, plus a short list of
kings.

Vindobonensis [ca. 1327]. This set of annals from
Vienna is also a continuation of Mellicenses [ca. 1564]. It
is in one hand until 1267 and in many hands for later years.
Thus it is presumably original for years later than 1267.
For some reason there are no entries between 1303 and 1312.

Zwetlensis [ca. 1329]. This is the last continuation
of Mellicenses with which we shall have to deal, at least in
this work. There were two other continuations written in
Zwettl that I used in MCRE [pp. 256-260]; the editor identifies this one as Continuatio Zwetlensis III.

4.   Sources from Poland and Russia

Colbazienses [ca. 1568]. This source consists of a
manuscript written in the 12th century, to which annals running to 1568 have been added. The entries seem to be contemporaneous from about 1124 on. The annals are described
as coming from the monastery at Colbaz in Pomerania, on the
Ina River near Stargard. I have not been able to find Colbaz,
so I shall use the coordinates of Stargard instead. Since
the Ina is a short river and Stargard is near its midpoint,
this cannot be a serious error.

Cracoviensis [ca. 1331]. These annals begin with the
death of Bede, which is put in 730,† and they are original
from 965 until their close in 1331. Many references to the
affairs of Krakow prove that the annals were kept in that
city.

---

†The correct year is 735.

Heinricus [ca. 1227]. All we know about Heinricus is that he was probably a priest, and that he wrote a chronicle of Livonian affairs from 1186 to 1227. We must assume that the place for eclipse observations can be anywhere in Livonia, which lies in the modern soviet republics of Estonia and Latvia. We may take it to be the rectangle lying between latitudes 57° and 59° and between longitudes 24° and 26°.

Igor [ca. 1185]. I use the name Igor in citing this work not because he wrote it but because it is about him; the author is unknown. Igor was a prince of Novgorod-Seversk† who led an expedition against a nomadic tribe called the Kumans. The campaign ended in Igor's defeat and capture.

Lubenses [ca. 1281]. This set of annals has only six entries, for years between 1241 and 1281, although some of the entries give events occurring in two different years. The manuscript was found in the library of S. Mark's in Venice, where it is labelled as being from Lubensia in Silesia. I imagine that this is the town near Wroclaw that is now called Lubin.

Mechovienses [ca. 1434]. This set of annals starts with records of early events that are obviously derivative, but it is an original source from about 1295 to its close in 1434. The place (Miechovia) seems to be a monastery about 5 Polish miles from Krakow, but I do not know the direction. Hence I shall use the coordinates of Krakow.

Povest' Vremennykh Let [ca. 1110]. This title is translated as The Tale of Bygone Years, although the title of the cited translation is given as The Russian Primary Chronicle. It is a chronicle of Russian affairs from 852, which is considered to be the dawn of Russian written history, to 1110, and the chronicle itself is preceded by a short outline of the history of the world. The oldest text that we have dates from the late 15th century, more than 3 centuries after the last event given in the chronicle.

Scholars of the subject differ considerably about the writing of the chronicle. The translators of the edition cited conclude [p. 21] that it is the work of a single author. Zenkovsky [1963, p. 11], on the other hand, concludes that it was written by at least six different annalists from about the year 1040 on. The events described take place over a wide area; I noticed events in Novgorod in northern Russia and in Kiev in the Ukraine. This means that there is no way to know where the single eclipse in the chronicle was observed. Hence I shall not use this eclipse record, but I discuss the source so that the reader will know it has not been overlooked.

---

†Novgorod-Seversk is in the northern Ukraine. It is not the same as Novgorod, which is about 150 kilometers south of Leningrad.

Sambiensis [ca. 1352]. This is actually 13 independent sources which the editor has published sequentially under a single title. The sources are concerned with the affairs of Prussia and Livonia, which between them cover a wide area even if we use only the area of Prussia in the 14th century. Hence I shall not use the eclipse records contained in Sambiensis, but I shall list them for reference.

5. Scandinavian Sources

The Scandinavian records to be used in this work are all records after the year 1300 from sources that were used in MCRE [Section XIII.2]; I used only records before 1300 in MCRE. In fact, the records used here are all from either Iceland or Norway. There is a serious problem that arises with these records. They cover events in both Norway and Iceland with about equal emphasis, and only rarely does a record in one of these sources say whether an eclipse was observed in Iceland or Norway. Hence most records cannot be used for lack of knowing the place of observation.

6. Sources from Spain and Portugal That Have Not Been Used Before

Alcobacense [ca. 1111]. The work of M called my attention to this chronicle. It gives a sparse summary of events from the early 4th century, when the Goths left their northern territories, to 1111, when King Cirus captured the town of Santarem on May 22. In fact, most of the events listed are the capturing of places from Muslims by Christians, and I see no reason to conclude that the chronicle is specifically associated with Alcobaca, Portugal. Thus I believe that M is somewhat optimistic when he concludes [p. 8.51] that the eclipse observation in the chronicle is "almost certainly . . . from Alcobaca." The uncertainty in the place is well exemplified by the fact that this chronicle was associated for a long time with Alcala de Henares, which is a short distance northeast of Madrid. However, there is a concentration of events in places fairly close to Alcobaca, and I believe, on the basis of the events that are chronicled, that we may safely say that the eclipse was observed somewhere in the triangle whose corners are Alcobaca, Santarem, and Sintra, which are all in Portugal.

Clavius [1593]. I discussed this source in Section X.2. Clavius observed an eclipse in 1560 when he was in Portugal and he observed another in 1567 when he was in Rome.

ibn Hayyan [ca. 1070]. My attention was called to this work because both M and MS use it. The translator of the cited edition says that the part cited is the last part of the work. Thus it is probable that earlier parts of the work were published elsewhere, but neither the translator, M, nor MS give an indication that I could find about the location of the earlier parts. The part that I have seen is a history of the Islamic region whose capital was Cordoba.

7. Eclipse Records from These Parts of Europe

In this section, I shall present the eclipse records from the relevant parts of Europe that have not been used before. In addition, I shall discuss some records used in MCRE whose characteristics will be changed. Finally, I shall mention a few records from MCRE that will be used without change but that need more discussion.

464 July 20 E,SP. In this record [MCRE, p. 510], the annalist said that the sun had the form of a moon on its 5th night. I shall take this to mean a magnitude of 0.85, with a standard deviation of 0.075.

718 June 3 E,SP. This record [MCRE, p. 512] gave several contradictory clues to the date. I tentatively accepted 718 June 3 as the correct date, but calculation showed [MCRE, p. 592] that the eclipse of 720 October 6 was also a possibility. Therefore this eclipse cannot be identified, and I did not use it in MCRE nor shall I use it here.

810 July 5 E,CE. As I explained in Section IX.3 above, I found two records in MCRE [pp. 263 and 395] which seemed to be independent records of this eclipse, one from Germany and one from Austria. It turned out [p. 595] that the eclipse was not visible in either place. Hence the records cannot be used.

840 May 5. The entry for the year 840 in Juvavenses [ca. 956] reads: "DCCCXL. The emperor Louis died. This year the sun was eclipsed." I have not found any other record with just this wording, but it is not safe to use such a brief notice of a well-reported eclipse.

891 August 8a E,CE. A record of this eclipse from Switzerland was designated as 891 August 8 E,CE in MCRE [p. 287], so I use an a after the date in designating this one, which is from Juvavenses [ca. 956]. The first part of the entry for 891 reads: "DCCCLXXXI. The sun was obscured from the third until the sixth hour." I can find no eclipse that was visible in Austria in 881, and the hours are reasonable for the eclipse of 891 August 8. Further, it is easy to drop an X from DCCCLXXXXI, thus accidentally changing the date from 891 to 881. For these reasons, I believe that this record refers to the eclipse of 891. The only objection to this conclusion is that the event is out of order if the year is 891, but this is not a serious objection. The reliability will be taken as 0.5, the place is Salzburg, the value of η is 0, and the standard deviation of the magnitude is 0.18.

912 June 17 E,SP. ibn Hayyan [ca. 1070, p. 321] has an interesting account of this eclipse: "In the same year, on a Wednesday, the last day of the month of Sawal, there was a

total eclipse of the sun. The darkness covered the earth
and the stars appeared. The greater part of the people believed that the sun had set below the horizon and raised
themselves in order to perform the prayer of Magrib. Presently the shadow disappeared and the sun appeared; but
after half an hour it hid itself at the usual time behind
the wide horizon." Perhaps one should replace "wide" by
"distant".

ibn Hayyan has already identified the year as 299 of
the Muslim era, and Sawal, the 10th month in the Muslim calendar, has 29 days. The Julian day number corresponding to
this date [APO, Section II.5] is 2 054 334, which is the
date 912 June 17 in the Julian calendar. This day was Wednesday, so the dating of the eclipse is precise and accurate.
Nonetheless, I find a problem with the record.

Calculation with provisional accelerations puts the
maximum of this eclipse at Cordoba at 31 minutes before sunset. In line with this, M [p. 8.48] says that "the description seems to be trying to convey" the impression that the
eclipse was only a few minutes before sunset. The description does not give me this impression, and it seems instead
to be incompatible with the calculated circumstances. First,
we find that many people "raised themselves" (se levantó),
which M translates as "got up", for the prayer of Magrib,
which is the prayer of sunset. If they got up for the prayer
of sunset, it was presumably because it was growing dark, and
they presumably got up from a siesta, which is usually over
before the ordinary sunset. Second, the "shadow disappeared"
half an hour before sunset, which contradicts the calculations.
Maximum eclipse is hardly described as a disappearance of
the darkness.

Since the record gives the date exactly, this difficulty
is not enough to make us question the identification. Perhaps
there is no real problem, and we have an apparent problem
only because we are working with a translation. It is certain
that we are working with an account at second hand, because
ibn Hayyan wrote a century and a half after the event. I
shall take the reliability to be 0.5. The place is Cordoba.
Although the description of the darkness is vivid, I believe
that it is consistent with an eclipse that is not necessarily
total; we cannot take the word "total" literally. Hence I
shall use 0.054 for the standard deviation of the magnitude.

1009 March 29b E,BN. This is one of the records [MCRE,
p. 238] for which I erroneously took the place of observation
to be Blandain rather than Ghent. See Section II.1.

1023 January 24 E,BN. This is another record for which
the place should be Ghent rather than Blandain. I used it
on page 239 of MCRE.

1030 August 31 E,Sc. This is the eclipse which several
modern scholars claim happened at the death of S. Olaf in
battle at the place called Stiklestad, near Trondheim in

Norway. I studied this alleged record at length in AAO [pp. 81-86] and MCRE [pp. 499-501]. As I said in MCRE, it is almost inconceivable that this eclipse was seen at the time and place of S. Olaf's death. It is beyond all but the most frivolous question that S. Olaf died on 1030 July 29, and the idea of an eclipse at his death was not even mentioned until at least a century later. In MCRE, I admitted the possibility that the eclipse was observed somewhere in the Scandinavian regions and that its date was assimilated to the death of S. Olaf. On further reflection, I see no reason to use this possibility as an astronomical observation, although it is a possibility that cannot be disproved. For observations, we want statements that can be proved, not ones that cannot be disproved. In the future, I shall ignore this record, and let us hope that the "eclipse of Stiklestad" will vanish from the mythology of eclipses.

1033 June 29 E,BN. See the record 1037 April 18 E,BN which follows.

1037 April 18 E,BN. This record, the preceding one, and the following one are all records for which the place should be Ghent rather than Blandain. The records are found on pages 240 and 241 of MCRE.

1039 August 22 E,BN. See the record 1037 April 18 E,BN, which precedes this entry.

1079 July 1 E,SP. This record is from Alcobacense [ca. 1111, p. 150]: "In the year 1117† on the first day of July at the 6th hour of the day the sun was darkened and it stayed in that darkness for 2 hours while stars appeared in the sky and it was as if midnight had come." M [p. 8.51] takes "2 hours" to apply to the time that stars were visible, and remarks that it is common to exaggerate the duration of totality. I believe that "2 hours" refers to the time of darkening, that is, to the duration of the entire eclipse. I believe that the implication of totality is greater here than it is in the record 912 June 17 E,SP, and I shall take 0.034 for the standard deviation of the magnitude. The reliability is 1, and the place is the triangle whose vertices are Alcobaca, Santarem, and Sintra in Portugal.

1091 May 21 E,PR. On page 173 of the cited translation, under the year 1091, Povest' Vremennykh Let [ca. 1110] has a record of this eclipse: "In this year, there was a portent in the sun, which seemed about to disappear. At the second hour of the day, on May 21, only part of it remained visible, approximating the size of the moon." Unfortunately, as we saw in the discussion of this source in Section XI.4, the point of observation could have been almost anywhere in the

---

†This source uses the Spanish Era, which commemorates the conquest of Spain by Augustus Caesar in -38. Thus 1117 is the same as our 1079.

western part of Russia, and the record cannot be used.

1093 September 23 E,BN. The place of observation for this record [MCRE, p. 241] needs to be changed from Blandain to Ghent.

M [p. 8.51] uses a Russian record of the eclipse of 1124 August 11. However, he does not take the record directly from the annalistic source. Instead, he takes it from a study of eclipses in Russian annals and chronicles made in 1949, without studying the nature of the chronicle, its reliability, or other important matters. He does not specify the original chronicle involved, and the secondary source he cites is not available at the Library of Congress. I have not tried to locate the secondary source elsewhere and thus I shall not use this record. The record has an interesting feature; it says that during the eclipse the stars and the moon appeared.

1133 August 2c E,BN. This is yet another record [MCRE, p. 243] for which we must change the place from Blandain to Ghent.

1133 August 2b E,CE. In this record [MCRE, p. 266], the annalist says that the sun appeared to be covered over "like a round bag, ..." M [p. 8.52] writes: "This is precisely the subjective impression formed by an observer who has, previous to totality, been watching the waning crescent."† He then explains why the word "bag" is an unusual "but very cleverly adept" way of describing the corona. I believe that M is reading too much into the language. For example, in the records 1140 March 20a E,BN and 1147 October 26a E,BN [MCRE, pp. 244-246], which were both written by the same person, the writer says that the sun was surrounded by a misty cover. The same arguments that justify "round bag" as a description of the corona would also justify "misty cover" as such a description. However, "misty cover" in these records cannot be a description of the corona because the eclipse of 1147 October 26 was annular by a wide margin and the corona would not have been visible.

1133 August 2h E,CE. In MCRE [p. 280] I took the magnitude to be 0.97, with a standard deviation of 0.03. Experience has shown that the magnitude should usually be decreased and the standard deviation increased in such cases. Here I shall take the magnitude to be 0.92, with a standard deviation of 0.05.

1140 March 20b E,BN. I need to make two changes from MCRE [p. 245] in the treatment of this record. The first is that the place needs to be changed from Blandain to Ghent. The second concerns the standard deviation of the magnitude. I used 0.02 in MCRE because the record says that the eclipse was of a kind unknown before, but it gives no specific details

---

†The emphases in this quotation are M's.

such as the mention of stars. I believe we are justified
in taking the standard deviation of the magnitude to be less
than 0.18, but 0.02 seems too small. Hence I shall use the
value 0.068, which is the "new value" corresponding to an
"old value" of 0.02 in Table IV.4.

1140 March 20 E,PR. Colbazienses [ca. 1568] has the
only Polish record of this eclipse I have found: "1140, the
13th calends of April on the 4th day of the week after nones
there was darkness on the earth." The day stated is March
20, which was on a Wednesday that year. Maximum eclipse in
Poland was after midafternoon, so nones here apparently means
the 9th hour of the day. The reliability of the record is 1.
I shall use Stargard as the place since I cannot find Colbaz.
The value of η is 0 and the standard deviation of the magnitude is 0.18.

1147 October 26a E,BN. This record [MCRE, pp. 245-246]
says that stars became visible and it also says that the sun
was surrounded by a misty cover; see the discussion of the
record 1133 August 2b E,CE above. Since the eclipse was annular, it seems odd that stars were visible. I should have
taken the standard deviation to be 0.01 rather than 0 by the
rules of MCRE; this corresponds to using 0.054 here.

1185 May 1. In the cited translation of the poem about
Igor [ca. 1185], at the beginning of the section that deals
with Igor's departure, we find these words: "Then Igor
glanced up at the bright sun and saw that from it with darkness his warriors were covered." Then Igor makes a speech
to his warriors and begins to ride away on the campaign. At
this point: "The sun blocks his way with darkness. Night,
moaning ominously unto him, awakens the birds; . . ."†

The translator, in his foreword and his critical notes,
unequivocally takes these passages to be references to the
eclipse of 1185 May 1. For example, he says [p. 97]: "The
sudden night due to the eclipse causes the nocturnal birds
of prey to awake . . ." It seems to me that there are several objections to this conclusion.

For one thing, the date seems to be wrong. The translator says [pp. 73-74] that the campaign is recorded in contemporaneous annals and [pp. 1-2] that the events began on
Tuesday 1185 April 23, in Novgorod-Seversk. If this is correct, the darkness could not have been the eclipse, which
happened on Wednesday, 1185 May 1.

Next, we note that the poem in fact denies the existence
of an eclipse. Igor looked up at the "bright sun" and the
"sun blocks his way" as he began his march. In both cases,
Igor saw darkness when he saw the sun. This is probably a
poetic and romantic sign of his coming death, not the occurrence of an eclipse. The reference to the birds is ambiguous.

---

†I have not attempted to preserve the division into lines
in these quotations.

Darkness makes daytime birds become silent while the nocturnal birds awake to begin their activity. It seems to me
that the following is a reasonable interpretation of the
passage: Sunlight to others is darkness to Igor, because
of the omen just mentioned, so the night that awakens the
birds is in fact early morning.

Next, we should observe that the eclipse, both by the
map in Oppolzer's Canon [Oppolzer, 1887] and by my calculation with provisional values of the accelerations, was late
in the day. If the passages do refer to the eclipse, this
means that the departure began in late afternoon, close to
sunset. This is not the time of day when an army usually
begins a march that will take it several hundred kilometers
before it encounters the enemy. These events, from the military point of view, should have occurred in fairly early
morning, not in late afternoon, and this is consistent with
the interpretation suggested in the preceding paragraph.

Of course, it is true that the eclipse occurred at some
time during the campaign and subsequent defeat, and it is
conceivable that the poet was aware of the eclipse, that he
exaggerated its degree of totality, and that he rearranged
the course of history for poetic purposes. However, I do
not think we should take this possibility as astronomical
observation. Further, I think that the poet, while he may
have been influenced by the eclipse, is not in fact trying
to describe one.

M [pp. 8.57-8.58] uses a record of this eclipse from
the secondary source that I mentioned above in connection
with 1124 August 11. I shall not use this record for the
same reason that I did not use the record of 1124 August 11
from the same source.

1187 September 4 E,BN. The place for this record should
be changed from Blandain to Ghent. It appears on pages 246-
247 of MCRE.

1191 June 23c E,BN. Balduinus [ca. 1294] has an interesting record of this eclipse on page 537 of the cited
edition: "In the year of the Lord 1191 the sun underwent an
eclipse on the eve of John the Baptist." He goes on to say
that he remembers seeing this when he was a little boy.
(puerulus), mainly because he learned to watch it in water.
This sentence suggests that I was wrong in Section V.3 to
doubt that one can watch an eclipse by reflection in water,
but I definitely wonder how it is done except under conditions of very low wind.

The sentence also has an implication about the reliability of the record. The evidence of the text is that
Balduinus himself was still writing this chronicle in 1294.
On the other hand, as the editor points out, if Balduinus
could remember seeing the eclipse of 1191 June 23, he could
by no means have been writing in 1294. However, I can think
of two other explanations of the situation.

The first explanation is that Balduinus meant "this" to apply to the general phenomenon of a solar eclipse rather than to the specific eclipse of 1191 June 23. That is, he saw an eclipse when he was a little boy, and someone at that time taught him how to watch an eclipse in water, presumably by reflection. When he saw the account of the eclipse of 1191 June 23 in some source that he was using to write his chronicle, he remembered this learning and recorded his recollection.

The second explanation is that the original recorder learned to watch an eclipse in the water when he was a little boy, and that Balduinus uncritically copied this remark.

If the editor's interpretation were correct, it would mean that we do not have a consistent picture of the circumstances in which the chronicle was written, and hence we should be reluctant to use the record. If either of the other explanations is correct, it merely means that Balduinus used a source older than his own writing, which would have to happen if he wrote in 1294. This in turn means that the record in Balduinus is not original, although it certainly seems to be independent.

I shall assume that Balduinus did use a genuine record of the eclipse of 1191 June 23, and I shall give it the reliability of 0.5 that I frequently use when we have a record that is independent of others known but that cannot be original. The place is Ninove. The value of η is 0 and the standard deviation of the magnitude is 0.18.

We shall see in a moment that Balduinus has a record of the eclipse of 1230 May 14. It is possible that this is the eclipse which he saw as a boy. He also has a brief notice of the eclipse of 1207 February 28 that could have been abbreviated from any one of several sources and that is not safe to use.

1207 February 28 E,PR. Heinricus [ca. 1227] puts this eclipse in the 8th year of the bishop Albertus, whose first year is 1199: "This winter was an eclipse of the sun for a great hour of the day." Heinricus apparently begins his year at Easter, so the year and the season are correct. By a "great hour of the day" I presume that Heinricus means a time that is longer than an ordinary hour, but I do not know what he refers to. The record is contemporaneous and has a reliability of 1; the place is Livonia. The value of η is 0 and the standard deviation of the magnitude is 0.18.

1207 February 28 E,BN. The place for this record [MCRE, p. 247] should be changed from Blandain to Ghent.

1230 May 14 E,BN. This is the record by Balduinus [ca. 1294] that I mentioned a moment ago. It is found on page 542 of the cited edition: "In the year of the Lord 1233, Ferrandus, count of Flanders and Hainaut, died. In the space of three years before this year the sun had an eclipse." I

have used "in the space of three years" but I believe that
"at the interval of three years" is also a valid translation.
If so, we could translate the second sentence simply as:
"Three years before this year the sun had an eclipse." In
any case, I believe that there is no question about the identification of the eclipse. It is chronologically possible
that Balduinus saw this eclipse and that this is an original
record, but I think it more likely that the record is second-
hand even if Balduinus did see the eclipse as a small boy.
Hence I shall use a reliability of 0.5. The place is Ninove.
The value of η is 0 and the standard deviation of the magnitude is 0.18.

We saw in Section VIII.3 that the French writer Iohannes
Longus [ca. 1365] has a record of the eclipse of 1239 June 3,
and that it was partial in S. Omer where he wrote. He also
says that the eclipse was greatest in Navarre and particularly
at Pamplona. If we used this aspect of the record, we should
count it as a Spanish record. However, we have no idea how
Iohannes obtained his information about the eclipse as seen
in Spain, and we concluded in Section VIII.3 not to use the
Spanish aspect. M [p. 8.60] has two other Spanish records
of this eclipse. Since I have not directly consulted the
sources that M uses, I shall not use these records.

1240 April 23 E,PR. Lubenses [ca. 1281] has the only
record of this eclipse that I have found. It occurs on page
549 of the cited edition, after a discussion of the Easter
season in 1241: "Before this, at Easter in the year 1240,
there was an eclipse of the sun after midday." Easter in
1240 was on April 15, so it is consistent with medieval usage
to consider this as being at Easter, that is, in the Easter
season. Since the eclipse was penumbral, we cannot tell from
Oppolzer [1887] when its maximum occurred in Silesia. However, the time that he lists for the eclipse is $17^h 21^m.4$, so
that the maximum was almost surely after noon, as the record
says. The sparse entries in this set of annals were probably
recorded contemporaneously, so the reliability of the record
is 1, and the place will be taken as Lubin in Silesia. The
value of η is 0 and the standard deviation of the magnitude
is 0.18.

1241 October 6a E,BN. I found only one record of this
eclipse from the BN provenance in MCRE [p. 248], so that
record was designated simply as 1241 October 6 E,BN. Thus,
in spite of the letter "a" after the date, this is the second record from that provenance. It is from Menkonis [ca.
1273, p. 536]: "In the year of the Lord 1241 on the Lord's
day on the octave of Michael, the 27th of the moon, about
the hour of nones, there was an eclipse of the sun which according to the calculations of the philosophers occurred
naturally by the interposing of the moon between us and the
sun." Menkonis follows this by a long discussion of the
causes of eclipses, both lunar and solar. In spite of all
this, Menkonis takes the eclipse to be a prodigy because it

occurred close to the death of the pope.† This is an original record and it receives a reliability of 1. The place will be taken as Werm, Belgium. The value of η is 0 and the standard deviation of the magnitude is 0.18.

    <u>Balduinus</u> [ca. 1294, p. 543] has a brief notice of this eclipse that it is not safe to use.

    1241 October 6f E,CE. <u>Sancrucensis</u> [ca. 1310] records this eclipse under the correct year of 1241: "The sun was obscured in certain places, and there was darkness, such that stars were seen in the sky about the feast of S. Michael at the hour of <u>nones</u>." The eclipse had its maximum shortly after midday, so <u>nones</u> probably means our noon. The date is exactly a week after the feast of S. Michael. This is an original and contemporaneous record with a reliability of 1, made at Heiligenkreuz. I shall take the value of η to be 0, with the standard deviation of the magnitude equal to 0.054 because of the reference to the stars.

    <u>Zwetlensis</u> [ca. 1329, p. 655] has a brief notice of this eclipse, with no information except the year. It is not safe to use this record.

    1241 October 6g E,CE. <u>Praedicatorum Vindobonensium</u> [ca. 1283, p. 727] records this eclipse in its entry for 1241: "The sun was obscured and stars appeared at the hour of <u>nones</u> on the 2nd nones of October‡ on the Lord's Day." The eclipse was on Sunday, as the record says. Although this source frequently copies from other known sources, this record seems to be original and receives a reliability of 1. The place is Vienna. The value of η is 0 and the standard deviation of the magnitude is 0.054.

    1241 October 6h E,CE. <u>Frisacenses</u> [ca. 1300] records this eclipse in the following way: "Year of the Lord 1241. The Tartars devasted all Hungary with fire and the sword. The same year on the octave of S. Michael‡ the sun was eclipsed; that was on the Lord's day." Since the entries in this set of annals are quite intermittent, it is probable that they were made contemporaneously, so the record receives a reliability of 1; the place is Friesach, Austria. The value of η is 0 and the standard deviation of the magnitude is 0.18.

---

†Pope Gregory IX died on 1241 August 22 and Celestine IV died on 1241 November 10. I believe Menkonis is the only writer I have seen who attributes superstitious significance to this eclipse.

‡October 6.

‡That is, a week after S. Michael's day, and hence on October 6, which was a Sunday.

1241 October 6i E,CE. Stirensis [ca. 1346] has the
following brief record of this eclipse: "Item in the year
of the Lord 1241 there was an eclipse on the octave of S.
Michael." It does not say that the eclipse was solar. I
usually omit brief records of this sort because they could
have been copied easily from some unknown source. However,
Stirensis is a highly intermittent set of annals that has
only about 30 entries from 1220 to 1346, and such annals are
usually original. Hence I shall use this record, but with a
reliability of 0.5, and I shall take the place to be Graz.
The lowered reliability reflects both an uncertainty in the
place and in the originality. The value of η is 0 and the
standard deviation of the magnitude is 0.18.

1241 October 6 E,PR. This eclipse is recorded in Colbazienses [ca. 1568] in this short note: "1241, this year on
the 27th of the moon at the 8th hour on the 2nd nones of October (October 6) there was an eclipse of the sun." The records in this source are believed to be contemporaneous and
original, so this has a reliability of 1. I use the coordinates of Stargard because I cannot find those of Colbaz. The
value of η is 0 and the standard deviation of the magnitude
is 0.18.

Part (2) of Sambiensis [ca. 1352, p. 699] records this
eclipse as follows: "An eclipse of the sun, to wit a great
darkness, occurred on the Lord's day after the day of S.
Michael in the year of the Lord 1240." In spite of the error
in the year, it is likely that this is an original record, as
we judge from the intermittent nature of the annals. However,
I decided in Section XI.4 not to use the records from Sambiensis because of the great uncertainty in the place where they
were written.

1261 April 1 E,CE. The records of this eclipse are
few but widespread. There is a record from England [MCRE,
p. 178] and one from Germany [MCRE, p. 435], while this one
is from Austria. The day must have been cloudy over most of
Europe, with just a few clear patches. This record is found
in Sancrucensis [ca. 1310] under 1261: "This year there was
an eclipse of the sun about the third hour on the octave of
the Annunciation, which was on the 6th day of the week." The
date of the Annunciation is March 25, and its octave is a
week later on April 1, so the date given for the eclipse is
correct, and so is the weekday. The hour is reasonable if it
refers to the beginning of the eclipse. This seems to be an
original record and it receives a reliability of 1; the place
is Heiligenkreuz. I shall take the value of η to be 0, with
0.18 for the standard deviation of the magnitude.

1263 August 5 E,BN. The place for this record [MCRE,
p. 248] should be changed from Blandain to Ghent.

1263 August 5a E,CE. This record from Sancrucensis [ca.
1310] reads: "This year there was an eclipse of the sun on

-363-

Dominic the Confessor after midday." This entry appears
under the year 1263. "On Dominic the Confessor" presumably
means on the day dedicated to S. Dominic. His day is actually August 4, so the annalist has made an error of a day.
The reliability is 1, the place is Heiligenkreuz, the value
of η is 0, and the standard deviation of the magnitude is 0.18.

1263 August 5 E,PR. This record from Cracoviensis [ca.
1331] has the same error in date: "An eclipse of the sun on
the day of S. Dominic on the 27th of the moon." This appears
among the events of 1263. While the day of S. Dominic is
August 4, the 27th day of the ecclesiastical moon was on August 5, the correct date of the eclipse. This original record
has a reliability of 1 and it was written in Krakow. The value of η is 0 and the standard deviation of the magnitude is
0.18.

M [p. 8.64] refers to a Scandinavian record of this
eclipse that I have not consulted directly. The record is
interesting in that it seems to describe (correctly) an annular eclipse. M does not use the record, since he does not
use observations of annularity unless they were made by trained
astronomers. He also comments that the nature of the source
has not been studied adequately from the standpoint of the
originality of the records in it. I shall not use this record
either.

1267 May 25a E,CE. I found a single record of this
eclipse from Central Europe in MCRE [p. 275], for which I used
the designation 1267 May 25 E,CE. Thus I use the letter "a"
after the date for this record, even though it is the second
record I have found. It is from Zwetlensis [ca. 1329]: "In
the year of the Lord 1267 there was an eclipse of the sun on
the day of S. Urban near midday, and it lasted until vespers."
May 25 is the day of S. Urban, so the recorder has the correct date. However, he errs seriously about the time of day.
The eclipse began in Austria about midmorning and was over
about noon. Thus it did not last to any time near vespers.
I do not believe that a timing error this serious could have
happened if the record were contemporaneous or nearly so.
Thus I shall use a reliability of only 0.5 even though most
records in Zwetlensis seem to be original. The place is
Zwettl, the value of η is 0, and the standard deviation of
the magnitude is 0.18.

1267 May 25b E,CE. Sancrucensis [ca. 1310] dates the
eclipse in the same way but puts it at the right time of day:
"And on Pope Urban before midday there was an eclipse of the
sun." This occurs under the year 1267. We presume that "on
Pope Urban" means on his day. The reliability is 1, the place
is Heiligenkreuz, the value of η is 0, and the standard deviation of the magnitude is 0.18.

1267 May 25c E,CE. This record from Basileenses [ca.
1277] has the day wrong: "An eclipse of the sun on the eve
of Urban, predicted by Brother Gotfried, an astronomer of the

Praedicatorian order in Worms." The year is correctly given as 1267, but the record puts the eclipse on the day before May 25. The reliability of this record is 1, the place is Basel, and the value of η is 0. Since the eclipse had been predicted, the brothers in the monastery at Basel may have been watching for it particularly, and the magnitude might have been small. In spite of this, I shall use 0.18 for the standard deviation of the magnitude.

1270 March 23a E,CE. Claustroneoburgensis [ca. 1383, p. 648] has one of two Austrian records of this eclipse: "This year there was an eclipse of the sun on the 10th calends of April† about the first hour." The year has already been given as 1270. The eclipse was early in the morning in Austria, as the record says. The reliability is 1, the place is Klosterneuburg, the value of η is 0, and the standard deviation of the magnitude is 0.18.

1270 March 23b E,CE. Vindobonensis [ca. 1327, p. 703] has the other Austrian record of this eclipse under the year 1270: "Also in the same year there was an eclipse of the sun about the first hour, on the first Lord's day before the Annunciation to S. Mary." Annunciation Day is March 25, and the preceding Sunday in 1270 was indeed March 23. A footnote in the cited edition says that the writing changed hands at this point in the text. This is an original record with a reliability of 1, and it comes from Vienna. The value of η is 0 and the standard deviation of the magnitude is 0.18.

1310 January 31 E,BN. This record is from S. Iacobi [ca. 1393, p. 644], which is the final part of the source that I cited as S. Iacobi [ca. 1174] in MCRE [p. 223]; see also Section XI.2 above. The record appears under the year 1309: "In the same there was an eclipse of the sun on the last day of January." The preceding sentence deals with the winter of 1309-1310, and "the same" seems to refer to that winter. If so, the year given for the eclipse is really correct. In any case, there is no question about the identification of the eclipse. This seems to be an original source and the record has a reliability of 1; the place is Liège. The value of η is 0 and the standard deviation of the magnitude is 0.18.

1310 January 31a E,CE. Sancrucensis [ca. 1310, p. 735] has the following entry under the year 1310: "On the Sabbath before the Purification about vespers there was a solar eclipse." The Purification of the Virgin is February 2, and January 31 was the preceding Saturday in 1310. Thus Sabbath is being used in the Hebrew sense of the 7th day, and not as a synonym for the Lord's day, which is Sunday. Maximum eclipse in Austria was probably about the middle of the afternoon but the eclipse probably lasted until about vespers. The reliability is 1, the place is Heiligenkreuz, the value of η is 0, and the standard deviation of the magnitude is 0.18.

---

†March 23.

1310 January 31b E,CE. Florianensis [ca. 1310], which, like Sancrucensis [ca. 1310], is a continuation of Mellicenses [ca. 1564] (see Section XI.3), has a record of the same eclipse. Its record of this eclipse appears under the year 1309 and reads as follows: "About this time within one year, namely from the eve of S. Bartholomew in the year of the Lord 1309 until vespers on the following day of S. Lawrence, in the year of the Lord 1310, the moon was eclipsed and the sun once." I believe the writer accidentally left out a word and that he meant for the last part of the sentence to read "the moon was eclipsed twice and the sun once." The lunar eclipses of 1309 August 21 and 1310 August 11 were both visible in Austria, weather permitting, and these dates correspond closely but not exactly to the saints' days stated. The day of S. Lawrence is August 10 and the day of S. Bartholomew is August 24 in some churches and August 25 in others. I see no reason for him to mention both days unless he meant to refer to both eclipses.

Whether this is so or not, the solar eclipse must be that of 1310 January 31. This part of Florianensis is believed to be original, so the record has a reliability of 1; the place is S. Florian. The value of η is 0 and the standard deviation of the magnitude is 0.18.

1310 January 31c E,CE. This record is from S. Rudberti [ca. 1327], under the year 1310: "This year there was a solar eclipse about midday, on the 2nd calends of February." The date given is January 31, but the eclipse was closer to mid-afternoon than to midday. This source has at least two authors for years near this one, and this usually means that the records were being made contemporaneously. Hence I shall give this record a reliability of 1 in spite of the fact that such a short record could easily have been copied or condensed from some other source. The place is Salzburg. The value of η is 0 and the standard deviation of the magnitude is 0.18.

Part (12) of Sambiensis [ca. 1352] records this eclipse in the following way: "In the year of the Lord 1310, on the 2nd calends of February, there was an eclipse of the sun before the hour of vespers and it was over in the first stroke." I do not understand the use of pulsus, which I have translated as "stroke", unless it means the first ringing of a bell that followed the completion of vespers. I do not use this record for the reason given in the discussion of Sambiensis in Section XI.4.

1321 June 26 E,CE. Mellicenses [ca. 1564] has the only Austrian record of this eclipse, although there are several records from Germany and Poland. The record occurs among the events for 1321: "On the day of John and Paul there was an eclipse of the sun." June 26 is the day of S. John and Paul, so the dating is correct. The hand changes with almost every entry in this part of the annals, so the record is almost surely original. The reliability is 1, the place is Melk, the value of η is 0, and the standard deviation of the magnitude is 0.18.

1321 June 26a E,PR. <u>Mechovienses</u> [ca. 1434] has the
first of three Polish records of this eclipse: "1321. There
was a great eclipse of the sun on the day of John and Paul,
and it lasted from early morning until the 6th hour." The
hours are approximately correct, although the duration implied
seems somewhat long. The reliability is 1, and the place is
Krakow or very close to it. The value of η is 0. Since the
eclipse is described as great, it was probably larger than the
usual observed eclipse, and I shall use 0.1 rather than 0.18
for the standard deviation of the magnitude.

1321 June 26b E,PR. This record is from <u>Colbazienses</u>
[ca. 1568], and it is given under the year 1321: "An eclipse
of the sun occurred on the day of John and Paul at the third
hour, on the 6th day of the week." The weekday given is correct. The reliability is 1 and I shall use Stargard for the
place. The value of η is 0 and the standard deviation of the
magnitude is 0.18.

Part (12) of <u>Sambiensis</u> [ca. 1352] has the following
record of this eclipse: "Also in the new year of the Lord
1321 there was a solar eclipse on the day of the blessed
martyrs John and Paul at the 3rd hour." Although this record
is probably original, I do not use any records from <u>Sambiensis</u>
for the reason already explained. Part (9) of the same source
also has the following record of the eclipse of 1324 April 24
that I shall not use either: "Also in the same year on the
8th calends of May† was a solstice from sunrise to the 3rd
hour." Although the writer uses the word "solstice" rather
than "eclipse", I think there is little doubt that he was re-
referring to the eclipse. Since this eclipse was penumbral,
it is not surprising that I have found no other record of
this eclipse in any source.

The last eclipse record in <u>Sambiensis</u> is found in Part
(11): "Also there was an eclipse in the year of the Lord
1330, on the 17th calends of August." The date is 1330 July
16.

1330 July 16 E,Sc. <u>Lögmanns</u> [ca. 1430] has a short
notice of this eclipse under the year 1330: "A solar eclipse
was seen in Iceland." <u>Flatøbogens</u> [ca. 1394] has the same
entry under 1328. I shall count these as a single record of
the eclipse with a reliability of 1; no other eclipse near
this date was visible in Iceland. The place is Iceland, the
value of η is 0, and the standard deviation of the magnitude
is 0.18.

1339 July 7 E,CE. This record is found in <u>Novimontensis</u>
[ca. 1396] under the year 1339: "This year on the nones of
July (July 7) about the hour of vespers there was a total
(<u>universalis</u>) eclipse of the sun." I discussed the meaning

---

†This date is April 24. The annalist has already stated
that the year is 1324.

to be attached to <u>universalis</u> in Section IX.3 and decided
that it should be translated as total. However, I showed in
connection with records of the eclipse of 1321 June 26 that
the word was not used with rigor, and that we should not
take the use of the word as an indicator of strict totality.
It is unfortunate that we do not know where this eclipse was
total, because it was an annular-total eclipse, and the zone
of totality was therefore very narrow. Because of the use of
<u>universalis</u>, I take the standard deviation of the magnitude
to be 0.054, in accordance with previous practice. The reliability is 1, the place is Neuberg, and the value of η is 0.

1339 July 7 E,PR. <u>Mechovienses</u> [ca. 1434] has this record under the year 1338, but there is no entry for 1339 so
the error in year is probably an oversight. The record reads:
"This year there was a destruction (<u>destruccio</u>) of the sun on
the 4th day of the week after Peter, at the hour of vespers."
I have not seen "destruction" used before to describe an
eclipse. I do not understand the reference to Peter. The
nearest day to July 7 associated with Peter that I can find
is June 29, which is dedicated to Peter and Paul jointly.
Even if this were the day, 1339 June 29 was on a Tuesday and
the following Wednesday was therefore June 30, not July 7.
However, there is no doubt about the identification of the
eclipse. This source is original after about 1295, so the
record has a reliability of 1. The place is Krakow and the
value of η is 0. One might take the use of "destruction" to
indicate a particularly large eclipse, but I do not believe
that this is a safe reading, and I shall use 0.18 for the
standard deviation of the magnitude.

1339 July 7 E,Sc. There are several Scandinavian records of this eclipse but only <u>Lögmanns</u> [ca. 1430] tells where
it was observed. The record is almost identical with 1330
July 16 E,Sc: "There was a solar eclipse in Iceland." The
reliability is 1, the place is Iceland, the value of η is 0,
and the standard deviation of the magnitude is 0.18.

1344 October 7 ? Both codices of <u>Novimontensis</u> [ca.
1396] describe terrible events under the year 1344 that are
foretold by a solar eclipse, but the two accounts differ
considerably from each other. Neither codex says that an
eclipse was actually observed, nor does either mention any
details that we can associate specifically with an eclipse.
The eclipse of 1344 October 7 was probably visible in Neuberg, and it is possible that we have an implicit record of
the eclipse, but it would not be safe to make this assumption.

1364 March 4 E,BN. The entries in <u>Fossenses</u> [ca. 1384]
for 1363 and 1364 are arranged this way:

1363 . . . . And the following year, namely
1364 the sun underwent an eclipse on the 4th day of
the month of March, between midday and nones†; and
this year winter began at the feast of Nicholas, and
lasted for fourteen weeks with snow continuously on the
ground, and there was great want of animal pasture.‡

This is a clear and original record of the eclipse of 1364
March 4 and it receives a reliability of 1. The place is
Fosse, the value of η is 0, and the standard deviation of the
magnitude is 0.18.

1384 August 17 E,BN. I referred to this record in MCRE,
where I used it to assign the correct date to the source
identified as Fossenses [ca. 1384]. The last entry in this
source reads: "1389. On the 17th day of August the sun suffered an eclipse from the 13th hour to the 14th hour or beyond, and this to the extent of about a third part, according
to what could be judged by the spectators, with the rest remaining in the light." The year 1389 is clearly wrong. My
belief is that the number of the year was originally written
in Roman numerals, and the accidental insertion of a V changed
the year from 1384 to 1389. For this reason, I use 1384 in
citing this source, although the cited edition gives its year
as 1389. This is an original record of a partial eclipse of
the sun, with a reliability of 1, coming from Fosse.

There is a problem about the magnitude. The relevant
part of the text reads: "et hoc pro tercia parte quantitatis
ipsius, prout respicientibus spectari poterat, residuo
remanente in suo lumine." I believe this must be taken as
meaning that a third of the sun was eclipsed, with the rest
remaining lighted. However, a provisional calculation shows
that the magnitude was slightly greater than 0.7, and the
discrepancy seems too great to be an error in observation; it
is the difference between considerably less than half and
considerably more than half.

There seem to be only two possibilities. Either I am
wrong about the translation, or the writer made a simple slip,
saying one third when he meant two thirds. I believe it is
safe to take the indicated magnitude as 0.67, with 0.20 for
the standard deviation of the magnitude.

1408 October 19 E,CE. Claustroneoburgensis [ca. 1383]
records the eclipse in this way under the year 1408: ". . and
in that year occurred an eclipse of the sun in the morning

---

†Nones clearly means the 9th hour, that is, midafternoon,
in this context.

‡The actual entries are in Latin; the translation is mine.
The great want mentioned in the last column may have been
of pasture animals, but I believe that animal pasture is a
more likely translation.

about the seventh† hour." The only possible eclipse in 1408
is that of 1408 October 19, which was rather small in Austria
but which did come in the fairly early morning. The only
other eclipse near this time that came in the early morning
was 1406 June 16. Most sources are accurate about the year
at this time in history, in spite of the preceding example,
and I believe that identification of the eclipse as 1408
October 19 is safe. The reliability is 1, the place is
Klosterneuburg, the value of η is 0, and the standard deviation of the magnitude is 0.18.

The remaining eclipse records from Central Europe all
come from <u>Mellicenses</u> [ca. 1564], and all supply an explicit
statement about the magnitude. Thus it is convenient to consider them all together. The reliability is 1 and the place
is Melk for all these records.

1485 March 16 E,CE. This record is the last sentence
under 1485: "On the 16th day at the 4th hour in the month
of March there was a total eclipse of the sun." Since the
eclipse was in late afternoon in Austria, the writer is numbering the hours from noon.

1486 March 6 E,CE. This record is given under the year
1486: "On the 5th day at the 17th hour in the month of March
there was an eclipse, having nine points (<u>puncta</u>)." The
eclipse was in the early morning, and the hour is correct if
we assume that the writer numbers the hours from noon as he
does in the preceding record. In other words, the writer has
apparently adopted the practice that astronomers followed for
centuries of beginning the day at noon, so that the astronomical day with a particular number began at noon of the civil
day with the same number. However, the hour is wrong unless
it is a calculated one. 17 hours after noon at this time of
year is before sunrise, and the writer could not have seen a
solar eclipse begin when the sun was invisible.

1502 October 1 E,CE. This record occurs under the events
of the year 1502: "An eclipse of the sun with 10 points was
seen on the day of S. Egidius." This is probably S. Aegidius
of Assisi, a companion of S. Francis, but I do not know his
day. This record gives a specific estimate of the magnitude
using the same word "point" as the preceding record, but the
method of dating has nothing in common with the two preceding
records. Thus we cannot tell from the style whether the
writer is the same or not.

From the fact that the writer uses "total" (<u>plena</u>) in
referring to the eclipse of 1485 March 16, <u>M</u> [p. 8.66] concludes that ". . there is no reason to question the reliability of the magnitude despite the brevity." This conclusion

---

†This writer must have already adopted the custom of numbering the hours from midnight.

seems optimistic to me. We saw in Section IX.3† that annalists did not use the similar terms generalis and universalis with rigor, and we decided to use a standard deviation
of 0.054 for the magnitude for records that use these words.
Since this source gives an explicit estimate of the magnitude
for the next two records, the writer apparently paid more
attention to the magnitude than most annalists do, and we
might decide to take the eclipse as total for this reason.
However, I believe that it is more conservative to use 0.054
for the standard deviation of the magnitude, as we did with
the earlier similar records. The value of η is 0.

The other records provide apparently precise estimates
of the magnitude, but we do not know the unit involved. Since
the eclipse of 1502 had 10 points, all we can conclude with
certainty is that the unit is less than 1/10. It is quite
likely that the unit is the one that we call the digit, which
is 1/12 of totality, but I prefer not to make this assumption.
Instead, I shall merely take the statements of magnitude as
indications that the eclipses were definitely partial and not
total. That is, I shall give the records 1486 March 6 E,CE
and 1502 October 1 E,CE the conventional characteristics that
we have adopted for records that specify a partial eclipse
without specifying its magnitude.

1560 August 21 E,SP. In the preceding chapter we discussed the observation that Clavius [1593, pp. 508-509] made
of the eclipse of 1567 April 9 in Rome. Clavius also saw
the eclipse of 1560 August 21, and he records the observation
of both eclipses in the same passage. Apparently this happened by accident; it does not appear that he arranged his
life in order to see either eclipse.

After commenting that he has had the opportunity to observe two remarkable eclipses of the sun, Clavius writes:
". . . I observed the first in the year 1559 in Coimbra in
Lusitania about midday,‡ in which the moon interposed itself
directly between the vision and the sun,‡ such that it covered the whole sun for a not short space of time, and produced
a great darkness like night." We should notice that Clavius
gives the year as 1559. Since he does not give the day of
the year, he may have been writing from memory. Regardless
of the explanation of his error, it is interesting that a
trained astronomer joins many novices in giving the wrong
year for a solar eclipse.

In the eclipse of 1567 April 9, the apparent angular
diameters of the sun and moon were almost equal and almost
the entire chromosphere of the sun was visible even at the

---

†See the discussion of the record 1321 June 26c E,G.

‡Meridiem.

‡Clavius uses the word visus, which I have translated as
vision. We would probably say "between the eye and the
sun."

TABLE XI.2

ECLIPSE RECORDS FROM OTHER PARTS OF EUROPE THAT ARE
NEW OR THAT HAVE BEEN RE-INTERPRETED

Designation	Place	Standard deviation of the magnitude	Reliability
464 Jul 20 E,SP	Galicia	0.075[a]	1
891 Aug 8a E,CE	Salzburg	0.18	0.5
912 Jun 17 E,SP	Cordoba	0.054	0.5
1009 Mar 29b E,BN	Ghent	0.18	0.5
1023 Jan 24 E,BN	Ghent	0.18	0.5
1033 Jun 29 E,BN	Ghent	0.18	0.5
1037 Apr 18 E,BN	Ghent	0.18	0.5
1039 Aug 22 E,GN	Ghent	0.18	0.5
1079 Jul 1 E,SP	triangle[b]	0.034	1
1093 Sep 23 E,BN	Ghent	0.18	1
1133 Aug 2c E,BN	Ghent	0.18	1
1133 Aug 2h E,CE	Prague	0.05[c]	1
1140 Mar 20b E,BN	Ghent	0.068	1
1140 Mar 20 E,PR	Stargard	0.18	1
1147 Oct 26a E,BN	Gembloux	0.054	1
1187 Sep 4 E,BN	Ghent	0.18	1
1191 Jun 23c E,BN	Ninove	0.18	0.5
1207 Feb 28 E,PR	Livonia	0.18	1
1207 Feb 28 E,BN	Ghent	0.18	1
1230 May 14 E,BN	Ninove	0.18	0.5
1240 Apr 23 E,PR	Lubin	0.18	1
1241 Oct 6a E,BN	Werm	0.18	1
1241 Oct 6f E,CE	Heiligenkreuz	0.054	1
1241 Oct 6g E,CE	Vienna	0.054	1
1241 Oct 6h E,CE	Friesach	0.18	1

[a] The eclipse will be taken as partial, with a magnitude of 0.85.

[b] The triangle whose vertices are Alcobaca, Santarem, and Sintra, all in Portugal.

[c] The magnitude will be taken as 0.92.

TABLE XI.2 (Continued)

Designation	Place	Standard deviation of the magnitude	Reliability
1241 Oct 6i E,CE	Graz	0.18	0.5
1241 Oct 6 E,PR	Stargard	0.18	1
1261 Apr 1 E,CE	Heiligenkreuz	0.18	1
1263 Aug 5 E,BN	Ghent	0.18	1
1263 Aug 5a E,CE	Heiligenkreuz	0.18	1
1263 Aug 5 E,PR	Krakow	0.18	1
1267 May 25a E,CE	Zwettl	0.18	1
1267 May 25b E,CE	Heiligenkreuz	0.18	1
1267 May 25c E,CE	Basel	0.18	1
1270 Mar 23a E,CE	Klosterneuburg	0.18	1
1270 Mar 23b E,CE	Vienna	0.18	1
1310 Jan 31 E,BN	Liège	0.18	1
1310 Jan 31a E,CE	Heiligenkreuz	0.18	1
1310 Jan 31b E,CE	S. Florian	0.18	1
1310 Jan 31c E,CE	Salzburg	0.18	1
1321 Jun 26 E,CE	Melk	0.18	1
1321 Jun 26a E,PR	Krakow	0.1	1
1321 Jun 26b E,PR	Stargard	0.18	1
1330 Jul 16 E,Sc	Iceland	0.18	1
1339 Jul 7 E,CE	Neuberg	0.054	1
1339 Jul 7 E,PR	Krakow	0.18	1
1339 Jul 7 E,Sc	Iceland	0.18	1
1364 Mar 4 E,BN	Fosse	0.18	1
1384 Aug 17 E,BN	Fosse	0.20[d]	1
1408 Oct 19 E,CE	Klosterneuburg	0.18	1
1485 Mar 16 E,CE	Melk	0.054	1
1486 Mar 6 E,CE	Melk	0.14[e]	1
1502 Oct 1 E,CE	Melk	0.14[e]	1
1560 Aug 21 E,SP	Coimbra	0	1

[d]The magnitude will be taken as 0.67.
[e]The magnitude will be taken as 0.81.
Note: The eclipse is taken to be central if no estimate of the magnitude is given.

position of maximum eclipse. While there is little problem
in judging whether or not most eclipses are total, there is
a serious problem with eclipses of this sort, and Clavius
in fact reports the eclipse as annular although it was total.
Further, there is nothing in his record which shows that he
was within the zone of totality; he was merely within the zone
in which some of the chromosphere remained visible and bright.
For these reasons, I took 0.01 for the standard deviation of
the magnitude for 1567 April 9.

For 1560 August 21, on the other hand, the moon was
larger than the sun by a safe margin, and there would have
been no difficulty in deciding whether the eclipse was total
or partial. Since Clavius says explicitly that the moon covered the whole sun and that the covering lasted for a significant time, and since Clavius was a trained astronomer, I
believe that we may safely use 0 for the standard deviation
of the magnitude. By plotting the zone of totality for the
eclipse, I have determined that the value of η on the edge of
the zone was 0.0120. The standard deviation of the value of
η within the zone is therefore this divided by $\sqrt{3}$, or 0.0069.

That is, in analyzing the record, I shall take η to be
0, with 0.0069 for the standard deviation of η and 0 for the
standard deviation of the magnitude. The reliability is 1
and the place is Coimbra, Portugal.

MS [p. 487] and M [pp. 8.66-8.67] both take this to be
the record of a total eclipse, and I owe my knowledge of the
existence of the record to their writings.

The records from the parts of Europe that have not been
used in earlier chapters (Belgium and the Netherlands, Central
Europe, Poland and Russia, Scandinavian countries, and Spain
and Portugal), and those records whose characteristics have
been changed since an earlier use, are summarized in Table XI.2,
which has the same format as the preceding tables with a similar function.

ECLIPSE RECORDS FROM THE NEAR EAST

1. Preliminary Remarks

In this chapter I shall take up all the sources that
have not been used in previous chapters. All such sources
that will be used in this work come from the part of the
world that we call the Near East. Two sources that are important from the literary and historical viewpoints, but
which will not be used here, come from other parts of Asia,
namely India† and Japan. I shall discuss these sources, including the reasons why I shall not use them, in the next
section.

The sources that will be used come from the Byzantine
Empire, Egypt in the post-Hellenistic period, the Holy Land,
and Syria. The Holy Land and Syria were both parts of the
Byzantine Empire at one time. When I use them as separate
provenances, I use the Holy Land [MCRE, p. 559] only for reports written by writers who were in the Palestine area for
a Crusade or who were serving the Christian kings of Jerusalem. I use Syria as a separate provenance only for reports written in the area of Syria after it ceased to be part
of the Byzantine Empire.

The symbols used for these provenances are M,B for the
Byzantine Empire, Is,E‡ for Egypt, M,HL for the Holy Land,
and M,S for Syria.

There are no new sources from the Holy Land. A separate
section will be devoted to the sources that have not been
used before for each of the other three provenances.

Eclipse records from the Near East that have been used
before and that will be used here without change are listed
in Table XII.1. This table has the same format as the similar tables in earlier chapters. Note that Syria is listed
as the place of observation for the eclipse record 360 August
28 M,B, but that the record is assigned to the Byzantine
Empire because Syria was still part of the Empire in 360.

2. Sources from India and Japan That Will Not Be Used

I have not made a great effort to find sources from
either India or Japan because I have little acquaintance
with their literature. In the course of incidental reading,
I have come across one source from each area that contains

---

†I use India to mean the entire peninsula, not the modern
country with the same name.

‡Is stands for Islamic. I should perhaps use Is rather than
M in the designation for Syria also.

TABLE XII.1

RECORDS OF SOLAR ECLIPSES FROM THE NEAR EAST
THAT HAVE BEEN USED BEFORE

Designation	Place	Standard deviation of the magnitude	Reliability
346 Jun 6a M,B	Constantinople	0.054	0.2
348 Oct 9 M,B	Constantinople	0.18	0.2
360 Aug 28 M,B	Syria	0.054	0.5
418 Jul 19a M,B	Constantinople	0.054	1
418 Jul 19c M,B	Constantinople	0.18	0.5
484 Jan 14 M,B	Athens	0.054	0.5
590 Oct 4 M,B	Constantinople	0.068	1
644 Nov 5 M,B	Constantinople	0.18	0.5
693 Oct 5a M,B	Constantinople	0.054	0.5
693 Oct 5b M,B	Constantinople	0.054	0.5
760 Aug 15 M,B	Constantinople	0.18	0.5
787 Sep 16 M,B	Constantinople	0.068	1
812 May 14 M,B	Constantinople	0.110	1
813 May 4 M,B	Constantinople	0.18	1
891 Aug 8a M,B	Constantinople	0.054	0.75
891 Aug 8b M,B	Constantinople	0.054	0.75
891 Aug 8c M,B	Constantinople	0.054	0.2
968 Dec 22b M,B	Constantinople	0.054	0.5
968 Dec 22c M,B	Constantinople	0.18	1
1113 Mar 19 M,HL	Jerusalem	0.1[a]	1
1124 Aug 11 M,HL	Jerusalem	0.14[c]	1
1147 Oct 26 M,B	—[b]	0.14[c]	1
1187 Sep 4 M,HL	Jerusalem	0.18	1

[a]The magnitude is taken to be 0.8.

[b]The place is the point with latitude 41°, longitude 30° E.

[c]The eclipse will be taken as partial, with a magnitude of 0.81.

Note: The eclipse is taken to be central if no estimate of the magnitude is given.

eclipse records. However I shall not use these records for
reasons that will be explained in this section.

The source from India is Kalhana [1150]. The title in
the original language is given as Rajatarangini. The translator of the cited translation gives the title in English as
A Chronicle of the Kings of the Kasmir. If Rajatarangini is
an adequate representation of the title in the original language, I suspect that the English title has been invented; I
doubt that there was a single word which meant a chronicle
of the kings of Kasmir, even though that word seems to be a
compound one. The translator calls this a didactic work.
His reason for this is perhaps illustrated by a passage on
page 22 of volume 1, which reads " ; . . all the rulers of
the various regions showed eager haste to throw themselves
on the kingdom like vultures on the carrion." Although he
calls the work didactic, the translator feels that the writer
Kalhana is basically impartial.

The work was begun in 1148 and finished in 1150. The
first part of the work is a "history" of early times and is
clearly mythical. About 850 and after, however, Kalhana begins to draw upon written sources. We may probably accept
the work as being basically historical after this date, although there may still be some mythical elements and of course
there may be errors. One example of the basic historicity
after 850 is that many events after that year are assigned
dates that are precise to the day.

I found one passage in Kalhana that clearly refers to a
solar eclipse, if we may rely upon the language of the translation, and one passage that refers to a great darkness which
is probably not an eclipse. These passages will now be discussed.

1133 August 2 ? A passage on page 171 of volume 2
reads: "The prince who had come to Kuruksetra on occasion
of the solar eclipse, met there the Lavanya and abandoned .
. his enmity . ." I do not know the meaning of the underlined words. The first seems to be a place while the second
seems to be a person or group of people. A footnote to the
passage reads: "The great pilgrimages to the Tirthas of
Kuruksetra take place on solar eclipses; . . ." I gather
that Tirthas are holy places of some sort.

Shortly before this passage, we have an event that is
precisely dated in terms of a lunar calendar and that must
come sometime in the summer of 1133. There is therefore a
temptation to identify the eclipse with the famous eclipse
of 1133 August 2. However, the eclipse of 1134 July 23 had
the same lunar date and it had about the same magnitude as
the one of 1133 August 2 in Kasmir, and I do not believe
that we can trust the chronology well enough to favor one
year over the other. Thus this record cannot be dated.

1147 October 26 ? On page 210 of volume 2 we find this
passage: "Then the world became darkened by clouds . ., the
light being suppressed by [darkness]." The brackets are in

TABLE XII.2

A CONNECTED GROUP OF THREE JAPANESE ECLIPSE RECORDS

Date as stated	Page number	Date using equivalent year of Suiko	Date if 36 III 2 = 628 Apr 10	Visible in Japan?
Suiko 36 III 2[a]	155	36 III 2	628 Apr 10	Yes
Jomei 8 I 1	b	44 I 1	636 Apr 11	Possibly
Jomei 9 III 2	168	45 III 2	637 Apr 1	Yes

[a]The record says that this eclipse was total.
[b]I inadvertently failed to record this page number.

TABLE XII.3

A CONNECTED GROUP OF EIGHT JAPANESE ECLIPSE RECORDS

Date as stated	Page number	Date using equivalent year of Temmu	Date if 20 X 1 = 691 Oct 27	Visible in Japan?
Temmu 9 XI 1	348	9 XI 1	680 Nov 27	Yes
Temmu 10 X 1	353	10 X 1	681 Nov 16	Possibly
Jito 5 X 1	404	20 X 1	691 Oct 27	No
Jito 7 III 1	411	22 III 1	693 Apr 11	No
Jito 7 IX 1	413	22 IX 1	693 Oct 5	Possibly
Jito 8 III 1	414	23 III 1	694 Mar 31	No
Jito 8 IX 1	417	23 IX 1	694 Sep 24	Possibly
Jito 10 VII 1	420	25 VII 1	696 Aug 3	No

the printed translation, and I imagine that "darkness" was supplied by the translator in order to complete the meaning. This passage refers to some year between 1144 and 1149, but we cannot date it more closely by means of the text alone. If it does refer to an eclipse, the only likely eclipse I can find is that of 1147 October 26. However, I see no reason to doubt the statement that the darkness was caused by clouds. The event is probably a dark day of the kind discussed under the heading "-647 April 6 ?" in Section VI.1.

The Japanese source that I have studied is Nihongi [720]. It is probably fair to describe this as a national chronicle of Japan. The translator (W.G. Aston) of the cited translation says [volume 1, p. 1] that Nihon, Nippon, and Niphon are all the same word and that "Japan is merely a Chinese pronunciation of this word, modified in the mouths of Europeans."

According to the translator, a commentary on Nihongi written about 824 says that it was completed and laid before the empress in 720 by Prince Toneri and Yasumaro Futo no Ason. This statement furnishes my basis for using the specific year 720 in citing Nihongi. It is possible that the work was written by the people named, but it is also possible that they were in the nature of sponsors rather than authors.

The part of Nihongi before about 500 must certainly be considered as mythology. After about that date, however, it begins to resemble genuine annals or chronicles, and much of the work from then on is probably based upon older written works and is hence basically historical. The calendar used is a mixed lunar-solar calendar, in which the months are lunar, and the years have either 12 or 13 months, with the number of months being chosen to keep the average year equal to the solar year. It is interesting that almost all the solar eclipses recorded in Nihongi took place on the first day of the month. This suggests that the Japanese intended to begin their month with the true conjunction of the sun and moon, not with the first or last visibility as many peoples did, but that they made an error sometimes. The year began at about the same season as ours.

A year is identified as the year Y of a certain ruler. The year in which a reign ended is considered as belonging entirely to the outgoing (or dead) ruler, and the first year of his successor starts at the beginning of the year after his accession.

I found records of eleven solar eclipses in Nihongi, which occur in two clusters, both in our 7th century. All the records are found in volume II of the cited translation. Table XII.2 gives the three eclipse records that occur in the earlier cluster and Table XII.3 gives the eight that occur in the later one. The first column in each table gives the date by means of the ruler's name, the number of his (or her) year, the number of the month, and the day of the month, which is usually 1. The words Temmu, Jito, Suiko, and Jomei are the names of the rulers. Jito and Suiko were empresses. The others were not identified, so far as I noticed, and they are probably emperors. The second column gives the page number in volume II.

In order to tie the years together conveniently, it is helpful to refer all years in each cluster to the first ruler mentioned. The third column in each table does this. The year in this column in Table XII.2 is the regnal year that Suiko would have had if she had lived that long; the rest of the date is the same as in the first column. The corresponding year in Table XII.3 is the regnal year that Temmu would have had if he had lived long enough.

The translator, apparently without using the eclipse, says that the first year in Table XII.2 is 628. An eclipse in the third month of that year would therefore have to be the eclipse of 628 April 10. It is interesting that the record says that this eclipse was total, and Oppolzer's map [Oppolzer, 1887] shows that it was indeed very large and perhaps total in most of Japan. Thus we seem to have excellent agreement between calculation and the record.

Since the month given in the records is the lunar month, we cannot identify a date exactly unless we know all the intercalary months. If we assume that all the dates in the records must be dates of solar eclipses, the second date in Table XII.2 must be 636 April 11. The eclipse of this date was penumbral, with the maximum eclipse coming in late afternoon, Japanese time. Thus the eclipse could have been seen in Japan so far as the hour is concerned, but the moon's shadow almost missed the earth completely, and it is doubtful that the eclipse was visible in Japan. Nonetheless, for the sake of argument, let us assume that it was.

The third eclipse in Table XII.2 must be 637 April 1 by the assumed chronology, and the eclipse of that date was very large in Japan. Thus, if we assume that the second eclipse was visible, we seem to have a cluster of three eclipse records, all of which can be dated. However, before we decide to use these records, we should look at Table XII.3.

The translator says that the record dated Jito 5 X 1, which is equivalent to 20 X 1 of Temmu, comes in the year 691. The only eclipse in 691 which was visible in Japan was the eclipse of 691 May 3, which was probably visible as a small eclipse about the middle of the morning. However, an eclipse in May cannot be in the 10th lunar month, as the record requires. The only eclipse that could have been in the right month and year, if the year is 691, is the eclipse of 691 Oct 27. This eclipse was hardly visible outside of Antarctica, so we cannot identify this record as one of a visible eclipse.

For the moment, let us accept the translator's dating and see where it leads us for the remaining records. The relevant dates and conditions of visibility are given in the last two columns of Table XII.3. Four of the eight eclipses were certainly not visible in Japan, one was certainly visible, and three were possibly visible; their visibility cannot be determined without detailed calculation. Thus the eclipses do not agree at all well with the translator's assignment of date.

Nonetheless, the translator had sound reasons for his assignment of dates, and we should not discard his dates

lightly. I suspect that the correct answer to our puzzle is
the following: The Japanese of this period, who were under
considerable cultural influence from China, calculated the
dates of solar eclipses and put them in the records, just as
their contemporaries in the Tarng dynasty of China did.† In
my opinion, this conclusion should neither be accepted nor
rejected until the matter has been studied thoroughly by
persons who are expert in the relevant literature.

If this conclusion is correct, we cannot accept a record for use in this work unless it has wording that definitely
indicates observation rather than calculation. The only record in Tables XII.2 and XII.3 that contains a statement beyond
the mere statement that there was a solar eclipse on a given
date is the record dated 628 April 10. This record says that
the eclipse was total. This could be the result of observation, but there are no details, such as the visibility of
stars, that confirm totality, and the statement could also be
the result of calculation. Thus no record from Nihongi can
be used as the record of an observed solar eclipse without
further research.

3.   Sources from the Byzantine Empire That Have Not Been
      Used Before

Acropolita [ca. 1262]. Acropolita was known as an unsuccessful military commander, but he was successful in his
career as chancellor, which he became in 1244. His work is
a history of the Byzantine empire from the capture of Constantinople by the Crusaders in 1204 to its recovery by the
Greeks in 1261.

Cosmas [ca. 550]. Cosmas is generally known as Cosmas
Indicopleustes; the second part of his name apparently refers to India. Cosmas was well travelled, but he probably
never travelled to India in spite of his name. I do not
know whether Cosmas is his own name or whether it was applied
to him in reference to the universality of his subject. He
wrote this work in an attempt to prove that the earth is flat
and not spherical as the astronomers maintained.

In his travels one year he was in Ethiopia south of the
tropic at the summer solstice, so that the shadow fell toward
the south at noon. In Alexandria, on the other hand, the
shadow is toward the north at noon at all times of year. It
is interesting to see Cosmas try to reconcile these observations with a flat earth. de Camp [1970, p. 19] describes
Cosmas's work as a "monument of unconscious humor," and I
cannot think of a better description.

Gregoras [ca. 1359]. Nicephorus Gregoras lived from
about 1295 to 1360. He attracted attention at an early age
on account of his learning and was appointed court archivist

---

†See Section V.5.

by Andronicus II at about the age of 30. When Andronicus II
was deposed by his grandson Andronicus III in 1328, Gregoras
shared in his downfall, but he was able to regain favor after
the death of Andronicus II in 1332. He wrote many still extant works on many subjects, most of which are still unpublished. His works include writings on astronomy, and in 1326
he proposed a calendar reform. I do not see how his reform
could have been successful. His basis for the reform is the
statement in Section VIII.13 of his cited history that the
vernal equinox "in our days" comes on March 17.† Picking
1332 as a typical year "in our days", I find that the equinox came at about noon, Constantinople time, on 1332 March
12, so the main datum that Gregoras used is wrong by 5 days.

Malalas [ca. 563]. Ioannis Malalas was born in Antioch
in about 491 and he died in about 578. Since his history is
centered on Antioch when it is not centered on Constantinople,
I imagine that he spent his life in Antioch except, perhaps,
when he was travelling. This is a history of the world from
early Egyptian times through the year 563. It is considered
to be a work of very poor historical quality, whose main
interest is that it is the earliest chronicle we have that
was written for the "common man" rather than for the learned.

Michael I [ca. 1195]. I classify eclipse records according to the place where the observation was made rather
than according to the place where the author lived or wrote.
If we use this basis, we must class Michael I as both a
Byzantine and a Syrian source. I shall discuss this source
in Section XII.5, which deals with Syrian sources.

Panaretos [ca. 1426]. When the Crusaders captured
Constantinople in 1204 and set up the so-called Latin empire,
Alexius I Comnenus, a relative of the legitimate emperor, escaped and set up an independent government. Since his capital was at Trebizond‡, this government is usually called the
Empire of Trebizond, although it seems to me rather small to
merit the title of empire. Regardless of its size, it long
remained independent and it held out against the Muslim conquerors longer than Constantinople. It was not conquered
until 1461 but Constantinople fell in 1453. Panaretos is a
fairly short history of the Empire of Trebizond, going to
the year 1426. Since a small part near the end of the manuscript has been lost, it is possible that the original went
to a slightly later date. Although Trebizond was not part
of the Byzantine Empire during the period covered by the
history, I count this as a Byzantine source because it was
part of that culture.

---

†This section of his history is apparently an abstract of
his longer work on calendar reform. The date appears as
March 17 in both the published Greek text and the Latin
translation.

‡Now Trabzon, near the northeast corner of modern Turkey.

Zosimus [ca. 475]. This is a history of the period of
the Roman (or Byzantine) emperors, which deals with them
rather sketchily down to Constantine. It is written from a
non-Christian point of view.

In addition to these sources and those used in MCRE, I
have studied at least thirty Byzantine sources without finding any records of solar eclipses.

4. Egyptian Sources

Many uninformed writers on the history of science say
that ancient Egypt† had a vast store of astronomical knowledge
and understanding. Egyptian records of solar eclipses show
that this is far from the truth. Except for the allusion to
the "eclipse of Hipparchus"‡ by Ptolemy [ca. 142], the first
known eclipse record from Egypt is dated 601 March 10.

Djeme [ca. 601]. E. B. Allen studies a set of annals
written in Coptic that apparently come from the town of Djeme,
which is now called Madinat, near Thebes. He does not give
an ancient title for the work, nor does he give the period
that the annals cover. Thus I have arbitrarily called the
work the Annals of Djeme, and for its date I have used the
date of the only event that Allen discusses. This event is
a solar eclipse. Because the reader would not find anything
in the literature listed under Djeme, I am also listing this
source as Allen [1947] in the references.

I have not been able to locate Madinat in any reference,
even though it is presumably the modern name of the place.
The only clue that Allen gives is that it is near Thebes.
The only place near Thebes with a name at all like Madinat
that I can find in the Times Atlas [1955] is Astun el Mata'na,
about 40 kilometers from Thebes. It does not seem likely
that this is the place, so I shall use the coordinates of
Thebes.

ibn Iyas [ca. 1522]. Apparently we know nothing about
ibn Iyas except the obvious facts that we gather from this
source, which is almost a diary. Some obvious facts are
that he lived in Cairo at least from 1468 January 26, when
the source begins, to 1522 November 19, when it ends. The
cited French translation, which is the only form of the
source I have seen, is published in four volumes. The translation calls the first two Histoire des Mamlouks Circassiens
and the last two Journal d'un Bourgeois du Caire. The reason for the second title is clear but the reason for the
first is not so obvious. The translator also calls the entire work Chronique d'ibn Iyas, and this is the title I have
used in the list of references.

---

†That is, Egypt before the Alexandrian conquest.

‡See the discussion dated -128 November 20 in Section VI.1.
The eclipse is also mentioned by two other ancient writers.

Most of this work is a chronicle arranged by months, with about one page in the cited translation for each month. One interesting feature is that ibn Iyas sometimes records the date when the Nile reached its highest stage, and the reading of the Nilometer† on that date.

These are the only sources from Egypt with eclipse records that I have found, except for the allusion by Ptolemy and others that was already mentioned.

5.   Syrian Sources

Chalipharum Liber [ca. 636]. A subtitle to this work that appears in the printed edition calls this a chronicle of the 8th century, but the latest date that I found in it is 636. However, the events in the work are far from being in chronological order, and it is possible that I missed some events from the 8th century.

The work has three main parts. The first part is an account of the emperors from Constantine through Heraclius. The second is a disordered chronicle of imperial affairs which first gives events from 456 to 570, then from 511 to 631, then from 367 to 629, then 623 to 636, and finally 32 to 529. The third part gives accounts of the church councils through the Council of Chalcedon. The work seems to be a compilation made by one person.

The source has only one record of a solar eclipse, and that one cannot be dated. I ignore most sources that have only unidentifiable eclipse records, and I include this one only because the translator's discussion of the record needs considerable amplification.

Michael I [ca. 1195]. The translator of the cited edition describes the author of this chronicle as Michael the Syrian, Patriarch of the Jacobite Church of Antioch from 1166 to 1199. He was one of the last people to write in the Syriac language. For a long time his chronicle was known only in an abridged Armenian translation, but the Syriac manuscript‡ was found in a library in Edessa near the end of the 19th century. It is a chronicle of the history of the church and empire down to his own time.

This source also has no usable records of solar eclipses, and I include it only in order to correct some errors in M's [pp. 8.46 and 8.56] use of it. It should be considered as a Byzantine source for years before about 630 and as a Syrian source in later years.

---

†This is the instrument that measured the height of the river, but I do not know the basis for choosing the reference point. Use of the Nilometer goes back to ancient times, I believe.

‡I do not believe that this is intended to mean Michael's holograph.

I have also studied the famous source called the Syriac
Chronicle [569], but it contains no useful records of solar
eclipses. Thus there are no known useful records from
Syrian sources.

6. Records from the Near East

In this section I shall discuss records of solar eclipses
from the Near East that have not been used before, as well as
a few records that have been used before but that will be
used here with changed characteristics. In addition, I shall
discuss a few records that will be used without change from
earlier work, but that require some further comment.

295 March 3 ? In Book XII, on page 306 of the cited
edition, Malalas [ca. 563] writes: "Under Diocletian there
was great darkness for the entire day." Nothing about this
says that the phenomenon was an eclipse. In fact, if the
darkness lasted an entire day, it was not an eclipse but a
"dark day". If we assume that the record does refer to an
eclipse, the date 295 March 3 is the most likely one, but it
is not the only possibility. Thus this record is not usable
even if it should refer to an eclipse.

360 August 28 M,B. I used this record in AAO [p. 118]
and in MCRE [p. 537], and M [p. 8.39] also discusses it. All
these sources cite the original source, and M further cites
a discussion of the record in Stephenson's unpublished dissertation that I have not seen. Referring to Stephenson, M
writes: "The latter† giving particular detail and relying
directly on the original source, concludes that the place of
observation is unknown. Presumption that it occurred near
Amida, a town supposedly under siege at the time, as done by
Newton, is unwarranted since the eclipse occurred a full
year after‡ the battle, and there is no evidence in the original source to suggest a connection with any other event:
Stephenson (1972). This is an example of a case where there
is no substitute for consulting the original‡ source as
Stephenson did."

This is a serious and unjustified misrepresentation of
what I did. I specifically pointed out that there was no
reason to assume that the eclipse observation coincided with
the battle or siege in either place or time. I identified
the eclipse by using the year 360, which the annalist clearly
identifies, and I assumed that the observation could have
been made anywhere in an extended region. I used slightly
different regions in AAO and MCRE, but the region in both
cases was intended to be an approximation to the Roman province of Syria. This is reasonable, since the annalist was
living in Syria or serving in the army on its frontier at
the time [MCRE, p. 537].

---

†Stephenson, that is, in his dissertation.

‡The emphasis is M's.

I did write in AAO that the eclipse was probably assimilated to the battle, and that the amount of assimilation
was probably small,† but I decided against even this assumption in MCRE. Judging from what M says, Stephenson wrote
his conclusion before MCRE was published. If so, this provides some justification for Stephenson's conclusion. M,
however, ignored what I wrote in MCRE and he did not study
what I wrote in AAO. He relied upon an outdated quotation
from an unpublished work instead of consulting the original
source.

I shall modify slightly the region used in MCRE, and
I shall take Syria to be the rectangle bounded by latitudes
34° and 37° North and by longitudes 36° and 42° East.

393 November 26 ? Section 4.58 of Zosimus [ca. 475]
deals with a battle between the emperor Theodosius and a
rival Eugenius that lasted for several days. On the first
day of this battle, which Zosimus dates precisely as the
equivalent of 394 September 5, there was such an eclipse of
the sun that for some time it was considered to be night
rather than day. However, there was no eclipse on 394 September 5 or on any day that I can get by postulating reasonable errors in writing the date. For this reason, we expect
this to be a literary eclipse, except that Zosimus does not
seem to make the eclipse play any part in the conduct or outcome of the battle. The eclipse of 393 November 26 is the
only eclipse that was large in the right part of the world
anytime near the specified date, but I do not feel that it
is safe to use this identification. Thus I shall not use
this record.

Michael I [ca. 1195] has records of several eclipses
during the period that we consider him to be a Byzantine
source rather than a Syrian one. In Chapter VIII.11 he says
that there was darkness over all the earth on the day Marcian
was crowned, which was in 450. There was no eclipse visible
in the right part of the world at that time, although the
eclipse of 449 May 8 was probably large in Syria at sunrise.
In Chapter IX.7 he says that there was an eclipse of the sun
in the first year of Anastasius, which should be 491 or 492.
The eclipse of 492 January 15 may have been visible in Syria
and perhaps in Constantinople, and the eclipse of 493 January 4 was large at sunrise in Syria. In Chapter IX.11 he
says that in the time of Anastasius there was an eclipse of
the sun on the 6th feria (Friday) from the 3rd to the 9th
hour. The eclipse of 512 June 29 is the only one I can find
during the reign of Anastasius that came on Friday during
the middle of the day, although the eclipse of 511 January
15 was very large at sunset in much of the Near East. It
came on a Saturday, however. Finally, in Chapter IX.26 he
says that the sun was darkened for about 1½ years around the
year 536.‡ He finds this hard to believe but says that he

---

†Meaning a year or so in time and a few degrees in position.

‡A similar statement is found in sources much older than
Michael I. See MCRE, pp. 458-459.

finds it stated in many reliable sources. We are probably safe in considering that this is not an eclipse record.

The date of 512 June 29 is the only one we can find that is unique for the circumstances stated in any of these records, and it is the only one that we can even consider using. However, Michael's source for this may well be the Syriac Chronicle [569]. This chronicle says in Chapter VII.9 that a wonderful sign occurred during the reign of Anastasius, namely an eclipse of the sun which produced darkness from the 6th to the 9th hour. This record does not give the week day, and it gives a different initial hour, but it shows that Syrian sources could be among those which Michael used. Thus we cannot safely assign a place of observation, or even a region of reasonably small size, and this record is not safe to use.

512 June 29 M,B. This record was quoted in MCRE [p. 542]; the original writer identifies the year as being near 512. The eclipse of 512 June 29 was very large if not total at Constantinople, and I tentatively adopted this date on page 542. At the same time, I said that this identification would have to be abandoned if the eclipse of 511 January 15 was prominent in Constantinople. I found on page 592 that the eclipse of 511 January 15 was large in Constantinople at sunset, so that it was likely to be observed. Hence the eclipse cannot be identified, and I did not use it in MCRE nor shall I use it here.

547 February 6 ? I mentioned in Section XII.3 that Cosmas [ca. 550], in Book VI of his work, uses meridian elevation angles of the sun in an attempt to prove that the earth is flat in accordance with the Scriptures. In the same book, in column 321 of the text and column 322 of the Latin translation, he describes two successful eclipse predictions which he intends to use for the same purpose. One prediction was for a solar eclipse and the other was for a lunar eclipse in the same Egyptian year; the year is identified only as being 10 of the Indiction. The solar eclipse was predicted for the 12th of the Egyptian month Mechir and the lunar eclipse was predicted for the 24th of the month Mesor,† and indeed both eclipses happened according to Cosmas. This means that both eclipses were observed, so we may take this as a record of an observed solar eclipse.

If we give or take a day according to the details of when a leap year occurred, we may say that the solar eclipse came on February 6 and the lunar eclipse came on August 17.

---

†Mechir is the 6th month of the Egyptian year and Mesor is the 12th month. The Indiction was a cycle of 15 years used by the Romans in connection with taxation and tax assessment. In terms of years of the common era, year 10 of the Indiction must have the form 532 + 15N for years in the time of Cosmas, with N being an integer. The year of the Indiction began on September 1, and this formula applies only to dates before September 1.

According to the footnote, the year must have the form
532 + 15N. Indeed, for N = 1, we find a solar eclipse visible in the Near East on 547 February 6 and a lunar eclipse
visible there on 547 August 17. Thus there seems to be no
doubt about the dates involved.

In spite of this, I feel some malaise about the dates,
perhaps because of the nature of Cosmas's work. However,
even if we accept the date, we cannot use the record. Cosmas
was famous for the extent of his travels and we cannot even
guess where the observation was made.

601 March 10 Is,E. Judging from the discussion by <u>Allen</u>
[1947], this record is an isolated fragment that does not
come from a chronicle or an extended set of annals, and thus
there is no orthodox ancient title to use in its citation.
Since the fragment came from the town of Djeme, I arbitrarily
used the citation <u>Djeme</u> [ca. 601] in Section XII.4. As I
noted there, Djeme is the town now known as Madinat, near
Thebes.

In the translation offered by Allen, the record reads:
"On the fourteenth of Phamenoth† of the fourth indiction, the
sun was eclipsed in the fourth hour of the day and in the
year in which Peter, son of Palu, was made village official
in Djeme." The original is in Coptic. Since we are given
the month and day of the month in the Egyptian calendar, and
since we are given the year of the Indiction, we know the
date exactly except that the year is uncertain by a multiple
of 15 years.‡ Thus we need to see what eclipses come on a
possible date that began in Egypt in or before the 4th hour;
we should not rely upon the statement about the hour any
more than this.

In order to identify the eclipse, Allen searched through
all the eclipses listed in <u>Oppolzer</u> [1887] between the year
297 and 1580 for eclipses that meet the following conditions:
(a) The date must be of the required form, with an ambiguity
of 15 years, (b) the eclipse must be total (as opposed to
annular), and (c) the path of totality must pass near mid-
Egypt. He found that only the eclipse of 601 March 10 satisfies all these conditions. Finally, he calculated the time
associated with the eclipse at Djeme and found that the
greatest obscuration occurred during the 4th hour of the day,
as "required by the record."

Allen was lucky that his dating of the eclipse was correct, because two of the conditions he used are incorrect.
We know enough about eclipse records to know that a stated
hour does not need to be the hour of greatest obscuration.
It can just as well be the beginning hour, or, more probably,
the time when the occurrence of an eclipse was first noticed.
All we can safely say in this regard is that the eclipse

---

†The 7th month in the Egyptian calendar.

‡See the discussion of the Indiction under the date 547
February 6 above.

probably began before the end of the 4th hour. Further,
there is absolutely nothing in the record to suggest that
the eclipse was total. As to the path of the eclipse passing near mid-Egypt, the validity of this assumption depends
upon how "near" it needed to be. All that is needed is that
the eclipse be reasonably observable in mid-Egypt.

The only conditions that we may reasonably impose are
the following: (a) The date must satisfy the required form,
(b) the eclipse must have been reasonably observable in central Egypt, and (c) it must have begun there in or before
the 4th hour of the day. The only two eclipses that satisfy
the first condition are those of 601 March 10 and 1141 March
10. The eclipse of 1141 was not visible in Egypt, and hence
the eclipse of 601 March 10 is the correct one. Thus we do
not need most of the conditions that Allen used, because the
conditions on the date and the zone of visibility are so restrictive in themselves.†

Thus we are safe in taking the passage from Djeme [ca.
601] to be a record of the eclipse of 601 March 10. There
is no reason to question the originality of the record that
I know of, so I shall take the reliability to be 1. The
place is Madinat and the standard deviation of the magnitude
is 0.18. Since I cannot find the coordinates of Madinat,
I use those of Thebes.

627 October 15 ? On page 115 of Chalipharum Liber
[ca. 636] we find the sentence: "Year 938, day 15 of September,‡ both the sun and the moon were eclipsed." The sentence
says that Chosroes (Khosrau) died in February of the year 934.
These years are counted from -311 October 1 as the beginning
of the first year, so the day of the eclipses is 627 September 15 and the month of Khosrau's death is 623 February. This
must refer to Khosrau II, who died in 628; this shows that
the chronology is not accurate.

The editor of the cited edition notes that there cannot
be a solar and a lunar eclipse on the same day, and he sup-
poses that the writer meant to say the same month. He fails
to note that the confusion is much more extensive than this,
starting with the large error in the date of Khosrau. The
year 938 runs from 626 October 1 to 627 September 30. Within
this interval, the only lunar eclipse was on 626 November 9,
and its maximum was at $11^h 32^m$, Greenwich time, according to
Oppolzer [1887]. This was still in the middle of the day in
the Near East, so this eclipse was not observable. There was
a solar eclipse on 626 October 26, but it was not visible in
the Near East either. The eclipse of 627 October 15 is

---

†To satisfy the reader's probable curiosity, I add that
maximum eclipse at Djeme came at the end of the 4th hour of
the day, according to a calculation made with the final
values of the accelerations. The calculated magnitude is
0.87.

‡The original has the Syrian names of the months, which I
have changed into their English equivalents.

tempting because its date is off by exactly a month, but it was not visible either. These eclipses are unidentifiable.

MS [pp. 482-483] and M [pp. 8.43-8.44] both use a record of the eclipse of 693 October 5 from a Syrian chronicle that I have not read. In my opinion, they have not studied the text of the chronicle enough to justify their use of the record.† Since I have not studied the source either, I shall not use this record.

812 May 14 M,S. On page 26, volume 3 of Michael I [ca. 1195] we find this record: "In the year 1123, the 14th of May,‡ there was a total eclipse of the sun, from the 9th to the 11th hour, and the darkness was as deep as night; one saw the stars and people lighted torches. Afterward the sun reappeared for about an hour." The eclipse is dated accurately, and the chronicler has stated that the eclipse was total. The problem concerns the place of observation.

M [p. 8.46] assigns the place to be either Edessa or Harran, with a high probability (0.80), and he justifies the choice by the following argument: "This can be inferred with reasonable probability from the fact that the only events described by Michael during the time around the eclipse took place in Edessa or Harran. Presumably he was working from manuscripts in his possession, and his information at this epoch was apparently limited to the region bounded by Edessa and Harran."

I do not understand the basis for this conclusion. To test it, I chose a few lines in the text on either side of the eclipse record, and I started to count the places referred to in these few lines. I quit counting when I had reached 12 places, which range from Constantinople to Baghdad. In fact, within the limited span of this count, there are more events recorded at either Constantinople or Baghdad than at either Edessa or Harran. Clearly the observation could have been made anywhere over a wide region, and the record is unusable.

891 August 8. On page 119 of the same volume, Michael I [ca. 1195] writes: "Shortly after, the same year, there was an eclipse of the sun at the middle of the day, so that one saw the stars in all the sphere of the heavens." He has already given the year as 1200 with respect to the era of -311 October 1, so the year specified runs from 888 October 1 to 889 September 30. The only eclipse visible in the Near East during this year was that of 888 October 9, which was a small eclipse visible at sunrise in parts of the Near East.

---

†See particularly the discussion of the record dated 1176 April 11 below, which shows the dangers of studying only one record from an extensive source.

‡Again I have given the English equivalent of the Syrian name. The date given in the chronicle is equivalent to 812 May 14.

However, the path of the eclipse of 891 August 8 goes through the near East, and no other path near this time does so. Hence this must be the eclipse meant. However, the eclipse was rather annular, with a maximum magnitude of about 0.95, so the statement about the stars is unlikely. This fact reflects upon the reliability of Michael in the preceding record. It is not safe to use the record that apparently relates to 891 August 8 because we do not know the place of observation.

968 December 22a M,B. I used this record in MCRE [p. 549], and I have already discussed my reasons for changing its characteristics in Section II.1 of this work. I shall change the standard deviation of the magnitude to 0.034.

1109 May 31 ? On page 197 of volume 3, in the year 1419, Michael I [ca. 1195] says that a thick darkness like dust covered the sun from the 1st hour to the 3rd on April 4; from the 3rd hour to the 6th hour the sun glowed feebly; then for three more hours the sun's globe was like fire and it did not shine at all. This kind of behavior lasted for 12 days. Finally, on May 5, the sun was darkened for 3 hours. The year specified runs from 1107 October 1 to 1108 September 30,† so the dates specified are 1108 April 14 and 1108 May 5.

There was no eclipse visible in the Near East during the year stated, and the nearest eclipse I can find is that of 1109 May 31, which would have been a small eclipse in the middle of the afternoon. The record does not indicate that we are dealing with a solar eclipse, and the dates involved confirm this conclusion.

1133 August 2. Michael I [ca. 1195] records this famous eclipse on page 235 of volume 3: "On the 2nd of August,‡ there was an eclipse of the sun ..." At the time when the eclipse happened, he tells us that 40 Knights Templar were killed, along with 400 other Christians, and the deacon Bar Qorya. We have learned earlier from Michael's chronicle that Bar Qorya was deacon at Amida, but it is not safe to assume that this is the place. The deacon could well have been fighting away from home. Further, although we are now almost in Michael's own time, we may not safely assume that the place of observation is Antioch, which was Michael's home. The reasons for this will appear in the discussion of the record dated 1176 April 11 below. Thus the record cannot be used.

---

†The editor says that the year is equivalent to 1109, but I do not see how he reached this conclusion. Perhaps it is a typographical error.

‡He has already said that the year was 1444, so the date is 1133 August 2.

1176 April 11. This record is also from <u>Michael I</u>
[ca. 1195, <u>volume 3</u>, p. 367]: "In this year 1487, on New
Monday,[†] the 11th of April, at the beginning of the morning,
at the end of the office, that is after the reading of the
Gospel, the sun was totally obscured: night came, and the
stars appeared; the moon itself was seen near the sun. It
was a sad and terrifying sight, which made many people lament
and weep; the sheep, the cattle, and the horses mingled one
with the other from fright. The darkness lasted 2 hours,
then the sun returned." Immediately after this, Michael says
that there was a lunar eclipse 15 days later, at the end of
a Monday, and at the moment of the evening, in the same part
of the sky where the solar eclipse had been.

The statement that the moon was seen near the sun is
puzzling, but otherwise we clearly have an attempt to describe a total eclipse. Since we are definitely in Michael's
own time, we seem entitled to assume that Michael saw the
eclipse himself in Antioch, where he was probably living,
and $\underline{M}$ [p. 8.56] assumes that this is so with a probability
of 0.90.

This illustrates vividly one danger in studying only a
limited set of records from a particular source. We must
study this record along with Michael's records of 1133 August
2, which was quoted above, and with his records of 1187 September 4 and 1191 June 23, which will be quoted below. When
we do so, we see that the conclusion of the preceding paragraph cannot be sustained.

The records of 1187 September 4 and 1191 June 23 both
say that stars could be seen around the sun, and the record
of 1191 June 23 even repeats that the moon could be seen.
It is not possible that stars could have been seen in Antioch
during either of these eclipses. Further, the eclipse of
1191 June 23 was annular by a considerable margin, and it is
doubtful that stars could have been seen "all about the sun"
even within its central zone. From these facts, we know
either that stars were not seen during the eclipses of 1187
September 4 and 1191 June 23 or that the observations were
not made in the place we expect, namely Antioch. Thus we
cannot assume that the record of 1176 April 11 refers to an
eclipse that was total there. Similarly, we cannot assume
that Michael's observation of the eclipse of 1133 August 2
was made in Antioch.

The record of 1191 June 23 in fact tells us that we may
not rely upon Michael's statements about seeing stars and,
by implication, upon his other statements that indicate totality. The record of 1191 June 23 is quite interesting,
because Michael distinctly implies that the eclipse was partial and not very large, in spite of the statement that stars
were seen all around the sun. In sum, we must conclude that
all Michael's eclipse records are unreliable, and we cannot
use any of them.

[†]In spite of its name, this means the first Sunday after
Easter, so that the date given for the eclipse is Sunday,
1176 April 11. I continue to give the English equivalents
of the Syrian month names given in the text.

1187 September 4. On pages 397-400, volume 3 of the
cited edition, Michael I [ca. 1195] has a long passage which
says that the false prophets, that is, the astronomers, predicted that there would be a solar eclipse on 1186 September
14, and a conjunction of all the planets, followed by a wind
that would destroy all life. When the day came, many hid in
caves or other places, but the day was clear and there was
no eclipse and no wind. Then "the people praised the Lord
and the kings scorned the astronomers, whose art is decep-
tion." I hope that by astronomers he means what we call
astrologers.

A few pages later, on page 403, Michael records the
eclipse of 1187 September 4: "In the year 1498, on Friday
the 4th of September,† at the 8th hour, there was an eclipse
of the sun, and one saw the stars around the sun." This
record is unusable for reasons that were discussed above
under the date 1176 April 11.

1191 June 23. Michael I [ca. 1195] records this eclipse
on page 410 of volume 3 of the cited edition: "In the same
year,‡ on the 23rd of Haziran, there was an eclipse of the
sun. More than half of its globe was darkened, and all about
the sun one saw the stars, and even the moon in its vicinity."
Note the distinct implication that the magnitude of the
eclipse was not much greater than 0.5, but stars were seen
all around the sun. This record cannot be used for reasons
that were discussed above under the date 1176 April 11.

M [p. 8.60] gives a record of the eclipse of 1239 June
3 from a source that he identifies as Synaxarium. However,
he does not list this source among his references, and he in
fact uses a secondary source rather than the primary source.
I shall ignore this record.

1245 July 25 ? Acropolita [ca. 1262, pp. 67-69] says
that there was an eclipse of the sun when the sun was in
Cancer during the reign of the emperor John III Ducas. This
caused the emperor to ask Acropolita to explain the cause of
eclipses, which Acropolita proceeds to do. This was apparently a short time before the death of the emperor, which
happened in 1254, and it was apparently after Acropolita became chancellor in 1244. However, I can find no eclipse
that was significant in or near Constantinople with the sun
in Cancer between 1191 June 23 and 1293 July 5. The sun on
1245 July 25 could perhaps be described as being in Cancer
by someone who was not being careful, and the eclipse of that
date was possibly visible in Constantinople; I cannot tell
without detailed calculation. It is clear that this eclipse
cannot be identified.

---

†This date is 1187 September 4, which was a Friday.

‡Michael has already given the year as 1502, with reference
to the era -311 October 1. The date he gives for the
eclipse is thus 1191 June 23.

1267 May 25 M,B. <u>Gregoras</u> [ca. 1359, Section IV.8]
records this eclipse as follows: "At that time the moon
obscured the sun, as it was passing the 4th degree of Gemini,
at the 3rd hour before noon, on the 25th day of May, year
6775.† The eclipse was 12 digits, as closely as possible.
At the center time of the eclipse darkness appeared on the
earth, such that many stars could be seen." The record seems
to say that the eclipse was total, and thus we use 0.034 for
the standard deviation of the magnitude. The place is probably Constantinople. Since Gregoras was not born until about
1290, the record cannot be original, so it will receive a
reliability of 0.5. <u>M</u> [p. 8.64] and <u>MS</u> [pp. 486-487] accept
this as a record of a total eclipse seen in Constantinople,
but with a reliability less than 1.

1331 November 30 M,B. <u>Gregoras</u> [ca. 1359, Section IX.14]
notes that many omens preceded the death of the emperor Andronicus II, who had abdicated in 1328 and who died on 1332
February 12. First was an eclipse of the sun, which preceded
his death by as many days as he had lived in years. A lunar
eclipse followed the solar eclipse. Then came an earthquake
on the night of 1332 January 17. Finally, on the day he died,
there was a tidal wave accompanied by a strong wind, which
did much damage to Constantinople and its walls. Although
these events are used in a magical way, they seem to be duly
recorded and not invented. In particular, there are 74 days
from 1331 November 30 to 1332 February 12 and Andronicus was
72 years old. In view of the uncertainty about which days
were included in the counting or not, the agreement is good.
Since we are in Gregoras' own time, the reliability is 1,
and the place is Constantinople. The value of η is 0 and
the standard deviation of the magnitude is 0.18.

1337 March 3a M,B. <u>Gregoras</u> [ca. 1359] records this
eclipse in his Section XI.3. He says that the "Romans" had
been hurt in a war with people whom he calls Scythians, and
eclipses of the sun and moon, coming 16 days apart, foretold
this damage. The lunar eclipse came when the sun was in the
1st degree of Pisces and the solar one when it was in the
15th degree. These positions, along with the fact that the
year has already been identified, tell us that the eclipses
were on 1337 February 15 for the moon and 1337 March 3 for
the sun. This record receives the same characteristics as
the preceding one.

1337 March 3b M,B. The text of this record occurs on
page 17 and the translation occurs on page 46 of the cited
edition of <u>Panaretos</u> [ca. 1426]: "In the month of March,
the 2nd day, in the great holy forty-day fast, an eclipse of
the sun occurred from the 4th to the 7th hour, and the people
rose against the emperor, gathered outside the castle, and
threw stones at him." Although the chronology of this record

---

†The year was reckoned from the time that the Byzantines
assigned to the Creation, which was -5508 September 1
[<u>MCRE</u>, Section XV.1]. Thus the date is 1267 May 25.

is careless, there is little doubt that it refers to the eclipse of 1337 March 3. The eclipse is placed between two events that occurred in 1336 July and 1336 October, so the year is wrong. Further, the eclipse was on the 3rd day of the month rather than the 2nd day, and neither day was in Lent, which began on March 5 that year. The date is confirmed by the fact that the hours given look reasonable. The record is independent of any others known, but it cannot be the original, so it receives a reliability of 0.5. The place is Trebizond, the capital involved. The value of η is 0 and the standard deviation of the magnitude is 0.18.

1341 December 9 M,B. This is the last eclipse record in Gregoras [ca. 1359] and it comes in Section XII.15. It says there was an eclipse of the sun and the moon at the same time, and that there was another lunar eclipse before 6 full months had passed. The sun was in Sagittarius at the time of the pair of eclipses. I thank my colleague S. M. Krimigis for reading the passage for me in the original. He says that the phrase which I have translated as "at the same time" should not be taken precisely, and it probably means only that the pair of eclipses was close together compared with the interval to the next eclipse. There is no doubt that the lunar eclipses are those of 1341 November 23 and 1342 May 21 and that the solar eclipse was on 1341 December 9. The reliability is 1, the place is Constantinople, the value of η is 0, and the standard deviation of the magnitude is 0.18.

1361 May 5 M,B. The text of this record is found on page 28 and its translation is found on page 57 of Panaretos [ca. 1426]: "In the month of May on the 5th, on the 2nd day, Indiction 14, year 6869,† happened an eclipse of the sun such as never before happened in the memory of man; and it lasted 1 hour and 100 (sic) minutes. The emperor Lord Alexius and his mother Lady Irene and some of the archons and I found ourselves by accident in the Cloister Sumelas not far from Matzuka; we prayed much and called on all the saints." The phrase "on the 2nd day" probably means on the 2nd day of the week, but the eclipse was on the 4th day of the week; this is a minor error.

This record poses problems with regard to the standard deviation of the magnitude and with regard to the reliability. The eclipse must have been impressive, since it provoked prayer and since there had not been such an eclipse in the memory of man. However, there is no specific mention of great darkness, no reference to stars, nor any other indicator of a very large or total eclipse. M [p. 8.65] takes it to be total, with a probability of 0.90, but I cannot go this far. I shall take the standard deviation of the magnitude to be 0.054‡ because of the vigorous description, although the absence of specific indicators makes this choice perhaps

---

†The date stated is the equivalent of 1361 May 5.

‡This is the value used if a record says that stars were seen without saying that the eclipse was total.

somewhat optimistic.

The other question concerns the author of the record and hence its reliability. Panaretos presumably lived at least until 1426, since his history runs to this date in its surviving form and some of it seems to be missing. The eclipse occurred 65 years earlier. It is possible that Panaretos was old enough 65 years or more before his death to be making written records, but it is unlikely. Further, he would have been so young that he would probably not have been chosen as a member of such an august party. Thus it is probable that Panaretos took the record from some unknown source, preserving the first person "I" as he did so; there are precedents for such careless copying. Since it is unlikely that the record is original, I shall take a reliability of 0.5 rather than 1. M [p. 8.65], who worked only from a secondary source, did not notice this point and apparently assumed without question that "I" refers to Panaretos.

The cloister mentioned is on Mount Sumelas. M [Appendix 2] has found its coordinates. Its latitude is 40°.67 and its longitude is 39°.64.

1473 April 27 Is,E. This record and the remaining records from the Near East are from ibn Iyas [ca. 1522]. This record is on page 96 of volume 2, under the Islamic month that corresponds roughly to April of 1473: "There was a total eclipse of the sun: complete obscurity lasted for about 30 degrees." Here I must rely only upon a French translation of the source, and I do not know what is the original term that is translated as 30 degrees. I suppose this could mean 30 minutes, or it could be the time for the earth to rotate 30°, which would be 2 hours. Either way, the time of complete obscurity is an exaggeration. Since the eclipse was annular, there was no period of complete obscurity. However, I think it is safe to take this as an eclipse that was nearly central at Cairo, and I shall use 0.054 for the standard deviation of the magnitude. The reliability is 1.

1502 October 1 Is,E. This record is on page 38 of volume 3:† "A solar eclipse happened immediately after sunrise and lasted about an hour." The month corresponds approximately to October of 1502, and the eclipse of 1502 October 1 did begin fairly early in the morning in Cairo. The reliability is 1, the value of η is 0, and the standard deviation of the magnitude is 0.18.

1513 March 7 Is,E. This record is on page 276 of volume 3: "On Thursday the 26th, the day was very dark, a violent wind blew and the cold became intense; a disagreeable eclipse of the sun took place, 14 degrees before the middle of the

----

†I am referring to the volume numbers of the entire work. For some reason, the actual 3rd and 4th volumes were renumbered volumes 1 and 2, so that in the work as printed there are two volumes numbered 1 and two volumes numbered 2.

TABLE XII.4

ECLIPSE RECORDS FROM THE NEAR EAST THAT ARE NEW
OR THAT HAVE BEEN RE-INTERPRETED

Designation			Place	Standard deviation of the magnitude	Reliability
601 Mar	10	Is,E	Madinat	0.18	1
968 Dec	22a	M,B	Constantinople	0.034	1
1267 May	25	M,B	Constantinople	0.034	0.5
1331 Nov	30	M,B	Constantinople	0.18	1
1337 Mar	3a	M,B	Constantinople	0.18	1
1337 Mar	3b	M,B	Trebizond	0.18	0.5
1341 Dec	9	M,B	Constantinople	0.18	1
1361 May	5	M,B	Mt. Sumelas	0.054	0.5
1473 Apr	27	Is,E	Cairo	0.054	1
1502 Oct	1	Is,E	Cairo	0.18	1
1513 Mar	7	Is,E	Cairo	0.18	1

Note: The eclipses are all taken to be central within the
uncertainty indicated by the standard deviation of the
magnitude.

afternoon; it lasted about an hour." Again I do not know
the meaning of a degree in this context. The month corresponds to February/March of 1513, so there is no question
about the identification. However, the eclipse was on a
Monday rather than a Thursday. It did come about the middle
of the afternoon in Cairo. This record receives the same
characteristics as the preceding one.

The records from the Near East that I am using here for
the first time, or that I have used before and am using here
with different characteristics, are summarized in Table
XII.4. This table has the same format as the similar tables
in earlier chapters.

CHAPTER XIII

DETAILS OF THE ANALYSIS

1. The Standard Deviation of the Coordinate η

The standard deviation σ_η that will be attached to the value of η used for a record is the resultant of three components. The first component to be discussed, which I shall denote by σ_G, comes from the geometry of the central eclipse zone and the relation of the place of observation to it. If the place of observation is a region rather than a specific spot, I use an approximate center of area in the calculation σ_G.

In order to calculate σ_G, I begin with the standard deviation σ_μ of the magnitude that is assigned to each record in accordance with the principles of Section IV.1. I then calculate the values of μ and η for each eclipse at several points that are chosen to lie approximately on a straight line passing through the place of observation normal to the center line of the eclipse path. From these calculations, it is trivial to find η as a function of μ and hence to find the value of σ_G that corresponds to the assigned value of σ_μ.

TABLE XIII.1

THE RELATION BETWEEN σ_μ AND σ_G

Old σ_μ	New σ_μ	σ_G
0	0.034	0.0321
0.01	0.054	0.0437
0.02	0.068	0.0490
0.03	0.082	0.0554
0.04	0.096	0.0618
0.05	0.110	0.0681
0.06	0.124	0.0745
0.07	0.138	0.0809
0.08	0.152	0.0872
0.09	0.166	0.0936
0.10	0.180	0.1000
	0.140	0.0783
	0.190	0.1045
	0.360	0.2000

Experience soon showed that the value of σ_G for a given value of σ_μ hardly changes from one eclipse to another if σ_μ is greater than about 0.07. Accordingly, I calculated the average value of σ_G for several eclipses for each of several values of σ_μ that occur frequently. The values of σ_μ for which this was done were 0.068, 0.14, 0.18, 0.19, and 0.36. The results of this computation are shown in Table XIII.1.

The first column of Table XIII.1 gives what I call the old σ_μ. This is the value that I would have assigned to a record in MCRE, and the first column of Table XIII.1 is identical with the first column of Table IV.4. For each old σ_μ, the second column of Table XIII.1 gives the new σ_μ, which is the value that I assign in this work, as well as three additional values that are not made to correspond to any "old" σ_μ. Through the value 0.180, this column is identical with the second column of Table IV.4. Finally, the third column gives the value of σ_G determined in the way just described.

The values of σ_μ that are most frequently used are 0.180 and 0.360. The average values of σ_G found for these values of σ_μ were very close to 0.1 and 0.2, respectively, so I rounded to these values for simplicity. The reader may remember that the magnitude assigned to a record which says that an eclipse was partial but which gives no clue to the magnitude is 0.81 ± 0.14; this accounts for the presence of the values 0.14 and 0.19 in the second column of Table XIII.1. The value 0.068 for the new σ_μ corresponds to 0.02 for the old σ_μ. I obtained the values of σ_G listed opposite the values of the old σ_μ from 0.02 to 0.10 by linear interpolation.

This gives us two values of σ_G for the value 0.14 of the new σ_μ. One is the value 0.0783 that is tabulated in the last column of the table. The other is the value 0.0818 that we obtain by linear interpolation in the upper part of the table. The difference is not important, but I point it out in order to indicate the level of uncertainty or variability in the table.

The values of σ_G for σ_μ equal to 0.034 and 0.054 vary considerably from eclipse to eclipse, and in particular they change by large amounts in going from eclipses that were total somewhere to eclipses that were only annular. Further, the value of σ_G is much smaller for annular eclipses than for ones that were total somewhere. It is not reasonable to assign more accuracy to annular than to total eclipses, and thus it is not reasonable to use the values of σ_G that we actually derive from annular eclipses. Thus I have calculated the first two values of σ_G in the table by using only eclipses that were total somewhere.

Only one other comment is needed about σ_G. In earlier chapters, I sometimes assigned a value of 0.05 to σ_μ for eclipses in which the magnitude was specifically estimated. It is not quite reasonable to use this value, because it is less than σ_μ for eclipses in which stars were seen. In order to avoid this problem, I have decided never to use a value of σ_G for such eclipses that is less 0.0437, which is the value used with records which assert that stars were seen. An even larger minimum value of σ_G is probably justified in these cases.

The second component of σ_η comes from the uncertainty in the place of observation; I$^\eta$call this component σ_{p1}. It occurs only for records to which we must assign a region rather than a specific spot for the place of observation. In order to evaluate σ_{p1}, I calculate η for the center of area of the region and at its extremities, and take σ_{p1} to be equal to the largest difference found this way.

The resultant of σ_G and σ_{p1} is clearly $(\sigma_G^2 + \sigma_{p1}^2)^{\frac{1}{2}}$. This is the value that σ_η would have if there were no question about the data that we have inferred from the record. Uncertainty about the recorded data gives the third component of σ_η, and it is represented by the number called the reliability that I assign to each record. If I feel that there is no appreciable question about the data inferred from a record, I give it a reliability of 1. If I feel that the data are so uncertain as to be virtually worthless, I give the record a reliability of 0, and I give intermediate values of the reliability to records with intermediate properties.

In other words, the reliability P is my assessment of the probability that we have inferred the correct data from a record. The value of σ_η, reflecting all its three components, is then:

$$\sigma_\eta = \left[(\sigma_G^2 + \sigma_{p1}^2)/P\right]^{\frac{1}{2}}. \qquad \text{(XIII.1)}$$

In most cases, $\sigma_{p1} = 0$ and $P = 1$, so that σ_η reduces to σ_G. In all cases admitted to this work, I believe, σ_{p1} is appreciably smaller than σ_G, so that the uncertainties in place have little effect upon the value of the records or upon the inference of the accelerations.

The relative weight that will be assigned to a record is $1/\sigma_\eta^2$. Thus the weight is proportional to the reliability P.

2. The Eclipse Calculations

The core of the method of analysis used in this work is a computer program for calculating the circumstances of solar eclipses. The time inputs required for the program are the Julian day number of an eclipse and an estimated hour for the conjunction; these are taken from Oppolzer. In addition, the program needs the latitude and longitude of the place where the observation was made.

The coordinates of the places of observation used in this work are given in Appendix III.

Except for one important change that will be described in a moment, the program is still the one that was used in MCRE. The central part of the program is the computation of the solar and lunar positions at any specified time. In analyzing ancient and medieval observations, we do not need the full accuracy of modern orbital theories. Accordingly I use only about the largest third of the perturbations in the lunar

theory of Brown [1919] and in the solar theory of Newcomb
[1895].† However, I have tried to keep all secular effects
in the programs.

The root-mean-square error in the programs has been
determined by calculating solar and lunar positions at specific times and comparing with the positions tabulated in
the American Ephemeris and Nautical Almanac. The conclusion
is that the error is about 1 second of arc for each body in
each coordinate. A more direct test for our purposes is to
compare the central line of an eclipse path with that calculated by the Naval Observatory. The agreement is usually
within about 0°.1 in longitude measured on the earth, although a disagreement of about 0°.4 was found for one eclipse.

It takes the earth about 24 seconds to rotate through
0°.1, and during this time the moon moves about 12 seconds of
arc with respect to the sun. However, we expect only about
2 seconds of arc error because of the errors in the computed
longitudes of the sun and moon. The errors in the computed
paths seem to come mostly from errors in the calculated latitudes of the sun and moon. Since most paths lie nearly in
an east-west direction, they should be more sensitive to
latitude than to longitude errors in the calculated positions of the sun and moon.

Once the positions of the sun and moon are found for
appropriate times, I calculate the circumstances of an
eclipse using the methods described in Chapter 9 of the
Explanatory Supplement [1961], with one exception. The
Explanatory Supplement calculates the northern and southern
limits of a path by a method of successive approximations.
I have instead made some analytic approximations to certain
functions needed in calculating the limits, and I can then
find the limits by a closed process. The differences in
limits found this way are quite small except for points very
near the beginning and end of a path.

The program calculates both general and local circumstances for an eclipse; it can calculate local circumstances
for as many points as desired. The general circumstances
include what are called the Besselian elements, as well as
several other useful quantities, at different times during
the eclipse, and the coordinates of a number of points on
the central line of the eclipse path and on the limits of
the path. The local circumstances consist of the local times
of sunrise and sunset, the times when the eclipse began and
ended locally, the maximum magnitude, and the values of the
magnitude and of the quantity η at various times during the
local eclipse. If the eclipse was not visible at the specified place, the program still finds the times of sunrise and
sunset. It also estimates the time when the eclipse magnitude, though always negative, had its algebraically greatest
value, and it provides this value along with the time at
which it occurred. This gives a measure of the amount by
which the eclipse failed to be visible.

---

†This still amounts to 347 perturbations in the lunar
program and 73 in the solar program.

The change that has been made in the program relates to the way that time is handled in calculating the mean longitudes of the sun and moon. In earlier work, I used ephemeris time (ET) measured from the epoch 1900 January 0.5 ET as the independent variable, except that I allowed the lunar acceleration $\dot{n}_M$ with respect to ephemeris time to have an arbitrary value instead of the specific value -22.44. This meant that the eclipse path was given in terms of ephemeris longitude rather than geographic longitude, and thus I had to use the ephemeris longitude rather the geographic longitude of the observer's position. Since the ephemeris longitude is a quadratic function of time that could differ from the geographic longitude by amounts up to around 100° for ancient observations, this caused awkwardness in assigning the coordinates needed as inputs to the program.

In the present work I use solar time (UT)† rather than ET as the independent variable, and I take the fundamental epoch to be 1799 December 30.5 UT,‡ for reasons that were discussed in Section I.3.

Let us first change the time base in the ephemeris of the moon. The expression for the lunar mean longitude that is currently used in calculating the ephemeris is [Explanatory Supplement, 1961, p. 107]:

$$L_M = 270°.434\ 164 + 481\ 267°.883\ 1420T - 0°.001\ 133T^2$$
$$+ 0°.000\ 001\ 89T^3. \hspace{2cm} \text{(XIII.2)}$$

In this, T is ephemeris time measured from the beginning of 1900.

I have said that Brown's theory is now used in calculating the position of the moon, but this is not strictly correct. The theory currently used is that of Eckert, Jones, and Clark [1954]. This differs from Brown's theory in several ways, but the only difference important enough to concern us here is in the mean longitude. As I have noted, after Brown had constructed a purely gravitational theory, he found that there was still a discrepancy between his expression for the mean longitude and values obtained from observations. Accordingly he added (Section I.2) the "great empirical term" $10''.71 \sin (140°T + 240°.7)$ in order to improve the agreement with observation. In doing this, he was still using solar time as the time base; that is, T in Brown's original theory is solar time measured from the beginning of 1900.

---

†Properly speaking, this is the symbol for universal time rather than solar time. Since the difference is quantitatively trivial, I use the familiar symbol UT instead of introducing a new symbol. I hope the reader will not be misled.

‡The exact reason for this peculiar choice will appear in a moment. I usually call this epoch the beginning of 1800 for brevity.

When ephemeris time was adopted as the time base in 1960, the great empirical term was removed and replaced by the quadratic [Explanatory Supplement, 1961, p. 87] - 8".72 - 26".74T - 11".22T^2; this corresponds to an acceleration of -22.44"/cy^2. In terms of degrees, this quadratic is -0°.002 422 - 0°.007 4278T - 0°.003 1167T^2. If we add the negative of this to the expression for L_M in Equation XIII.2, we get G_M, the mean longitude of the moon in a strictly gravitational theory:

$$G_M = 270°.436\ 586 + 481\ 267°.890\ 5698T + 0°.001\ 9837T^2$$
$$+ 0°.000\ 001\ 89T^3. \hspace{2cm} (XIII.3)$$

Since this is strictly a gravitational theory, T may be interpreted as either solar time or ephemeris time.

The data of Brouwer [1952] and Martin [1969], supplemented by a few additional observations as described in Section I.4, give us observed values of the lunar mean longitude as a function of solar time from 1627 to 1960. A fit to these data gives us

$$-3".79 - 14".48T - 6".31T^2 = -0°.001\ 053 - 0°.004\ 022T$$
$$-0°.001\ 7528T^2$$

for the difference between G_M and the observed longitude. Adding this to the right member of Equation XIII.3 gives

$$L_M = 270°.435\ 533 + 481\ 267°.886\ 5478T + 0°.000\ 2309T^2$$
$$+ 0°.000\ 001\ 89T^3 \hspace{2cm} (XIII.4)$$

for the best representation we can make of the lunar mean longitude as a function of solar time.

Before we change the time origin to the beginning of 1800, let us derive the corresponding expression for the sun. The expression for the mean longitude of the sun that is currently used in calculating its ephemeris is [Explanatory Supplement, 1961, p. 98]:

$$G_S = 279°.696\ 678 + 36\ 000°.768\ 9250T + 0°.000\ 3025T^2.$$
$$(XIII.5)$$

This is already the expression that results from a strictly gravitational theory, so it corresponds to Equation XIII.3 for the moon. The best estimate we can make of the effect of the non-gravitational forces is, from Section I.4:

$$0''.38 + 0''.99T + 0''.59T^2 = 0^c.000\ 106 + 0°.000\ 2750T$$

$$+ 0°.000\ 1639T^2.$$

Thus the best representation we can make for the mean longitude L_S of the sun as a function of solar time is

$$L_S = 279°.696\ 784 + 36\ 000°.769\ 2000T + 0°.000\ 4664T^2.$$

$$(XIII.6)$$

These results have been determined by averaging data whose average epoch is close to 1800, so 1800 is really the correct epoch of Equations XIII.4 and XIII.6 in spite of the fact that they are referred to the beginning of 1900. Thus we want to change the time origin to the proper epoch. If we let τ denote solar time referred to the beginning of 1800 rather than of 1900, we let

$$T = \tau - 1.\qquad (XIII.7)$$

When we substitute from Equation XIII.7 into Equations XIII.4 and XIII.6, we find

$$L_M = 322°.549\ 214 + 481\ 267°.886\ 091\tau + 0°.000\ 225\ 23\tau^2$$

$$+ 0°.000\ 001\ 89\tau^3,\qquad (XIII.8)$$

$$L_S = 278°.928\ 050 + 36\ 000^\theta.768\ 267\tau + 0°.000\ 4664\tau^2.$$

As I have already said, the epoch for Equations XIII.8 is 1799 December 30.5 UT. This peculiar epoch arises from using the simple time transformation in Equation XIII.7. The unit of time is the Julian century of 36525 days, but the century running from 1800 January 0 to 1900 January 0 had only 36524 days, since 1800 was not a leap year and there was no date 1800 February 29. Hence 36525 days before 1900 January 0.5 was a day earlier than 1800 January 0.5, namely 1799 December 30.5 UT.

We should note carefully that Equations XIII.8 do not correspond to $\nu_M' = 0$ and $\nu_S' = 0$. Instead, they are based upon the accelerations

$$\nu_M' = -12.62, \qquad\qquad \nu_S' = +1.18.$$

3. The Method of Inference

The basic method of inference is the method of weighted least squares, which I have already discussed in general terms in earlier parts of this work. I shall give a more detailed description here, for the principal purpose of explaining the notation.

The data will be divided into samples containing all the eclipse observations that were made during a relatively short interval of time. For each eclipse in a sample, I assume that the coordinate η is a linear function of the parameters D'' and W,[†] and I write this function in the form

$$\eta = \eta_o + \eta_D D'' + \eta_W W. \qquad \text{(XIII.9)}$$

The coefficients η_D and η_W are clearly the partial derivatives of η with respect to D'' and W, respectively. They are calculated numerically.

In calculating η_D, I vary D'' while holding ν_S' and the nodal perturbation $\delta\Omega$ constant. There are two obvious ways to calculate η_W, in both of which D'' is held constant. In one method, I let $\partial\eta/\partial W = \partial\eta/\partial(\delta\Omega)$. That is, I vary the nodal perturbation while holding ν_S' and D'' constant. In the other, I let $\partial\eta/\partial W = -(2/\tau^2)(\partial\eta/\partial\nu_S')$, and I calculate $\partial\eta/\partial\nu_S'$ by varying ν_S' while holding D'' and the nodal perturbation constant. The first method is the one used in this work.[‡]

Thus, by using the eclipse program described in the preceding section, I find values of the quantities η_o, η_D, and η_W for each eclipse. Now let η_{obs} denote the value of η implied by an observation and define R by

$$R = \eta_{obs} - \eta_o. \qquad \text{(XIII.10)}$$

If there were no error in the value η_{obs}, the quantity $\eta_D D'' + \eta_W W$ would equal R. As it is, this quantity equals R within an error estimate E:

$$\eta_D D'' + \eta_W W = R \pm E,$$

or

$$\eta_D D'' + \eta_W W - R = \pm E.$$

We want to find the values of D'' and W that minimize

[†] W is the parameter introduced in Section IV.6. It is the difference between the perturbation in the lunar node and the perturbation in the mean longitude of the sun.

[‡] As a test, I calculated η_W and inferred the parameters D'' and W using both methods for a few data samples. The differences between the results were always small compared with the estimated standard deviations. The differences presumably arise from neglecting the solar eccentricity and other small effects in defining W.

the weighted sum of the squares of the E's, where the sum
is taken over all eclipse observations in a given sample.
We decided in Section XIII.1 to give each observation a
relative weight proportional to $1/\sigma_\eta^2$, with σ_η being defined
by Equation XIII.1. That is, the estimates of D'' and W in
the method of weighted least squares are those which minimize
the function F:

$$F = \tfrac{1}{2}\sum\left[(\eta_D D'' + \eta_W W - R)/\sigma_\eta\right]^2.$$

We can simplify the discussion slightly if we introduce
new variables η_d, η_w, r and ε defined by

$$\eta_d = \eta_D/\sigma_\eta, \qquad \eta_w = \eta_W/\sigma_\eta, \qquad r = R/\sigma_\eta, \qquad \varepsilon = E/\sigma_\eta.$$

$$(XIII.11)$$

In terms of these variables, F becomes

$$F = \tfrac{1}{2}\sum(\eta_d D'' + \eta_w W - r)^2 = \tfrac{1}{2}\sum\varepsilon^2. \qquad (XIII.12)$$

In other words, when we use the variables η_d, η_w, r, and ε,
all observations are on the same footing in calculating F.

It is easy to show that D'' and W, considered as components of a vector (D'', W), are the solution of the matrix
equation

$$\begin{pmatrix} \sum\eta_d^2 & \sum\eta_d\eta_w \\ \sum\eta_d\eta_w & \sum\eta_w^2 \end{pmatrix} \begin{pmatrix} D'' \\ W \end{pmatrix} = \begin{pmatrix} \sum\eta_d r \\ \sum\eta_w r \end{pmatrix}. \qquad (XIII.13)$$

Let M_{11}, M_{12}, M_{21}, and M_{22} be the coefficients in the matrix
that is the inverse of the one in the left member of Equation
XIII.13.† Then

$$\begin{pmatrix} D'' \\ W \end{pmatrix} = \begin{pmatrix} M_{11} & M_{12} \\ M_{21} & M_{22} \end{pmatrix} \begin{pmatrix} \sum\eta_d r \\ \sum\eta_w r \end{pmatrix}. \qquad (XIII.14)$$

In the remaining discussion, D'' and W will denote the values
found from Equation XIII.14.

It only remains to find the error estimates (standard
deviations) to be attached to the values D'' and W. Let these
standard deviations be denoted by $\sigma(D'')$ and $\sigma(W)$. These are

---

†Obviously $M_{12} = M_{21}$ for this case.

related to the matrix coefficients M_{11} and M_{22} in Equation
XIII.14 by means of the errors in the individual observations.
We see from Equation XIII.12 that the error ε for an individual observation equals $\eta_d D'' + \eta_w W - r$. Let $\sigma(\varepsilon)$ be the standard deviation of the ε's for a given sample. Then

$$\sigma(D'') = \sigma(\varepsilon) M_{11}^{\frac{1}{2}}, \qquad \sigma(W) = \sigma(\varepsilon) M_{22}^{\frac{1}{2}}. \qquad \text{(XIII.15)}$$

If the values of σ_η for various classes of eclipses have
been realistically chosen, $\sigma(\varepsilon)$ is close to unity. If this
happens, we can replace the values of $\sigma(D'')$ and $\sigma(W)$ in Equations XIII.15 simply by $M_{11}^{\frac{1}{2}}$ and $M_{22}^{\frac{1}{2}}$, respectively.

If we have a moderately large body of data, the sums in
Equation XIII.13 are proportional to N, the number of observations used. The coefficients in the inverse matrix in
Equation XIII.14 are then inversely proportional to N. Hence
the standard deviations $\sigma(D'')$ and $\sigma(W)$ in Equations XIII.15
are proportional to $N^{-\frac{1}{2}}$.

4.   Values of D'' and W at Various Epochs

I use the term "coefficients of condition" to denote
the quantities η_d, η_w, and r that appear in Equations XIII.12.
These coefficients, along with the standard deviation σ_η that
determines the weight of each observation, are listed in
Tables XIII.5 through XIII.21. In order not to break up the
main text, I have placed these tables at the end of the
chapter. If the reader wants the values of the variables
η_D, η_W, and R, he can find them by multiplying the tabulated
variables by σ_η. If he wants the value of η_o that I have
calculated for any observation, he must first go to the discussion of the record in Chapters V through XII to find the
conclusion that has been reached about the value of μ or η.
For most records, the ones that say nothing about the magnitude μ, I take the eclipse to be central, which means
$\eta_{obs} = 0$. If a record says that an eclipse was partial
without stating the magnitude, I take μ to be 0.81. In all
other cases, a specific numerical value of μ is given. The
value of η_{obs} corresponding to a magnitude of 0.81. is taken
to be 0.1045, in accordance with Table XIII.1. The value of
η_{obs} for any other value of μ is also to be taken from Table
XIII.1, using $1 - \mu$ as the value for the quantity called the
new σ_μ. Finally, to find η_o, the reader subtracts R from
η_{obs}.

It should be pointed out again that the dates listed in
Tables XIII.5 through XIII.21 are in Greenwich time. For
Chinese records, the local date is sometimes a day later
than the Greenwich date. For records other than Chinese, I
believe that the local date and the Greenwich date are always the same.

I have listed σ_η to a number of significant figures
that seems absurd. The reason is that σ_η is involved in
transforming from the variables η_D, η_W, and R to η_d, η_w, and
r, and η_D and η_W do have several significant figures. It is

important to preserve the correct relation between the variables in making the transformation, and hence it is necessary to use σ_η as a precise quantity even though it is not.

A few records that were discussed in earlier chapters do not appear in the tables. To start with, there were substantial questions about the correct dates of the record designated -644 August 28 C in Table V.2 and the record designated -34 November 1 C in Table V.3, and I decided in Chapter V not to use these records. Calculation shows that the eclipse of -644 August 28 failed to be visible by a considerable margin. The eclipse of -34 November 1, on the other hand, was visible everywhere in heartland China as a small eclipse at sunset. Since even a small eclipse at sunset is readily seen, it is quite likely that the record in question is actually a record of the eclipse of -34 November 1. However, we cannot establish the date without the use of computation, and it is safest not to use this record.

There were no apparent textual problems with the records designated -552 August 31 C in Table V.2 and 1240 April 23 E,PR in Table XI.2, but these eclipses were not visible at the place of observation used. The most likely explanation is that reports of these eclipses were brought from elsewhere, but other explanations, such as eclipse prediction, are possible.

The other four records which appear in the appropriate tables in earlier chapters, but which do not appear in Tables XIII.5 through XIII.21, are -557 May 30 C in Table V.2, -441 March 11 C in Table V.2, 756 October 28 C in Table V.5, and 360 August 28 M,B in Table XII.1. The three Chinese records share a common problem. The eclipse of 756 October 28, for example, had its maximum magnitude shortly before sunset for some values of the parameters used in calculating the coefficients of condition, but its magnitude was still rising at sunset for other values.† Thus the calculation did not yield valid values of the partial derivatives needed. It would be possible to devise special methods of calculating the partial derivatives for these eclipses, but it does not seem worth the trouble.

With the record 360 August 28 M,B, there was no eclipse for some values of the parameters and a small eclipse at sunrise for others. Again the calculations did not yield valid partial derivatives, so this record is also omitted.

For three records, namely -15 November 1 C in Table V.3, 808 July 27 C in Table V.5, and 1181 July 13 E,F in Table VIII.2, the coefficients of condition are based upon the magnitude μ rather than the coordinate η. This situation arises from a peculiarity of the eclipse program as it is currently written. The program finds the times when the eclipse began or ended locally by a method of successive approximations. It is instructed to quit searching when it has made a certain number of attempts without adequate convergence. When this happens, the program in its present

---

†The eclipses of -557 May 30 and -441 March 11 present the symmetrical problem for eclipses occurring near sunrise.

TABLE XIII.2

VALUES OF D″ AND W AT VARIOUS EPOCHS

Epoch	D'' $''/cy^2$	$\sigma(D'')$ $''/cy^2$	W $'$	$\sigma(W)$ $'$
- 660	+ 6.55	3.48	+15.53	38.92
- 551	+ 9.18	5.27	-23.56	33.54
- 398	+ 7.45	5.43	-16.50	28.13
- 166	+ 7.46	2.94	-15.26	19.31
- 44	- 2.35	4.35	+36.47	26.46
+ 122	+ 5.25	4.74	+ 0.53	21.96
415	+ 2.63	7.27	-15.60	15.71
602	+ 1.44	7.08	-10.88	15.89
772	+24.18	12.85	-12.87	8.74
878	- 4.35	11.69	+ 8.03	12.14
1005	+13.40	11.37	-17.93	8.13
1128	- 9.12	10.52	- 7.39	7.75
1174	-29.45	12.18	- 8.74	5.79
1248	-11.87	13.51	- 0.89	6.65
1354	- 4.05	28.80	+ 2.39	8.11
1446	+21.56	45.91	+ 6.31	9.07
1534	-43.66	30.64	- 2.51	3.21

form lists the maximum magnitude and the time when the maximum occurs,[†] but it does not list the coordinate η. Now it is valid to use μ rather than η for calculating the coefficients provided that η has the same sign for all values of the parameters used in the calculations, and this happened in all three cases mentioned. Thus, instead of revising the program and running it again, I simply used the magnitudes provided.

In MCRE, Tables XVIII.1 through XVIII.9, I grouped eclipse records according to time, putting together all records pertaining to a given century or half-century. In Tables XIII.5 through XIII.21 of this work, I have adopted instead the following principles:

[†]The program finds the local maximum magnitude before it finds the times of beginning and ending.

a. The time span of a particular group of records must be short enough that we can approximate W by a constant. We expect the largest component of W to be proportional to the square of the time from 1800.

b. Within the limits imposed by principle a, the spans are chosen so that $\sigma(D'')$, the standard deviation of the inferred value of D'', is approximately the same for all groups.

c. Within the limits of both preceding principles, the time span is chosen for convenience, such as a particular century.

Table XIII.2 gives the values of D'' and W, along with their standard deviations, that are inferred from each group in Tables XIII.5 through XIII.21. Each line in Table XIII.2 corresponds to a particular table, and the epoch listed in Table XIII.2 is the average of the first and last epochs in the corresponding table. This is not an accurate way of finding the average epoch, but it is as accurate as the situation demands.

As the footnotes indicate, the values of r in Tables XIII.5 through XIII.21 are referred to the values† $D'' = 6$ $''/cy^2$ and $W = -0'.025\tau^2$. However, the values of D'' and W in Table XIII.2 are referred to $D'' = 0$ and $W = 0$.

5. Time and Space Distribution of the Records

Tables XIII.5 through XIII.21 contain 631 records of 189 individual eclipses. This sample is large enough to support intensive statistical studies of various properties of the records. A large study of the records' statistics is beyond the scope of this work, but a few simple comments are interesting and useful.

A great deal has been written about the periods with which eclipses recur, the so-called eclipse cycles. The most famous of these is probably the cycle of 18 years plus 11 days. This is 223 lunar months. At the end of 223 months the mean longitude of the moon is $10°.8$ greater than it was at the beginning while the longitude of the node is $11°.3$ greater. Thus the moon and the node repeat their relative positions, to fairly high accuracy, with a period of 18 years plus 11 days. This is the period that is usually mis-called the saros.‡

---

†Note that the minute of arc rather than the second is taken as the basic unit in dealing with W, contrary to the usual choice of units.

‡Saros (a Babylonian word) means the number $3600 (= 60^2)$, and it has nothing to do with astronomical motions. See Neugebauer [1957, pp. 141-142] for an interesting discussion of how this error arose. It seems to be impossible to eradicate this error from the astronomical literature.

It is indeed correct that eclipses have a strong tendency
to recur at this interval, provided that we consider all
eclipses. If we consider only eclipses that are visible at
a particular spot, the interval is still fairly accurate for
lunar eclipses. That is, if a lunar eclipse was visible on
day D, there is a good chance that an eclipse was also visible
18 years plus 11 days after day D.

If a solar eclipse was visible at a particular spot on
day D, there is a good chance that there was another solar
eclipse 18 years plus 11 days after day D, but there is a poor
chance that the eclipse will be visible at the same spot. The
reason is that 223 months is 6585.321 days, about a third of
a day more than an integral number of days. Thus, when the
second eclipse occurs, its path on the average is shifted by
a third of a revolution, or 120° in longitude.

However, three such periods come to 19756 days, almost
exactly, which is 54 years plus 33 days. Thus, if there is
an eclipse on day D, there is a reasonably good chance that
another eclipse visible at the same spot will occur 54 years
plus 33 days later. We see an example of this interval in
Table XIII.5. The eclipses of -654 August 19 and -600 September 20 are separated by just 54 years plus 33 days. However, I find no other example of this interval in the first
two tables, which cover the years from -719 to -504.

Another interesting interval is 111 months, which equals
3277.90 days, or 9 years minus 9 days. If an eclipse occurs
when the moon is exactly at one node on day D, it will be
about 16° from the other node after 111 months, and there
will be another eclipse at this time. Further, since the
interval contains nearly an integral number of days, there
is a strong tendency for the second eclipse to be visible at
the same place. We see three examples of this interval in
Tables XIII.5 and XIII.6. They are the eclipse pairs -663
August 28 and -654 August 19, -558 January 14 and -549 January 5, and -519 November 23 and -510 November 14.

After two such intervals, the distance between the moon
and the node changes by about 32°, and a third eclipse is
possible but unlikely. To take a medieval example, there
were eclipses visible in Europe on 1124 August 11 and 1133
August 2, and we see from Table XIII.16 that both eclipses
were well recorded. 111 months before 1124 August 11, namely
1115 August 20, there was no eclipse anywhere (the nearest
eclipse was a month earlier, on July 23). Similarly, there
was no eclipse 111 months after 1133 August 2, on 1142 July
24, but there was one on 1142 August 22.

Johnson [1889, p. 64][†] writes: "When a great solar
eclipse happens at any particular place, it is frequently
followed by three at the space of half a Chaldaean period
(nine years) between them." He then lists three such sets
of four eclipses that were visible in England. Two of the

[†] I do not know where or when the interval of 111 months was
discovered. Johnson is the earliest writer I have seen who
mentions it, but I doubt that he discovered it.

TABLE XIII.3

DISTRIBUTION OF ECLIPSE RECORDS IN TIME AND LOCATION

Interval	Number of eclipses	Number of records from China	Mediterranean and Babylonia	Northern Europe	Total number of records
-719/-700	2	2			2
-699/-600	11	11			11
-599/-500	16	16			16
-499/-400	5	3	3		6
-399/-300	4	1	3		4
-299/-200	1	1			1
-199/-100	5	4	1		5
- 99/0	6	6			6
1/100	4	2	3		5
101/200	1	1			1
201/300	1	1			1
301/400	4	1	3		4
401/500	7	1	9		10
501/600	9	2	5	3	10
601/700	7		5	5	10
701/800	9	4	2	12	18
801/900	14	2	9	42	53
901/1000	5		8	12	20
1001/1100	12		6	48	54
1101/1200	17		15	160	175
1201/1300	15	1	24	111	136
1301/1400	15		18	29	47
1401/1500	14		21	9	30
1501/1567			4	2	6
Totals	189	59	139	433	631

sets came in 1706, 1715, 1724, and 1733 and in 1833, 1842, 1851, and 1860. The third set contains the eclipses of 1406 June 16, 1415 June 7, 1424 June 26, and 1433 June 17. We see from Table XIII.20 that the first and third of these are recorded in English sources but that the second and fourth are not. The second and fourth are, however, recorded in continental sources. These days must have been cloudy in

-413-

England because both eclipses were large though not total there.

I do not believe that one should use the word "frequently" in connection with such a set of four eclipses. The interval between the first and third and between the second and fourth eclipses is 223 months, and we have seen that eclipses separated by this interval are rarely visible at the same spot. To have this rare event happen twice in a set of four eclipses must be rare indeed.

Table XIII.3 shows us how the eclipse records are distributed in time and location. In preparing this table, I have divided the observer's positions into three regions only, instead of trying to show each individual provenance. One region is China. The region that I call "Mediterranean and Babylonia" includes Italy, Spain, and Portugal, as well as the Mediterranean provenance and the provenance of Babylonia and Assyria. "Northern Europe" contains all provenances not included in the first two regions.

TABLE XIII.4

ECLIPSES WITH TEN OR MORE RECORDS

Date	Number of records	Date	Number of records
840 May 5	12	1187 Sep 4	17
878 Oct 29	13	1191 Jun 23	31
968 Dec 22	10	1207 Feb 28	15
1033 Jun 29	13	1239 Jun 3	30
1093 Sep 23	12	1241 Oct 6	44
1133 Aug 2	38	1263 Aug 5	16
1140 Mar 20	17	1310 Jan 31	13
1147 Oct 26	13		
1178 Sep 13	22		
1185 May 1	12		

Table XIII.3 is misleading with regard to Chinese records in the medieval period. Chinese records are actually as frequent in the 10th and later centuries as they are in earlier centuries, but I have not used Chinese records later than the 9th century in this work and they are not included in the table, with a single exception.

It is interesting to see how the recording of eclipses peaked in the 12th century. There are more records in this century than in any other, and there are also more individual eclipses recorded.

The table shows many more records from northern Europe

than from the Mediterranean region. I do not know the extent
to which this is real. I tend to believe that it is fairly
real, and not merely a reflection of the interests of people
who have edited series of medieval documents. The reason is
that I have made heavy use of the Italian series called <u>Rerum</u>
<u>Italicarum Scriptores</u>, which is at least as large as any of
the series dedicated to German, French, and English history.

It is also interesting to see which eclipses were the
most recorded. Table XIII.4 lists the dates of the eclipses
for which I have used 10 or more records. There are 17 such
eclipses, and 7 of them come in the 12th century.   In <u>MCRE</u>,
the most frequently recorded eclipse was that of 1133 August
2, but that eclipse has now slipped to second place. First
place now belongs to 1241 October 6, for which I have used
44 records, while 1133 August 2 has only 38 records.

There is one final point of interest about Table XIII.3.
The earliest usable record is that of the eclipse of -719
February 22, in the -8th century. If we count centuries in
the usual way, there is at least one extant record of a solar
eclipse in every century since then.† However, there are
three intervals, from -299 July 26 to -197 August 7, from 120
January 18 to 243 June 5, and from 243 June 5 to 346 June 6,
which are longer than 100 years and from which we have no
existing records of solar eclipses. There are numerous observations of other astronomical phenomena in the interval
from -299 to -197, but there are none that I know of from
120 to 243 or from 243 to 346.‡ Thus there will remain two
gaps of more than 100 years since the -8th century from which
we have no extant observations, even when we consider astronomical observations of all kinds.

---

† Table XIII.3 goes only through the 16th century, but there
are obviously many records from every century since then.

‡ <u>Ptolemy</u> [ca. 142] quotes some observations which he attributes to other astronomers that have dates near 130 and that
may be genuine. However, it is not safe to assume that
these are genuine observations in the present state of
knowledge. See Table XIII.2 of <u>Newton</u> [1977].

TABLE XIII.5

COEFFICIENTS OF CONDITION FOR ECLIPSES
FROM -719 TO -600

Designation	$1000\eta_d$ $cy^2/''$	$1000\eta_w{}^a$	r^b	σ_η
- 719 Feb 22 C	- 24.03	- 4.97	-1.075	0.2830
- 708 Jul 17 C	-277.32	- 19.38	-0.280	0.0721
- 694 Oct 10 C	- 41.87	- 5.50	-0.907	0.2741
- 675 Apr 15 C	- 2.31	+ 5.20	-0.622	0.2817
- 668 May 27 C	+ 76.48	- 8.28	-0.169	0.1932
- 667 Nov 10 C	- 57.02	+ 5.42	+0.456	0.2754
- 663 Aug 28 C	- 69.76	+ 7.14	+0.386	0.1996
- 654 Aug 19 C	- 75.62	- 4.84	+0.136	0.2850
- 647 Apr 6 C	- 5.54	- 4.97	+1.378	0.2841
- 625 Feb 3 C	+ 34.71	- 5.67	-0.363	0.2729
- 611 Apr 27 C	+ 33.35	- 6.98	+0.454	0.2009
- 601 May 7 C	+ 26.06	+ 5.05	+1.150	0.2820
- 600 Sep 20 C	-209.24	- 20.06	-1.198	0.0693

[a] In units of reciprocal minutes of arc.
[b] Referred to $D'' = 6 ''/cy^2$ and $W = -0'.025_T{}^2$.

TABLE XIII.6

COEFFICIENTS OF CONDITION FOR ECLIPSES
FROM -598 TO -504

Designation	$1000\eta_d$ $cy^2/''$	$1000\eta_w{}^a$	r^b	σ_η
- 598 Mar 5 C	- 15.80	+ 7.56	-0.820	0.2073
- 574 May 9 C	+ 2.92	- 5.29	-0.116	0.2828
- 573 Oct 22 C	- 31.02	+ 5.03	+0.622	0.2828
- 558 Jan 14 C	+ 66.65	+ 5.20	+0.768	0.2828
- 551 Aug 20 C	- 52.15	+ 5.60	+0.510	0.2828
- 549 Jan 5 C	- 51.62	- 5.36	-0.205	0.2828
- 548 Jun 19 C	- 0.71	+ 20.22	-0.170	0.0707
- 545 Oct 13 C	- 27.93	- 5.05	-0.198	0.2828
- 534 Mar 18 C	+ 44.72	- 5.14	-1.203	0.2828
- 526 Apr 18 C	+ 80.79	+ 5.20	+0.228	0.2828
- 524 Aug 21 C	- 51.88	- 7.76	+0.348	0.2000
- 520 Jun 10 C	+ 48.75	- 7.52	+1.082	0.2000
- 519 Nov 23 C	- 32.53	+ 5.01	-0.814	0.2828
- 517 Apr 9 C	+ 27.14	- 5.31	+0.825	0.2828
- 510 Nov 14 C	- 69.56	- 4.76	+0.781	0.2828
- 504 Feb 16 C	+ 68.41	+ 5.24	+1.187	0.2828

[a]In units of reciprocal minutes of arc.
[b]Referred to $D'' = 6 ''/cy^2$ and $W = -0'.025_T{}^2$.

TABLE XIII.7

COEFFICIENTS OF CONDITION FOR ECLIPSES
FROM -497 TO -299

Designation	$1000\eta_d$ $cy^2/''$	$1000\eta_w$ [a]	r [b]	σ_η
- 497 Sep 22 C	- 48.35	+ 5.59	+0.122	0.2828
- 494 Jul 22 C	- 1.59	+ 5.07	-0.946	0.2828
- 480 Apr 19 C	+ 41.10	- 5.10	-0.331	0.2828
- 430 Aug 3a M	-138.05	- 18.87	+0.646	0.0791
- 430 Aug 3b M	- 36.99	- 4.96	+0.208	0.3007
- 423 Mar 21 M	+ 58.54	- 19.87	+0.522	0.0790
- 393 Aug 14 M	- 62.58	- 20.27	-0.888	0.0783
- 381 Jul 3 C	+ 64.57	+ 7.14	+0.698	0.1971
- 363 Jul 13 M	+ 28.49	+ 4.13	+0.537	0.3431
- 309 Aug 15a M	+ 42.87	+ 10.54	+0.056	0.1347
- 299 Jul 26 C	+ 42.24	- 7.49	+0.256	0.1917

[a] In units of reciprocal minutes of arc.
[b] Referred to $D'' = 6 ''/cy^2$ and $W = -0'.025_T{}^2$.

TABLE XIII.8

COEFFICIENTS OF CONDITION FOR ECLIPSES
FROM -197 TO -135

Designation	$1000\eta_d$ $cy^2/''$	$1000\eta_w$ [a]	r [b]	σ_η
- 197 Aug 7 C	+ 72.15	+ 28.29	+0.011	0.0561
- 187 Jul 17 C	-205.35	- 22.55	-1.216	0.0642
- 180 Mar 4 C	+144.52	- 22.87	+0.359	0.0626
- 146 Nov 10 C	- 97.11	+ 16.27	-0.009	0.0880
- 135 Apr 15 BA	+317.52	+ 45.24	-0.418	0.0311

[a] In units of reciprocal minutes of arc.
[b] Referred to $D'' = 6 ''/cy^2$ and $W = -0'.025_T{}^2$.

TABLE XIII.9

COEFFICIENTS OF CONDITION FOR ECLIPSES
FROM -88 TO -1

Designation	$1000\eta_d$ $cy^2/''$	$1000\eta_w$ [a]	r [b]	σ_η
- 88 Sep 29 C	- 47.29	+ 18.66	+0.559	0.0793
- 79 Sep 20 C	-215.83	- 21.09	+1.075	0.0661
- 27 Jun 19 C	+149.29	+ 21.75	-0.301	0.0645
- 15 Nov 1 C	+ 19.88	+ 13.32	+2.160	0.2000
- 14 Mar 29 C	+ 38.45	- 13.32	-1.578	0.1184
- 1 Feb 5 C	- 16.12	+ 18.99	+0.263	0.0775

[a] In units of reciprocal minutes of arc.
[b] Referred to $D'' = 6 ''/cy^2$ and $W = -0'.025_T{}^2$.

TABLE XIII.10

COEFFICIENTS OF CONDITION FOR ECLIPSES
FROM 2 TO 243

Designation	$1000\eta_d$ $cy^2/''$	$1000\eta_w$ [a]	r [b]	σ_η
2 Nov 23 C	-127.91	+ 23.14	+0.311	0.0627
59 Apr 30 E	+ 4.25	- 16.25	-0.882	0.1001
59 Apr 30 M	- 30.29	- 13.24	-1.164	0.1007
65 Dec 16 C	-142.65	- 22.11	+0.515	0.0627
118 Sep 3 E,I	- 7.70	+ 6.13	+0.387	0.2370
120 Jan 18 C	+ 86.41	- 24.06	+0.051	0.0587
243 Jun 5 C	+ 28.96	- 8.21	+1.106	0.1934

[a] In units of reciprocal minutes of arc.
[b] Referred to $D'' = 6 ''/cy^2$ and $W = -0'.025_T{}^2$.

TABLE XIII.11

COEFFICIENTS OF CONDITION FOR ECLIPSES
FROM 346 TO 484

Designation	$1000\eta_d$ $\mathrm{cy}^2/''$	$1000\eta_w$ [a]	r [b]	σ_η
346 Jun 6a M,B	+ 46.82	+ 14.53	−0.904	0.0977
348 Oct 9 M,B	− 18.11	− 6.26	+0.745	0.2236
360 Aug 28 C	− 15.87	+ 8.47	−0.209	0.1907
393 Nov 20 E,I	− 14.40	+ 6.20	+0.090	0.2256
402 Nov 11 E,SP	− 32.73	− 9.65	+0.369	0.1444
418 Jul 19a E,I	− 1.81	+ 9.64	−0.157	0.1518
418 Jul 19 E,SP	+ 9.79	+ 6.41	−0.107	0.2272
418 Jul 19a M,B	− 61.21	+ 33.48	−0.579	0.0437
418 Jul 19c M,B	− 18.92	+ 10.35	−0.179	0.1414
429 Dec 12 C	− 0.35	+ 19.41	+0.296	0.0722
447 Dec 23 E,SP	+ 31.02	+ 13.77	−0.108	0.1007
458 May 28 E,SP	+ 78.14	+ 21.74	−0.122	0.0666
464 Jul 20 E,SP	+ 46.41	− 28.30	−0.049	0.0555
484 Jan 14 M,B	− 44.90	+ 22.77	+0.136	0.0618

[a] In units of reciprocal minutes of arc.
[b] Referred to $D'' = 6\ ''/\mathrm{cy}^2$ and $W = -0'.025\tau^2$.

TABLE XIII.12

COEFFICIENTS OF CONDITION FOR ECLIPSES
FROM 512 TO 693

Designation			$1000\eta_d$ $cy^2/''$	$1000\eta_w$ [a]	r [b]	σ_η
512 Jun 29	E,I		+ 26.52	+ 10.25	-0.778	0.1414
516 Apr 18	C		+ 16.00	+ 9.08	-0.337	0.1734
522 Jun 10	C		+ 19.16	- 7.15	-0.417	0.1971
534 Apr 29	E,I		+ 12.80	+ 6.88	-0.732	0.2286
538 Feb 15	E,I		+ 3.53	+ 3.11	-0.288	0.4536
540 Jun 20	E,I		+ 8.50	- 5.07	-0.134	0.2796
563 Oct 3	E,F		- 12.23	+ 17.49	+1.302	0.0899
590 Oct 4	E,F		- 77.44	- 23.85	+1.220	0.0636
590 Oct 4	M,B		- 96.94	- 31.08	+0.069	0.0490
592 Mar 19	E,F		+ 17.76	+ 7.46	+0.133	0.1886
601 Mar 10	Is,E		+ 28.75	- 14.40	+0.500	0.1000
603 Aug 12	E,F		- 17.18	+ 9.87	-0.415	0.1426
644 Nov 5	M,B		- 24.57	- 10.94	-0.017	0.1414
655 Apr 12	E,SP		+ 15.32	- 18.55	+0.066	0.0783
664 May 1	B,E		- 6.14	+ 9.90	+0.170	0.1465
664 May 1a	B,I		- 3.94	+ 9.97	+0.180	0.1458
664 May 1b	B,I		- 1.25	+ 3.15	+0.057	0.4611
688 Jul 3	B,I		+ 5.85	- 8.34	+0.393	0.1838
693 Oct 5a	M,B		- 33.98	+ 22.98	-0.251	0.0618
693 Oct 5b	M,B		- 33.98	+ 22.98	-0.251	0.0618

[a] In units of reciprocal minutes of arc.
[b] Referred to $D'' = 6 ''/cy^2$ and $W = -0'.025_T{}^2$.

TABLE XIII.13

COEFFICIENTS OF CONDITION FOR ECLIPSES
FROM 702 TO 841

Designation		$1000\eta_d$ $cy^2/''$	$1000\eta_w{}^a$	r^b	σ_η
702 Sep 26	C	− 28.77	− 17.11	−1.104	0.0817
729 Oct 27	C	− 25.00	+ 16.53	−0.750	0.0860
733 Aug 14a	B,E	+ 2.24	+ 10.66	−0.092	0.1453
733 Aug 14b	B,E	+ 3.19	+ 19.74	+0.204	0.0784
753 Jan 9	B,E	− 2.03	− 10.78	−0.433	0.1475
753 Jan 9	B,I	− 3.01	− 7.07	−0.236	0.2245
754 Jun 25	C	+ 9.45	+ 18.03	+0.198	0.0820
760 Aug 15	E,CE	− 7.04	− 4.68	+0.235	0.3162
760 Aug 15	M,B	− 12.02	− 10.54	+0.383	0.1414
761 Aug 5	C	− 31.95	− 23.46	+0.023	0.0610
764 Jun 4	B,I	+ 6.69	− 10.82	−0.042	0.1457
764 Jun 4	E,F	+ 3.18	− 11.15	+0.744	0.1414
764 Jun 4	E,G	+ 0.71	− 11.15	+0.530	0.1414
764 Jun 4	E,CE	− 1.42	− 4.97	+0.293	0.3162
787 Sep 16a	E,G	− 4.00	+ 15.40	−1.496	0.1000
787 Sep 16b	E,G	− 3.36	+ 10.87	−1.307	0.1414
787 Sep 16	E,CE	− 2.13	+ 4.86	−0.440	0.3162
787 Sep 16	M,B	− 28.57	+ 31.37	−1.369	0.0490
807 Feb 11	B,W	+ 2.35	− 7.11	+0.168	0.2236
807 Feb 11	E,F	+ 6.36	− 11.24	+0.426	0.1414
807 Feb 11	E,G	+ 12.50	− 15.90	+0.863	0.1000
807 Feb 11	E,I	+ 12.20	− 11.22	+1.138	0.1414
807 Feb 11a	E,CE	+ 10.78	− 11.24	+0.823	0.1414
807 Feb 11b	E,CE	+ 5.06	− 5.02	+0.388	0.3162
808 Jul 27	C	− 13.75	− 20.64	+1.184	0.2000

[a] In units of reciprocal minutes of arc.
[b] Referred to $D'' = 6 \ ''/cy^2$ and $W = -0'.025_T{}^2$.

TABLE XIII.13 (Continued)

Designation	$1000\eta_d$ $cy^2/''$	$1000\eta_w$ [a]	r [b]	σ_η
809 Jul 16 B,E	+ 6.51	+ 10.69	+0.922	0.1459
810 Nov 30 E,G	- 19.75	- 14.03	+0.461	0.1000
810 Nov 30a E,CE	- 13.08	- 9.95	+0.264	0.1414
810 Nov 30b E,CE	- 5.38	- 4.46	+0.075	0.3162
812 May 14a E,G	+ 9.37	+ 10.32	-1.517	0.1414
812 May 14b E,G	+ 8.66	+ 10.32	-1.595	0.1414
812 May 14 M,B	- 0.73	+ 21.59	-0.675	0.0681
813 May 4 M,B	+ 10.75	+ 13.97	+0.279	0.1000
818 Jul 7a E,G	+ 11.31	- 11.12	+1.735	0.1414
818 Jul 7b E,G	+ 11.14	- 11.10	+1.694	0.1414
840 May 5a E,F	+ 4.60	- 10.04	-0.254	0.1414
840 May 5b E,F	+ 13.44	- 30.54	+0.090	0.0465
840 May 5c E,F	+ 4.25	- 14.20	-0.413	0.1000
840 May 5d E,F	+ 8.75	- 14.20	-0.042	0.1000
840 May 5a E,G	+ 1.57	- 6.36	-0.088	0.2236
840 May 5b E,G	+ 1.68	- 6.35	-0.088	0.2236
840 May 5c E,G	+ 6.86	- 32.49	-1.430	0.0437
840 May 5d E,G	+ 10.30	- 32.49	-1.895	0.0437
840 May 5e E,G	+ 5.51	- 31.28	-0.747	0.0454
840 May 5a E,I	+ 1.41	- 10.04	+0.132	0.1414
840 May 5b E,I	+ 4.59	- 28.98	+0.706	0.0490
840 May 5c E,I	+ 7.71	- 31.34	+0.267	0.0454
841 Oct 18 E,F	- 10.98	+ 15.21	-2.682	0.1002

[a] In units of reciprocal minutes of arc.
[b] Referred to $D'' = 6$ $''/cy^2$ and $W = -0'.025_T{}^2$.

TABLE XIII.14

COEFFICIENTS OF CONDITION FOR ECLIPSES
FROM 865 TO 891

Designation	$1000\eta_d$ $cy^2/''$	$1000\eta_w$ [a]	r [b]	σ_η
865 Jan 1 B,I	+ 4.40	− 10.11	+0.015	0.1422
878 Oct 29 B,E	+ 4.97	+ 14.12	+0.014	0.1006
878 Oct 29 B,I	− 4.92	+ 9.98	+0.068	0.1423
878 Oct 29 B,W	− 2.91	+ 6.35	+0.060	0.2236
878 Oct 29a E,F	− 2.34	+ 6.32	+0.141	0.2241
878 Oct 29b E,F	− 3.50	+ 14.17	+0.094	0.1000
878 Oct 29c E,F	− 6.28	+ 39.66	+0.684	0.0358
878 Oct 29d E,F	− 3.75	+ 14.17	+0.254	0.1000
878 Oct 29e E,F	− 6.42	+ 27.99	+0.152	0.0506
878 Oct 29f E,F	− 2.65	+ 10.02	+0.052	0.1414
878 Oct 29a E,G	− 0.53	+ 10.02	+0.146	0.1414
878 Oct 29b E,G	− 0.53	+ 10.02	+0.146	0.1414
878 Oct 29c E,G	− 2.34	+ 44.14	+0.081	0.0321
878 Oct 29d E,G	− 2.00	+ 14.17	+0.064	0.1000
885 Jun 16 B,I	+ 16.66	+ 14.58	+0.329	0.0960
885 Jun 16 B,SM	+ 9.72	+ 9.92	+0.105	0.1414
888 Apr 15 C	+ 51.47	+ 21.55	−2.073	0.0699
891 Aug 8 E,F	− 4.60	− 11.29	−0.330	0.1414
891 Aug 8 E,CE	− 7.96	− 11.22	−0.074	0.1414
891 Aug 8a E,CE	− 9.55	− 11.19	−0.206	0.1414
891 Aug 8a M,B	− 52.52	− 30.52	−0.424	0.0505
891 Aug 8b M,B	− 52.52	− 30.52	−0.424	0.0505
891 Aug 8c M,B	− 27.12	− 15.76	−0.219	0.0977

[a] In units of reciprocal minutes of arc.
[b] Referred to $D'' = 6 ''/cy^2$ and $W = -0'.025_T{}^2$.

TABLE XIII.15

COEFFICIENTS OF CONDITION FOR ECLIPSES FROM 912 to 1098

Designation	$1000\eta_d$ $cy^2/''$	$1000\eta_w{}^a$	r^b	σ_η
912 Jun 17 E,SP	− 24.27	− 22.65	+0.256	0.0618
939 Jul 19 E,G	+ 11.49	+ 18.10	−0.315	0.0783
939 Jul 19a E,I	+ 13.73	+ 18.05	−0.940	0.0783
939 Jul 19c E,I	+ 7.78	+ 9.99	−0.096	0.1414
939 Jul 19 E,CE	+ 10.25	+ 14.13	−0.695	0.1000
961 May 17a E,G	+ 16.75	+ 15.20	−0.145	0.1000
961 May 17b E,G	+ 16.75	+ 15.20	−0.145	0.1000
968 Dec 22 E,F	− 17.30	+ 31.47	−0.214	0.0448
968 Dec 22 E,G	− 3.89	+ 9.97	−0.185	0.1414
968 Dec 22a E,I	− 3.18	+ 9.95	+0.052	0.1414
968 Dec 22b E,I	− 3.36	+ 9.95	+0.076	0.1414
968 Dec 22 E,CE	− 5.50	+ 14.07	−0.251	0.1000
968 Dec 22a E,BN	− 4.01	+ 9.82	−0.244	0.1436
968 Dec 22b E,BN	− 1.29	+ 3.15	−0.078	0.4472
968 Dec 22a M,B	+ 10.90	+ 43.71	+0.084	0.0321
968 Dec 22b M,B	+ 5.66	+ 22.70	+0.044	0.0618
968 Dec 22c M,B	+ 3.50	+ 14.03	+0.027	0.1000
990 Oct 21a E,G	− 11.50	+ 15.87	+0.602	0.1000
990 Oct 21b E,G	− 11.50	+ 15.87	+0.625	0.1000
990 Oct 21c E,G	− 8.11	+ 11.19	+0.405	0.1418
1009 Mar 29a E,BN	+ 5.00	− 14.57	+2.854	0.1000
1009 Mar 29b E,BN	+ 3.18	− 10.30	+1.972	0.1414
1018 Apr 18 E,G	− 6.08	+ 13.29	−0.338	0.1111
1018 Apr 18 E,BN	− 1.82	+ 4.68	−0.896	0.3162
1023 Jan 24 B,E	+ 13.00	+ 13.80	+0.071	0.1000

^{a}In units of reciprocal minutes of arc.
^{b}Referred to $D'' = 6 ''/cy^2$ and $W = -0'.025\tau^2$.

TABLE XIII.15 (Continued)

Designation	$1000\eta_d$ $cy^2/''$	$1000\eta_w{}^a$	r^b	σ_η
1023 Jan 24 B,I	+ 8.06	+ 9.94	−0.164	0.1426
1023 Jan 24 E,F	+ 11.31	+ 9.71	+0.460	0.1414
1023 Jan 24 E,BN	+ 10.43	+ 9.74	+0.324	0.1414
1030 Aug 31 B,I	− 6.76	+ 10.25	+0.904	0.1441
1033 Jun 29a E,F^c	+ 4.14	+ 23.17	+0.621	0.0665
1033 Jun 29a E,F^c	+ 4.51	+ 23.17	+0.863	0.0665
1033 Jun 29b E,F	+ 1.50	+ 15.40	−0.459	0.1000
1033 Jun 29c E,F	+ 3.07	+ 15.79	−0.212	0.0977
1033 Jun 29d E,F	+ 7.44	+ 35.31	−1.620	0.0437
1033 Jun 29e E,F	+ 3.75	+ 15.43	−0.174	0.1000
1033 Jun 29f E,F	+ 2.47	+ 10.91	−0.209	0.1414
1033 Jun 29a E,G	+ 0.75	+ 15.43	−0.938	0.1000
1033 Jun 29b E,G	+ 1.00	+ 15.43	−0.247	0.1000
1033 Jun 29c E,G	+ 1.00	+ 15.43	−0.417	0.1000
1033 Jun 29a E,I	− 2.89	+ 22.27	+1.139	0.0693
1033 Jun 29b E,I	− 1.35	+ 13.93	−0.275	0.1107
1033 Jun 29c E,I	+ 0.88	+ 10.94	+0.078	0.1414
1033 Jun 29 E,BN	+ 2.12	+ 10.89	−0.562	0.1414
1037 Apr 18a E,F	+ 25.89	+ 24.39	+0.812	0.0618
1037 Apr 18b E,F	+ 26.29	+ 24.32	+0.806	0.0618
1037 Apr 18 E,BN	+ 10.78	+ 10.68	+0.679	0.1414
1039 Aug 22a E,F	− 17.50	− 14.70	−0.626	0.1000
1039 Aug 22b E,F	− 18.50	− 14.67	−0.250	0.1000
1039 Aug 22a E,G	− 7.94	− 6.57	−0.375	0.2236
1039 Aug 22b E,G	− 17.00	− 14.73	−1.050	0.1000
1039 Aug 22c E,G	− 18.25	− 14.70	−1.120	0.1000
1039 Aug 22 E,BN	− 11.31	− 10.44	−0.491	0.1414
1044 Nov 22a E,F	− 8.50	+ 15.83	−0.374	0.1000
1044 Nov 22b E,F	− 6.01	+ 11.17	−0.285	0.1414
1044 Nov 22c E,F	− 8.25	+ 15.83	−0.386	0.1000

^{a}In units of reciprocal minutes of arc.
^{b}Referred to $D'' = 6 \ ''/cy^2$ and $W = -0'.025_T{}^2$.
^{c}This is a record with two possible places of observation.

TABLE XIII.15 (Continued)

Designation	$1000\eta_d$ $cy^2/''$	$1000\eta_w{}^a$	r^b	σ_η
1044 Nov 22a E,G	− 5.75	+ 15.83	−0.834	0.1000
1044 Nov 22b E,G	− 5.50	+ 15.83	−0.804	0.1000
1044 Nov 22c E,G	− 5.25	+ 15.83	−0.811	0.1000
1044 Nov 22 E,I	− 6.75	+ 15.83	−0.297	0.1000
1079 Jul 1 E,SP	− 38.62	− 42.81	+0.503	0.0330
1087 Aug 1 E,CE	+ 3.01	+ 10.80	−2.320	0.1414
1093 Sep 23a B,E	− 4.06	− 5.04	+0.342	0.2957
1093 Sep 23b B,E	− 3.80	− 5.05	+0.219	0.2957
1093 Sep 23a E,G	− 14.00	− 14.83	+0.262	0.1000
1093 Sep 23b E,G	− 15.50	− 14.73	+0.607	0.1000
1093 Sep 23c E,G	− 10.96	− 10.42	+0.514	0.1414
1093 Sep 23d E,G	− 15.50	− 14.73	+0.727	0.1000
1093 Sep 23e E,G	− 15.00	− 14.80	+0.439	0.1000
1093 Sep 23 E,I	− 16.93	− 13.16	+0.060	0.1107
1093 Sep 23 E,BN	− 13.75	− 14.87	+0.797	0.1000
1093 Sep 23a E,CE	− 7.27	− 6.57	+0.187	0.2236
1093 Sep 23b E,CE	− 15.50	− 14.77	+0.168	0.1000
1093 Sep 23c E,CE	− 36.04	− 33.71	+1.938	0.0437
1098 Dec 25 E,F	+ 3.00	+ 15.70	−1.353	0.1000
1098 Dec 25a E,G	+ 6.75	+ 15.57	−1.103	0.1000

[a] In units of reciprocal minutes of arc.
[b] Referred to $D'' = 6 ''/cy^2$ and $W = -0'.025_T{}^2$.

TABLE XIII.16

COEFFICIENTS OF CONDITION FOR ECLIPSES
FROM 1109 TO 1147

Designation	$1000\eta_d$ $cy^2/''$	$1000\eta_w$ [a]	r [b]	σ_η
1109 May 31 E,BN	+ 2.75	+ 15.67	+1.315	0.1000
1113 Mar 19 M,HL	+ 9.43	+ 22.01	+0.223	0.0636
1118 May 22 E,BN	+ 8.25	− 15.70	+2.670	0.1000
1124 Aug 11a B,E	− 2.32	+ 5.79	+0.330	0.2476
1124 Aug 11b B,E	− 2.22	+ 5.79	+0.320	0.2476
1124 Aug 11c B,E	− 2.22	+ 5.79	+0.323	0.2476
1124 Aug 11a E,G	− 7.00	+ 14.27	+1.510	0.1000
1124 Aug 11b E,G	− 6.50	+ 14.30	+1.273	0.1000
1124 Aug 11 E,BN	− 8.22	+ 18.08	+0.954	0.0791
1124 Aug 11 E,CE	− 7.75	+ 14.30	+1.465	0.1000
1124 Aug 11 M,HL	− 14.05	+ 18.14	+1.724	0.0783
1133 Aug 2a B,E[c]	− 10.30	− 16.05	−0.229	0.0874
1133 Aug 2a B,E[c]	− 11.16	− 16.05	−0.110	0.0874
1133 Aug 2b B,E	− 14.30	− 22.29	+0.936	0.6293
1133 Aug 2c B,E	− 14.01	− 22.47	+0.617	0.6244
1133 Aug 2d B,E	− 15.10	− 22.29	+0.659	0.0629
1133 Aug 2 B,I	− 4.94	− 9.59	+0.382	0.1466
1133 Aug 2 B,SM	− 8.05	− 14.61	+0.082	0.0963
1133 Aug 2a E,F	− 22.88	− 32.04	+1.043	0.0437
1133 Aug 2b E,F	− 10.75	− 14.00	+0.555	0.1000
1133 Aug 2c E,F	− 24.03	− 32.04	+1.144	0.0437
1133 Aug 2d E,F	− 34.36	− 40.85	+2.480	0.0342
1133 Aug 2e E,F	− 10.50	− 14.03	+0.716	0.1000
1133 Aug 2f E,F	− 4.70	− 6.27	+0.320	0.2236
1133 Aug 2b E,G	− 24.60	− 32.04	−0.011	0.0437

[a] In units of reciprocal minutes of arc.

[b] Referred to $D'' = 6$ $''/cy^2$ and $W = -0'.025_T{}^2$.

[c] This is a record with two possible places of observation.

TABLE XIII.16 (Continued)

Designation	$1000\eta_d$ $cy^2/''$	$1000\eta_w{}^a$	r^b	σ_η
1133 Aug 2c E,G	- 8.49	- 9.85	+0.120	0.1414
1133 Aug 2d E,G	- 8.49	- 9.85	+0.257	0.1414
1133 Aug 2e E,G	- 33.49	- 43.61	+0.782	0.0321
1133 Aug 2f E,G	- 8.31	- 9.88	+0.053	0.1414
1133 Aug 2g E,G	- 25.17	- 31.97	-0.135	0.0437
1133 Aug 2h E,G	- 18.20	- 22.61	+0.453	0.0618
1133 Aug 2i E,G	- 8.49	- 9.88	+0.101	0.1414
1133 Aug 2j E,G	- 37.38	- 43.52	+0.445	0.0321
1133 Aug 2k E,G	- 36.60	- 43.52	+0.455	0.0321
1133 Aug 2ℓ E,G	- 8.13	- 9.88	-0.107	0.1414
1133 Aug 2m E,G	- 18.20	- 22.61	+0.471	0.0618
1133 Aug 2n E,G	- 12.25	- 13.93	+0.098	0.1000
1133 Aug 2 E,I	- 18.84	- 17.62	-0.275	0.0783
1133 Aug 2a E,BN	- 24.60	- 32.04	+0.838	0.0437
1133 Aug 2b E,BN	- 24.03	- 31.97	+0.888	0.0437
1133 Aug 2c E,BN	- 10.25	- 14.00	+0.364	0.1000
1133 Aug 2d E,BN	- 24.60	- 32.04	+0.705	0.0437
1133 Aug 2e E,BN	- 30.37	- 43.61	+0.358	0.0321
1133 Aug 2a E,CE	- 23.01	- 25.14	-0.054	0.0554
1133 Aug 2b E,CE	- 38.94	- 43.30	+0.542	0.0321
1133 Aug 2c E,CE	- 39.72	- 43.30	+0.315	0.0321
1133 Aug 2f E,CE	- 12.98	- 14.46	+0.094	0.0963
1133 Aug 2g E,CE	- 8.66	- 9.88	-0.126	0.1414
1133 Aug 2h E,CE	- 27.46	- 31.88	+0.915	0.0437
1133 Aug 2i E,CE	- 12.50	- 13.93	-0.265	0.1000
1140 Mar 20a B,E[c]	+ 4.67	- 14.67	+0.029	0.0963
1140 Mar 20a B,E[c]	+ 4.67	- 14.67	+0.252	0.0963
1140 Mar 20b B,E	+ 8.30	- 23.45	+0.181	0.0602
1140 Mar 20c B,E	+ 7.47	- 23.45	+0.403	0.0602
1140 Mar 20d B,E	+ 4.75	- 14.17	+0.026	0.1000

[a]In units of reciprocal minutes of arc.
[b]Referred to $D'' = 6$ $''/cy^2$ and $W = -0'.025_T{}^2$.
[c]This is a record with two possible places of observation.

TABLE XIII.16 (Continued)

Designation	$1000\eta_d$ $cy^2/''$	$1000\eta_w$ [a]	r [b]	σ_η
1140 Mar 20e B,E	+ 4.93	− 14.67	+0.141	0.0963
1140 Mar 20f B,E	+ 4.50	− 14.17	+0.150	0.1000
1140 Mar 20g B,E	+ 3.36	− 10.02	+0.018	0.1414
1140 Mar 20 B,SM	+ 3.18	− 10.02	−0.310	0.1414
1140 Mar 20 B,W	+ 3.71	− 9.99	−0.021	0.1414
1140 Mar 20 E,F	+ 3.01	− 10.02	+0.226	0.1414
1140 Mar 20 E,G	+ 3.75	− 14.17	+0.926	0.1000
1140 Mar 20a E,BN	+ 12.46	− 44.14	+1.358	0.0321
1140 Mar 20b E,BN	+ 8.16	− 28.92	+0.706	0.0490
1140 Mar 20 E,PR	+ 2.25	− 14.17	+0.242	0.1000
1140 Mar 20a E,Sc	+ 2.50	− 14.17	−0.113	0.1000
1140 Mar 20b E,Sc	+ 3.84	− 19.77	−0.230	0.0717
1140 Mar 20c E,Sc	+ 3.84	− 19.77	−0.230	0.0717
1147 Oct 26a E,F	− 12.75	− 14.87	+0.634	0.1000
1147 Oct 26b E,F	− 12.75	− 14.87	+0.513	0.1000
1147 Oct 26c E,F	− 13.25	− 14.83	+0.815	0.1000
1147 Oct 26b E,G	− 27.46	− 34.10	+0.748	0.0437
1147 Oct 26c E,G	− 12.25	− 14.90	+0.225	0.1000
1147 Oct 26d E,G	− 12.75	− 14.87	+0.214	0.1000
1147 Oct 26e E,G	− 12.25	− 14.87	+0.153	0.1000
1147 Oct 26f E,G	− 9.37	− 10.47	+0.262	0.1414
1147 Oct 26a E,BN	− 28.60	− 34.10	+0.776	0.0437
1147 Oct 26b E,BN	− 28.03	− 34.03	+0.627	0.0437
1147 Oct 26c E,BN	− 11.75	− 14.93	+0.147	0.1000
1147 Oct 26 E,CE	− 9.72	− 10.47	+0.311	0.1414
1147 Oct 26 M,B	− 17.24	− 18.99	+0.375	0.0783

[a] In units of reciprocal minutes of arc.
[b] Referred to $D'' = 6$ $''/cy^2$ and $W = -0'.025_T{}^2$.

TABLE XIII.17

COEFFICIENTS OF CONDITION FOR ECLIPSES
FROM 1153 TO 1194

Designation	$1000\eta_d$ $cy^2/''$	$1000\eta_w$ [a]	r [b]	σ_η
1153 Jan 26a E,F	+ 10.75	+ 15.27	-0.673	0.1000
1153 Jan 26b E,F	+ 10.25	+ 15.27	-0.847	0.1000
1153 Jan 26c E,F	+ 10.75	+ 15.23	-0.653	0.1000
1153 Jan 26a E,G	+ 12.25	+ 15.13	-0.126	0.1000
1153 Jan 26b E,G	+ 12.25	+ 15.13	+0.020	0.1000
1153 Jan 26c E,G	+ 14.69	+ 19.41	+1.060	0.0783
1153 Jan 26d E,G	+ 11.00	+ 15.20	-0.471	0.1000
1153 Jan 26 E,CE	+ 9.02	+ 10.68	+0.196	0.1414
1163 Jul 3 E,I	+ 9.00	+ 15.60	-1.566	0.1000
1178 Sep 13a B,E	- 6.50	+ 14.27	-0.754	0.1000
1178 Sep 13b B,E	- 8.30	+ 18.22	+0.340	0.0783
1178 Sep 13c B,E	- 4.42	+ 10.09	-0.521	0.1414
1178 Sep 13 B,SM	- 13.16	+ 32.65	-1.153	0.0437
1178 Sep 13a B,W	- 2.68	+ 6.37	-0.302	0.2242
1178 Sep 13b B,W	- 2.80	+ 6.38	-0.270	0.2236
1178 Sep 13a E,F	- 5.30	+ 10.06	-0.187	0.1414
1178 Sep 13b E,F	- 7.00	+ 14.23	-0.770	0.1000
1178 Sep 13c E,F	- 17.16	+ 32.56	+0.741	0.0437
1178 Sep 13d E,F	- 7.50	+ 14.23	-0.527	0.1000
1178 Sep 13e E,F	- 17.73	+ 32.56	+1.085	0.0437
1178 Sep 13f E,F	- 16.02	+ 32.56	+0.535	0.0437
1178 Sep 13g E,F	- 11.90	+ 20.49	+0.143	0.0693
1178 Sep 13h E,F	- 6.19	+ 10.06	-0.112	0.1414
1178 Sep 13i E,F	- 19.02	+ 30.93	+0.839	0.0460
1178 Sep 13 E,G	- 5.30	+ 10.04	-0.641	0.1414

[a] In units of reciprocal minutes of arc.
[b] Referred to $D'' = 6 \ ''/cy^2$ and $W = -0'.025_T{}^2$.

TABLE XIII.17 (Continued)

Designation	$1000\eta_d$ $cy^2/''$	$1000\eta_w{}^a$	r^b	σ_η
1178 Sep 13a E,I	− 9.00	+ 14.17	−0.328	0.1000
1178 Sep 13b E,I	− 11.49	+ 18.14	+0.917	0.0783
1178 Sep 13c E,I	− 5.66	+ 10.06	−0.476	0.1414
1178 Sep 13d E,I	− 6.36	+ 10.02	−0.216	0.1414
1178 Sep 13e E,I	− 19.45	+ 32.49	+0.410	0.0437
1178 Sep 13 E,Sc	− 4.21	+ 9.98	−1.141	0.1426
1180 Jan 28 E,F	+ 6.50	− 15.43	+0.687	0.1000
1181 Jul 13 E,F	+ 6.94	− 44.52	−0.056	0.1800
1185 May 1a B,E	+ 4.50	+ 14.10	+1.031	0.1000
1185 May 1b B,E	+ 5.11	+ 17.97	+0.065	0.0783
1185 May 1c B,E	+ 4.00	+ 14.10	+0.995	0.1000
1185 May 1d B,E	+ 2.01	+ 6.31	+0.474	0.2236
1185 May 1e B,E	+ 4.25	+ 14.10	+0.863	0.1000
1185 May 1a B,SM	+ 9.28	+ 30.72	+0.751	0.0458
1185 May 1b B,SM	+ 3.53	+ 9.95	+0.356	0.1414
1185 May 1a B,W	+ 3.14	+ 8.04	−0.126	0.1751
1185 May 1b B,W	+ 2.86	+ 8.04	−0.070	0.1751
1185 May 1a E,F	+ 3.36	+ 9.97	+1.204	0.1414
1185 May 1c E,F	+ 5.11	+ 18.01	+0.579	0.0783
1185 May 1 E,Sc	+ 1.13	+ 7.99	+0.378	0.1769
1186 Apr 21a B,E	+ 1.75	+ 14.20	+3.081	0.1000
1186 Apr 21b B,E	+ 2.23	+ 18.17	+2.623	0.0783
1187 Sep 4 B,E	− 12.45	− 17.66	+0.738	0.0783
1187 Sep 4 E,F	− 13.41	− 17.59	+0.992	0.0783
1187 Sep 4a E,G	− 11.50	− 13.73	+1.283	0.1000
1187 Sep 4b E,G	− 11.75	− 13.73	+1.255	0.1000
1187 Sep 4c E,G	− 10.50	− 13.80	+0.975	0.1000
1187 Sep 4d E,G	− 11.00	− 13.80	+0.903	0.1000
1187 Sep 4e E,G	− 10.50	− 13.80	+1.286	0.1000
1187 Sep 4f E,G	− 11.75	− 13.73	+1.283	0.1000

[a] In units of reciprocal minutes of arc.

[b] Referred to $D'' = 6\ ''/cy^2$ and $W = -0'.025_\tau{}^2$.

TABLE XIII.17 (Continued)

Designation	$1000\eta_d$ $cy^2/''$	$1000\eta_w{}^a$	r^b	σ_η
1187 Sep 4g E,G	− 14.35	− 18.38	−0.510	0.0749
1187 Sep 4a E,I	− 8.84	− 9.69	+1.194	0.1414
1187 Sep 4b E,I	− 8.84	− 9.69	+1.213	0.1414
1187 Sep 4 E,BN	− 10.25	− 13.80	+1.487	0.1000
1187 Sep 4a E,CE	− 11.75	− 13.73	+0.965	0.1000
1187 Sep 4b E,CE	− 11.75	− 13.73	+1.179	0.1000
1187 Sep 4c E,CE	− 19.01	− 22.17	+1.733	0.0618
1187 Sep 4a E,Sc	− 6.46	− 9.65	+0.396	0.1433
1187 Sep 4 M,HL	− 13.50	− 13.63	+0.666	0.1000
1191 Jun 23a B,E	− 1.35	− 14.42	−0.634	0.1107
1191 Jun 23b B,E	− 1.58	− 14.42	−0.656	0.1107
1191 Jun 23c B,E	− 0.67	− 7.14	+0.193	0.2236
1191 Jun 23d B,E	− 2.79	− 25.49	+0.415	0.0626
1191 Jun 23e B,E	− 1.60	− 20.34	−0.748	0.0783
1191 Jun 23f B,E	− 3.39	− 36.05	+0.490	0.0443
1191 Jun 23 B,SM	− 0.75	− 15.93	−0.230	0.1000
1191 Jun 23 B,W	− 0.34	− 7.14	+0.174	0.2236
1191 Jun 23a E,F	− 5.72	− 36.45	−0.043	0.0437
1191 Jun 23b E,F	− 1.50	− 15.93	+1.024	0.1000
1191 Jun 23c E,F	− 2.00	− 15.97	+0.957	0.1000
1191 Jun 23d E,F	− 3.36	− 21.38	+0.683	0.0745
1191 Jun 23e E,F	− 3.25	− 15.87	+0.670	0.1000
1191 Jun 23f E,F	− 2.25	− 15.93	+0.762	0.1000
1191 Jun 23g E,F	− 1.59	− 11.26	+0.337	0.1414
1191 Jun 23a E,G	− 3.18	− 11.22	+0.517	0.1414
1191 Jun 23b E,G	− 7.34	− 23.25	+1.007	0.0681
1191 Jun 23c E,G	− 3.54	− 11.19	+0.453	0.1414
1191 Jun 23d E,G	− 3.50	− 15.90	+0.374	0.1000
1191 Jun 23e E,G	− 3.61	− 14.36	−0.197	0.1107
1191 Jun 23f E,G	− 6.07	− 25.68	+0.879	0.0618

[a] In units of reciprocal minutes of arc.
[b] Referred to $D'' = 6 ''/cy^2$ and $W = -0'.025_T{}^2$.

TABLE XIII.17 (Continued)

Designation	$1000\eta_d$ $\text{cy}^2/''$	$1000\eta_w$ [a]	r [b]	σ_η
1191 Jun 23g E,G	− 4.50	− 15.87	+0.273	0.1000
1191 Jun 23a E,I	− 3.89	− 11.17	+0.759	0.1414
1191 Jun 23b E,I	− 3.89	− 11.17	+0.885	0.1414
1191 Jun 23a E,BN	− 3.00	− 15.93	+0.497	0.1000
1191 Jun 23b E,BN	− 3.25	− 15.93	+0.446	0.1000
1191 Jun 23c E,BN	− 1.77	− 11.26	+0.312	0.1414
1191 Jun 23a E,CE	− 6.00	− 15.80	+0.544	0.1000
1191 Jun 23b E,CE	− 5.25	− 15.80	+0.689	0.1000
1191 Jun 23c E,CE	− 4.07	− 11.17	+0.431	0.1414
1191 Jun 23d E,CE	− 5.25	− 15.80	+0.598	0.1000
1194 Apr 22 B,E	+ 0.64	− 18.10	+1.858	0.0783

[a] In units of reciprocal minutes of arc.
[b] Referred to $D'' = 6 \ ''/\text{cy}^2$ and $W = -0'.025_T{}^2$.

TABLE XIII.18

COEFFICIENTS OF CONDITION FOR ECLIPSES
FROM 1207 TO 1290

Designation		$1000\eta_d$ $cy^2/''$	$1000\eta_w{}^a$	r^b	σ_η
1207 Feb 28a	B,E	+ 6.89	+ 10.68	−0.694	0.1414
1207 Feb 28b	B,E	+ 6.89	+ 10.70	−0.663	0.1414
1207 Feb 28c	B,E	+ 7.07	+ 10.70	−0.672	0.1414
1207 Feb 28d	B,E	+ 6.89	+ 10.70	−0.663	0.1414
1207 Feb 28	E,F	+ 7.78	+ 10.66	−0.529	0.1414
1207 Feb 28a	E,F	+ 14.05	+ 19.25	+0.534	0.0783
1207 Feb 28b	E,F	+ 15.33	+ 19.07	+0.885	0.0783
1207 Feb 28c	E,F	+ 11.25	+ 15.03	−0.358	0.1000
1207 Feb 28a	E,G	+ 11.50	+ 15.00	−0.045	0.1000
1207 Feb 28b	E,G	+ 11.50	+ 15.00	+0.119	0.1000
1207 Feb 28c	E,G	+ 10.50	+ 15.07	−0.429	0.1000
1207 Feb 28d	E,G	+ 11.75	+ 15.00	+0.149	0.1000
1207 Feb 28e	E,G	+ 11.50	+ 14.97	+0.035	0.1000
1207 Feb 28	E,PR	+ 8.40	+ 15.06	+0.145	0.1012
1207 Feb 28	E,BN	+ 10.50	+ 15.07	−0.664	0.1000
1221 May 23	C	+ 11.00	+ 28.19	−0.524	0.0500
1230 May 14	B,E	+ 3.50	− 14.07	+0.055	0.1000
1230 May 14	E,G	+ 3.83	− 18.01	−0.959	0.0783
1230 May 14	E,BN	+ 2.47	− 9.95	+0.159	0.1414
1230 May 14	E,Sc	+ 2.99	− 14.03	+0.124	0.1003
1232 Oct 15	E,G	− 2.70	+ 12.81	−0.905	0.1111
1236 Aug 3	E,F	− 3.25	+ 15.97	+2.379	0.1000
1236 Aug 3	E,Sc	+ 0.74	+ 15.80	+0.339	0.1008
1239 Jun 3a	B,E	+ 3.75	+ 14.17	−1.536	0.1000
1239 Jun 3b	B,E	+ 4.00	+ 14.17	−1.617	0.1000

[a] In units of reciprocal minutes of arc.
[b] Referred to $D'' = 6\ ''/cy^2$ and $W = -0'.025\tau^2$.

TABLE XIII.18 (Continued)

Designation	$1000\eta_d$ $\mathrm{cy}^2/''$	$1000\eta_w$ [a]	r [b]	σ_η
1239 Jun 3a E,F	+ 3.01	+ 10.02	-0.164	0.1414
1239 Jun 3b E,F	+ 13.24	+ 44.14	-0.028	0.0321
1239 Jun 3c E,F	+ 3.84	+ 12.82	-0.246	0.1107
1239 Jun 3d E,F	+ 3.50	+ 14.20	-0.070	0.1000
1239 Jun 3e E,F	+ 9.73	+ 32.43	-0.092	0.0437
1239 Jun 3a E,G	+ 2.23	+ 18.14	+0.238	0.0783
1239 Jun 3b E,G	+ 1.75	+ 14.23	-0.784	0.1000
1239 Jun 3a E,I	+ 1.24	+ 10.06	-0.211	0.1414
1239 Jun 3b E,I	+ 4.41	+ 31.34	-0.346	0.0454
1239 Jun 3c E,I	+ 5.45	+ 44.33	-1.016	0.0321
1239 Jun 3d E,I	+ 6.23	+ 44.33	+0.115	0.0321
1239 Jun 3e E,I	+ 1.75	+ 14.23	-0.298	0.1000
1239 Jun 3f E,I	+ 2.57	+ 20.90	-0.438	0.0681
1239 Jun 3g E,I	+ 2.00	+ 14.20	-0.274	0.1000
1239 Jun 3h E,I	+ 4.58	+ 32.49	-0.627	0.0437
1239 Jun 3i E,I	+ 2.25	+ 14.23	-0.375	0.1000
1239 Jun 3j E,I	+ 3.67	+ 20.90	-0.395	0.0681
1239 Jun 3k E,I	+ 5.72	+ 32.56	+0.458	0.0437
1239 Jun 3ℓ E,I	+ 5.15	+ 32.56	-0.501	0.0437
1239 Jun 3m E,I	+ 6.23	+ 44.33	-0.620	0.0321
1239 Jun 3n E,I	+ 1.50	+ 14.27	+0.368	0.1000
1239 Jun 3o E,I	+ 1.75	+ 14.23	+0.145	0.1000
1239 Jun 3p E,I	+ 5.45	+ 44.33	+0.452	0.0321
1239 Jun 3q E,I	+ 3.24	+ 23.03	-0.254	0.0618
1239 Jun 3r E,I	+ 1.41	+ 10.06	-0.111	0.1414
1239 Jun 3s E,I	+ 2.43	+ 23.03	+0.183	0.0618
1239 Jun 3a E,CE	+ 1.25	+ 14.23	-0.676	0.1000
1239 Jun 3b E,CE	+ 1.25	+ 14.23	-0.718	0.1000
1241 Oct 6a B,E	- 8.75	- 13.67	+0.722	0.1000
1241 Oct 6b B,E	- 11.17	- 17.50	-0.280	0.0783

[a] In units of reciprocal minutes of arc.

[b] Referred to $D'' = 6 \ ''/\mathrm{cy}^2$ and $W = -0'.025_T{}^2$.

TABLE XIII.18 (Continued)

Designation	$1000\eta_d$ $cy^2/''$	$1000\eta_w$ [a]	r [b]	σ_η
1241 Oct 6c B,E	− 8.75	− 13.70	+0.815	0.1000
1241 Oct 6d B,E	− 6.36	− 9.69	+0.541	0.1414
1241 Oct 6a E,F	− 6.54	− 9.67	+0.321	0.1414
1241 Oct 6b E,F	− 9.00	− 13.67	+0.430	0.1000
1241 Oct 6a E,G	− 17.86	− 27.96	+0.412	0.0490
1241 Oct 6b E,G	− 8.75	− 13.70	+0.192	0.1000
1241 Oct 6c E,G	− 26.48	− 42.77	+0.240	0.0321
1241 Oct 6d E,G	− 20.02	− 31.35	+0.302	0.0437
1241 Oct 6e E,G	− 8.75	− 13.70	+0.136	0.1000
1241 Oct 6f E,G	− 24.14	− 42.77	−0.069	0.0321
1241 Oct 6g E,G	− 27.26	− 42.77	+0.302	0.0321
1241 Oct 6h E,G	− 19.82	− 30.18	+0.388	0.0454
1241 Oct 6i E,G	− 28.04	− 42.68	+0.548	0.0321
1241 Oct 6k E,G	− 14.16	− 22.17	+0.456	0.0618
1241 Oct 6ℓ E,G	− 8.25	− 13.73	+0.026	0.1000
1241 Oct 6m E,G	− 18.88	− 31.42	+0.066	0.0437
1241 Oct 6n E,G	− 27.26	− 42.77	+0.162	0.0321
1241 Oct 6o E,G	− 18.72	− 30.18	+0.090	0.0454
1241 Oct 6p E,G	− 19.45	− 31.35	+0.059	0.0437
1241 Oct 6q E,G	− 9.00	− 13.70	+0.203	0.1000
1241 Oct 6r E,G	− 9.00	− 13.67	+0.249	0.1000
1241 Oct 6s E,G	− 6.19	− 9.69	+0.188	0.1414
1241 Oct 6t E,G	− 18.88	− 31.42	−0.259	0.0437
1241 Oct 6u E,G	− 20.02	− 31.35	+0.382	0.0437
1241 Oct 6v E,G	− 14.15	− 21.54	+0.615	0.0636
1241 Oct 6w E,G	− 13.75	− 22.22	+0.015	0.0618
1241 Oct 6x E,G	− 27.26	− 42.77	+0.025	0.0321
1241 Oct 6y E,G	− 19.45	− 31.35	+0.162	0.0437
1241 Oct 6 E,I	− 9.25	− 13.70	+0.245	0.1000
1241 Oct 6 E,BN	− 8.75	− 13.67	+0.480	0.1000

[a] In units of reciprocal minutes of arc.
[b] Referred to $D'' = 6 ''/cy^2$ and $W = -0'.025_T{}^2$.

TABLE XIII.18 (Continued)

Designation		$1000\eta_d$ $cy^2/''$	$1000\eta_w$ [a]	r [b]	σ_η
1241 Oct	6a E,BN	− 8.75	− 13.70	+0.399	0.1000
1241 Oct	6a E,CE	− 20.02	− 31.35	+0.185	0.0437
1241 Oct	6b E,CE	− 20.02	− 31.35	+0.000	0.0437
1241 Oct	6c E,CE	− 26.48	− 42.68	+0.031	0.0321
1241 Oct	6d E,CE	− 27.26	− 42.68	+0.084	0.0321
1241 Oct	6e E,CE	− 16.84	− 28.02	−0.273	0.0490
1241 Oct	6f E,CE	− 19.45	− 31.42	−0.304	0.0437
1241 Oct	6g E,CE	− 18.88	− 31.42	−0.359	0.0437
1241 Oct	6h E,CE	− 8.75	− 13.70	+0.040	0.1000
1241 Oct	6i E,CE	− 6.01	− 9.69	−0.021	0.1414
1241 Oct 6	E,PR	− 7.50	− 13.77	−0.332	0.1000
1241 Oct 6	E,Sc	− 5.09	− 9.67	−0.135	0.1425
1245 Jul 25	E,G	+ 3.50	− 15.77	+2.374	0.1000
1255 Dec 30	B,E	+ 1.00	− 15.57	−1.796	0.1000
1255 Dec 30	E,CE	+ 3.50	− 15.57	−1.372	0.1000
1261 Apr 1	B,E	+ 7.25	+ 15.17	−2.695	0.1000
1261 Apr 1	E,G	+ 6.19	+ 10.61	−1.269	0.1414
1261 Apr 1	E,CE	+ 9.25	+ 14.97	−1.454	0.1000
1263 Aug	5a B,E	− 8.25	− 15.30	+1.242	0.1000
1263 Aug	5b B,E	− 8.00	− 15.27	+1.215	0.1000
1263 Aug	5c B,E	− 8.25	− 15.30	+1.173	0.1000
1263 Aug 5	E,G	− 8.00	− 15.33	+0.601	0.1000
1263 Aug	5a E,G	− 7.23	− 15.32	+0.358	0.1003
1263 Aug	5b E,G	− 7.25	− 15.37	+0.365	0.1000
1263 Aug	5c E,G	− 7.00	− 15.37	+0.422	0.1000
1263 Aug	5a E,I	− 8.00	− 15.27	+0.770	0.1000
1263 Aug	5b E,I	− 8.50	− 15.27	+0.921	0.1000
1263 Aug 5	E,BN	− 8.00	− 15.30	+0.959	0.1000
1263 Aug 5	E,CE	− 9.90	− 19.58	−0.628	0.0783
1263 Aug	5a E,CE	− 7.50	− 15.33	+0.350	0.1000

[a] In units of reciprocal minutes of arc.
[b] Referred to $D'' = 6 ''/cy^2$ and $W = -0'.025_T{}^2$.

TABLE XIII.18 (Continued)

Designation			$1000\eta_d$ cy^2/″	$1000\eta_w$ [a]	r [b]	σ_η
1263 Aug	5	E,PR	− 7.00	− 15.40	+0.031	0.1000
1263 Aug	5a	E,Sc	− 6.00	− 15.47	+0.302	0.1000
1263 Aug	5b	E,Sc	− 7.00	− 15.37	+0.379	0.1000
1263 Aug	5c	E,Sc	− 4.77	− 10.89	+0.185	0.1414
1267 May	25	E,G	+ 4.25	− 15.07	−1.646	0.1000
1267 May	25a	E,G	+ 3.49	− 15.02	−1.979	0.1004
1267 May	25	E,CE	+ 4.00	− 15.07	−1.580	0.1000
1267 May	25a	E,CE	+ 3.75	− 15.07	−1.617	0.1000
1267 May	25b	E,CE	+ 3.75	− 15.07	−1.504	0.1000
1267 May	25c	E,CE	+ 4.75	− 15.03	−1.763	0.1000
1267 May	25	M,B	+ 5.51	− 33.19	−0.128	0.0454
1270 Mar	23	E,G	+ 0.71	− 10.70	−0.453	0.1414
1270 Mar	23a	E,CE	+ 1.25	− 15.13	−0.241	0.1000
1270 Mar	23b	E,CE	+ 1.25	− 15.13	−0.234	0.1000
1270 Mar	23	E,Sc	+ 0.75	− 15.13	−0.702	0.1000
1288 Apr	2	B,E	+ 1.25	− 15.13	+3.613	0.1000
1288 Apr	2	B,W	+ 2.54	− 19.17	+3.081	0.0788
1290 Sep	5a	E,F	+ 0.75	+ 15.97	+0.814	0.1000
1290 Sep	5b	E,F	+ 0.96	+ 20.40	−0.295	0.0783
1290 Sep	5	E,G	0.00	+ 20.40	−1.285	0.0783
1290 Sep	5	E,I	− 0.22	+ 14.37	−0.703	0.1111

[a] In units of reciprocal minutes of arc.
[b] Referred to $D'' = 6$ ″/cy^2 and $W = -0'.025_T{}^2$.

TABLE XIII.19

COEFFICIENTS OF CONDITION FOR ECLIPSES
FROM 1310 TO 1399

Designation	$1000\eta_d$ $cy^2/''$	$1000\eta_w$ [a]	r [b]	σ_η
1310 Jan 31a B,E	+ 4.00	− 15.67	−0.209	0.1000
1310 Jan 31b B,E	+ 3.75	− 15.67	−0.364	0.1000
1310 Jan 31a E,F	+ 4.50	− 15.63	+0.128	0.1000
1310 Jan 31b E,F	+ 5.75	− 19.96	−1.166	0.0783
1310 Jan 31c E,F	+ 5.75	− 19.96	−1.166	0.0783
1310 Jan 31a E,G	+ 5.75	− 19.96	−0.890	0.0783
1310 Jan 31b E,G	+ 7.47	− 24.58	+0.954	0.0636
1310 Jan 31a E,I	+ 2.62	− 7.80	+0.524	0.2000
1310 Jan 31b E,I	+ 2.62	− 7.80	+0.560	0.2000
1310 Jan 31 E,BN	+ 4.25	− 15.63	+0.125	0.1000
1310 Jan 31a E,CE	+ 4.75	− 15.60	+0.903	0.1000
1310 Jan 31b E,CE	+ 4.75	− 15.63	+0.796	0.1000
1310 Jan 31c E,CE	+ 5.00	− 15.60	+0.775	0.1000
1312 Jul 5 E,G	+ 5.75	+ 18.77	+1.799	0.0783
1321 Jun 26a E,G	+ 9.73	− 34.55	+0.215	0.0437
1321 Jun 26c E,G	+ 8.01	− 34.62	−0.810	0.0437
1321 Jun 26d E,G	+ 3.75	− 15.13	−0.224	0.1000
1321 Jun 26 E,CE	+ 4.00	− 15.13	+0.309	0.1000
1321 Jun 26a E,PR	+ 5.90	− 23.79	+0.461	0.0636
1321 Jun 26b E,PR	+ 3.50	− 15.13	−0.288	0.1000
1330 Jul 16 B,E	− 3.75	+ 14.90	+0.430	0.1000
1330 Jul 16a E,I	− 3.60	+ 13.41	−1.937	0.1111
1330 Jul 16b E,I	− 3.01	+ 10.51	+0.386	0.1414
1330 Jul 16c E,I	− 3.60	+ 13.38	−1.944	0.1111
1330 Jul 16 E,Sc	− 2.47	+ 14.80	−0.150	0.1011

[a] In units of reciprocal minutes of arc.
[b] Referred to $D'' = 6$ $''/cy^2$ and $W = -0'.025_\tau{}^2$.

TABLE XIII.19 (Continued)

Designation			$1000\eta_d$ $cy^2/''$	$1000\eta_w{}^a$	r^b	σ_η
1331 Nov	30	M,B	− 6.25	− 13.60	+2.325	0.1000
1333 May	14	E,I	− 4.47	+ 19.67	+0.845	0.0783
1337 Mar	3a	M,B	+ 7.25	+ 15.03	−1.413	0.1000
1337 Mar	3b	M,B	+ 5.66	+ 10.51	−0.356	0.1414
1339 Jul	7	B,E	− 9.58	− 28.54	−1.372	0.0522
1339 Jul	7	E,I	− 7.98	− 18.90	−0.336	0.0783
1339 Jul	7	E,CE	− 13.16	− 33.94	+0.247	0.0437
1339 Jul	7	E,PR	− 5.25	− 14.87	−0.348	0.1000
1339 Jul	7	E,Sc	− 2.22	− 14.89	−0.102	0.1014
1341 Dec	9	M,B	− 1.50	+ 14.13	+2.576	0.1000
1354 Sep	17	E,I	− 4.52	− 13.16	+0.126	0.1107
1361 May	5	B,E	+ 4.25	− 14.33	−3.371	0.1000
1361 May	5	E,I	+ 4.00	− 14.30	−1.973	0.1000
1361 May	5	M,B	+ 5.26	− 23.19	−0.304	0.0618
1364 Mar	4	E,BN	+ 3.50	− 15.70	+1.862	0.1000
1384 Aug	17a	B,E	− 3.00	+ 15.13	−1.360	0.1000
1384 Aug	17b	B,E	− 3.00	+ 15.13	−1.360	0.1000
1384 Aug	17	E,I	− 5.90	+ 23.69	+0.808	0.0636
1384 Aug	17	E,BN	− 2.93	+ 13.59	+0.347	0.1111
1386 Jan	1a	E,I	− 2.21	− 20.46	−0.286	0.0681
1386 Jan	1b	E,I	− 3.43	− 31.88	+1.030	0.0437
1399 Oct	29	E,I	− 0.71	+ 10.44	+0.981	0.1414

[a] In units of reciprocal minutes of arc.

[b] Referred to $D'' = 6\ ''/cy^2$ and $W = -0'.025_T{}^2$.

TABLE XIII.20

COEFFICIENTS OF CONDITION FOR ECLIPSES
FROM 1406 TO 1486

Designation	$1000\eta_d$ $cy^2/''$	$1000\eta_w$ [a]	r [b]	σ_η
1406 Jun 16 B,E	+ 2.75	+ 14.17	-0.429	0.1000
1406 Jun 16 E,G	+ 3.25	+ 14.13	+0.427	0.1000
1406 Jun 16 E,I	+ 4.06	+ 16.37	-0.766	0.0863
1408 Oct 19a E,I	- 3.54	- 10.11	-1.402	0.1414
1408 Oct 19b E,I	- 4.52	- 12.91	-0.847	0.1107
1408 Oct 19 E,CE	- 4.75	- 14.33	-2.752	0.1000
1409 Apr 15a E,I	+ 0.35	+ 11.17	+1.013	0.1414
1409 Apr 15b E,I	+ 0.45	+ 14.27	+0.349	0.1107
1415 Jun 7a E,G	+ 8.57	- 44.14	-0.146	0.0321
1415 Jun 7b E,G	+ 2.75	- 14.17	-0.049	0.1000
1415 Jun 7a E,I	+ 3.51	- 18.10	-0.808	0.0783
1415 Jun 7b E,I	+ 3.51	- 18.05	-0.429	0.0783
1415 Jun 7c E,I	+ 2.75	- 14.17	+0.115	0.1000
1424 Jun 26 B,E	- 2.00	+ 14.30	+0.840	0.1000
1431 Feb 12a E,I	+ 2.65	+ 9.81	-0.175	0.1414
1431 Feb 12b E,I	+ 3.75	+ 13.87	-0.247	0.1000
1431 Feb 12c E,I	+ 6.07	+ 22.44	+0.812	0.0618
1431 Feb 12d E,I	+ 3.75	+ 13.87	+0.115	0.1000
1431 Feb 12e E,I	+ 10.90	+ 43.08	-0.075	0.0321
1433 Jun 17 E,F	- 3.21	- 12.82	-1.617	0.1090
1448 Aug 29a E,I	- 6.29	- 24.26	+1.415	0.0636
1448 Aug 29b E,I	- 3.61	- 13.93	+0.013	0.1107
1460 Jul 18 E,I	+ 1.81	+ 12.82	-0.329	0.1107
1465 Sep 20a E,I	0.00	- 24.06	+0.200	0.0636
1465 Sep 20b E,I	- 0.25	- 15.30	+1.496	0.1000

[a]In units of reciprocal minutes of arc.
[b]Referred to $D'' = 6\ ''/cy^2$ and $W = -0'.025_T{}^2$.

TABLE XIII.20 (Continued)

Designation			$1000\eta_d$ $cy^2/''$	$1000\eta_w$ [a]	r [b]	σ_η
1473 Apr 27	Is,E		+ 5.15	− 36.32	−0.606	0.0437
1478 Jul 29	E,I		− 2.25	+ 14.20	−1.239	0.1000
1485 Mar 16	E,I		+ 1.57	+ 22.22	+0.781	0.0636
1485 Mar 16	E,CE		+ 1.72	+ 32.43	+0.005	0.0437
1486 Mar 6	E,CE		0.00	+ 18.48	−0.084	0.0783

[a] In units of reciprocal minutes of arc.
[b] Referred to $D'' = 6 \ ''/cy^2$ and $W = -0'.025_T{}^2$.

TABLE XIII.21

COEFFICIENTS OF CONDITION FOR ECLIPSES
FROM 1502 TO 1567

Designation			$1000\eta_d$ $cy^2/''$	$1000\eta_w$ [a]	r [b]	σ_η
1502 Oct 1	E,CE		− 1.92	− 20.14	−0.295	0.0783
1502 Oct 1	Is,E		− 3.50	− 15.20	+2.474	0.1000
1513 Mar 7	Is,E		+ 1.50	− 14.93	+0.468	0.1000
1544 Jan 24	E,G		+ 2.34	+ 44.86	−0.137	0.0321
1560 Aug 21	E,SP		− 28.99	−203.33	+1.725	0.0069
1567 Apr 9	E,I		+ 15.87	−239.21	−0.190	0.0063

[a] In units of reciprocal minutes of arc.
[b] Referred to $D'' = 6 \ ''/cy^2$ and $W = -0'.025_T{}^2$.

CHAPTER XIV

DISCUSSION OF THE RESULTS

1. The Time Dependence of D''

Values of D'', the acceleration coefficient in the lunar elongation, are listed in Table XIII.2 for various epochs from -660 to +1534. We see from the table that the standard deviations $\sigma(D'')$ for the epochs 1446 and 1534 are much larger than for other epochs. These values come from using eclipses with dates from 1406 to 1486 and from 1502 to 1567, respectively. We could decrease the standard deviation by combining these two samples were it not for the problem with W. We remember that we must be able to approximate W by a constant for a given sample of eclipses, and that the largest component of W is a quadratic function of time as measured from 1800. If we tried to use the data from 1406 to 1567 as a single sample, we would find that the square of the time varies from about 16 to about 5.4. This variation is by a factor of about 3, so approximating W by a constant is not very accurate. Thus we are forced to use two separate samples between 1406 and 1567.

However, once we have found D'' and W separately, we may combine the two estimates of D'', which depends almost linearly on time, into a single estimate. Thus I estimate that

$$D'' = -23.56 \pm 25.49 \qquad (XIV.1)$$

at the epoch 1486.

The values of D'' from Table XIII.2, after replacing the values for 1446 and 1534 by the single value from Equation XIV.1, are plotted in Figure XIV.1. The central estimate for each epoch is plotted as a circle, and the distance from a circle to either end of the associated vertical bar is the standard deviation $\sigma(D'')$. The figure also shows the value of D'' at the epoch 1800, as obtained from modern data.

It is clear from the figure that D'' has not been constant between -700 and the present. The confidence in this conclusion is so high that I shall not bother to make a formal calculation of the confidence level.

Equation I.9 gives the time behavior of D'' that was obtained by combining the results of <u>AAO</u>, <u>MCRE</u>, <u>Brouwer</u> [1952], and <u>Martin</u> [1969]†; it presents D'' as a set of three straight lines. I repeat Equation I.9 here for the convenience of the reader:

---

† After allowing for the fact that some data were used in more than one source.

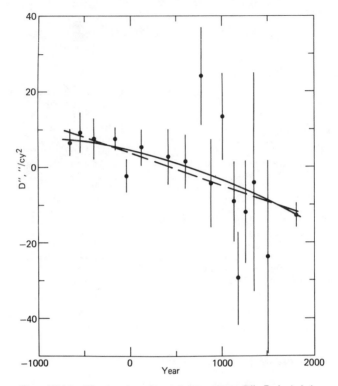

Figure XIV.1. The time dependence of the parameter D''. Each circle is a value of D'' that has been inferred for a particular epoch, and the half-length of the associated bar is the standard deviation of the inferred value. The dashed line is the best-fitting straight line and the solid curve is the best-fitting quadratic. The difference between the two functions is statistically significant at about the "1σ" level.

$$D'' = \begin{cases} -14 + T , & - 6 \le T \le 0 , \\ -44 - 4T , & -12\frac{1}{2} \le T \le - 6 , \\ 9.125 + \frac{1}{4}T, & -26 \le T \le -12\frac{1}{2}. \end{cases} \qquad (XIV.2)$$

Equation XIV.2 gives a rather good fit to the values plotted in Figure XIV.1. This shows, among other things, that changing the epoch from the beginning of 1900 to the beginning of 1800 has at most a small effect upon the results.

However, there is an interesting difference between Figure XIV.1 and Equation XIV.2. The data which lead to Equation XIV.2 are plotted in Figure 4 of Newton [1972c]. That figure shows a small but definite positive slope to the values from -700 to + 700 that is not seen in Figure XIV.1. There are

-446-

two main reasons for this difference:

      a. Figure XIV.1 is based upon many more solar eclipse observations than Figure 4 of the reference. In particular, the present work uses many more eclipses before -500.

      b. Figure 4 of the reference was based upon both eclipse observations and many other types of observation. Many of the other types of observation were based upon the records of Ptolemy [ca. 142] that are now known to be fraudulent [Newton, 1977].

Thus Figure XIV.1 is based upon the only valid body of data that has been analyzed in the current literature. The relation between the data from solar eclipses and from other types of observation will be the subject of a later work.

Since the data in Figure XIV.1 show a slight negative slope instead of a positive one between -700 and +700, there is now no reason for using three straight lines, and we should fit the data using a simpler function. The data suggest to the eye that there may be a quadratic term in the behavior of D''. To test the matter, I have found both the straight line and the quadratic that best fit the values of D''. Since the year 600 is approximately the average value of the time for the data, I refer the functions to the year 600. The best linear fit is

$$D'' = -1.54 \pm 1.44 - (0.854 \pm 0.155)C, \qquad \text{(XIV.3)}$$

in which C denotes time in Julian centuries from the year 600. The best quadratic function is

$$D'' = +0.58 \pm 2.55 - (0.844 \pm 0.156)C - (0.0244 \pm 0.0243)C^2.$$

$$\text{(XIV.4)}$$

The coefficient of C^2 in Equation XIV.4 is almost exactly equal to its standard deviation, so the presence of the term in C^2 is statistically significant at only a marginal level.

The functions in Equations XIV.3 and XIV.4 are plotted in Figure XIV.1. They differ from each other by about 2 $''/\text{cy}^2$ and the uncertainty in the value of either function at any time is about the same size. Thus we are entitled to say that we know D'' at any time since -700 within about 2 $''/\text{cy}^2$ standard deviation.

The coefficient of C in either Equation XIV.3 or XIV.4 is more than five times its standard deviation. Thus the coefficient is highly significant and there is no serious question that D'' has changed by a large amount within historic times.

Since the difference between the best straight line and

the best quadratic is only marginally significant, I shall use the straight line for simplicity in most of the remaining discussion. I shall use the quadratic when I wish to explore specifically the effect of the difference between the two functions, and when I explore the origin of the accelerations.

2. The Time Dependence of W

The function W was defined in Equation IV.22 as the perturbation in the lunar node minus the perturbation in the sun's mean longitude. The values of W listed in Table XIII.2 are plotted in Figure XIV.2, along with their standard deviations. Since the perturbations are to be calculated from the fundamental epoch at the beginning of 1800, W equals zero at 1800 by definition.

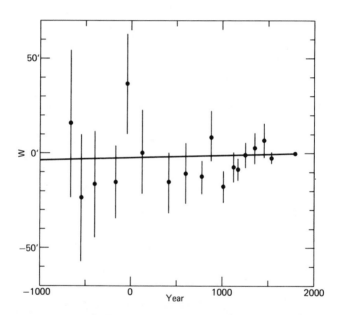

Figure XIV.2. The time dependence of the function W. W is the perturbation in the lunar node minus the perturbation in the solar mean longitude. Note that W is in minutes rather than seconds of arc in the figure. W equals zero at the epoch 1800 by definition. The straight line is the function $W = 0'.1367\tau$, in which τ is time in centuries from 1800. This function is drawn arbitrarily and is not intended to be the best linear fit to W.

We expect the biggest perturbation in the lunar node to be a linear function of time, and <u>M</u> [p. 12.9] finds that this perturbation $\delta\Omega$ is

$$\delta\Omega = (8''.2 \pm 3''.6)\tau. \qquad\qquad (XIV.5)$$

This function, after being converted from seconds to minutes of arc, is plotted as the straight line in Figure XIV.2. The immediate reaction to this line is to say that it is "lost in the noise" and that a nodal perturbation of a plausible size cannot be found from the eclipse data. However, it is possible that the significance of the straight line is greater than the eye realizes, and the matter is an important one. Hence I shall study the matter further by explicit calculations in the next section.

I ask the reader to let me anticipate these calculations and to assume that we cannot find the nodal perturbation from Figure XIV.2. This means that the solar perturbation that we find from the figure is not affected by the nodal perturbation at a level that is statistically significant. In other words, we can find the solar perturbation by taking the nodal perturbation to be zero. I shall find the solar perturbation in this way in the next section.

3.   The Time Dependence of the Solar Acceleration $\nu_S{}'$

If we use δL_S to denote the perturbation in the sun's mean longitude, we define the solar acceleration $\nu_S{}'$ by

$$\delta L_S = \tfrac{1}{2}\, \nu_S{}'\, \tau^2.$$

Thus $\nu_S{}'$ at the epoch τ is a kind of average acceleration between the epoch and 1800; it is not the instantaneous acceleration that applies at the epoch. D'' is a similar kind of average. See <u>Newton</u> [1972c] for a detailed discussion of this matter.

Since we are taking the nodal perturbation to be zero in studying Figure XIV.2, W reduces to

$$W = -\tfrac{1}{2}\, \nu_S{}'\, \tau^2. \qquad\qquad (XIV.6)$$

It is now simpler to use $\nu_S{}'$ rather than W in the analysis, and we find a value of $\nu_S{}'$ at each epoch in Figure XIV.2 by using Equation XIV.6. The resulting values of $\nu_S{}'$, along with their standard deviations $\sigma(\nu_S{}')$, are listed in Table XIV.1. They are also plotted in Figure XIV.3, with one change. Because of the large standard deviations at the epochs 1446 and 1534, I have replaced the values at these epochs by a single value at the epoch 1486:

TABLE XIV.1

VALUES OF ν_S' AT VARIOUS EPOCHS

Epoch	ν_S' $''/\text{cy}^2$	$\sigma(\nu_S')$ $''/\text{cy}^2$
− 660	− 3.08	7.72
− 551	+ 5.12	7.28
− 398	+ 4.10	6.99
− 166	+ 4.74	6.00
− 44	−12.87	9.34
+ 122	− 0.23	9.36
415	+ 9.76	9.83
602	+ 9.10	13.29
772	+14.61	9.92
878	−11.34	17.14
1005	+34.04	15.44
1128	+19.64	20.59
1174	+26.76	17.73
1248	+ 3.50	26.19
1354	−14.42	48.92
1446	−48.78	70.11
1534	+42.57	54.44

$\nu_S' = +8.21 \pm 43.00$ at 1486.

The standard deviations of the ν_S' values are about twice those of the D'' values in Table XIII.2. We expect them to be larger, for reasons that have been frequently discussed, and I am somewhat surprised that the differences are not even larger than they are.

Figure XIV.3 also shows a value of ν_S' at the epoch 1800. This value, whose standard deviation is 0.36, is taken from Table I.4 in Section I.4.

The straight line drawn in Figure XIV.3 is the best linear fit to the values of ν_S'. Since ν_S', unlike D'', is determined with high accuracy at the epoch 1800, I took 1800 as the base epoch in fitting to ν_S'. The best linear fit is

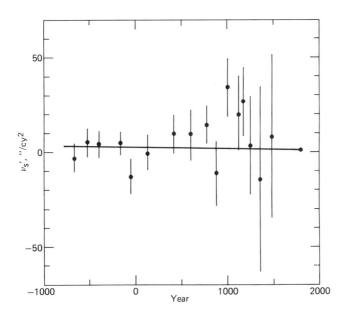

Figure XIV.3.   The time dependence of the solar acceleration ν_S'. The solid line is the best-fitting straight line to the individual values of ν_S'.

$$\nu_S' = 1.21 \pm 0.36 - (0.087 \pm 0.138)\tau. \qquad (XIV.7)$$

Since the coefficient of τ is less than its standard deviation, we do not know from the eclipse data whether ν_S' has increased or decreased within historic times.   In fact, the matter is worse than this.   If we calculate the value of ν_S' from Equation XIV.7 at the year -700, when $\tau = -25$, we find

$$\nu_S' = +3.38 \pm 3.48 \qquad \text{at } -700.$$

Thus we cannot even tell the sign of ν_S' in -700 with high confidence from eclipse data alone, although we can say that it was positive with odds of about 5 to 1.

In order to see whether the inference of ν_S' is affected significantly by using M's value for $\delta\Omega$ (Equation XIV.5), I took $W = (8.2\tau - \frac{1}{2}\nu_S'\tau^2)/60$ (in minutes of arc) and repeated the analysis for ν_S'.   The new result is

$$\nu_S' = 1.21 \pm 0.36 - (0.041 \pm 0.138)\tau.$$

To the accuracy quoted, the constant term is not affected by including the nodal perturbation. The coefficient of τ is changed by 0.046, which is just a third of the standard deviation. Thus the nodal perturbation has no statistically significant effect on $\nu_S{}'$.

It is also interesting to find the level of significance of the linear coefficient in W that we find when we fit to the data in Figure XIV.2. When I do this, I take advantage of the fact that the value of $\nu_S{}'$ in 1800 is determined almost entirely by the modern data, and that its value is not affected by introducing the nodal perturbation. Hence I assume that δL_S has the form $\frac{1}{2} \times 1.21\tau^2 + \frac{1}{2} A\tau^3$, so that W has the form[†]

$$W = \dot{\Omega}\tau - 0.0101\tau^2 - \tfrac{1}{2} A\tau^3.$$

I then find the parameters $\dot{\Omega}$ and A by fitting to the data. The result, when converted to units of seconds and centuries, is

$$\dot{\Omega} = 54.2 \pm 25.2 \text{ ''/cy}, \quad A = +0.217 \pm 0.197 \text{ ''/cy}^3.$$

Let us look first at the standard deviation of $\dot{\Omega}$. It is three times the value of 8.2 ''/cy that $\underline{M}$ quotes, and it is seven times the standard deviation of 3.6 that he finds for his estimate of $\dot{\Omega}$. Thus it is clear that any value of $\dot{\Omega}$ consistent with modern results is too small to find by analyzing the values of W. In other words, we cannot make a useful contribution to the study of the lunar node by using the total collection of untimed solar eclipse data that is currently available. All the more, then, we cannot make a contribution by using only the subset of data that $\underline{M}$ uses.

We must also look at the central value of $\dot{\Omega}$, namely 54.2 ''/cy. This quantity, which is the coefficient of τ in the function W, can come only from a combination of a perturbation in the nodal precession rate and an error in the mean motion of the sun. A value of 54.2 is far too large to be compatible with modern results. Thus this value cannot be significant in spite of the fact that it equals about two standard deviations. That is, the value must be an error of some sort. Since an error of this size occurs by chance about 1 time in 20, it is not utterly implausible that the error is the result of a sampling accident in the data. It is more likely, I suspect, that I have underestimated the standard deviations and that the inference of $\nu_S{}'$ from the data is even more difficult than it has seemed.

In the next section I shall estimate the coefficient A by combining modern data with ancient solar observations. By assuming this value of A, I could use the values of W to

---

[†]The coefficient of τ^2 is divided by 60 in order to convert to minutes.

estimate the single parameter $\dot{\Omega}$. However, it is not likely
that the standard deviation of $\dot{\Omega}$ would be significantly
lowered by doing this, and thus it is not likely that the
resulting value of $\dot{\Omega}$ would be significant. Hence I shall
not bother with making this estimate.

4. The Time Dependence of $\dot{n}_M$

Recent studies about the time dependence of $\dot{n}_M$, the acceleration of the moon with respect to ephemeris time, illustrate vividly the principle often called the innate cus-
sedness of the inanimate world. Spencer Jones [1939], we
remember, found $\dot{n}_M$ = -22.44 from modern data. However, in
AAO [p. 272] I found $\dot{n}_M$ = -41.6 ± 4.3 at the epoch -200 and
-42.3 ± 6.1 at the epoch +1000. These values suggest that
$\dot{n}_M$ has been constant within historic times, at a value near
-40, and that the variation found in D'' must be the result
of a variation in y, the acceleration of the earth's spin
with respect to ephemeris time.

This conclusion seemed to be confirmed almost immediately by two independent studies. van Flandern [1970], using
lunar occultations timed by cesium clocks from 1955 to 1968,
found $\dot{n}_M$ = -52 while Oesterwinter and Cohen [1972], using a
more extensive body of observations timed with cesium clocks,
found $\dot{n}_M$ = -38. Thus it seemed that $\dot{n}_M$ had a value around
-40 and that Spencer Jones's value was seriously in error.

More recently, however, Morrison and Ward [1975] have
made a careful new analysis of the modern data, including
both the data that Spencer Jones used and some additional
data, and they find $\dot{n}_M$ = -26 ± 2. In Section I.4 of this
work, using more general methods of analysis but confining
myself to the data of Spencer Jones, I find (Equation I.23)
$\dot{n}_M$ = -28.38 ± 5.72. These results at first seem to conflict
with the results mentioned in the preceding paragraph.

However, the conflict is only apparent. The acceleration found by van Flandern and by Oesterwinter and Cohen is
with respect to atomic time rather than ephemeris time. If
the gravitational constant G is decreasing slowly with time
[van Flandern, 1974], the acceleration with respect to atomic
time is more negative than $\dot{n}_M$, the acceleration with respect
to ephemeris time. A change of 1 part in 10^8 per century in
G is enough to make a difference of about 40 between the two
accelerations. Thus it is not necessary for the acceleration
with respect to atomic time to agree with the results of
Spencer Jones or Morrison and Ward.

This leaves us with an apparent conflict between the
modern value of $\dot{n}_M$ and the value found from the ancient data.
For simplicity, in referring to the AAO results, I shall use
only the value -41.6 ± 4.3 for the epoch -200. This differs
from the value -22.44 by about $4\frac{1}{2}$ standard deviations, so
that the difference seems to be highly significant. This difference is what M [p. 6.2] calls the Spencer Jones anomaly.
The anomaly is confirmed by the results of MS, who find
[p. 529] $\dot{n}_M$ = $-37\frac{1}{2}$ ± 5 from their analysis of ancient and
medieval solar eclipses. M [Chapter XII] claims to have

removed the Spencer Jones anomaly by introducing the perturbation in the lunar node.

However the Spencer Jones anomaly is not real and there is in fact no anomaly to remove. The anomaly is simply the result of the tendency, which is almost universal, to underestimate the errors in measurements or observations. Three points need to be made in this regard.

a. In saying above that the anomaly amounted to $4\frac{1}{2}$ standard deviations, I tacitly took the value -22.44 to be exact. This is common with many writers on the subject. However, there is a nonzero standard deviation attached to the value -22.44, and it should be combined with the standard deviation of 4.1 attached to the ancient value. Thus, the anomaly is not as large as $4\frac{1}{2}$ standard deviations.

b. The analysis in Section I.4 gives $\dot{n}_M = -28.38 \pm 5.72$. This is closer to the AAO value by about 6 $''/cy^2$, and this immediately removes a good fraction of the anomaly. In fact, the difference between this value and the AAO value is not much more than the sum of the standard deviations, and thus it is not highly significant.

c. The standard deviation of the AAO value was seriously underestimated and the central value -41.6 itself is probably wrong. I underestimated the errors in the solar eclipse observations by a factor of 3.† Further, of the observations in AAO other than eclipses, the ones with the smallest indicated standard deviations tend to be those derived from Ptolemy [ca. 142]. We know now that these values were fabricated, and this is probably the reason why they have small apparent errors; the "errors" are mostly errors in computation that Ptolemy made when fabricating his "data". Hence, when we remove the fabricated data, we may well change the inferred value of $\dot{n}_M$.‡

Thus, pending a new study of the observations other than the solar eclipses, we have no values of $\dot{n}_M$ in ancient times to use except those found here. The best estimate of D'' is that in Equation XIV.3, if we restrict ourselves to linear functions of time. The best estimate of ν_S' is the one in Equation XIV.7, and $\dot{n}_M$ is related to D'' and ν_S' by (Equation I.29):

$$\dot{n}_M = D'' - 12.3683\nu_S'. \qquad\qquad (XIV.8)$$

---

† I used 0.06 for the standard deviation of the magnitude of ancient eclipses in AAO but I use 0.18 here.

‡ I have said elsewhere that removal of these fraudulent data does not affect my main results, but this was said explicitly with respect to the parameter D''. The more sensitive parameter $\dot{n}_M$ may well be affected.

When we combine all these relations, we find:

$$\dot{n}_M = -26.75 \pm 4.68 + 0.222\tau \pm 0.156C \pm 1.706\tau \qquad (XIV.9)$$

In this, τ is measured in centuries from 1800 and C is measured in centuries from 600. The standard deviation $\sigma(\dot{n}_M)$ of $\dot{n}_M$ has three independent components which must be combined at any epoch by taking the square root of the sum of the squares.

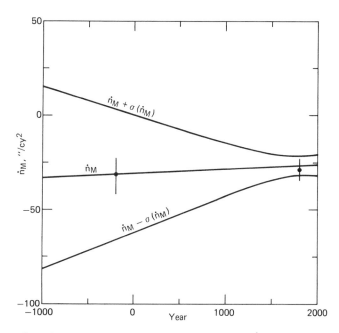

Figure XIV.4. The time dependence of the acceleration $\dot{n}_M$ of the moon with respect to ephemeris time. The straight line is the best estimate we can form by using only observations of solar eclipses. The curves represent $\dot{n}_M$ plus or minus one standard deviation, as inferred from the solar eclipses alone. The circle and associated error bar shown for the year 1800 represent the best estimate we can make from modern lunar observations. To find the circle and error bar shown at the epoch −200, we combine the estimate of D″ from the eclipse observations with the estimate of ν_S' from the equinox observations of Hipparchus. There is no observable variation of $\dot{n}_M$ with time.

The acceleration $\dot{n}_M$ and its "1σ" limits $\dot{n}_M + \sigma(\dot{n}_M)$ and $\dot{n}_M - \sigma(\dot{n}_M)$ are plotted as functions of time in Figure XIV.4. $\dot{n}_M$ has a positive slope, but the linear term 0.222τ is only about an eighth of the component 1.706τ of the standard deviation and thus it has no significance.

The curves in Figure XIV.4 show us that we can tell almost nothing about $\dot{n}_M$ from the untimed observations of solar eclipses. However, we can find reliable values of $\dot{n}_M$ at two different epochs. We have the value -28.38 ± 5.72 at the epoch 1800 from the analysis of the modern data; this value is plotted in the figure. Further, we can make a reliable estimate of ν_S' from the equinox observations made by Hipparchus. This estimate is [APO, p. 397]:

$$\nu_S' = 2.98 \pm 0.71 \quad ''/cy^2 \qquad (XIV.10)$$

at the epoch -140, approximately. This corresponds to $\tau = -19.4$, which I shall round to -20 for simplicity.

The value of D'' at $\tau = -20$ ($C = -8$) from Equation XIV.3 is 5.29 ± 1.91. When we combine this with the value from Equation XIV.10 by using Equation XIV.8, the result is -31.57 ± 8.98 for $\dot{n}_M$ and its standard deviation. This value is plotted at the epoch -200 in Figure XIV.4. The values plotted at -200 and $+1800$ show that a sizeable variation of $\dot{n}_M$ with time is possible but not probable.

Martin and van Flandern [1970] have estimated that the lunar nodal rate needs a correction of $+4.31 \pm 0.37$ $''/cy$. If we use this, we eliminate almost all of the small time derivative of $\dot{n}_M$ that is indicated in Figure XIV.4. We found in the preceding section that adding 8.2 $''/cy$ to the nodal rate $\dot{\Omega}$ adds 0.046τ to the value of ν_S' that we infer from the eclipse data.[†] Thus adding 4.31 to $\dot{\Omega}$ adds 0.0242τ to ν_S', and this, by Equation XIV.8, adds -0.299τ to $\dot{n}_M$. This changes the linear term in Equation XIV.9 from $+0.222\tau$ to -0.077τ.

To find the point that is plotted at the epoch -200 in Figure XIV.4, we combine the value of D'' from the solar eclipses with the value of ν_S' from the equinox observations (Equation XIV.10). Neither value is affected by introducing a perturbation in the lunar node. Thus, if we accept the nodal perturbation found by Martin and van Flandern, the point at -200 will no longer show the exact agreement with the straight line that it shows in the figure. However, the difference will be less than the standard deviation associated with the point, and thus it will not be significant.

All things considered, the data indicate no appreciable change in $\dot{n}_M$ within historic times, whether we use the nodal correction of Martin and van Flandern or not. If we assume that $\dot{n}_M$ has been a constant, the best estimate of the constant

---

[†] Changing the nodal rate does not affect the estimate of ν_S' in Equation XIV.10, which comes only from solar observations.

we can find is close to the best estimate that we have found
for the epoch 1800. Further, this estimate is found by using data that are not affected to first order by a nodal
perturbation. In summary, our tentative conclusion is that
$\dot{n}_M$ has been constant in historic times and equal to†

$$\dot{n}_M = -28.38 \pm 5.72 \quad ''/cy^2. \tag{XIV.11}$$

This conclusion may be subject to correction when we combine
the observations of solar eclipses with other kinds of observation that involve the moon. It is not affected by any reasonable change in the rate of precession $\dot{\Omega}$ of the lunar node.

5. The Standard Deviations of the Magnitude

In Section IV.1 I established two main categories of
record for Chinese records of solar eclipses and three main
categories for records from other provenances. For the
Chinese records, we have those which assert that the eclipse
was central (jih) and what we may call a general Chinese record. For the other provenances, we have records which state
explicitly that the eclipse was total, those which state
that stars could be seen without asserting totality, and
what we may call a general record.

I calculated standard deviations of the magnitude for
each of these five categories in Section IV.1 by direct inspection of the magnitudes calculated for each eclipse using
provisional values of the astronomical parameters. These
values, given now to three decimal places, appear in the
second column of Table XIV.2. I have repeated the calculations using the final values of the parameters. Specifically,
I use the linear form for D'' from Equation XIV.3 and the form
for ν_S' from Equation XIV.7, and I take the nodal perturbation to be zero. The resulting values of the standard deviation of the magnitude appear in the last column of Table
XIV.2.

As they should, all estimates of the standard deviation
decrease in going from the provisional to the final values
of the parameters, with the exception of the general Chinese
records. That standard deviation rises slightly, but not by
an amount that is important for such a small sample.

The biggest change is for the Chinese records that report the eclipse as central. For the records from other
provenances, the changes in the standard deviation occur
only in the third decimal place, and I give the deviations
to three decimal places in order to demonstrate that the

---

†I have used only the estimate from Section I.4 in writing
Equation XIV.11, and I have not used the value -26 ± 2 obtained by Morrison and Ward [1975]. The reason is that
Morrison and Ward have used some assumptions that are plausible but that are not used in Section I.4. I have preferred
to use only the less restrictive analysis of Section I.4,
even though it is based upon a smaller body of data.

TABLE XIV.2

STANDARD DEVIATION OF THE MAGNITUDE

Class of Record	Standard deviation of the magnitude	
	With provisional parameters	With fitted parameters
Chinese eclipses reported as central	0.071	0.058
General Chinese records	0.360	0.366
Eclipses reported as total	0.034	0.030
Eclipses in which stars were seen	0.054	0.051
General records other than Chinese	0.179	0.177

fitting did improve the standard deviation. The change does not look large, but it is important when we remember that there are altogether 631 records used in this study.

The changes in the standard deviations are not large enough to require another inference of the parameters. I shall continue to use the previously quoted parameters and the associated error estimates in future studies. However, when it is necessary to quote values for the standard deviation of the magnitude, I shall quote the values from the last column of Table XIV.2.

There are two categories of eclipse records that do not appear in Table XIV.2. One category contains the records which make it clear that an eclipse was partial but which give no indication of the magnitude. In Section IV.1 I found that the average magnitude for such records was 0.81, with a standard deviation of 0.14. Since Table XIV.2 shows that provisional estimates of the magnitudes were quite accurate, it does not seem necessary to repeat the calculations for this category.

The final category contains the records which make it clear that an eclipse was partial and which give an indication of the magnitude. The records in this class which give a quantitative statement about the magnitude were discussed in a preliminary manner in Section IV.1 and I shall complete the discussion in the next section.

I shall close this section with a reference to the interesting record designated as 1239 June 3b E,F in Section VIII.3. The record says that at the height of the eclipse a crescent could still be seen that was red as fire. On this basis, I took the magnitude to be 0.995. However, calculation with the final parameters yields a magnitude of 1.003, and I believe that the uncertainty in this calculation is about 0.002. Thus we have a total eclipse that was reported as partial, in parallel with the record 1567 April 9

E,I, in which the observer Clavius reported a total eclipse as annular. The explanation is probably the same in both cases. Apparently the observer saw part of the chromosphere and thought it was part of the sun's disc.

6. Records That Specifically Estimate the Magnitude

Each record that specifically estimates the magnitude poses a special problem, and it is not possible to establish a meaningful standard deviation of the magnitude for this class taken as a whole. We can only study carefully what the recorder wrote and attempt to make a reasonable estimate of the accuracy of his statement about the magnitude.

TABLE XIV.3

RECORDS CONTAINING AN ESTIMATE OF THE MAGNITUDE

Designation	Estimated magnitude	Estimated standard deviation	Calculated magnitude
458 May 28 E,SP	0.80	0.10	0.76
464 Jul 20 E,SP	0.85	0.075	0.81
563 Oct 3 E,F	0.80	0.10	0.60
590 Oct 4 E,F	0.80	0.10	0.71
592 Mar 19 E,F	0.67	0.17	0.77
1018 Apr 18 E,G	0.50	0.20	0.41
1033 Jun 29a E,F	0.88	0.06	a
1037 Apr 18a E,F	0.94	0.05	0.77
1113 Mar 19 M,HL	0.80	0.10	0.85
1178 Sep 13c E,F	0.90	0.05	0.95
1178 Sep 13e E,F	0.90	0.05	0.98
1178 Sep 13i E,F	0.94	0.06	0.97
1178 Sep 13e E,I	0.90	0.05	0.92
1181 Jul 13 E,F	0.30	0.18	0.29
1187 Sep 4g E,G	0.75	0.125	0.86

[a]The magnitude is 0.958 if the observation was made in Beze and 0.988 if it was made in Cluny.

TABLE XIV.3 (Continued)

Designation	Estimated magnitude	Estimated standard deviation	Calculated magnitude
1191 Jun 23a E,F	0.94	0.05	0.89
1239 Jun 3e E,F	0.94	0.05	0.93
1290 Sep 5 E,I	0.67	0.20	0.79
1330 Jul 16a E,I	0.50	0.20	0.90
1330 Jul 16c E,I	0.50	0.20	0.90
1339 Jul 7 B,E	0.75	0.075	0.89
1384 Aug 17 E,I	0.71	0.10	0.79
1384 Aug 17 E,BN	0.67	0.20	0.73
1448 Aug 29a E,I	0.67	0.10	0.80
1465 Sep 20a E,I	0.75	0.10	0.71

The situation regarding the records which give a quantitative estimate of the magnitude is summarized in Table
XIV.3. The first column in the table gives the designation
of the eclipse, the next column gives the magnitude estimated
by the observer, the third gives the standard deviation that
I have attached to the observer's estimate, and the last column gives the magnitude that I calculate with the final values of the parameters.

As I noted in Section IV.1, $\underline{M}$ [p. 5.25] emphasizes
strongly that one cannot use these estimates because of irradiation in the eye. The irradiation makes the sun appear
larger than it really is, he writes, which is the same as
saying that it makes the magnitude appear too small. Thus,
if $\underline{M}$ is correct, the estimated magnitudes should be system-
atically too small.

A number of records, which are not specifically identified in Table XIV.3, state the estimate of the magnitude by
saying that the sun looked like the new moon a certain number
of days after the crescent first became visible. This may
seem like a crude procedure to us, but we must remember that
medieval people, particularly those who lived in a Church
establishment, were accustomed to using a lunar calendar,
and many of them followed the phases of the moon closely
because of the connection between the lunar calendar and the
date of Easter. Thus many medieval people were intimately
acquainted with the appearance of the moon at a certain age
of the crescent, and their statements about the appearance
of a crescent should be reasonably accurate.

$\underline{M}$ particularly objects to this procedure and writes
[p. 5.25]: "The vast majority of estimates in the form of
item 17† from the medieval records come from magnitudes in

---

†This is the way in which $\underline{M}$ refers to records that compare
the partly eclipsed sun to a crescent moon.

excess of .98!† This is understandable, as few eclipses of lesser magnitude (statistically speaking) will even be noticed."

I am not able to corroborate these statements. We see from Table XIV.3 that the record 1178 September 13e E,F is the only record for which the magnitude is as large as 0.98, unless the eclipse of 1033 June 29 was observed at Cluny rather than Beze; we do not know where this observation was made. Thus either 23 or 24 eclipses out of this sample of 25 had magnitudes less than 0.98, and the magnitudes range down to 0.29. I see no basis for saying that few eclipses with magnitudes less than 0.98 "will even be noticed." Further, I see no basis for saying that the "vast majority" of the records in question correspond to eclipses with a magnitude greater than 0.98.

The straightforward way to see if irradiation has affected the estimates of magnitude is to compare the average estimated magnitude with the average calculated magnitude. The average of the estimates is 0.753 and the average of the calculated values is 0.799. Thus we see that the estimates are in fact rather accurate. This almost surely means that the people who made the estimates used optical aids to block irradiation in most cases.

In summary, estimates of the magnitude contained in medieval records make a useful contribution to the inference of the astronomical accelerations, and the estimated magnitudes are not affected significantly by irradiation.

7. Some Matters Held for Further Discussion

At several points in the earlier chapters, I promised to review a certain topic when more information became available, and in one case I held the decision about using a record in abeyance pending further investigation. This is a useful place to take up some of these matters.

In Section V.3 I discussed the record -15 November 1 C, with a promise to return to the question of its magnitude. The record says explicitly that the eclipse was observed at Charng-an, and the magnitude calculated for Charng-an is only 0.056. I believe that this is the smallest magnitude for any eclipse which we have reason to believe was actually recorded in ancient or medieval times. I calculate that the maximum of the eclipse came about an hour before sunset, and this fact may have helped to see it.

In Section V.6 I promised to review the three eclipses from the Annals of Lu that were reported as central; all three eclipses were in fact total if they were central. The dates of the eclipses are -708 July 17, -600 September 20, and -548 June 19, and the magnitudes that I calculate for them at Kufow are 0.94, 0.93, and 1.02 respectively. I can

---

†This exclamation point and the parenthesis in the next sentence are M's.

find no set of accelerations that make all three eclipses total at the same place.

M and MS apparently found no trouble with the eclipse of -708 July 17, but they were unable to make the eclipse of -600 September 20 be total at Kufow. They thereupon stated, with no historical basis that I can find, that the observation of that date was made in Ying, a point perhaps 1000 kilometers southwest of Kufow.

I found in Section V.6, from a study of the historical situation, that Luoh-yang was the most likely place if the observation of -600 September 20 was not made at Kufow. In pursuit of this question, I have calculated the magnitude of the eclipse at Kufow, Luoh-yang, and Ying. The calculated magnitudes are 0.928 at Kufow, 0.998 at Luoh-yang, and 0.985 at Ying. Thus there is no basis, either astronomical or historical, for taking Ying to be the place. Luoh-yang is the most likely place if Kufow is not correct.

M and MS worked under the plausible assumption that a statement of centrality in a Chinese record was rather accurate. However, we have found that this is not so, and that the standard deviation of the magnitude for eclipses reported as central (and that were in fact total somewhere) is 0.058. The deviation for -600 September 20 if the observation was made at Kufow is only 0.07, and thus it is not large enough to cause concern. That is, there is no reason to doubt that the observation was made in Kufow.

M and MS also assume that an observation found in an astronomical treatise was made at the capital unless there is an explicit statement to the contrary. This leads them to conclude that the eclipse of 120 January 18 was almost but not quite total at Luoh-yang, the capital at the time. From a study of the text, I showed in Section IV.5 that the observation of near-totality was probably not made at Luoh-yang. I find that the magnitude at Luoh-yang was 0.996 when I use the best linear fit to D'' and that it was 1.013 when I use the best quadratic fit. Thus the eclipse was probably total at Luoh-yang, contrary to the conclusion of M and MS, but the observation does not provide a truly critical test of their conclusion.

I decided that we can conclude only that the eclipse of 120 January 18 was reported as total, meaning that the magnitude exceeded 0.942, somewhere in heartland China. This conclusion is correct, and no stronger one seems warranted.

The record 1033 June 29a E,F is one of those for which the observer estimated the magnitude by comparing the crescent sun with a crescent moon. I conclude from his comparison that the magnitude (Table XIV.3) was 0.88 ± 0.06. We have the further fact that we cannot tell from the record whether the observation was made in Beze or Cluny. I decided in Section VIII.3 to calculate the circumstances for both places, and that I would not use the record if the magnitude differed significantly at the two points.

I find that the magnitude is 0.958 at Beze and 0.988 at Cluny.† The difference is 0.03, which is just half of the assigned standard deviation. Thus I use this record. I treat it as two records, one from Beze and one from Cluny, and I divide the total weight assigned to the record equally between these two hypothetical records. The theoretical justification for this procedure is discussed in Section III.7.

The record 1310 January 31a E,F says that the eclipse (observed at S. Denis) occurred at 1 hour and 24 minutes after noon. The precision of the stated time suggests calculation, and I decided to compare the stated time with the time calculated from the final parameters. I find that the eclipse began 16^m after noon, that the middle came at 1^h 52^m after noon, and that the end came at 3^h 18^m after noon. Thus the chronicler seems to be referring to the middle time of the eclipse, but his time is not particularly accurate. If he were trying to make an accurate reading of the time, I believe he could have done better than this, and I suspect that the time is indeed one calculated from rather poor ephemrides. However, I see no reason to doubt the general reliability of the record.

I also noted that the words might be interpreted to mean that the observer was within the central zone of this annular eclipse, but I did not feel warranted in adopting this interpretation. Calculation puts S. Denis almost exactly on the southern edge of the central zone, so the suggested interpretation is possible. However, I still do not feel warranted in concluding that S. Denis was in the central zone.

In Section IX.2 I hesitated to use the source Saxonici [ca. 1271] because the place of observation for its eclipse records could not be localized more closely than the region of Saxony. In retrospect, I feel that I have been too conservative in demanding a small region of observation. The contribution which the uncertainty in place makes to σ_η for the records in Saxonici is far smaller than the contribution made by the uncertainty in the magnitude. Records from a region ten or more times the size of Saxony would in fact be useful.

The record 840 May 5d E,G from the town in Xanten in Germany says that stars could be seen, but the magnitude of the eclipse there was only 0.88. When I believed that a reference to stars implied a standard deviation of the magnitude of 0.01, this seemed like a serious error. However, we know now that a reference to stars only implies a standard deviation of 0.051, and the error is no longer a matter of concern.

In Section X.3, under the heading "1084 October 2 ?", I discussed a record of what seems to be a total eclipse. It

---

†The average magnitude is 0.973, which differs from the observer's estimate by about $1\frac{1}{2}$ standard deviations. This is not large enough to raise any serious question.

is dated 1084 February 6, and it purportedly came between
the 6th and 9th hours of the day, that is, between noon and
mid-afternoon. The record seems to be genuine, but the
dating has been thoroughly garbled. The two most plausible
dates are 1084 October 2 and 1086 February 16. The observation was made in Sicily. I find that the eclipse of 1084
October 2 at a point in the center of Sicily reached a magnitude of 0.95 at $3^h.3$ after noon, which agrees with the
record in neither time nor magnitude. The eclipse of 1086
February 16 reached totality, but again with the eclipse
center coming $3^h.3$ after noon. Thus the magnitude agrees
for 1086 February 16 but the hour does not. The eclipse
cannot be identified.

8.   The Accuracy of Calculating Ancient Eclipses, and
      Some Historical Matters

     Within the period for which we have data, that is from
about -700 to the present, we know the parameter D'' with an
accuracy of about 2 $''/cy^2$. When we go to dates from -700,
however, we are extrapolating beyond the data and the uncertainty in D'' necessarily increases.

     Equation XIV.3 gives the best linear fit of D'' to the
data, while Equation XIV.4 gives the best quadratic fit. The
eclipse data give no strong basis for choice between them,
and we may use either in extrapolating backward. The difference between them is then the uncertainty in evaluating
D'' at any epoch before -700, and this uncertainty limits the
accuracy with which we can calculate the circumstances of
ancient eclipses.

     In the year -1400, for example, when the variable C in
Equations XIV.3 and XIV.4 equals -20, the linear function
gives $D'' = 15.65$ and the quadratic function gives 7.70. The
difference is 7.95 $''/cy^2$. The effect of D'' is calculated
from the year 1800, and -1400 is 32 centuries before 1800.
Thus the uncertainty in the lunar elongation in -1400 is
$\frac{1}{2} \times 7.95 \times 32^2 = 4070''$. It takes the elongation about 2.2
hours to change this much. An uncertainty of 2.2 hours in
the time of an eclipse may be the difference between a total
eclipse and one that is not even visible.

     Just this happens with the eclipse of -1374 May 3,
which we studied in Section III.4. We remember that Stephenson [1970] assumed that he could calculate the time of an
eclipse at this period of history with an accuracy of 5 minutes in time and 0.01 in the magnitude. Using this assumption, he assigned the date of a solar eclipse record from
Ugarit[†] as -1374 May 3, and he found that the magnitude of
the eclipse there was 0.99. M [p. 13.2] finds that the
eclipse was total there, and writes: "It can be said with
high confidence that the -1374 date and event is correct,
thereby establishing the event as the earliest independently
verifiable Ugaritic date, ..."

_____

[†]Ugarit is at latitude 35°.62 and longitude 35°.78.

When we calculate the circumstances of the eclipse at
Ugarit using the quadratic formula, we find that the magnitude was 1.0006, meaning a total eclipse as M says. However,
when we use the linear formula, the maximum magnitude we
find is -0.73, meaning that the eclipse was not visible in
Ugarit. Thus we do not know whether the eclipse was visible
in Ugarit or not, and we have no idea of the magnitude if it
was visible. There is no basis for taking -1374 May 3 to
be the date of the record.

In the next section, we shall find a semi-empirical
basis for extrapolating D'' to years before -660, which is the
earliest year for which we have an astronomical estimate of
D''. This method of extrapolation gives values of D'' that are
about the same as those given by the quadratic formula for
years back to -1400, but the standard deviation in the estimate is at least 8.0. This is greater than the difference
between the values given by the linear and quadratic formulas.
Thus, if 8.0 is correct for the standard deviation, the probability is slightly greater than 0.5 that the eclipse of
-1374 May 3 was visible at Ugarit, but there is almost no probability that it was total there.

The sensitivity of the -1374 eclipse to a change in D''
is more than average, and matters are not so bad with some
other eclipses. For example, MS claim that a certain Chinese
inscription is a record of a solar eclipse and that the date
is -1329 June 14. M [pp. 13.2-13.3] says that the eclipse
was total in An-yang, the presumed place of observation, and
that the dating is established with a very high confidence.†

We saw in Section III.5 that the inscription in question
does not refer to a solar eclipse, and that we could not date
it if it did. Nonetheless, it is interesting to calculate
the circumstances of the eclipse at An-yang. When we use the
linear extrapolation formula, we find $D'' = 14.75$ and a magnitude of 0.56. When we use the quadratic formula, we find
$D'' = 7.78$ and a magnitude of 0.97. Further, An-Yang was
north of the eclipse path for both values of D'', which means
that the eclipse was not total for any case between these
two. Thus there is no basis for assigning the date -1329
June 14 to the passage.

In Section II.2 I discussed the claim of Sawyer [1972a]
that the Biblical reference to the sun's standing still was
actually a record of a total solar eclipse seen at Gibeon‡
and that the date was -1130 September 30. M [p. 8.14] takes
this claim to be a genuine astronomical observation and it
in fact plays an important part in his inference of the
parameters.‡ Again the passage is almost surely not a record
of an eclipse at all, and it could not be dated if it were.
The date has been assigned only on the basis of a highly optimistic assumption about the accuracy of calculating ancient
eclipses.

---

† An-yang is at latitude 36°.07 and longitude 114°.33.

‡ At latitude 31°.85 and longitude 35°.20.

‡ See Section IV.3.

In Section II.2 I pointed out that -1123 May 18, -1083 March 27, -1062 July 31, and -1040 November 23 have just as much justification as -1130 September 30. When I wrote that, I intended to calculate the circumstances of all these eclipses at Gibeon, but it turns out not to be necessary. When I calculate the circumstances at Gibeon on -1130 September 30, I find $D'' = 13.06$ and a magnitude of 0.74 using the linear formula for D''. Using the quadratic formula, I find $D'' = 7.88$ and a magnitude of 0.97. Gibeon was south of the eclipse path for both values of D'' and thus the eclipse was not total for any value of D'' between the two limits. Thus there is no basis for assigning the date -1130 September 30.

MS [p. 529] give -17.7 ± 1.0 for the parameter they write as $\dot{e} - 1.97\dot{n}$. This parameter is $-1.97D''$, so this corresponds to $D'' = +9.0 \pm 0.5$. M [p. 11.1] finds $D'' = 9.1 \pm 0.3$. Even when we use all the eclipse records available, the standard deviation of D'' is 2 or more. Thus MS are optimistic about their accuracy by a factor of at least 4 and M is optimistic by about an order of magnitude.

Let us look at the uncertainty in the magnitude of the -1374 May 3 eclipse using the error estimates of MS and M. Using 0.5 for the uncertainty in D'', the uncertainty in the magnitude is about 0.11. Using 0.3 for the uncertainty in D'', the uncertainty in the magnitude is about 0.064. Thus, even if we use the highly optimistic error estimates of MS or M, there is still no basis for assigning the date -1374 May 3 to the Ugaritic record.

It is interesting to calculate the circumstances of a few other eclipses in the class that I have called "wrong but romantic" [Newton, 1969]. The "eclipse of Plutarch" is a famous passage in Plutarch's writing that many people have tried to interpret as a record of a solar eclipse. I discussed this eclipse in Section VI.1 under the heading "71 March 20 ?". The most recent writing about this eclipse that I know of is by M [p. 13.6], who writes: "If Plutarch was at home in Chaeronia on March 20, 71AD, then he witnessed a total solar eclipse (weather permitting)." He also writes later on the same page that Plutarch's account is the first clear reference to the corona, and he takes this to be consistent with the totality of the eclipse.

In Oppolzer [1887], the path of the eclipse of 75 January 5 is almost identical with that of 71 March 20 in the neighborhood of Chaeronea. If this is correct, there is no way to choose between the two eclipses, and the eclipse cannot be dated. By calculation I find that the magnitudes of the eclipses at Chaeronea were 0.95 for 71 March 20 and 0.83 for 75 January 5. Thus Oppolzer's paths for these eclipses are not particularly accurate even relative to each other. However, we can be almost sure that neither eclipse was total at Chaeronea.

As we saw in Section VI.1, the known facts of Plutarch's life make it unlikely that he was in Chaeronea in either 71 or 75. If he saw either eclipse as total, it must have been somewhere else and, since we do not know where, we have no hope of identifying the eclipse even if we assume that Plutarch saw it as a total eclipse.

There is another interesting point. The maximum magnitude that either eclipse had anywhere is about 1.004. If an observer is anywhere in the zone of totality for an eclipse this small, he sees most of the bright chromosphere, and I do not believe that he can see the corona in this situation. Those who wish to maintain that Plutarch based his passage upon seeing either eclipse must give up the idea that he makes a reference to the corona. It is possible that this reference to a "kind of light" around the eclipsed sun is instead a reference to the chromosphere.†

In Section VI.1, I discussed two possible references to solar eclipses by the Roman historian Livy. M [pp. 8.28-8.30], apparently relying upon unpublished work by Stephenson, confidently dates the eclipses as -189 March 14 and -187 July 17. He does not use the first eclipse because he concludes from the text that the eclipse was only partial at Rome, but he concludes from the text that the second eclipse was total there and he admits it to his corpus of observations. M later verifies these magnitudes by using his final parameter values and writes [P. 13.5]: "Stephenson's confidence in Livy at this epoch is apparently well founded."

I show in Section VI.1 that neither date can be correct and further that the second passage by Livy probably does not refer to an eclipse at all. Nonetheless it is interesting to calculate the circumstances of the two eclipses in Rome. I find that the eclipse of -189 March 14 had a magnitude of 0.92, so that it was readily visible. The eclipse of -187 July 17 was not total by my calculations; its maximum magnitude was only 0.98. Thus it appears that M's and Stephenson's confidence in Livy is not justified after all.

There is a further consideration. The event in the second passage, whatever it was, happened between the 3rd and 4th hours of the day. I find that the eclipse of -187 July 17 ended at about 1¾ hours of the day, before the event in question happened. Thus we have independent grounds for concluding that the event was not the eclipse of -187 July 17.

Cicero [ca. -53, Section I.16] quotes a passage from an earlier writer that clearly describes an eclipse visible in Rome near the year -400. M [p. 8.25], taking an unpublished work by Stephenson as his basis, concludes that the eclipse was large or total just at sunset in Rome and thus he assigns the date -399 June 21. I showed in Section VI.1 that this conclusion is based upon an error in translation. The record, if genuine, describes an eclipse that was large or total‡

---

† Of course it is possible that the reference is to the corona. All I am saying is that, if the passage does refer to the corona, Plutarch based it upon something other than seeing the eclipse of either 71 March 20 or 75 January 5.

‡ It had to be large enough to justify comparing the darkness to night.

that could have come at any time of the day, provided that
maximum eclipse was well separated from either sunrise or
sunset. The only eclipse that meets these conditions is the
eclipse of -401 January 18.

Calculation confirms these conclusions that were reached
from a study of the text. The eclipse of -399 June 21 began
about 35 or 40 minutes before sunset and the magnitude, still
rising, reached 0.74 at sunset. Thus this eclipse does not
meet the conditions of the record. The sun rose eclipsed on
-401 January 18 at Rome, but with a magnitude of only 0.37,
and the eclipse reached totality about 50 minutes later. This
gives an appreciable interval of bright daylight between sunrise and totality and the eclipse satisfies the record in
this regard.

However, as I pointed out in Section VI.1, the record
says that the eclipse came in June, so this eclipse is not
right either. No eclipse satisfies the passage. This is
not surprising. The eclipse is magical and the passage is
probably a literary forgery committed perhaps two centuries
before the time of Cicero that Cicero accepted in good faith.

I discussed the so-called eclipse of Archilochus at
considerable length in Section VI.1. Most people who have
tried to use this eclipse have said that the eclipse of -647
April 6 was total on the island of Paros and have thence concluded that the poet was on Paros on that date. I calculate
that the eclipse of -647 April 6 had a magnitude of 1.004
there, and changing D'' by 2 $''cy^2$ changes the magnitude by
about 0.04. Thus the probability is only about 0.5 that the
eclipse of -647 April 6 was total on Paros. It seems clear
to me that attempts to locate Archilochus in place and time
by astronomical calculations of eclipses are hopeless, even
if we make the dubious assumption that he personally witnessed
a total eclipse.

The three synoptic Gospels in the Christian New Testament
all say that there was a great darkness beginning at noon and
lasting for 3 hours on the day that Jesus was crucified. Many
people have tried to rationalize this statement into a record
of a total solar eclipse. The eclipse of 29 November 24 is
the nearest eclipse in time that was total somewhere in the
generally right part of the world. Hence all people who have
tried to rationalize this account have taken this to be the
date, undeterred by the fact that this eclipse was on a Thursday in late autumn while the Crucifixion, if it were a real
event, was on a Friday in early spring that came on the day
after a full moon.

Most people who have used this eclipse have tried to
make it total in Jerusalem or Nicaea. Sawyer [1972b], noting
that the word "eclipse" occurs in one of the Gospels but not
in the others, concludes, by an argument that I personally
cannot accept, that the eclipse was total in Antioch.†

I find that the center of the eclipse path on 29 November 24 went almost exactly halfway between Jerusalem and

---

†See Section II.3.

Antioch and that it was total in neither place. I find a
magnitude of 0.96 in Jerusalem and 0.97 in Antioch. Thus
Sawyer's conclusion is not consistent with the astronomical
results.

By inadvertence I failed to calculate the circumstances
of the eclipse at Nicaea. I do not believe that the repair
of this omission is worth the trouble.

I conclude this section with comments on a few valid
historical records. Thucydides (see Section VI.1) says explicitly that the sun took on the form of a crescent on -430
August 3 and that some stars could be seen. I concluded
from the historical circumstances that the observation was
made somewhere between Athens and Thasos. The greatest magnitude in this region was 0.90 at Thasos, and it is doubtful
that stars (in the plural) could be seen at this magnitude.
The most likely explanation is that Thucydides combined an
observation made at Thasos with one coming from farther east
that reached him somehow.

The record 418 July 19a M,B [MCRE, p. 538] says that a
cone-shaped object could be seen during the eclipse; the observation was presumably made in Constantinople. All students
of the subject that I know of have taken the object to be a
comet. I find that the magnitude was only 0.94 at Constantinople. Since some comets can be seen in the daytime, it is
also possible that there was a comet that could not be seen
in ordinary daylight but that could be seen when about 94
per cent of the sun's light was removed by the eclipse. Since
the eclipse was total at many places in Greece and Asia Minor,
it is also possible that the report reached the capital from
elsewhere.

The record 360 August 28 M,B (Section XII.6 and MCRE,
[p. 537]) says that stars could be seen during the eclipse.
The writer was somewhere in the Roman province of Syria at
the time. However, the eclipse was only a small eclipse
seen at sunrise everywhere in Syria, and it is doubtful that
stars could be seen during the eclipse anywhere east of Iran.
This record remains a mystery.

The record 968 December 22a M,B gives what sounds like
a description of the corona, and both M [pp. 8.48-8.49] and
MS [pp. 483-484] take the eclipse to be total in Constantinople on this basis. In Section II.1 I expressed some doubts
about the accuracy of this assumption, and my doubts may be
justified. I find that the magnitude was only 1.002 at
Constantinople, and it is doubtful that the corona could be
seen during an eclipse that reaches only this magnitude.

The records 1133 August 2b B,E and 1140 March 20b B,E
furnish an interesting illustration of the human tendency
toward exaggeration. In both records, the writer William of
Malmesbury gives vivid eyewitness accounts of what are made
to sound like total eclipses. The accounts are quoted on
pages 161 and 163-164 of MCRE. I find that William was justified on 1140 March 20, when the magnitude reached 0.995,
but the magnitude on 1133 August 2 was only 0.939. Since
William's work was not written until 1142, he may have gained

the knowledge needed to write his exaggerated account of the 1133 eclipse from his experience in the 1140 eclipse.

The record 1153 January 26c E,G in Section IX.3 contains a drawing of the appearance of the sun at the time of the greatest eclipse that the observer saw. As I noted there, M [pp. 8.55-8.56] finds that the eclipse was annular and central at the place of observation, in disagreement with the record. He suggests that clouds filtered the sun by just the amount needed to see the sun clearly until the eclipse was nearly central, but that increasing cloud cover prevented seeing the central phase.

I find that the eclipse was indeed central, and it seems likely, as M says, that clouds prevented seeing the central phase. The rest of M's explanation does not follow, however. As I showed in Section XIV.6, there is strong reason to believe that medieval observers used aids that prevented irradiation and they did not need to rely upon thin cloud cover in order to follow the development of an eclipse.

Also in Section IX.3 I decided to use three German records of the eclipse of 1321 June 26. Two records say explicitly that the eclipse was total while the third says nothing about the magnitude. I find that the magnitudes for the first two records were 0.97 and 0.94, while the magnitude for the record that does not state totality was 0.97. This furnishes another illustration of the tendency toward exaggeration.

In the record 1567 April 9 E,I in Section X.3, the astronomer Clavius reported the eclipse as annular, although it was total if he were within the central zone. Because he reported the eclipse as annular, both MS [p. 487] and M [p. 8.67] conclude that he actually saw the eclipse as total. I pointed out in Section X.3 that this is not necessarily so. I calculate that the magnitude was 1.0002 where Clavius was, and the last figure is not significant. Thus the probability is only about 0.5 that the eclipse was actually total where Clavius saw it.

The observer for the record 1486 March 6 E,CE (Section XI.7) estimated the magnitude as 9 points. The observer for the record 1502 October 1 E,CE, who may or may not have been the same, estimated the magnitude of that eclipse as 10 points. I pointed out that the "point" was probably the same unit as the digit, which means a twelfth part, but I did not use this assumption. Calculation shows that the assumption is highly probable and that the estimates were rather accurate. If the point is the digit, the recorded magnitude for 1486 March 6 is 0.75 while the calculated magnitude (using the final values of the parameters) is 0.68. Likewise, the recorded magnitude for 1502 October 1 is 0.83 and the calculated magnitude is 0.83. I actually used 0.81 for both eclipses, and thus I did not make a serious error in making this choice.

This gives further evidence, if more evidence is needed, that medieval estimates of the magnitude were not seriously affected by irradiation. See the discussion in Section XIV.6.

9. The Source of the Accelerations

In this section, I shall discuss the accelerations with respect to ephemeris time, and what we can conclude from them about the force system that causes the accelerations.

The best estimate we can make of $\dot{n}_M$, the acceleration of the moon with respect to ephemeris time, is that it has been constant within historic times and equal (Equation XIV.11) to -28.38 ± 5.72 in the customary units. Within the accuracy that concerns us here, the only known source of a lunar acceleration is friction in the lunar tide. This friction also decelerates the earth's spin, and we can calculate the contribution of tidal friction to y, the acceleration parameter of the earth's spin with respect to ephemeris time, from the value of $\dot{n}_M$ by using the conservation of angular momentum. I shall return to this point in a moment.

We can identify the following force systems that contribute to the value of y:

a. The first contribution is the one from tidal friction that was just mentioned. Elsewhere [Newton, 1972c, Equation 3] I have shown that this contribution equals $1.165\dot{n}_M$. Hence the contribution of tidal friction to y, which I shall call y_{tidal}, is

$$y_{tidal} = -33.06 \pm 6.66. \tag{XIV.12}$$

b. The second contribution comes from magnetic coupling between the core and mantle. I have been speaking of the acceleration of the earth's spin, but it would be more accurate to call it the angular acceleration of the earth's crust, on which all the places of observation are located. Crustal motions are trivial in this context, so this is also the acceleration of the mantle.

The core may have a different angular velocity from the mantle. If so, there is likely to be an interaction between them which accelerates both, and I shall assume that the interaction arises from the magnetic field generated in the core. Specifically, I shall assume that there is a contribution y_{mag}, which I shall call the magnetic contribution, that is a function of the earth's dipole moment $\mathcal{M}$. I shall return in a moment to the problem of relating y_{mag} to $\mathcal{M}$.

c. A brief item in the Quarterly Journal of the Royal Astronomical Society[†] says that Professor Lyttleton of the Cambridge University Institute of Astronomy has been studying the effect of the accretion of the earth's core on the apparent accelerations of the sun and moon, meaning the accelerations that I call $\nu_S{}'$ and $\nu_M{}'$. The item says specifically:

[†]The item appears on page 267 of volume 18 (1977). I do not list this item in the list of references because I do not know who wrote it.

"Allowance for a decreasing moment of inertia of the Earth at the rate required by the phase-change hypothesis for the nature of the terrestrial core . . . points to a value for the lunar secular acceleration of about $-30''$ cy^{-2} for the non-dynamical part, . . ."

I shall use the symbol y_C to denote the contribution to y that comes from the accretion of the core. Since the accretion of the core is decreasing the moment of inertia of the earth, y_C must be positive. The "non-dynamical" part of the lunar acceleration means the term $-1.7373y$ in ν_M' in Equation I.26. Hence a contribution of -30 to ν_M' means a contribution of $-30/-1.7373$ to y, so that

$$y_C = +17.27. \qquad (XIV.13)$$

There is undoubtedly an uncertainty in this estimate of y_C, but the journal item does not say what it is.

d. Let G denote the gravitational constant and let $\dot{G}$ denote its rate of change in a time scale that is independent of gravitation. We have two such time scales. One is atomic time and the other is solar time. (Solar time is defined by the earth's spin and not by its orbital motion.)

Suppose that a body of unit mass is in orbit around a massive body of mass m with a mean motion n and an orbital radius r. Then

$$n^2 r^3 = mG.$$

If G changes with time, n and r also change, but the change takes place at constant angular momentum. The angular momentum P equals nr^2, or $r = \sqrt{(P/n)}$. If we substitute this value of r into the preceding equation and square the result, we find

$$nP^3 = m^2 G^2.$$

From this, if P and m remain constant,†

$$\dot{n}/n = 2\dot{G}/G. \qquad (XIV.14)$$

---

† I do not know who first obtained this result. Some theories give a coefficient in the right member of Equation XIV.14 that is different from 2, but I shall use 2 for the sake of definiteness. The interested reader may consult van Flandern [1976].

If we let n be the mean motion of the sun, or, rather, of the earth in its orbital motion, the resulting value of ṅ is the orbital angular acceleration of the earth, which is the same as an apparent acceleration of the sun. In other words, the value of ṅ calculated this way is a contribution to ν_S'. The corresponding contribution to y, which I shall denote by y_G, is found by using Equation I.27 for the relation between ν_S' and y. The result is

$$y_G = -1.994 \times 10^9 (\dot{G}/G). \qquad (XIV.15)$$

e. There are many other contributions to y which are discussed extensively by Munk and MacDonald [1960, Chapter 11]. The total of these does not seem to be within an order of magnitude of the observed value of y. The largest one that is usually mentioned comes from the torque exerted by the sun on the mass distribution in the atmospheric tide. This contribution to y amounts to about 2.0 in the usual computation. However, this computation neglects the fact that the atmospheric tide loads the earth and oceans, and there is little if any tidal mass distribution associated with the totality of earth, oceans, and atmosphere.† In other words, the contribution of the atmospheric tide to y is much less than 2.0. The contribution has not been calculated accurately, so far as I know.

I shall neglect all contributions to y except those enumerated in items a, b, c, and d above.

It is likely that Ġ has kept the same value within historic times, to the accuracy that concerns us here. It is also likely that y_C has been substantially constant within historic times. Further, the astronomical data indicate that $ṅ_M$ and hence y_{tidal} have been rather constant within historic times. This means that the change in y that has occurred within historic times must come almost entirely from y_{mag}, and this allows us to estimate the relation between y_{mag} and $\mathcal{M}$.

I start by using the estimate of D'' from Equation XIV.4. For a reason that will appear in a moment, I shall use the symbol $\overline{D}''$ for this quantity in the rest of this work. If we omit the uncertainties in the estimated parameters, we have

$$\overline{D}'' = 0.58 - 0.844C - 0.0244C^2.$$

In this, we remember, C denotes time in centuries from the epoch 600, and we next want to change the reference epoch to 1800, which is the standard reference epoch for the lunar

---

†This statement applies, of course, only to the total effect of the atmospheric tide.

and solar ephemerides. If we use τ to denote time in centuries from 1800, we have

$$\overline{D}'' = -13.06 - 1.4296\tau - 0.0244\tau^2. \qquad (XIV.16)$$

We remember from earlier discussion that this value of $\overline{D}''$ is the average value† between any time τ and the epoch when $\tau = 0$. The reason for this is that we took the total relative displacement of the moon (with respect to the sun) and equated it to $\frac{1}{2}\overline{D}''\tau^2$. In studying the force system, however, we need the value of an acceleration that applies at any instant rather than the average over an extended period of time. Henceforth I shall use the familiar symbol D'' only to denote the instantaneous value of the corresponding parameter.‡ It is clear that D'' defined in this way is the second derivative of $\frac{1}{2}\overline{D}''\tau^2$, since the latter quantity gives a displacement. Thus

$$D'' = -13.06 - 4.2888\tau - 0.1464\tau^2. \qquad (XIV.16a)$$

Since $D'' = \dot{n}_M - 1.6073y$, and since $\dot{n}_M = -28.38 \pm 5.72$ from Equation XIV.11,

$$y = -9.53 + 2.6683\tau + 0.09108\tau^2.‡ \qquad (XIV.17)$$

We must next ask in general terms about the nature of the interaction between the dipole moment $\mathcal{M}$ and the mantle (which will be used to include the crust in this discussion) which can give rise to the contribution y_{mag}. If the interaction is between $\mathcal{M}$ and material in the mantle that is permanently magnetized, y_{mag} is proportional to $\mathcal{M}$. On the other hand, if the magnetization in the mantle is induced, the magnetization itself is proportional to $\mathcal{M}$, and the interaction is proportional to $\mathcal{M}^2$.

As I understand current ideas about the mantle, most of the mantle is probably too hot to allow permanent magnetization. Only the cool upper part (including the crust), which

---

† This is the reason for putting a bar over the D in $\overline{D}''$. I shall use a similar notation for other average parameters.

‡ More precisely, I use this to mean the quantity that I have called the "running average" in other work [Newton, 1973a, Section 5]. This is an average over an interval that is long compared with the time scale of the fluctuations (Section I.4) but that is short compared with the scale of historic time.

‡ This value of y is also an instantaneous or "running average" value. When we need to use the average between any value of τ and the reference epoch ($\tau = 0$), I shall use the symbol $\overline{y}$ for it. The problem of averaging does not arise with $\dot{n}_M$, since it seems to be a constant. I shall also use y_{mag} (without a bar) to denote an instantaneous value.

is relatively small, can support permanent magnetization. Thus we expect that y_{mag} arises mostly from an interaction that is proportional to $\mathcal{m}^2$. However, for the sake of completeness, I shall test the consequences of assuming both that y_{mag} is proportional to $\mathcal{m}$ and that it is proportional to $\mathcal{m}^2$, and I shall then see if we can draw any conclusions from the results.

In both cases, we have $y = y_{tidal} + y_{mag} + y_C + y_G$. In both cases, we assume that $y_{tidal} + y_C + y_G$ is a constant which we can estimate. We have independent estimates of y_{tidal} and y_C, so we use the estimate of the constant part of y to yield an estimate of y_G and hence of $\dot{G}/G$. For both cases, we use the work of Smith [1967] to furnish the needed values of $\mathcal{m}$. Smith [p. 344] plots values of $\mathcal{m}$, and the associated standard deviations, at six epochs from about -1100 to 1960. The error in the value of $\mathcal{m}$ at the epoch 1960 is negligible compared with the errors at the other epochs, and I take the error to be zero for 1960 in the calculations.

Case I. In this case, we assume that y_{mag} is proportional to $\mathcal{m}$, so that y has the form

$$y = A + B\,\mathcal{m}.$$

When I evaluate A and B by the obvious statistical procedures, I find

$$y = 30.89 \pm 2.62 - (4.49 \pm 0.17)\mathcal{m}. \qquad \text{(XIV.18)}$$

In using this, we take the unit of $\mathcal{m}$ to be 10^{25} gauss-cm^3. The uncertainty in the constant term includes the effects of uncertainties in both the values of y and the values of $\mathcal{m}$.

The constant part of this equals $y_{tidal} + y_C + y_G$, and we have numerical estimates of y_{tidal} and y_C in Equations XIV.12 and XIV.13, respectively. When we use these values, we get

$$y_G = 46.7 \pm 8.0, \qquad 10^9 \times \dot{G}/G = -23.4 \pm 4.0. \qquad \text{(XIV.19)}$$

Case II. In this case we assume that y_{mag} is proportional to $\mathcal{m}^2$, so that we take y to have the form $A + B\,\mathcal{m}^2$. When I proceed as before, I find

$$y = 9.37 \pm 2.72 - (0.225 \pm 0.005)\mathcal{m}^2,$$

$$y_G = 25.2 \pm 7.2, \qquad \text{(XIV.20)}$$

$$10^9 \times \dot{G}/G = -12.6 \pm 3.6.$$

Since the value of $\dot{G}/G$ in Case II is only about half the value in Case I, we naturally ask whether we have any information about $\dot{G}/G$ that will help us to distinguish between the two cases. We first look to cosmology. Astronomical observations indicate that every cosmological distance in the universe is increasing at the same relative rate H_O, which is called the Hubble constant. The reciprocal of H_O gives us an estimate of the age of the universe that we may call the Hubble age.

In 1937, P. A. M. Dirac devised a set of fundamental units (mass, length, and time) derived from fundamental parameters such as the charge on the electron, Planck's constant, and so on.[†] In this set of units, it turns out that the product of G and the Hubble age is about 1. If this is a law and not an accident, G must be decreasing at a relative rate numerically equal to H_O. That is, $-\dot{G}/GH_O = 1$. Hence we should compare the values of $-\dot{G}/G$ from Equations XIV.19 and XIV.20 with H_O.

Although some observers claim to have measured H_O with considerable accuracy, many cosmologists do not seem to accept these results, and it is an open question [Tinsley, 1977] whether H_O is as small as 3 parts in 10^9 (3×10^{-9}) per century or as large as 12×10^{-9}. In discussions, most cosmologists whom I have read recently use $H_O = 10 \times 10^{-9}$, probably because it is a convenient number that lies in the possible range. The main uncertainty in H_O comes from the uncertainty in the distance scale of the universe, not in measuring the speed of recession of distant objects.

If H_O does lie in the range from 3 to 12×10^{-9}, the value of $-\dot{G}/GH_O$ ranges from 1.95 to 7.8 in Case I and from 1.05 to 4.2 in Case II. If we insist that $-\dot{G}/GH_O$ should be 1 or less, we can thus reject Case I as the dominant mechanism, although we could admit that some of the interaction producing y_{mag} comes from permanent magnetism.[‡]

The smallest value of $-\dot{G}/GH_O$ for Case II is 1.05, which is larger than 1. However, the uncertainty in the value 1.05 on the basis of Equations XIV.20 is about 0.3, so that the excess of 1.05 over 1 is not statistically significant.

[†] I refer the reader to the survey by van Flandern [1976]. I have not directly consulted the article by Dirac (Nature, 139, p. 323, 1937), which is called "The Cosmological Constants". Dirac's speculation about the change in G is sometimes called his law of large numbers.

[‡] Most cosmologists seem to insist in fact that $-\dot{G}/GH_O$ should be small compared with 1. So far as I am aware, there is no observational basis for this insistence; there is merely a "hunch". I can readily imagine a cosmology in which G is inversely proportional not to the radius of the universe, as Dirac's speculation has it, but to the volume of the universe. This would allow $-\dot{G}/GH_O$ to be as large as 3 and would thus allow Case I to be correct. Tentatively, however, I accept the insistence that $-\dot{G}/GH_O$ should be 1 or less.

Further, the statistical uncertainty quoted in Equations XIV.20 is too small, because I assumed in deriving those equations that the value of y_C is exact, although this is certainly not the case.

The choice between various theories of gravitation, including both those in which G is constant and those in which it varies, must ultimately rest upon observation. Thus we need to ask if there is an independent measurement of $\dot{G}$. The only one I know of that has been published is by van Flandern [1974]. van Flandern has analyzed a series of precise lunar occultations timed by means of cesium clocks from 1955 to 1974, and he has found that the acceleration of the moon with respect to atomic time is -83 ± 10 $''/cy^2$. The difference between this and the value of $\dot{n}_M$ (-28.4 ± 5.7) can be attributed to $\dot{G}/G$. This leads to $\dot{G}/G = (-16 \pm 3) \times 10^{-9}$,† which is consistent with either Case I (Equations XIV.19) or Case II (Equations XIV.20).

Recently, however, van Flandern (private communication) has increased the time span of his data from 19 to 22 years, and he now finds that the lunar acceleration is only -38 ± 5. This changes $\dot{G}/G$ to $(-2.9 \pm 1.4) \times 10^{-9}$, which is much smaller than the value from either Case I or Case II, but which is consistent with the insistence that $-\dot{G}/GH_O$ should be much smaller than 1.

However, if the estimate of the lunar acceleration changed from -83 ± 10 to -38 ± 5 as a consequence of adding only 3 years to the time span, I think that the standard deviation must be underestimated. Since we can be sure that the standard deviation of $\dot{G}/G$ in Equations XIV.20 has been underestimated, it is probably the situation that van Flandern's measurement is consistent with Case II, but it would be hard to make it consistent with Case I.

Thus we have concluded from three different arguments that Case II probably represents the correct situation, at least in its dominant features. The arguments are: (a) Case II is more plausible on the basis of what we think we know about the mantle, (b) Case II yields a value of $-\dot{G}/GH_O$ near to or less than 1 while Case I does not, and (c) Case II yields a value of $\dot{G}/G$ that is consistent with the only measurements of $\dot{G}/G$ that I know of, while Case I does not. Each of these arguments is weak by itself, but the sum of them supports the conclusion strongly enough that we may accept it on a tentative basis. However, we should certainly not take Case II as a matter that has been demonstrated strongly.

It is interesting to see how well the first of Equations XIV.20 represents m^2 when we take the value of y from Equation XIV.17. The values of m^2, using the values of Smith

---

†We find this value by letting $\dot{n}$ and n in Equation XIV.14 apply to the moon. That is, we take n = 1.732×10^9 $''/cy$.

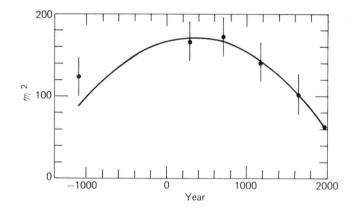

Figure XIV.5. The square of the reduced dipole moment of the earth plotted as a function of time. The reduced dipole moment $\mathcal{M}$ is expressed in units of 10^{25} gauss-cm^3, so that the value marked 200 on the vertical axis, for example, corresponds to $\mathcal{M}^2 = 200 \times 10^{50}$ gauss^2-cm^6, or to $\mathcal{M} = 14.14 \times 10^{25}$ gauss-cm^3. The solid circles show values of $\mathcal{M}^2$ at various epochs; the half-length of a vertical bar is the estimated standard deviation of a value. The standard deviation at the epoch 1960 is too small to plot. The curve is the function $(9.37 \cdot y)/0.225$, in which y is an estimate of the acceleration parameter of the earth's spin.

[1967] for $\mathcal{M}$, are plotted as the circles in Figure XIV.5. The half-length of the bar associated with each value is the estimated standard deviation of that value. The standard deviation of the value at the epoch 1960 is too small to plot on the scale used. The standard deviations of all earlier values are almost the same, and I have taken them to be the same for simplicity in drawing the figure. The earliest epoch shown in the figure is -1100, but the earliest epoch for which we have an estimate of y is -660. Hence I did not use the value of $\mathcal{M}^2$ at -1100 in deriving Equations XIV.20, nor did I use the value of $\mathcal{M}$ at -1100 in deriving Equation XIV.18.

The curve plotted in Figure XIV.5 is obtained by substituting the form of y from Equation XIV.17 into the first of Equations XIV.20 and solving for $\mathcal{M}^2$. The agreement between the curve and the points is excellent except for the value at -1100, which lies above the curve. This simply means that $\mathcal{M}^2$ has not followed a quadratic throughout the historical period, and there is no reason that it should.

We now have a semi-empirical basis for estimating the accelerations at epochs before -660. For times after -660, I recommend using Equation XIV.17 for y, and for earlier times I recommend using the first of Equations XIV.20. In order to estimate $\mathcal{M}^2$ at times before -660, I recommend using the straight line that connects the values of $\mathcal{M}^2$ for the years -1100 and +280. Doing this will make a slight discontinuity in y at the epoch -660, but this discontinuity is not large enough to be important. After finding y in this way, we must find $\overline{y}$ by the suitable averaging process. We can then find $\overline{D}''$ or ν_M' using $\dot{n}_M = -28.38 \pm 5.72$.

The findings of this section have an impact upon the historical review in the preceding section. We saw there that the uncertainty in the value of $\overline{D}''$ for years preceding -660 is quite large.† For example, it is so large that we cannot tell whether the eclipse of -1374 May 3 was even visible in Ugarit or not, much less whether it was total there. The use of m^2 to extrapolate for years before -660 gives us the soundest basis we have for an extrapolation, but it does not lead to a small uncertainty in the value of $\overline{D}''$. If we extrapolate back to, say, -1350 by the method of the preceding paragraph, we get $\overline{D}'' = +8.1 \pm 9.6$, so that we cannot even tell the sign of $\overline{D}''$ at an epoch this remote with assurance.

Altogether we have now made three estimates of $\overline{D}''$ at the epoch -1350, namely:

$$\text{best linear fit:} \qquad \overline{D}'' = +15.1,$$

$$\text{best quadratic fit:} \quad \overline{D}'' = + 7.8, \qquad\qquad \text{(XIV.21)}$$

$$\text{using } m^2: \qquad\qquad \overline{D}'' = + 8.1 \pm 9.6.$$

The value obtained by using m^2 agrees fairly well with the value from the best quadratic fit, but the standard deviation of this value is greater than the difference between the linear and quadratic fits.‡ It is certainly true that there is no present basis for identifying any solar eclipse record in the literature earlier than -719 February 22. Specifically, there is no basis for identifying the eclipse in the Ugaritic record as that of -1374 May 3, and we cannot even tell whether the eclipse of that date was visible or invisible at Ugarit.

10. A Final Summary

In this work, I have used mainly qualitative observations of solar eclipses made in ancient and medieval times, coming down approximately to the invention of the telescope. In referring to qualitative observations, I mean observations in which an observer at a known place on a known date

---

† I remind the reader that I used D'' in earlier sections for the quantity that I now denote by $\overline{D}''$.

‡ If our main interest had been in finding an estimate of $\overline{D}''$ that had the smallest standard deviation, we could have found D'' from $\overline{D}''$ (Equation XIV.16a) and then fitted this form of D'' directly to m^2. However, the largest single contribution to the standard deviation of $\overline{D}''$ comes from the standard deviation of m^2, and I estimate that the standard deviation of $\overline{D}''$ would still be about 8 if we had done this.

says that he saw an eclipse of the sun. If he gives information which indicates that the eclipse was total or very large, or if he makes a remark such as saying 2/3 of the sun was eclipsed, I have used this information, but I have not used careful quantitative measurements of the magnitudes or times of solar eclipses made by trained astronomers. Further, I have not used other types of ancient or medieval observations that involve the moon, but I have used some ancient solar observations. I reserve observations not used here for later work.

I have also made a new analysis of the data used by Spencer Jones, avoiding some assumptions that he made and also avoiding a serious theoretical error in his analysis. The result is to find that $\dot{n}_M$, the acceleration of the moon with respect to ephemeris time, is -28.4 ± 5.7 rather than the value -22.44 that is currently used in calculating the ephemeris of the moon.

The main differences between this work and other works that deal with the acceleration of the moon are three: (a) I now have many more eclipse observations, 631 to be specific. (b) I have eliminated the effects of Ptolemy's fraudulent observations. (c) I have eliminated the effects of an error in the theories from which the ephemerides of the sun and moon are currently calculated.

The results show that D'', the second derivative of the lunar elongation with respect to solar time, is a strong function of time but that $\dot{n}_M$ is constant. Hence y, the acceleration parameter of the earth's spin with respect to ephemeris time, is a strong function of time[†] that can be calculated from the results for D'' and $\dot{n}_M$.

It is likely that there are only four effects that make large contributions to y. They are tidal friction, core-mantle coupling acting through the earth's magnetic field, the accretion of the core, and a time dependence of the gravitational constant G. The only effect that seems to be time dependent is the coupling through the magnetic field. We are able to conclude tentatively that this coupling is proportional to the square of the earth's magnetic dipole moment.

Since we can estimate the contributions of tidal friction, the magnetic field, and of the accretion of the core, we can estimate the contribution made by a changing G and thence we can estimate the value of $\dot{G}/G$. Our estimate is that $\dot{G}/G$ equals -12.6 ± 3.6 parts in 10^9 per century.[‡] Most cosmologists seem to prefer a value of $\dot{G}/G$ that is substantially less than 10 in magnitude, but the real uncertainty

---

[†] I am using D'' and y to denote the quantities that I called "instantaneous values" or "running averages" in the preceding section.

[‡] A correct estimate of the uncertainty is larger than the number quoted, because I have not, for lack of information, included an uncertainty in the contribution to y from core accretion.

in the value found is large enough to make the value tentatively consistent with this preference.

Our lack of knowledge about many geophysical mechanisms is so large that we probably cannot make an important contribution to the study of $\dot{G}$ by using the earth's spin acceleration. The best methods of estimating $\dot{G}$ are probably van Flandern's method and the method of radar ranging of the planets. Within two decades, say, it is highly likely that these methods will tell us the value of $\dot{G}$ within rather narrow limits.

We have also found from the equinox observations made by Hipparchus about −140 that $\bar{\nu}_S'$ at that time was 2.98 ± 0.71, and modern data give $\bar{\nu}_S' = 1.18 \pm 0.36$ at the epoch 1800. The corresponding values of $\bar{y}$ are −22.9 ± 5.5 at −140 and −9.1 ± 2.8 at 1800. These values are consistent with the value of $\dot{n}_M$ and the time dependence found for D'' or $\bar{D}''$.

The results found in this work are independent of a possible correction to the rate of precession of the lunar node. In fact, as I have shown, it is not possible to make a significant estimate of this correction by using ancient and medieval data on solar eclipses, at least not at the level of accuracy currently available from such data.

It is gratifying to find values of the accelerations of the sun and moon that are consistent with each other and with modern data, and to find a plausible explanation of the time dependence of y, the acceleration of the earth's spin. It is also gratifying that we can make an estimate of $\dot{G}/G$ that, although rather crude, is consistent with current cosmological thinking and with an independent measurement. However, we should not put too much stress on these findings. Many students of the accelerations have found gratifying consistency in their results, only to have this gratification destroyed by the addition of more data.

The next stage in the study of the astronomical accelerations is to combine the qualitative observations of solar eclipses with other kinds of ancient and medieval astronomical observations. This will be the subject of the later volume or volumes of this work.

Note added in proof: In a recent paper, J. G. Williams, W. S. Sinclair, and C. F. Yoder (Tidal acceleration of the moon, Geophysical Research Letters, 5, pp. 943–946, November, 1978) use radar ranging of the moon to find the acceleration of the moon with respect to atomic time. They find −23.8 ± 4.0 $''/\text{cy}^2$, while I find −28.38 ± 5.72 in Equation XIV.11 for the acceleration with respect to ephemeris time. These results, if confirmed, deny the possibility that G is decreasing, in contradiction to the result I find here.

APPENDIX I

ROMAN CALENDARS

Many modern sources such as encyclopedias, or the
Explanatory Supplement [1961, p. 410], give brief discussions
of the Roman calendar before the calendar reform introduced
by Julius Caesar. Actually, there were at least three Roman
calendars that were used before the time of Caesar. Statements found in the sources mentioned are generally true for
one of these calendars, but statements that apply to different calendars are often presented as if they all applied to
the same calendar. Thus these sources may give the reader
an incorrect impression of the situation.

So far as I know, the most recent extensive discussion
(written in English) of the Roman calendar before Caesar is
by Michels [1967]. Except for incidental remarks, it deals
only with the calendar that immediately preceded the Julian
calendar. Further, even for that calendar, Michels is not
primarily interested in the principles on which it was constructed, and its relation to astronomical motions. Instead,
most of her attention is given to the special days in the
calendar, such as the days on which private law suits could
be initiated in the courts or the days on which various officials could be elected.

I know of only two sources [Ideler, 1826 and Mommsen,
1859] that give an adequate discussion of all the Roman calendars we know of before Caesar. Both these sources are now
rather old and inaccessible, and neither is in English. Accordingly, it seems useful to summarize in this appendix
what I have learned about the early Roman calendars in a
rather limited study, and to see whether this knowledge helps
us to judge some of the eclipse records in Chapter VI that
come from Roman sources.

Ideler seems to me to be both more inclusive and easier
to read than Mommsen. Therefore I shall take most of the
explicit citations from Ideler, citing Mommsen only where he
expresses a different point of view from Ideler.

The earliest Roman calendar of which we have any knowledge is sometimes called the calendar of Romulus.† We can
be rather certain that the first month in this calendar was

---

†According to tradition, Romulus founded Rome in -752. This
does not mean the settling of a group of people there. It
means instead what we may call the formal incorporation of
a well-settled area into a city under legal procedures of
long standing. Archaeological exploration indicates that
the traditional date is fairly accurate. We should not
take the association of the calendar with Romulus too literally, although Romulus may well have some historical
reality.

March and that the months then followed their present order through December,[†] but this is the only thing that we can say with considerable certainty about the calendar of Romulus. We do not know with certainty whether it had only 10 months or whether it had the full 12 months.

Ideler [p. 29] cites several ancient Roman customs which were based upon a fundamental period of 10 months. These include the duration of mourning and the period of credit granted to the purchaser of crops. Since the mourning period particularly is one that rests heavily on tradition, this indicates strongly that a period of 10 months was a dominant period in Roman timing at some point in their history, and hence that their earliest calendar had 10 months. Thus it is probable but not certain that the calendar of Romulus had 10 months.

If we grant this, there is still an important question. Did the 10-month year have only about 300 days, or did it have about 365 days? I cannot review all the discussion on this point, and I shall merely state the conclusion that most students of the subject now seem to accept: The year was a normal year, containing 365 days, more or less. It contained 10 months that were named, followed by a "dead time" of about 60 days or so in which there was no official calendar. Of course the early Romans must have had some way of keeping track of the days during the dead period, but we do not know what it was.

There is also the question of whether the months were conventional months, as ours are, or whether they were lunar months. I shall leave this question aside until we have described the second Roman calendar.

The second Roman calendar is the one that Ideler calls the calendar of Numa. According to tradition, Romulus was the first king of Rome and Numa was the second, out of seven kings altogether. There is no particular reason to doubt that there were about seven kings, but all the writing about them belongs to a much later stage in Roman history, and we must be cautious in what we accept about them.

There are two important features that must be noted about the calendar of Numa. The first is that it definitely

---

[†]As the reader probably knows, the names September through December simply mean 7th month through 10th month; these are the numbers that they have in a calendar that begins in March. Through the time of Julius Caesar, the two preceding months also lacked individual names and were simply called 5th month and 6th month. Augustus caused "5th month" to be renamed July [Ideler, 1826, p. 134] in honor of Julius, because it was the month in which Julius was born. For the month that was named after himself, however, he did not choose his birth month, which was September. Instead he chose "6th month", which is now called August, because that was the month in which some of his most important victories had occurred.

contained 12 months with names. These are the same months
that we use today, but the year still began with March, so
that January and February were the 11th and 12th months,
rather than the 1st and 2nd as they are with us. According
to several old writers, March, May, July, and October had
31 days. All the other months had 29 days except February,
which, then and now, had 28. This adds up to a year of 355
days.

The second feature to be noted is that this originated
as a lunar calendar. According to ancient writers, a specified Roman priest had the responsibility to watch for the
crescent new moon at sunset. When he saw it, he informed
another priest, who called out on the next day that the new
month had begun, and he also called out the number of days
that would elapse before the nones of the new month.[†] Hence
the first day of the month was known as the calends, from
which the word calendar comes.

This feature of the calendar seems to conflict with the
statements that the months had fixed numbers of days. Before
I discuss this point, however, I wish to discuss some other
interesting features of the situation.

One interesting feature is that the word calends comes
from the verb meaning to call. This verb is apparently a
very old word in the Indo-European family of languages, for
it has essentially the same form in Greek, Latin, and many
Germanic languages. However, the Latin word mensis (month)
is not derived from the Latin word for the moon but comes
instead from the Greek. Thus it may be that the calendar of
Numa, including the use of calends for the first day of the
month, was obtained from the Greeks.

Another interesting feature concerns the division of
the month. Each month had three days with specific names.
One such day was the calends, which was the first day, as
we have just seen. Another was the ides, which came on the
15th day in the long months (March, May, July, and October)
and on the 13th day in the other months. Finally, the nones,
which the priest announced on the calends, was the 9th day
before the ides, in inclusive counting. Thus the nones came
on the 7th day in the long months and on the 5th day in the
others.

The name calends, as we have seen, comes from the fact
that it was the day on which the priest called the new moon.
The name nones comes from the fact that it was the 9th day
before the ides. We now wonder what was the meaning of the
name ides.

Ideler [p. 43ff] mentions several etymologies of the
word ides that were given by various ancient writers on the
subject. The most common one, and the one that appears in
my Latin dictionary, is that it has the same root as divide,
and hence that this day is called the ides because it divides

---

[†]See Michels [1967, p. 20] for more details.

the month. Whether this etymology is correct or not, the ides does divide the month, but it does so unevenly. We should remember that the full moon comes at the midpoint (within a few hours) between two conjunctions of the moon with the sun. However, if we reckon the monthly interval from one visible new moon to the next, the full moon comes about 2 days earlier than the midpoint of this interval. Since the ides has just this property, it seems almost certain that the ides means the time of the full moon, whatever its etymology may have been.

Thus the three days with names come from a time when the month was a lunar month. The calends was the day that followed the first visibility of the crescent new moon in the west at sunset. The nones then came at the first quarter and the ides came at the full moon, approximately. There is little doubt that the months in the calendar of Numa, which we might better call the calendar of early republican Rome even though Numa was a king, were lunar months.

Michels [1967, pp. 131ff] raises a minor objection at this point. She believes that the nones did not mark the first quarter during the period when the months were lunar. Part of her reason is that it is easier to judge when the quarter moon occurs (when the disk is bisected) than it is to judge the full moon [p. 13]. Her main reason, however, is that the interval from the first quarter to the full moon varies from about 6.5 days to slightly more than 8 days, "so that it would be impossible to assume that the day of the first quarter would always be the ninth day† before the Ides. The name Nonae could be used only when the lunar months had been abandoned and the Ides had been assigned to a fixed day in each month."

The purpose of the calling was to let the people know in advance when the nones (Nonae) and ides would occur, because many important events in Roman life were based upon the nones and ides. Once the beginning of the month was determined by direct observation, it was undesirable to determine any other day by observation; to do so would be to make it unknowable in advance. In other words, it was important to adopt conventional intervals from the calends to the nones and from there to the ides and not to let the intervals be found from observation. An interval of 8 days is a good conventional value to choose for the interval from the quarter to the full moon, and its use seems to me entirely consistent with the use of lunar months.

Let us now return to the calendar of Romulus for a moment. It is not likely that the calendar first used in Rome was altered seriously within a few years, as we would have to conclude if we assume that the first calendar was instituted by Romulus and the second by his successor Numa. Even if we assume no more than that the calendar of Numa was introduced early in the period of the Roman kings, it is still unlikely that the calendar used at the founding of the city

---

† This means the ninth day in inclusive counting, or an actual interval of 8 days.

was displaced within a short time. It may be that the calendar of Romulus was already being used by the people who became Romans before the formal establishment of the city.

If we knew that the names calends, nones, and ides were already used with the calendar of Romulus, this would be fairly good evidence, in my opinion, that the months in this calendar were lunar ones. However, so far as I know, all the testimony to this effect was written long after the fact, and it probably reflects oral tradition and not written documentation that has since been lost.

However, considerations about the rapidity and direction of calendrical change do seem to me to have some validity. It is much easier for people to realize that a month is a cycle with a fixed number of days than to realize that the solar year is such a cycle.† Hence, in most regions, the first calendar that any people uses is a lunar one.‡ As agricultural and commercial considerations become important in the life of a people, the solar year tends to be introduced. This may be done by using a mixed lunar-solar calendar or by going to a calendar that makes no attempt to incorporate a lunar period. Once this is done, there is no obvious incentive to go back to a lunar calendar, and we do not expect an urban people that has adopted a solar calendar to abandon it and go back to a lunar one.

Since the calendar of Numa was definitely lunar in origin, it is unlikely that the earlier calendar of Romulus was solar. However, people have done things with their calendars that seem strange to us and we cannot categorically deny the possibility that the Romans went from a solar calendar under the earliest kings to a lunar calendar under later ones.

To me, the most plausible conclusion is that both the calendars of Romulus and Numa were lunar, and that the only change between the two was the naming of the months that occur during the original "dead period" from December to March. I believe that this suggestion is consistent with everything we know about the early Roman calendar, and it is consistent with plausible changes and advances in calendrical technique.

I mentioned above that the idea of having fixed numbers of days in each month seems to conflict with the idea that

---

†I use the term solar year in order to avoid specifying whether I mean a tropical year or a sidereal year. For the moment, at least, a solar year means simply a year determined by the motion of the sun.

‡Specifically, all the early calendars used in Mediterranean regions and Babylonia/Assyria were lunar, so far as I know. In fact, all calendars that I know of were lunar ones at some period of their development, with the possible exception of those used in Central America, of which the Mayan calendar [Thompson, 1974] is the prototype. The Mayan calendar as known has no element based upon the moon. However, it is a very elaborate calendar that obviously has much unknown history behind it, and it may have had an early lunar phase.

the months were basically lunar. This conflict seems so
strong that Mommsen [pp. 14-15] concludes that the calendar
was not really lunar. He agrees that it was originally
lunar. However, as he points out, the discrepancy between
the average length of a calendar month and the synodic month
would soon make the calendar cease to have any relation to
the lunar phases unless the calendar were corrected occasionally by means of direct observation. The Greeks were
flexible enough to do this, but the Romans were too rigid
and too much bound by rules to make the necessary adjust-
ments, according to Mommsen.

There is no evidence to show that Mommsen's conclusion
is correct, so far as I know, and Mommsen cites no evidence
to support it. It seems to rest only upon a doubtful assumption about the Roman national character. Further, even
if the assumption were true, the conclusion would not follow.

Let us first consider the nature of the assumption,
which is that the Romans were rigid followers of rules. As
we shall see in a moment, the Romans developed a solar calendar long before the time of Julius Caesar. If they had in
fact been bound by rules, this calendar would have stayed
accurate and there would have been no need for Caesar's calendar reform.[†] Instead, they were bound by the rules so
little that the calendar was in error by 80 days when the
Julian calendar was introduced. Thus Mommsen's assumption
about the Roman character, like most such assumptions, seems
to be invalid.

Now let us consider the relation between the assumption
and Mommsen's conclusion. Here the basic question is what
in fact the rule might have been. For example, we might
have the following rule: At the beginning of the last month
in every third year, look for the crescent moon and adjust
the length of the last month, if necessary, to keep the calendar in phase with the moon. In this case, rigidly following the rule would preserve the agreement of the calendar
with the moon.

Michels [p. 123] expresses an opinion similar to Mommsen's, saying that a month of 31 days is impossible in a
lunar calendar. Whether this is correct or not depends upon
what we mean by a lunar calendar. If we mean a calendar in
which the length of every month is governed directly by the
motion of the moon, the statement is correct. If we mean a
calendar in which the calendar month equals the lunar month
when averaged over a reasonable period, it is not correct.
The ordinary year in the calendar in question had 355 days
while 12 lunar months are slightly more than $354\frac{1}{3}$ days.
Thus the calendar gave a good approximation to the lunar
month, and the discrepancy could easily be removed by ad-
justing the length of the intercalary month as needed.

[†] This statement requires a modification that I shall state
after I give the rules for this solar calendar. The modification does not affect the argument of this paragraph.

Ideler [p. 45] reaches a quite different conclusion
from Mommsen. He concludes that the months in the calendar
of Numa did not have fixed lengths, in spite of explicit
statements by various ancient writers which assign a specific length to each of them. Instead, he contends, the
beginning of each month was determined by the direct observation of the crescent moon, just as it was in the early
Babylonian and Jewish calendars.† Ideler's contention
would be valid if it were not for the matter of intercalation.

Twelve lunar months have a duration of slightly more
than 354 days plus 8 hours, while the total duration assigned
to the 12 months in the Roman calendar was 355 days; this
fact is strong evidence that the calendar had a lunar origin.
Now the Romans wanted to have a particular month come always
at about the same season. For example, there was a festival
called Palilia which had to come in the spring but which also
had to come on the 11th calends of May.‡ In other words,
the average length of the year had to equal the solar year.
In order for this to happen, most years had to have 12 months,
but about 1 year in 3 had to have 13 months. By varying the
length of the extra or intercalary month according to need,
the beginning of each month could be kept close to the appearance of the new moon even though all months had a specified duration except the intercalary month.

This does not prove that the months (except the intercalary one) had specified durations, of course; it only shows
that they could have had without losing the essential quality
of being lunar months. The question of whether the named
months had fixed numbers of days or whether each calends was
determined by the sighting of the new moon does not matter
for our purposes.

The idea that the beginning of each month was determined
by the actual sighting of the new moon (weather permitting)
seems to conflict with a requirement that was stated earlier:
The priest had to announce on the calends whether the coming
month would be long (nones on the 7th day) or short (nones on
the 5th day). If the length of the coming month was in fact
going to be determined by when the next new moon would be
seen, about 30 days from the current calends, how could he
know what its length would be? For that matter, since the
days after the ides were identified by counting backward
from the next calends, how could this be done if the next
calends was to come on the still unknown day when the new moon
would next be seen?

Ideler [p. 46] suggests an ingenious answer to this
question. The new moon can sometimes be seen within about

---

† Of course there must be some provision about what to do if
weather prevents the observation of the moon.

‡ This was April 21 in later times when April had 30 days.
It was on April 20 when April had 29 days. However, the
Romans would not consider that the date had changed between the two calendars. To them, the true designation
was the 11th calends.

24 hours of the true conjunction, although such occasions are rare, and sometimes the new moon is not seen until about three days after the conjunction.[†] Thus the size of the crescent varies widely from one new moon to another. Further, if the moon is particularly large when it is first seen, we can expect the interval to the following new moon to be shorter than the average. Thus, in Ideler's suggestion, the responsible priest judged the size of the new moon and announced the length of the coming month accordingly, as either 29 or 31 days, except for the intercalary month whose length could be adjusted if necessary. If he announced a month of 31 days, it would sometimes happen that the next new moon was seen before the end of the month. In this case, the priest would delay his calling until the announced month of 31 days had elapsed, and the next month would have only 29 days.

This differs from a suggestion that I made earlier in only one significant way. In my suggestion, the priest adjusted the length of the month only once every few years. In Ideler's suggestion, he could choose the length of every month if necessary.

The next stage in the Roman calendar is sometimes called the calendar of the decemvirs. There were two set of decemvirs, who ruled Rome in two successive years about the year -450. Their principal accomplishment was to write down the laws for the benefit of the plebs, so that the plebs could know what the law was. Before this happened, the law was preserved only by oral means and its knowledge was the property of the patricians only. I shall mention the alleged connection of the decemvirs with the calendar in a moment. Since the connection of the decemvirs with the calendar seems doubtful to me, I shall call this the last republican calendar rather than the calendar of the decemvirs.

In this calendar, the 12 months with names had the lengths that were enumerated a while ago. That is, March, May, July, and October had 31 days, February had 28 days, and the other months had 29 days. Some late Roman writers say that the early Romans considered even numbers to be unlucky and that they adopted this arrangement in order to avoid having a large number of months with 30 days. This collection of 12 months contains 355 days, and the number of days in a full year was increased from time to time by introducing an intercalary month.

If we decide that the months in the calendar of Numa had specified numbers of days, the only difference between the calendar of Numa and the last republican calendar lay in the treatment of the intercalary month. In the calendar of Numa, the intercalary month was used to keep both the average month and year equal to their astronomical counterparts, at least in principle. In the last republican calendar, only the average year was preserved and no attempt

---

[†]These statements assume perfect visibility. Bad visibility can of course delay the first sight of the new moon by an unknown amount.

was made to keep the average month equal to the astronomical month. In other words, the calendar became solar.†

This was accomplished by having an intercalary month every second year. Much of the literature on the subject says that the intercalary month was called Mercedonius. However it seems that this name was used only by late writers. Writers who actually lived under the last republican calendar seem to have called it simply the intercalary month.

The length of the intercalary month was chosen in about the most awkward way that one can imagine. Since the total length of the 12 named months was 355 days, adding 20 days in 2 years would bring the average length of the year up to 365 days. When it was realized that the year was $365\frac{1}{4}$ days, it would be necessary to add 41 days in 4 years. There are several ways to do this, but the simplest is probably to institute a cycle of 4 years. In any such cycle, years 1 and 3 would have only the 12 named months and hence a length of 355 days. Year 2 would have an intercalary month of 20 days and year 4 an intercalary month of 21 days.

The Romans did not do this. In the last republican calendar, years 1 and 3 in a cycle of 4 years did have only 355 days with no intercalary month. In year 2 the intercalary month had 22 days and in year 4 it had 23 days. This made the average year equal to $366\frac{1}{4}$ days, approximately 1 day too long. In 20-odd years, the excess almost equalled the length of an intercalary month. Consequently the Romans introduced a cycle of 24 years. Within such a cycle, all the years except years 20 and 24 followed the rule of alternating intercalary months of 22 and 23 days; in both these years, the intercalary month would have 23 days according to the ordinary rule. In year 20, however, the intercalary month had 22 days instead of 23, and in year 24 it was omitted entirely. Thus 24 days were dropped in 24 years, and this brought the average length of the year down to $365\frac{1}{4}$

---

† The naming of these various calendars is confusing, and I may be contributing to the confusion. Ideler believes that this solar calendar was introduced by the decemvirs and thus he calls it the calendar of the decemvirs. Because I doubt that the decemvirs introduced it, I prefer not to use this name and I call it the "last republican calendar" for want of a better name. Mommsen, on the other hand, calls this calendar the calendar of Numa. This name (the calendar of Numa) is the name that Ideler and I use for the preceding lunar calendar, which Mommsen calls the "oldest" Roman calendar.

I do not understand Mommsen's nomenclature. The calendar just mentioned cannot properly be called the oldest Roman calendar because it was preceded by the calendar of Romulus. With regard to the solar calendar that I have just brought up, Mommsen [p. 30] says that it was not introduced until two centuries or more after the time of Numa. Thus I do not see why he names it after Numa.

days, the correct amount to high accuracy.†

Michels [1967, p. 169] dissents from the idea that the
Romans used such a 24-year cycle. She says that Macrobius
is the only ancient writer who mentions it. Since Macrobius
wrote around the year +400, she believes that the 24-year
cycle was devised by a writer under the Empire who thought
that the last republican calendar had to have an average
year of 365¼ days and thought of this way to generate such
a year. Actually, according to Michels, republican Rome
never lived under a calendar with this 24-year cycle.

Further, Michels says [p. 17], the number of interca-
lated days in 4 years was always either 44 or 45; this,
added to 4 × 355 days, makes an average year of either 366
or 366¼ days, about 1 day too long. Thus, if the priesthood
had rigorously followed the rules of intercalation that she
believes to be correct, the calendar year would have accumulated error at the rate of about 1 day per year, which quickly amounts to a serious amount. Note that the date in the
calendar would be earlier than the correct date after a year
or so, until it had lapped a full year.

Whether Macrobius or Michels is right about the 24-year
cycle, it is clear that the Romans had a solar calendar long
before the time of Julius Caesar, and that the months in
this calendar were no longer intended to keep in step with
the moon. If Macrobius is right, the length of the year was
the same as the Julian year, at least in principle. I shall
explain why I say "in principle" later on. If Michels is
right, the calendar lagged behind the Julian calendar by almost 1 day per year, again at least in principle.

Before I explain why I write "in principle" I want to
take up the question of when the last republican calendar
was introduced. Because the rules of intercalation in this
calendar are not certain, and because the rules of the calendar may have been modified during the time the calendar
was being used, I shall change the question to read: When
did the Romans go from a lunar calendar to a solar calendar?

As a preliminary to this question, let me describe the
Athenian calendar briefly. It was a lunar calendar, or more
accurately a mixed lunar-solar calendar, using lunar months,
but with the year kept at the right length on the average
by the intercalation of a lunar month about 1 year in every
3. In -431, the astronomer Meton introduced a cycle of 19
years, based upon the fact that 19 solar years are almost
exactly equal to 235 lunar months. Before this, the Athenians
had a cycle of 8 years called the octaeteris,‡ based upon the
approximation that 8 solar years almost equal 99 lunar months.
The error in the octaeteris is about 1½ days in 8 years,
while the error in the cycle of Meton is only about 2 hours
in 19 years.

_____

† It is possible that a cycle of 20 years may have been used
at times. See Mommsen [p. 44].

‡ This is an Anglicization of the Greek for eight years.

Now we can turn to the question of when the Romans adopted a solar calendar. Ideler says that the last republican calendar was one based upon an 8-year cycle, and he says that we can determine the time when it was adopted from this fact, with high probability. The decemvirs, he says [p. 66], sent an embassy to Athens to study their laws. At this time, about -450, the Athenians were still using the octaeteris. Ideler considers it highly probable that the Romans learned the octaeteris from the Athenians at this time and hence that this is the time when they adopted the solar calendar. Mommsen [p. 30] concurs with this account. There are many things wrong with this conclusion, it seems to me.

First, there is no evidence that such an embassy was ever sent and it is improbable on the face of it. On the other hand, the Etruscans traded widely and came into frequent contact with the Greeks, there were extensive Greek colonies in southern Italy, and the Romans could have learned something about Greek calendars indirectly through either of these intermediaries. However, there is no way to date when such learning occurred, if it ever did.

Second, the octaeteris dealt with a mixed lunar-solar calendar, and with a way of adapting to solar years while keeping lunar months. However, one fundamental principle of the last republican calendar is that it made no attempt to preserve the lunar month. Hence the makers of this calendar had nothing relevant to learn from the Greek octaeteris.

Third, the last republican calendar was not based upon a cycle of 8 years, in spite of Ideler's statement that it was. If Macrobius is right about the rules of intercalation, it was based upon a cycle of 24 years. If Michels is right, the cycle was 4 years. In neither case did the calendar use a cycle of 8 years.

Thus the last republican calendar was not related to any known Greek calendar, and we cannot date its introduction by means of any presumed contacts between Greeks and Romans. We can only guess at the date of its introduction, but we can be guided in our guess by the state of Greek astronomy at various times.

Let us first suppose that Macrobius is right about the 24-year cycle with its average year of $365\frac{1}{4}$ days. We need to ask when knowledge of the year became this accurate. I have already mentioned the calendar of Meton, introduced in -431, in which 235 months were taken equal to 19 years. The average length of the year in this calendar, however, was $365\frac{1}{4}$ days plus 1/76 of a day. About a century later, in -329, the astronomer Callippus modified Meton's calendar by omitting 1 day in each 76 years, so that the average year became $365\frac{1}{4}$ days exactly. This is the earliest use of the "Julian" year I know of, and it is unlikely that the Romans, with their little interest in astronomy, adopted such a calendar before the Athenians did. In other words, if the last republican calendar had $365\frac{1}{4}$ days, it was probably not adopted before -329.

As we have already noted, Michels rejects the testimony of Macrobius about the length of the year. Her reason is that Macrobius wrote about four centuries after the event (that is, after the last use of the republican calendar), and his sources may have been inaccurate. In contrast, she accepts the testimony of Cicero. In the discussion under the heading "-399 June 21 ?" in Section VI.1, I quoted Cicero about an eclipse that happened around the year -400, according to Cicero's sources. In this passage, Cicero [ca. -53, Section I.16] quotes the poet Ennius, who wrote around the year -200, and who in turn quoted an unknown source presumably written around the year -400. Cicero also quotes another and totally unknown source in the same passage.

In this passage, Cicero refers to two eclipses of the sun. One was on the nones of June in a year around -400. Michels [p. 126] says that the solar calendar must have already been in use at this time, since an eclipse cannot come on the nones of any month if the month is lunar. Hence she concludes that the time of the decemvirs (around -450) is the most likely time when the solar calendar was introduced, just as Ideler and Mommsen do, but for a different reason.

It is clear that the passage from Cicero is no more reliable than the passage from Macrobius, and there is no reason to accept one and reject the other. Further, as I pointed out in Section VI.1, the actual source for this eclipse is probably a late forgery, although Cicero was unaware of this.

If we accept this passage as reliable, it leads us to an absurdity. The passage also refers to an eclipse on the nones of July in the time of Romulus, in fact on the day that Romulus died. We cannot choose to accept one eclipse in this passage because it agrees with some theory and reject another in the same passage because it disagrees. We must accept both or reject both, because both have equal reliability. Now there could not have been a solar eclipse on the nones of July if the months were lunar, so the calendar must have been a solar one in the time of Romulus, if Michels is justified in her argument. Hence we have an absurdity, because it is almost certain that the months were lunar in the time of Romulus, a view that even Michels [Chapter 7] accepts. Thus we cannot accept either eclipse account in Cicero's passage as reliable, and it gives us no information about the introduction of a solar calendar.

In Michels' view, the year in the last republican calendar had about 366 days.† Hence we cannot argue much from the length of the year. Further, since the Athenians never adopted a solar calendar until they came under the influence of Rome, we see that Rome anticipated Athens in this regard.

---

† I am not sure whether Michels thought that the length of the calendar year (when averaged over 4 years) was variable or whether she thinks that we do not have enough data to tell its length exactly.

Hence we cannot argue from the state of Greek astronomy at any time. All we can say is that several ancient authors indicate that the months were lunar in the time of the kings, and that they ceased to be lunar well before the end of the Republic. It is not likely that the change came before -431, when Meton devised his famous cycle, but we cannot say that this is impossible.

In principle, as I said a moment ago, the average year had $365\frac{1}{4}$ days if Macrobius is right and about 366 days if Michels is right. In either case, the calendar either changed very slowly with respect to an accurate calendar or the calendar date fell behind the date in an accurate calendar at a rate of about 1 day per year. I say "in principle" because neither possibility just mentioned actually happened. When we come to the time of Julius Caesar, we find that the calends of January in -45 actually fell on the date that we would write as -46 October 13. That is, the calendar was ahead by 80 days. An error of this size and in this direction means that several intercalary months had been omitted, and this implies that the appropriate officials had retained control over the process of intercalation. In other words, the rules of intercalation never acquired the force of law. Whether the officials in question omitted the intercalary months from ignorance, from carelessness, or from political motives is something that we cannot tell.

Various writers, including Michels [p. 102], have tried to use the eclipse described by Livy, the one that I discussed under the heading "-189 March 14 ?" in Section VI.1, in order to study the errors in the last republican calendar. As we saw there, Livy gives the date corresponding to July 11 for the eclipse, but the only eclipse visible in Rome in -189 came on March 14, a discrepancy of about four months. However, as I said in Section VI.1, the eclipse reports transmitted by Livy are quite unreliable, and it is not safe to base any conclusions on them. We simply have no idea what the calendar error was in -189, but an error of four months exceeds any other calendar error I know of and it is thus unlikely.

In fact, as I showed in Section VI.1, there is no reliable basis for saying that the eclipse in question happened in -189, even if the record is genuine. The eclipse could well be the one of -187 July 17. If this is so, the calendar error was quite small.

We can now turn to the Julian calendar. We noted a moment ago that the date called -45 January 1 in the calendar actually in use came on the date that would be called -46 October 13 in the Julian calendar, an error of 80 days. In order to correct this error, Caesar decreed that the year -45 should have an intercalary month of 23 days, inserted in the usual place, plus two extraordinary intercalary months totalling 67 days which were placed between November and December. The total of the intercalary months is 90 days, which were added to the 355 days in the twelve regular months to give a year of 445 days. This brought Caesar in a position to inaugurate the calendar year of $365\frac{1}{4}$ days in the following year, -44, which was decreed to be a leap year in the modern sense of having a single intercalary day.

It is interesting to note where the intercalation was made in the various Roman calendars. When intercalation was employed with the last republican calendar, the intercalary month was usually placed after February 23. Then when the intercalary month was over, the remaining days of February nominally followed. I say nominally because the Roman method of stating dates wiped out the distinction between the intercalary month and the last days of February. After the ides of the intercalary month, the Romans had no way to specify a date except by saying that it was so many days before the next calends. However, the next calends was that of March, so the last days of the intercalary month were described as being so many days before the calends of March. This means that the last day of the intercalary month was the 7th calends of March. The next day, which nominally belonged to February, was the 6th calends of March. Thus, in effect, in an intercalary year, February had 23 days and the intercalary month had either 27 or 28 days.

When Caesar replaced the intercalary month by the single intercalary day, he still kept it in the same place, between the 7th calends of March (February 23) and the 6th calends of March (February 24). It was called the bissextus before the calends of March, that is, the second 6th before the calends. From this name, presumably, comes the fact that leap years were known for a long time as bissextile years.†

We do not know whether the intercalary month in the lunar "calendar of Numa" came after February 23 or not. However, this is such an odd place to insert the intercalary month that it almost surely had a very ancient origin. In other words, the intercalary lunar month in the calendar of Numa probably came after February 23, just as its later counterparts did.

We are accustomed to saying that the Julian calendar was adopted beginning with the year -44, a leap year. If we mean by this that the Romans actually used the Julian calendar in and after -44, the statement is not correct. It seems that the directions for the new calendar were not understood and that the addition of a leap day was made irregularly or every 3rd year instead of every 4th year, until about -7 or -8. At this point, someone finally understood the situation, and in consequence Augustus had all leap days omitted until the year was back in synchronization. The next year in which a leap day was added was +8, and the Julian calendar was followed from then without a break until the reform of the Gregorian calendar in 1582. Thus +8 is the year in which the use of the Julian calendar actually began.

---

†There is evidence [Mommsen, p. 22] that the intercalary month or the intercalary day sometimes came after February 24 rather than February 23. In the case of the last republican calendar, a change in position affected the lengths that February and the intercalary month seems to have. In the Julian calendar, it affected only the matter of which 6th calends of March was the regular day and which was the intercalary day.

Most historians ignore the period of irregularity between -44 and +8 and correct dates to what they would have been in the Julian calendar.

Julius Caesar changed the lengths of the twelve months so that they add up to 365 days rather than 355 days in ordinary years and to 366 days in leap years. He did this by leaving unchanged the months that already had 31 days, namely March, May, July, and October, and also by leaving February unchanged at 28 days in ordinary years. This left 213 days to be allocated to the 7 months that originally had 29 days each. The smoothest way to do this was to assign 31 days to 3 of them and 30 days to the others. He distributed the ones with 31 days as evenly as possible, making them January, August, and December. This distribution of the days is still the one we use today.

However, Caesar left the positions of the nones and ides unchanged even for the months that were lengthened to 31 days. Thus three months that now have 31 days, namely January, August, and December, still have their nones on the 5th day and their ides on the 13th day.

Many modern sources say that Caesar originally assigned 31 days to the alternate months January, March, and so on through September and November, 29 days to February in ordinary years and 30 in leap years, and 30 days to the remaining 5 months. This would have made an easily remembered scheme. Augustus Caesar, however, they go on to say, wanted his month of August to have as many days as Julius's month of July. In order to do this, he gave one of February's days to August and then interchanged the lengths of the months after August so that 30 and 31 days would alternate. Thus, they say, we owe our awkward arrangement of the lengths of the months to the vanity of Augustus.

However, this is not supported by the ancient sources. The ancient sources that mention the matter give the same distribution of the days in each month that we still use. Augustus does not seem to have done anything about the lengths of the months. His only contribution to the calendar seems to have been the restoration of the scheme of leap years that had been devised under Julius.

We have mentioned that the Roman year originally began with March. The beginning of the year was changed to January 1 about -150 and apparently remained there during the remaining life of the republic and of the Western Empire. However, the successor states to the Roman empire did not necessarily keep it on January 1. During the medieval period, Christmas, March 1, March 25, and September 1, as well as January 1, were all used as the beginning of the year in various places and times. The reader of medieval history can never take it for granted that the year was considered to begin on January 1.

In Section VI.1 we have four eclipse records that come from Rome before the imperial period. For three of them, the day of the month is given, and the days are impossible for solar eclipses if the months are lunar months. These days are the nones of July (July 7) for a year around -725, the nones of June (June 5) for a year around -400, and the 5th ides of July (July 11) for a year around -190. It is ironic that we cannot say with certainty whether the calendar was lunar or solar for any of these years. In my opinion, however, the probability is high that the months were lunar for the first two of these and that they were not for the last one. Thus it is unlikely that the first two records are genuine. They are probably dates that were calculated in late republican times by someone who did not know what the calendar was at the times claimed for the eclipses, or who did not realize that an eclipse cannot happen on the nones of a lunar month.

NOTE: Since I wrote this appendix in final form, I have had considerable correspondence with Dr. Pierre Brind-'Amour of the Department of Classical Studies, University of Ottawa (Ottawa, Ontario, Canada K1N 6N5) about the Roman calendar and about eclipse records that occur in Latin literature. Dr. Brind'Amour is writing a lengthy treatise about the Roman calendar, which should be the most important work on the subject that has appeared in a long time. I should point out that he disagrees with what I have said here on some important points.

He agrees that there was a correction to the calendar made under Augustus Caesar, but he believes that the rest of the traditional story (that the correction was needed because the pontiffs did not understand the method of intercalation) is not true. He believes that there was an error made in synchronizing the calendar in -44 under Julius Caesar, and that Augustus corrected this error.

Dr. Brind'Amour further believes that the "records" which I discussed in Section VI.1 under the headings "-189 March 14?" and "-187 July 17 ?" are indeed records of the eclipses on the indicated dates. He further believes that he has evidence that the second of these occurred in October in the Roman calendar. Thus both records, in his opinion, confirm that the Roman calendar was in error by three to four months at this stage in history.

Finally, Dr. Brind'Amour advances evidence independent of these eclipse "records" which indicates that the calendar was in error by three months at this stage. For example, a certain event which is dated in March is described as occurring in the winter.

I feel that it is important to let the reader know Dr. Brind'Amour's views on these matters, but I should like to close by reviewing my own. I have no firm opinion about the reason why Augustus chose to rectify the calendar during his reign. Dr. Brind'Amour and I concur that Julius, with the astronomical expertise available to him, could have given the calendar any relation to the seasons that he wanted. Dr. Brind'Amour believes that he chose some particular relation, but that Augustus, some years later, chose a relation that differed by several days, and that he achieved this relation by omitting the prescribed leap years for a number of leap years in a row. In deciding whether Dr. Brind'Amour is correct, we need to know whether there is a "pre-Julian" Roman tradition that puts the vernal equinox, for example, on March 25, which is the date that it had in the "Julian" calendar times when Julius and Augustus exercised the supreme authority.

I still firmly believe that one should not use the eclipse "records" just mentioned in research, either astronomical or calendrical. I see no reason to believe that they are genuine records at all. However, if we do accept them as genuine historical records, I can think of three possible interpretations of them:

1. The first record, discussed under the conventional date of −189 March 14, really refers to the eclipse of −187 July 17, and the other record refers to a "dark day" that happened in an unknown year, but one that is close to −187. I regard this as by far the most likely interpretation, if we assume that the records are genuine.

2. Both records refer to the eclipse of −187 July 17. There is ample precedent in the literature for a compiler's finding two records of the same eclipse and believing that they referred to two different eclipses. I regard this as the next most likely interpretation.

3. The first record refers to the eclipse of −189 March 14 and the second record refers to the eclipse of −187 July 17. I regard this as the least likely interpretation.

Thus we have at least three mutually exclusive interpretations of the records, each with some strength of argument behind it. In this circumstance, the only conclusion we can reach, in my opinion, is that we can reach no conclusion.

THE 60-DAY CYCLE IN THE CHINESE CALENDAR

In addition to using a calendar date, the Chinese for many millenia assigned to each day a position within a cycle of 60 days. I have used the term "cycle number" to designate this position.

A cycle number is specified by writing a pair of characters. The first character, called the stem, is chosen in order from a set of 10 and the second, called the branch, is chosen in order from a set of 12. Since 60 is the least common multiple of 10 and 12, the pair of characters begins to repeat after the 60th pair. Since there are 120 possible pairs, only half of the pairs are ever used.

The system of numbering is shown in Table A.II.1. The first character in a pair is chosen in order from the set of stems that appears in the left hand column while the second character is chosen from the set of branches that appears in the row along the top. The numbering starts in the upper left hand corner and proceeds diagonally as shown until we reach 10. We must then start over in the set of 10 while we continue onward in the set of 12. This brings us to the 11th character in the top row for number 11, and so on. The table shows the pair of characters used for every number from 1 to 60, which we reach in the lower right hand corner. Since we must now start over in both sets, we go back to the upper left hand corner.

The cycle number is simply related to the Julian day number. Add 50 to the Julian day number and divide by 60. The remainder is the cycle number, with one additional provision: If the remainder is 0, the cycle number is called 60. The relation can be written formally as

$$C = 1 + (J + 49)(\text{mod } 60).$$

In this, C denotes the cycle number and J denotes the Julian day number.

For examples, we encountered the designations ping-shen and ting-yu in Section III.5. We see from Table A.II.1 that these are cycle numbers 33 and 34, respectively.

This system of representing the numbers from 1 to 60 could be and was used for numbering entities other than days, and the entities did not have to refer to time. With enough practice, a person might memorize Table A.II.1 well enough that he could immediately think of the number that goes with any stem-branch pair, so that the pairs would just form a way of "spelling out" the numbers from 1 to 60. Cohen [1976] gives a simplified way of calculating the number represented which does, however, require memorizing the order in which

NAMES OF THE DAYS IN THE 60-DAY CYCLE

	tzu	ch'ou	yin	mao	ch'en	szu	wu	wei	shen	yu	shu	hai
chia	1		51		41		31		21		11	
i		2		52		42		32		22		12
ping	13		3		53		43		33		23	
ting		14		4		54		44		34		24
wu	25		15		5		55		45		35	
chi		26		16		6		56		46		36
keng	37		27		17		7		57		47	
hsin		38		28		18		8		58		48
jen	49		39		29		19		9		59	
k'uei		50		40		30		20		10		60

the stems and branches appear in the table.†

Let a stem be represented by a digit from 1 to 10 and a branch by a digit from 1 to 12. Cohen's rule is as follows: Add to the branch a multiple of 12 chosen so that the sum ends in the digit that corresponds to the stem. This sum is then the cycle number.

For example, take the combination ting-yu, which we have already used. The digit for ting is 4 and the digit for yu is 10. In order to obtain a sum ending in 4, we must add 2 × 12 = 24 to 10, getting 34. This is the number represented by ting-yu.

---

†Cohen says that he learned the method from one of his teachers who apparently never published it. He also cites a more complex method published by George A. Kennedy in 1952, that I have not seen.

APPENDIX III

PLACES AND REGIONS OF OBSERVATIONS

For most eclipse observations that are used in this work, we may conclude with reasonable confidence that the eclipse was observed at a specific place. Sometimes we have an explicit statement in the record about the place of observation; more often we infer the place from the history of the source and the circumstances under which it was written.

For some records, we can only conclude that the observation was made somewhere within a known geographical region. When the region is fairly small, such as England, the uncertainty in knowing the exact place is unimportant, and I have used many such records. For others, the possible region might have been as large as the Roman empire, and I have not used records for which the region is this large. So far it has not been necessary to formulate an exact criterion for accepting or rejecting a record on the basis of the size of the region.

The regions can be divided into two classes. In the more numerous class, the region can be approximated with considerable accuracy by an area that forms a rectangle on a Mercator projection. Table A.III.1 gives the coordinates of the corners that I have used for such regions. Four regions do not appear in the table. Three of these are lines, and the coordinates that I have used for the ends of the lines are given in Table A.III.2. The remaining region is the one that I have called Italy,† which cannot be approximated well by a rectangle, but can be represented rather well by a more general quadrilateral. The corners of this quadrilateral are also listed in Table A.III.2.

Table A.III.3 gives the coordinates of the specific places of observation. It is desirable to make a few remarks about choices of spelling and about the way the coordinates were found.

When a place (usually a town or city) has a familiar spelling in English, I have generally used this spelling even if it does not agree with the official spelling. An example is the use of Munich rather than München. Otherwise I have commonly used the spelling given in the Times Atlas. If a name does not appear in the Times Atlas, I use the spelling found in the source where I encountered the place in question.

The coordinates are those found in the Times Atlas for all places listed there, with two exceptions. I believe that the coordinates given there for Salzburg, Austria and for Heiligenkreuz, Austria are in error, and I have used instead coordinates obtained by measurements on a large-

---

†As I explain in Section X.1, this really means the part of Italy that is farther northwest than Naples.

TABLE A.III.1

RECTANGULAR REGIONS OF OBSERVATION

Name	Latitude range, degrees	Longitude range, degrees
Anjou (France)	46.7N/48.3N	1.3W/0.3E
Armenia	40.4N/41.6N	42.8E/45.2E
Belgium	50.3N/51.3N	3.0E/5.5E
Burgundy	45.5N/47.0N	4.3E/7.2E
Campania (Italy)	40.5N/41.0N	14.0E/15.0E
China	32.91N/35.21N	109.94E/117.08E
Denmark	55.0N/56.5N	9.0E/12.0E
England	51.3N/54.5N	3.0W/0.0
Galicia (Spain)	41.5N/43.5N	6.5W/8.5W
Iceland	64.0N/66.0N	18.0W/22.0W
Ireland	52.0N/55.0N	9.0W/6.0W
Livonia	57.0N/59.0N	24.0E/26.0E
Man, Isle of	54.1N/54.3N	4.6W/4.4W
Saxony	50.5N/51.25N	12.5E/14.5E
Spain	37.0N/43.0N	2.0W/8.0W
Syria	34.0N/37.0N	36.0E/42.0E

TABLE A.III.2

OTHER REGIONS OF OBSERVATION

Name	Description[a]
Italy	Corners at 41.0N, 14.0E, at 42.0N, 15.0E, at 45.0N, 9.0E, and at 46.0N, 10.0E
Kerulen River (Mongolia)	48.7N, 116.25E to 47.5N, 112.00E
Sicily, east coast of	37.0N, 15.2E to 38.0N, 15.2E
Western Wales	51.9N, 4.5W to 53.1N, 4.5W

[a]Latitudes and longitudes are in degrees.

SPECIFIC PLACES OF OBSERVATION

Place	Latitude, degrees	Longitude, degrees
Aachen, Germany	50.77N	6.10E
Admont, Austria	47.58N	14.47E
Alcobaca, Portugal	39.53N	8.98W
Angers, France	47.48N	0.53W
Angoulême, France	45.67N	0.17E
Antioch, Turkey	36.20N	36.17E
Arras, France	50.28N	2.77E
Athens, Greece	38.00N	23.73E
Augsburg, Germany	48.35N	10.90E
Autun, France	46.97N	4.30E
Auxerre, France	47.80N	3.58E
Babylon	32.55N	44.42E
Bamberg, Germany	49.90N	10.90E
Basel, Switzerland	47.55N	7.60E
Beauvais, France	49.43N	2.08E
Benevento, Italy	41.13N	14.77E
Bergamo, Italy	45.70N	9.67E
Bergen, Norway	60.38N	5.33E
Beze, France	47.47N	5.27E
Bicester, England[a]	51.90N	1.15W
Bologna, Italy	44.50N	11.33E
Bourbourg, France	50.95N	2.20E
Brauweiler, Germany	49.80N	7.60E
Brechin, Scotland	56.73N	2.67W
Brunswick, Germany	52.25N	10.50E
Burton-on-Trent, England	52.82N	1.60W
Bury S. Edmunds, England	52.25N	0.72E
Cairo, Egypt	30.05N	31.25E
Cambrai, France	50.17N	3.23E
Canterbury, England	51.28N	1.08E

[a]Used in place of Osney, England. See the discussion of Wykes [ca. 1289] in Section VII.2.

Place	Latitude, degrees	Longitude, degrees
Chaeronea, Greece	38.35N	22.97E
Charng-an, China	34.25N	108.97E
Cluny, France	46.42N	4.65E
Coggeshall, England	51.87N	0.68E
Coimbra, Portugal	40.20N	8.42W
Colbaz, Poland	see Stargard, Poland	
Collegio Romana, Rome, Italy	41.90N	12.48E
Colmar, France	48.08N	7.35E
Cologne, Germany	50.93N	6.95E
Constantinople	41.03N	28.98E
Cordoba, Spain	37.88N	4.77W
Corvei, Germany	51.80N	9.50E
Cremona, Italy	45.13N	10.02E
Diessen, Germany	47.95N	11.10E
Dijon, France	47.33N	5.03E
Disibodenberg, Germany	49.78N	7.82E
Dore, England	51.97N	2.89W
Douai, France	50.37N	3.08E
Durham, England	54.78N	1.57W
Egmond aan Zee, Netherlands	52.62N	4.62E
Eichstätt, Germany	48.90N	11.22E
Eisenach, Germany	50.98N	10.32E
Engelberg, Switzerland	46.82N	8.42E
Ensdorf, Germany	49.45N	12.02E
Erfurt, Germany	50.97N	11.03E
Essenbek, Denmark	56.45N	10.13E
Faenza, Italy	44.28N	11.88E
Farfa, Italy	42.22N	12.73E
Farnham, Surrey, England[b]	51.22N	0.82W
Ferrara, Italy	44.83N	11.63E

[b]Used in place of Waverly, England. See the discussion of Waverley [ca. 1291] on page 138 of MCRE.

Place	Latitude, degrees	Longitude, degrees
Flavigny, France	47.53N	4.52E
Florence, Italy	43.78N	11.25E
Foligno, Italy	42.95N	12.72E
Forli, Italy	44.22N	12.03E
Fosse, Belgium	50.40N	4.70E
Frankfurt-am-Main, Germany	50.10N	8.68E
Freiberg, Germany	50.92N	13.35E
Freising, Germany	48.40N	11.75E
Friesach, Austria	46.95N	14.42E
Fulda, Germany	50.55N	9.68E
Gembloux, Belgium	50.57N	4.70E
Ghent, Belgium	51.03N	3.70E
Gottweih, Austria	48.37N	15.63E
Graz, Austria	47.08N	15.37E
Gubbio, Italy	43.35N	12.58E
Hamburg, Germany	53.55N	10.00E
Harsefeld, Germany	53.45N	9.48E
Heiligenkreuz, Austria[c]	48.05N	16.13E
Heilsbronn, Germany	49.35N	10.83E
Herzogenrath, Germany	50.85N	6.10E
Hildesheim, Germany	52.15N	9.97E
Hradisch, Czechoslovakia	49.08N	17.50E
Ingolstadt, Germany	48.77N	11.45E
Jarrow, England	54.98N	1.48W
Jerusalem	31.78N	35.22E
Klosterneuburg, Austria	48.32N	16.33E
Krakow, Poland	50.05N	19.92E
Kremsmünster, Austria	48.08N	14.13E
Kufow, China	35.67N	117.02E
La Cava, Italy	40.70N	14.70E

[c]The Times Atlas gives latitude 46°30'N, longitude 16°16'E.

Place	Latitude, degrees	Longitude, degrees
Lambach, Austria	48.10N	13.87E
Laon, France	49.57N	3.62E
Liège, Belgium	50.63N	5.58E
Limoges, France	45.83N	1.25E
London, England[d]	51.52N	0.10W
Lorsch, Germany	49.65N	8.58E
Lübeck, Germany	53.87N	10.67E
Lubin, Poland	51.38N	16.17E
Lund, Sweden	55.70N	13.17E
Lyon, France	45.77N	4.83E
Madinat, Egypt	see Thebes, Egypt	
Magdeburg, Germany	52.13N	11.62E
Mainz, Germany	50.00N	8.27E
Malmesbury, England	51.60N	2.10W
Mantua, Italy	45.17N	10.78E
Margan, Wales	51.57N	3.73W
Marseilles, France	43.30N	5.37E
Melk, Austria	48.23N	15.35E
Melrose, Scotland	55.60N	2.73W
Memmingen, Germany[e]	47.98N	10.40E
Merseburg, Germany	51.37N	12.00E
Metz, France	49.12N	6.18E
Milan, Italy	45.47N	9.20E
Monte Cassino, Italy	41.48N	13.83E
Mont S. Michel, France	48.63N	1.50W
Mt. Sumelas, Turkey	40.67N	39.64E
Munich, Germany	48.13N	11.58E
Naples, Italy	40.83N	14.25E
Neresheim, Germany	48.77N	10.37E
Neuberg, Austria	47.67N	15.58E

[d]These are the coordinates of S. Paul's Cathedral.

[e]The point whose coordinates are given is 16 kilometers east of Memmingen.

Place	Latitude, degrees	Longitude, degrees
Nevers, France	47.00N	3.15E
Nieder-Alteich, Germany	48.76N	13.04E
Ninove, Belgium	50.83N	4.03E
Ober-Alteich, Germany	48.92N	12.67E
Orléans, France	47.90N	1.90E
Orvieto, Italy	42.72N	12.10E
Oslo, Norway	59.93N	10.75E
Osney, England	see Bicester, England	
Oxnead, England	52.72N	1.43E
Paderborn, Germany	51.72N	8.73E
Padua, Italy	45.40N	11.88E
Paris, France	48.87N	2.33E
Parma, Italy	44.80N	10.32E
Passau, Germany	48.58N	13.47E
Pegau, Germany	51.17N	12.25E
Peterborough, England	52.58N	0.25W
Piacenza, Italy	45.05N	9.68E
Pisa, Italy	43.72N	10.40E
Pohlde, Germany	51.67N	10.45E
Pont-à-Mousson, France	48.92N	6.05E
Prague, Czechoslovakia	50.08N	14.42E
Prum, Germany	50.20N	6.42E
Quedlinburg, Germany	51.80N	11.15E
Ravenna, Italy	44.42N	12.20E
Regensburg, Germany	49.02N	12.12E
Reggio nell'Emilia, Italy	44.70N	10.62E
Reichenau, Germany	47.70N	9.08E
Reichersberg, Austria	48.33N	13.40E
Rome, Italy	41.88N	12.50E
Rouen, France	49.43N	1.08E

Place	Latitude, degrees	Longitude, degrees
Rye, Germany	54.83N	9.55E
S. Albans, England	51.77N	0.35W
S. Amand-les-Eaux, France	50.45N	3.43E
S. Benoit-sur-Loire, France	47.80N	2.30E
S. Blasien, Germany	47.77N	8.13E
S. David's, Wales	51.90N	5.27W
S. Denis, France	48.93N	2.35E
S. Evroul-en-Ouche, France	48.78N	0.47E
S. Florian, Austria	48.22N	14.38E
S. Gall, Switzerland	47.42N	9.38E
S. Georgio d. E., Italy	46.68N	11.63E
S. Omer, France	50.75N	2.25E
S. Trudperti, Germany	47.87N	7.87E
Salerno, Italy	40.67N	14.77E
Salzburg, Austria[f]	47.80N	13.05E
Santarem, Portugal	39.23N	8.67W
Saumur, France	47.27N	0.08W
Sazava, Czechoslovakia	49.88N	14.90E
Schaffhausen, Switzerland	47.70N	8.63E
Scheftlarn, Germany	47.87N	11.57E
Sens, France	48.20N	3.30E
Siena, Italy	43.32N	11.32E
Sintra, Portugal	38.80N	9.37W
Stade, Germany	53.60N	9.47E
Stargard, Poland[g]	53.35N	15.02E
Stuttgart, Germany	48.78N	9.20E
Terracina, Italy	41.28N	13.25E
Tewkesbury, England	51.98N	2.15W
Thasos, Greece	40.77N	24.70E
Thebes, Egypt[h]	25.68N	32.67E

[f]The _Times Atlas_ gives the latitude as 47° 54′N.

[g]Used in place of Colbaz, Poland.

[h]Used in place of Madinat, Egypt.

Place	Latitude, degrees	Longitude, degrees
Thebes, Greece	38.32N	23.32E
Toulouse, France	43.62N	1.43E
Tours, France	47.38N	0.70E
Trebizond	41.00N	39.72E
Trier, Germany	49.75N	6.65E
Trois-Fontaines, France	48.67N	7.13E
Ursperg, Germany	48.27N	10.53E
Vendôme, France	47.80N	1.07E
Venice, Italy	45.43N	12.33E
Vicenza, Italy	45.55N	11.55E
Vienna, Austria	48.22N	16.37E
Vigeois, France	45.37N	1.50E
Volterra, Italy	43.40N	12.87E
Vormezeele, Belgium	50.82N	2.87E
Waverley, England	see Farnham, Surrey, England	
Weimar, Germany	50.98N	11.33E
Weingarten, Germany[i]	47.7 N	9.1 E
Weltenburg, Germany	48.89N	11.83E
Werm, Belgium	50.83N	5.47E
Westminster, England	51.50N	0.13W
Winchester, England	51.07N	1.32W
Wissembourg, France	49.03N	7.95E
Worcester, England	52.18N	2.22W
Worms, Germany	49.63N	8.38E
Würzburg, Germany	49.80N	9.95E
Xanten, Germany	51.67N	6.45E
Zwettl, Austria	48.62N	15.18E
Zwiefalten, Germany	48.23N	9.45E

[i] I have not been able to find more precise coordinates.

scale map. I have also used coordinates obtained this way
for some points not listed in the atlas; I have noted these
places in the discussion of the sources involved.

When I could not find a place either in the Times Atlas
or on a large-scale map, I turned to Ginzel [1884]. Ginzel
wrote before the meridian of Greenwich was adopted as the
standard reference meridian, and he used the meridian of the
Paris Observatory as his reference. He also lists 2° 26' W
as the longitude of London. Hence I first added 2° 26' to
his longitudes. These longitudes are too large, since Greenwich is east of London. Since I did not know which specific
spot Ginzel meant by London, I compared longitudes found this
way with those in the Times Atlas for several places, and decided that the agreement was best if I subtracted 5' from
the longitude referred to London. The places whose coordinates were found this way are Reichersberg, Austria, Rye,
Germany, and Zwettl, Austria.

Finally, there are a few points that I could not find
at all. For them, I could only substitute some place that
is known to be nearby. These places are noted in the table
and the reasons for the choice of substitute are given in
the discussions of the sources involved.

It is almost certain that there are some errors in a
table as large as Table A.III.3. I should appreciate being
informed of any errors that the reader may discover. It is
possible, of course, that a value listed in the table is in
error even though I used accurate coordinates in the calculations.

AGRICULTURAL SUCCESS IN ENGLAND FROM 1273 TO 1391

In his Historica Anglicana, Walsingham [ca. 1422] frequently closes his account of the year with a sentence such
as: "Transit annus iste frugifer et fructifer." We could
translate this as: "This year passed (or went out) fruitful
and profitable." However, it is clear from Walsingham's
usage that frugifer and fructifer are both used to refer to
products of agriculture and that they are not synonyms. Our
first task is to decide just what he meant by these words,
and my Latin dictionary is not much help.

The ending fer presumably comes from the word meaning
to bear or carry, so a year that was frugifer and fructifer
was one that bore "frug" and "fruct" and probably bore them
in an adequate quantity. The difficulty is that the roots
frug and fruct both refer to fruit, either literally or
figuratively, in general usage.

I found only one passage that helps decide the meanings
that Walsingham attached to the roots. In this passage, if
I may use the roots frug and fruct for the moment as if they
were words, he says that frug was abundant but that there
was almost no fruct of the trees. From this I infer that
fruct means the fruit of trees, but I suspect that it may
include other kinds of fruit such as strawberries as well.
In the rest of this appendix I shall translate fruct in this
sense, that is as the fruit of both trees and vines or bushes.

Since frug is also an important agricultural product,
the most likely meaning for it is grain, although there is a
perfectly good word frumentum that Walsingham could have used
for grain. I shall translate frug as grain in the rest of
this appendix. If I am wrong about the meanings of frug and
fruct, the reader can easily restore the correct translations
if they ever become known.

When Walsingham says only that a year was frugifer et
fructifer, the implication seems to be that the year was
neither particularly good nor particularly bad. I have
translated this condition by saying that both the fruit and
grain were successful.

Walsingham's summaries of the agricultural year are
listed in Table A.IV.1. The first column in the table gives
the year. The second column gives the page in the cited
edition where Walshingham's summary is found, and the third
column gives a translation of his summary. Pages for years
through 1379 are from volume 1 and later pages are from volume 2.

At one point Walsingham gives entries for two successive years but he labels both year 1308. In addition, he
has entries for all consecutive years from 1273 through 1308.
The only explanation seems to be that he gave all years
through 1307 a number that is 1 year too high. Thus, through

RESULTS OF AGRICULTURE IN ENGLAND
FROM 1273 TO 1391

Year	Page	Nature of agricultural results
1273	10	Both fruit and grain successful.
1274	12	Rich in both fruit and grain.
1275	14	Rich in both fruit and grain.
1276	15	Both fruit and grain successful.
1277	16	Fruit and grain successful in England but poor for Romans.
1278	18	Both fruit and grain successful.
1279	19	Both fruit and grain successful.
1280	20	Both fruit and grain successful.
1281	20	Both fruit and grain successful.
1282	21	Both fruit and grain successful.
1283	24	Copious fruit and grain.
1284	26	Fruit and grain rather good.
1285	26	Fruit and grain rather abundant.
1286	28	Fruit and grain successful for most people, but there was such a drought that some people died.
1287	28	Both fruit and grain successful.
1288	29	Both fruit and grain successful.
1289	31	Fruit and grain marvelously successful.
1290	32	The year started well but ended destitute.
1291	33	Food was dear but it was not a famine.
1292	38	Grain barely sufficed for the people.
1293	43	Very good grain and not a shortage.
1294	45	Grain was not productive and fruit was not fertile.
1295	49	So destitute in grain and fruit that paupers died of hunger.
1296	53	Troublesome for the rich and unbearable for the poor because of a want of grain and a great dearth.
1297	63	Hard on the people because of a lack of grain.

Year	Page	Nature of agricultural results
1298	73	A lack of grain for the lowly.
1299	77	Not abundant in grains but not altogether lacking in fruit.
1300	80	Neither good nor bad for grain.
1301	84	A joyful year for Christians because of victories in the Holy Land; he forgot to tell us about the crops.
1302	97	Grain not rich but not lacking.
1303	99	Mediocre in both fruit and grain.
1304	105	Hard for the rich, made want for the poor.
1305	107	Fruit and grain neither lacking nor abundant.
1306	108	Produced fruit and grain.
1307	111	Brought forth grain and fruit but not outstanding.
1308	120	Sufficient but not great for crops.
1308	122	Both fruit and grain successful.
1309	124	Very poor in grain and fruit.
1310	128	Mediocre in grain and fruit.
1311	135	Both fruit and grain successful.
1312	137	Abundant in neither fruit nor grain.
1313	143	Good in neither grain nor fruit.
1314	145	Cruel ... famine ..., and a pestilence that resulted from the famine.[a]
1315	148	Brought pestilence and death to many because of hunger, thirst, dysentery, and high fevers.
1316	153	Sterile.
1317	155	Insufficient in grain and fruit for our people and people in other lands.
1318	157	Poor in grain, lacking in fruit; ruinous.
1319		No entry.
1320	163	Ruinous.
1321	167	Ruinous in England.

[a]Several words are missing from the text.

Year	Page	Nature of agricultural results
1322	171	Little rejoicing over grain, but very rich in fruit.
1323	175	Unpraised in its production of fruit and grain.
1324	179	Moderately cheering in its production of fruit and grain.
1325	185	Both fruit and grain successful for the first year in a long time.
1326		Walsingham does not have this year in his history.
1327	190	Happy and fertile.
1337	222	Copious grain and fruit.
1338	223	Both fruit and grain successful.
1339	226	Rich in grain and abundant in fruit.
1341	253	Both fruit and grain successful; plentiful supplies for sale.
1377	345	Happy time, abundant grain, but the fruit of trees a little wanting.
1379	427	Successful grain and ample fruit.
1383	109[b]	Both fruit and grain successful.
1384	119	Fertile grain, and indeed the fruit was not altogether sterile.
1385	141	Quite rich in fruit.
1386	153	Ample in grain but mediocre in fruit.
1387	170	Successful in fruit and grain, and moderately healthy.
1388	179	Abundant fruit and grain.
1389	195	Moderate in grain but abundant in fruit.
1390	198	Year brought neither ample fruit nor grain.
1391	203	Very hard to many poor and middle class[c] because of dearth and disease.

[b]The page numbers come from volume 1 up to this point. The remaining page numbers are from volume 2.

[c]The word used is mediocris. In this context, I believe it is equivalent to "middle class" in our terminology.

the first appearance of 1308 in Table A.IV.1, we should
subtract 1 from the number of each year. From 1308 on,
the years that Walsingham gives (which are the ones that
appear in the table) are usually correct, although there
may be isolated errors.

With only a few exceptions, which are probably accidental, Walsingham gives an agricultural summary for every
year from 1273 through 1327. At that point, his sources
seemed to drop the agricultural summary as a regular matter,
because we find no additional entry for 10 years, and there
are only 16 entries after 1327 to the end of the history.
In fact, the history goes to 1422 but the last agricultural
summary is for the year 1391.

Walsingham refers to England specifically in his summaries for several years, and in two years (1277 and 1317)
he compares conditions at home with conditions abroad.† For
this reason, I have called Table A.IV.1 the results of agriculture in England for the years covered.

The brevity of Walsingham's summary causes difficulty in
translation for several years. However, there is only one
year for which I am concerned about the general accuracy of
the translation in the table. This year is 1286. If there
was such a drought that year that some people died in consequence, it seems unlikely that the agricultural year was
successful for most people. In spite of this concern, I
believe that the translation represents what Walsing'.am wrote.
Perhaps he ran together information pertaining to two different years.

---

†He compared English conditions with conditions abroad in
a few other years, but I do not give these in the table.

APPENDIX V

THE FLOODING OF THE NILE

For every year from 1469 to 1522, ibn Iyas [ca. 1522]
gives some information about the flooding of the Nile, upon
which the life of Egypt depends. Further, in connection
with the year 1520, when the flood was unusually late, he
gives information about three different years well before
his own time when the flood was also very late, and in connection with the year 1516, when the flood was extraordinarily early, he tells us it had not been this early since
1441. ibn Iyas usually refers to the Nile as "the blessed
Nile".

For most years, he gives the dates of two events. One
event is the event that I have translated as "flood",† and
the other is the event called the opening of the dike.

The dike, always identified in the singular, is opened
by a high official of the court, whom ibn Iyas frequently
identifies. It seems unlikely to me that a single dike
could have actually controlled the entire irrigation system.
In fact, since both banks of the Nile are irrigated, as I
understand the matter, it seems impossible that a single
dike could suffice. Therefore it is likely that the dike in
question was one maintained for ceremonial purposes and that,
once it was opened, the "working" dikes could be opened to
begin the actual irrigation.

Occasionally ibn Iyas refers to the "maximum" of the
river level rather than to the flood. I suspect that he
either used the word maximum by accident or that he used it
in ignorance of its correct meaning.‡ If the Egyptians
waited until the level was a maximum, that means that they
had to wait at least until a day when the level did not increase. In most years, this happens some time after the
level is high enough to allow irrigation. Waiting for the
maximum would waste water and perhaps lessen the area that
could be irrigated. Thus I suspect that the dike was opened
when the level reached a certain point that had been determined by experience or by calculation.

In a number of years, ibn Iyas gives us information
about the level of the river. Since ancient times there has
been an instrument called the Nilometer which measures the
river level; the one to which ibn Iyas refers is located at
Cairo. The level is measured by means of cubits and digits.
The cubit is related to the average length of the forearm
and is about half a meter. The digit means a twelfth part
in most ancient and medieval contexts, but it clearly does
not do so here because the number of digits is more than 12

---

†The word used in the French translation is crue.

‡So far as I can tell, if there is erroneous usage, it may
have been due to the translator rather than to ibn Iyas.

in most measurements. Since the largest number of digits I
noticed is 49, I suspect that 60 digits equal 1 cubit in
this work. This makes the digit slightly shorter than a
centimeter, and it may be that it was related to the thick-
ness of a finger.

ibn Iyas makes some remarks which indicate that the
dike was in fact opened when the Nilometer read a specific
level. On 1517 August 16 he says that the level was 16
cubits plus 1 digit, which was 4 digits short of the flood.
On 1518 August 20, on the other hand, he says that the river
reached the prescribed level of 16 cubits, and the dike was
opened. While there is a slight confusion here, it seems
likely that the dike could be opened as soon as the river
reached some level close to 16 cubits.

The record for 1504 August 18, however, contradicts this
conclusion. This record says that the Nile reached a maximum
on this date and that the dike was opened. It also says that
the level was 17 cubits plus 5 digits. The inference that I
drew from the record is that this was the level on the day
the dike was opened. It is possible, I suppose, that this
was indeed the maximum level reached and that ibn Iyas confused the day of maximum level with the day when the river
reached flood stage.

In sum, we cannot tell from the records of ibn Iyas
exactly what the flood meant, nor the conditions that determined the opening of the dike. Further study would be needed
to settle the question.

Several times when the level was not approaching flood
in a normal manner, the sultan read (or caused to be read)
the entire Koran at the Nilometer. When this happened, the
Nile resumed its normal approach to flood within a short
time, and this doubtless proved to the court and people that
the reading accomplished its purpose.

The information about the Nile flooding is summarized in
Table A.V.1. This table gives the year and the date (converted to the Julian calendar) in the first two columns. The
third column gives the page in the cited translation where
the record is found. The Roman numeral before the period
gives the volume number and the following Arabic numeral
gives the page number in that volume. The fourth column
tells what happened on that date.

In some years, ibn Iyas records only the month in which
the flood stage was reached. I have omitted these years from
the table.

We note that the dike was opened on the day that flood
was reached about half the time and that the dike opening
was deferred about half the time. In a few cases ibn Iyas
tells why the delay occurred, but in most cases he does not
indicate a reason.

The average date of the flood in Table A.V.1 is August 7.3, and the standard deviation from this date is 8.9 days.[†] This is probably somewhat less regular than the blooming of the Japanese cherry trees in Washington, but not outstandingly so. The Nile flooding certainly does not have the high regularity that some writers claim for it. The total range of the dates in the table is from July 23 (in 1479) to August 29 (in 1295), a range of 37 days.

This range does not count the years 1484 and 1520, when the Nile came close to or above flood stage out of season in addition to having its usual flood.

The reader should be able to find more information about the Nile flooding if he wants to. In particular, the translator of ibn Iyas refers several times to Histoire de Nil by Omar Toussoun and to The Cairo Nilometer by Popper. He does not give any further bibliographical information about these books, but they should not be hard to locate. Since the matter is a side issue to this work, I have not tried to find them.

[†] These calculations omit the first four dates, which ibn Iyas gives only because they were extremes.

NILE FLOODING FROM 1469 TO 1522
AND IN CERTAIN ISOLATED YEARS

Year	Date	Page	Event
1295	Aug 29	IV.339	Flood
1298	Aug 28	IV.339	Flood
1313	Aug 28	IV.339	Flood
1441	Jul 20	IV.55	Flood
1469	Aug 17	II.41	Flood; dike opened
1470	Aug 15	II.57	Flood; dike opened
1471	Aug 19	II.68	Flood; dike opened
1472	Aug 14	II.84	Flood; dike opened
1473	Jul 29	II.100	Flood; dike opened
1474	Aug 13	II.109	Flood; dike opened
1475	Aug 5	II.123	Flood; dike opened
1476	Jul 27	II.136	Flood; dike opened
1477	Jul 24	II.148	Flood; rarely this early
	Jul 25	II.148	Dike opened
1478	Jul 28	II.161	Flood; dike opened
	Jul 30	II.161	Height reached 17 c.[a]
1479	Jul 23	II.171	Flood
	Jul 24	II.171	Dike opened; an extraordinary event (presumably because it was so early)
	Jul 26	II.171	Height reached 17 c. 6 d.; very rare.
1481	Aug 8	II.202	Flood; dike opened
1483	Aug 11	II.225	Flood; dike opened
1484	Aug 15	II.233	Flood; dike opened
	?	II.233	Level fell rapidly after reaching 17 c. 22d.; most areas were not irrigated.
	~ Oct 1	II.235	Nile rose rapidly to great height out of season, too late to help the crops.
1485	~ Jun 1	II.243	Level at low stage was 8 c. 20 d., an unheard-of level (we do not know whether this was unusually high or low)
	Aug 13	II.245	Flood; dike opened

[a]"c." denotes a cubit and "d." denotes a digit. The digit here is probably 1/60 of a cubit.

Year	Date	Page	Event
1486	Aug 11	II.258	Flood; dike opened. Level was 16 c. 20 d.
	Aug 14	II.258	Level reached 16 c. 49 d., an extraordinary level.
1487	Aug 5	II.271	Flood; dike opened
1488	Aug 4	II.284	Flood; dike opened
1489	Jul 27	II.299	Flood
	Jul 28	II.299	Dike opened
1490	Jul 27	II.308	Nile rose 33 d. on this single day, a rarety.
	Jul 28	II.308	Flood; dike opened
1491	Aug 6	II.320	Flood; could not open dike because it was a fast day.
	Aug 8	II.320	Dike opened
1493	Aug 5	II.333	Flood; dike opened
1494	~Aug 29	II.343	After reaching a maximum, the Nile then fell, and then rose again to flood height on about August 29, and it reached a good height after all.
1497	Aug 20	II.409	Flood
	Aug 21	II.409	Dike opened. River then fell rapidly, so irrigation was poor.
1498	Aug 12	II.438	Flood; dike opened
1499	Jul 27	II.462	Nile rose 30 d.
	Jul 28	II.462	Nile rose 40 d.
	Jul 29	II.462	Nile rose 20 d. to flood stage.
	Jul 30	II.462	Dike opened
1500	Aug 1	II.486	Flood
	Aug 2	II.486	Dike opened
1501	Aug 2	III.16	Flood; dike opened
1502	Jul 28	III.32	Nile rose 40 d. on this single day.
	Jul 29	III.32	Nile rose 20 d.
	Aug 1	III.32	Flood
	Aug 2	III.32	Dike opened. Level was 11 d. above average.
1503	Aug 2	III.52	Maximum; dike opened

Year	Date	Page	Event
1504	Aug 18	III.63	Maximum; dike opened. Level was 17 c. 5 d.
1505	Aug 2	III.78	Flood; dike opened. Later the level rose 3 more digits, but we are not told the actual level. It was a good year.
1506	Aug 13	III.92	Maximum; dike opened.
	Aug 14	III.93	River rose still more.
1507	Aug 3	III.113	Flood; river had risen 70 digits between July 30 and August 1.
	Aug 4	III.114	Dike opened
1508	Aug 4	III.130	Nile rose 50 digits on this day, and a total of 90 digits from August 4 to August 6. Such a rise had happened only three times since the Hijra.
	Aug 7	III.131	Dike opened
1509	Aug 7	III.156	Flood; dike opened
1510	Aug 9	III.182	Sultan read the complete Koran at the Nilometer because the river had risen poorly. The river resumed rising.
	Aug 13	III.182	Flood
	Aug 14	III.182	Dike opened. The implication is that it was a good year.
1511	Aug 6	III.217	Nile ceased to rise, so the complete Koran was read at the Nilometer.
	Aug 8	III.218	Flood; dike opened; normal level
1512	Jul 25	III.255	Flood
	Jul 26	III.255	Dike opened. Level was 16 c. 10 d., an exceptional height, but it rose 28 more digits in next two days, thus attaining a total of 21 digits (sic).
1513	Aug 7	III.304	Flood; 5 digits above normal of 16 cubits
	Aug 8	III.304	Dike opened
1514	Aug 15	III.361	Flood; level was 16 c. 2 d.
	Aug 16	III.361	Dike opened. Three days later the level reached 17 c. 16 d., an extraordinary phenomenon.
1515	Jul 29	III.426	Flood
	Jul 30	III.426	Dike opened

Year	Date	Page	Event
1516	Jul 20	IV.55	Flood; had not been this early since 1441.
	Jul 21	IV.55	Dike opened
1517	Aug 11	IV.189	River level dropped from August 11 to August 15.
	Aug 16	IV.190	Level was 16 c. 1 d., 4 d. short of flood.
	Aug 17	IV.190	Dike opened
1518	Aug 20	IV.256	River reached the prescribed level of 16 c. and the dike was opened.
1519	Aug 12	IV.297	Flood 2 d. above required 16 c.
	Aug 13	IV.297	Dike opened
1520	Aug 16	IV.338	River fell before reaching flood.
	Aug 20	IV.338	Koran was read because river was still dropping.
	Aug 24	IV.339	Flood
	Aug 25	IV.339	Dike opened; very late but not a record. See years 1295, 1298, and 1313.
	~ Nov 16	IV.357	Flood out of season at 16 c. less 16 d.
1521	Aug 3	IV.387	Flood; level was 16 c. 6 d.
	Aug 4	IV.387	Dike opened
1522	Aug 6	IV.453	Flood; level was 16 c. 3 d.
	Aug 7	IV.453	Dike opened

ARE RECORDED SOLAR ECLIPSES CORRELATED WITH
OTHER PORTENTS IN CHINESE RECORDS?

In a famous paper, Bielenstein [1956] decided that the reports of solar eclipses found in the records from the Former Han dynasty are strongly correlated with the reports of other portents such as meteors, comets, novae, and the like. He also believed that eclipses or other portents were rarely fabricated, although two or three eclipses clearly were. From these considerations, he concluded that the appropriate court officials used the eclipses and other portents as a method of criticizing the emperor indirectly. That is, under a popular emperor, the officials often refrained from reporting many portents (including solar eclipses), but under an unpopular emperor the officials might report all portents that were observed. In this connection, the popularity of the emperor meant his popularity with the officials involved. This might be quite different from his popularity with the people at large.

In order to reach his conclusion, Bielenstein considered the number of solar eclipses recorded under the reign of each emperor and the number of other portents reported under the same emperor. Let X_i denote the number of eclipses reported in the reign of the ith emperor, let Y_i denote the number of other portents reported in the same reign, and let R_i denote the number of years in this reign. Bielenstein correctly observed that, other things being equal, more eclipses tend to be reported in a long reign than in a short one, and so do the other portents. Thus the variability in the lengths R_i may introduce a false correlation between the X_i and the Y_i.

In order to avoid this possibility, Bielenstein divided the number X_i of solar eclipses reported in the ith reign by R_i, obtaining the number of solar eclipses reported per year in the ith reign.† Let U_i denote this value, which equals X_i/R_i. Bielenstein similarly calculated V_i $(= Y_i/R_i)$, which is the number of other portents reported per year in the ith reign. He found that U_i and V_i are strongly correlated, and from this he concluded [p. 132] that "all the portents‡ form a homogeneous material, affected by one and the same power."

---

† Bielenstein actually subtracted the number of reported eclipses in each reign from the number of eclipses that could have been observed during that reign. In other words, Bielenstein actually dealt with the unreported eclipses instead of the reported eclipses when he made his calculations. I do not understand the advantage of this procedure.

‡ That is, the solar eclipses plus the other portents. I remind the reader that lunar eclipses were not reported at all, except one or two that were accidentally reported as solar eclipses.

He went on to reach the conclusion, already mentioned, that the reporting of portents was a means of expressing indirect criticism of the emperor.

I have pointed out elsewhere [Newton, 1977a] the error in mathematical statistics that Bielenstein makes in this procedure, and I shall only summarize the matter here. If the X_i and the Y_i are uncorrelated, the process of dividing both by the same set of R_i tends to introduce a correlation between the quotients U_i and V_i. This correlation is an artifact of the mathematical processes used and has no physical importance. On the other hand, suppose that the U_i and V_i are uncorrelated. Since $X_i = U_i R_i$ and $Y_i = V_i R_i$, the X_i and Y_i have a tendency to be correlated.† Since Bielenstein divided X_i and Y_i by R_i before he tested for a correlation, he would have removed this correlation.

The important point is that the process of dividing by the R_i to find the U_i and V_i may either increase or decrease the correlation, depending upon the circumstances. One cannot learn anything about the possibility of a true correlation between the underlying physical entities if one divides an original set of data (X_i, Y_i) by a common set of divisors R_i before testing for a correlation. Similarly, if X_i and Y_i are likely to be proportional to R_i, one cannot learn anything by testing X_i and Y_i for correlation before dividing.

A way to solve the problem correctly is the following (it is probably not the only correct way): Find the best straight-line fit between the X_i and the R_i. Then let the residual x_i be X_i minus the corresponding value on the straight line. In a similar way, construct residuals y_i from the values of the Y_i and R_i. Now examine the x_i and the y_i for correlation. If they are correlated, we may safely assume a correlation between the physical entities represented by the X_i and Y_i.

The considerations outlined so far in this appendix are important for many problems beyond the one that Bielenstein addressed. For this reason, I did not attempt to solve Bielenstein's problem in the paper cited.

Bielenstein and Sivin [1977] make two calculations in an attempt to show that the considerations of this appendix do not apply to Bielenstein's paper, and to show that there is a strong correlation between the reporting of eclipses and of other portents. Further, according to them, "the use of portents for indirect criticism is by now too well documented to need further elaboration." They do not say, however, where any of this documentation can be found. Actually, their calculations show just the opposite of what they state. The correlation that they find between the reporting of eclipses and of other portents is not particularly strong, nor does it have much statistical significance.

In one calculation, Bielenstein and Sivin [p. 187]

---

†In all the considerations of this appendix, I tacitly assume that the values of the R_i are not all the same.

calculate the correlation coefficients between the X_i and the Y_i and between the U_i and the V_i, and compare the two coefficients. They find that the correlation coefficient between the X_i and the Y_i is 0.89 while that between the U_i and the V_i is 0.50. They describe this as "more conclusive information" than that obtained from the procedure outlined in the third paragraph above this one.

We remember that Bielenstein in effect used the correlation between the U_i and the V_i as the basis for his conclusion about the use of portents (including solar eclipses) for indirect criticism. Actually, he never stated what the correlation coefficient was; he merely asserted that the correlation was strong. Further, he did so without examining its statistical significance, and <u>Bielenstein and Sivin</u> make the same mistake.

The method of finding the statistical significance of a correlation coefficient is given by <u>Kreyszig</u> [1970, p. 343]. What we do is to find the probability that the coefficient could have a particular value by chance, if the numbers involved were random. Suppose that we have N pairs of numbers and that the correlation coefficient between them is r. We first calculate a number t:[†]

$$t = |r| [(N - 2)/(1 - r^2)]^{\frac{1}{2}}. \qquad (A.VI.1)$$

We then look in a table on page 454 of <u>Kreyszig</u> [1970] in which t is tabulated as a function of two variables. One variable is $N - 2$ and the other is unity minus the probability sought.

In the case of the Han portents, $r = 0.50$ for the U_i and the V_i, and $N = 10$. For these values, we find $t = 1.63$. We find this value of t in the table opposite the value 0.92 in the column for which $N - 2 = 8$. This means that the probability is 0.08, about 1 chance in 12, that $|r|$ was as large as 0.50 purely by chance.

If the probability had been as large as about 0.3, I believe that most scientists would say that the correlation is insignificant. I also believe that most scientists do not consider that a correlation is established, even tentatively, unless the probability is less than 0.05, and many insist on a probability less than 0.01 before they consider that a correlation is well established.

Let us forget for the moment that the correlation between the U_i and the V_i does not necessarily tell us anything about the true correlation, and let us assume on the contrary that the correlation in question is a meaningful one if it is statistically significant. We see that the correlation is in an intermediate range. The probability of chance occurrence

---

[†]We use $|r|$ rather than r in front of the square root because r can be either positive or negative. In testing its significance, only the absolute value of r matters.

is not large enough to let us dismiss the correlation immediately as being insignificant. The probability of chance occurrence is not small enough to let us say that the correlation is significant, even on a tentative basis. In other words, Bielenstein was not entitled to take the correlation as one that is strong and well established. The most one can say is that one can neither accept nor reject the correlation without further study, and that the correlation is probably not strong even if it is real.

In their other calculation, Bielenstein and Sivin use the method that I outlined above. That is, they find the residuals of the X_i and the Y_i from the appropriate straight lines and then calculate the correlation coefficient between the residuals. They find that the coefficient r is 0.56, but again they do not test its significance.

In this case, the residuals have been found by fitting a straight line to 10 values of X_i as a function of the reign lengths R_i, and similarly fitting a straight line to 10 pairs of Y_i and R_i. Two parameters are needed in specifying the line in each case, so only 8 of the 10 residuals (x_i or y_i) in each case are linearly independent. Thus, when we test the correlation between the x_i and y_i, the number N of independent residuals is 8 rather than 10. That is, N = 8 in Equation A.VI.1. Again we find that the probability that $|r|$ is as large as 0.56 by chance is 0.08. Thus the correlation between the reporting of solar eclipses and the reporting of other portents is not particularly significant.

However, let us assume for the sake of argument that the coefficient is significant at a useful level of confidence. There is still another consideration which neither Bielenstein [1956] nor Bielenstein and Sivin [1977] address, and that is the strength of the correlation. If the reporting of other portents were governed strongly by the desire to express indirect criticism, or by any other single factor, we should expect the correlation coefficient to be 0.8 or larger for a sample of the size we have here. We find only 0.56. A value around 0.56 is what we expect to find if the reporting of eclipses is governed by several factors, most of which affect eclipses and other portents differently. That is, the correlation is weak even if we assume it to be statistically significant.

In sum, the reporting of solar eclipses and the reporting of other portents are probably governed by several factors, most of which affect eclipses and other portents differently. The eclipses and other portents may be correlated weakly by a desire on the part of the appropriate officials to express indirect criticism of the emperor, so far as we can tell by a statistical analysis, but there is no significant evidence that either confirms or denies this idea.

Until the writing of the present work, it has been assumed in the literature that the eclipse reports in the Han records represent actual observations, except perhaps for two or three fabricated eclipses. We concluded in Chapter V that this is probably not so. For example, the annals from the state of Lu, which run from about -720 to about -480,

contain reports of 37 eclipses, of which 33 clearly correspond to eclipses that were readily visible in Lu. There are
only 4 "impossible" cases, which probably result from scribal
errors that we cannot unscramble. In contrast, there are 55
reports of solar eclipses in the Former Han records, among
which I count 13 "impossible" cases.†

If we assume that all the reports are based upon observed
eclipses, except perhaps in a few cases, how do we explain a
rate of "impossibility" of about 1 in 4 for the Han records
when we have a rate of only about 1 in 10 for the older Lu
records? I do not see a plausible explanation, and hence I
conclude that the "impossible" cases were actually calculated
ones. It follows, as I show in Chapter V, that many of the
"possible" cases were also calculated.

Something that can be calculated cannot be a portent.
However, if the emperor did not know that eclipses could be
calculated, he would believe that they were portents. Hence
the eclipses could serve the purposes of criticizing the emperor whether they were observed or calculated.

Thus the presence of calculated eclipses in the Han records does not directly affect Bielenstein's argument. However, it does affect the argument strongly in an indirect
manner.

In order to use calculated eclipses as portents, which
is equivalent to fabricating the eclipses, only one or two
people could know the secret, or it would soon cease to be
a secret. Probably only the chief astronomer and a successor to whom he imparted the secret would know that eclipses
were being calculated for use as portents, and would also
know how to do the calculations.‡ By Bielenstein's hypothesis,
however, the decision about the number of portents to report
to the emperor's court was made by a set of bureaucratic officials. If a sizeable group of people were "in on" the conspiracy to criticize the emperor by using calculated eclipses,
the secret could not have been kept for the duration of the
dynasty, and bureaucratic heads would have rolled long before
the dynasty fell. Hence the calculated eclipses could not
have been used for indirect criticism by anyone except the
chief astronomer and perhaps one trusted associate.

Bielenstein discusses the popularity of the emperor with

---

† The number of "impossible" causes may be slightly greater
or slightly less than 13. It is hard to decide in a few
cases whether there has been a scribal error in writing the
date or whether the report constitutes an "impossible"
case. I have counted some of these cases as visible
eclipses and some of them as impossible. I doubt that the
number 13 can be in error by more than 1.

‡ We remember that the "impossible" dates were not mere
random choices. They are dates of new moons at which an
eclipse nearly happened, which is what we expect if the
eclipses were calculated from a theory based upon inaccurate parameters.

the court officials and concludes that the historically known popularity or unpopularity of the emperor closely parallels the number of portents, including eclipses, reported per year. However, if I am right about the use of calculated eclipses, the only official who mattered for the expression of indirect criticism was the chief astronomer (and perhaps an associate or a designated successor). I do not believe that the historical sources that Bielenstein used can tell us whether the emperor was or was not popular with the chief astronomer.

There is another consideration. If calculated eclipses were being used as fabricated portents, their occurrence should parallel closely the unpopularity of the emperor, if Bielenstein's main hypothesis is correct. However, the distribution of the "impossible" eclipses is unusual. The dynasty came to power in -201 and 10 of the "impossible" eclipses were reported between then and -134, an interval of 67 years. There were about 160 years left to the dynasty, but only 3 impossible cases occur during its remaining time.

According to Bielenstein, the last two emperors were highly unpopular with the court officials, so the reporting of "impossible" eclipses should have been a maximum under them. Instead, it is a minimum. In fact, the last "impossible" eclipse I found in the Han records is the one that Dubs dates -34 November 1.

The point about the conspiracy seems valid to me even if the eclipse reports were all based upon observations. Eclipses, novae, comets, and other types of astronomical "portents" could have been observed in many places and the observations reported to the office of the chief astronomer. However, the transmission of an observation to the emperor, which is often called its memorializing, must have been done by the chief astronomer. Under Bielenstein's hypothesis, the chief astronomer must have been in a conspiracy with other officials to decide upon the number of astronomical portents to memorialize each year. If anyone has had to work in an area which requires the preservation of a secret, he knows that it is impossible to keep one long, especially if many people know it. I do not believe that the conspiracy to criticize the emperor by the memorializing of portents could have been kept up for the duration of the Former Han dynasty, which was more than two centuries.

In summary, Bielenstein's hypothesis is that a set of court officials used the reporting of eclipses and other portents as a means of expressing indirect criticism of the emperor, and that this procedure was maintained for the two centuries or more that the dynasty ruled. As a corollary, this requires the maintenance of a successful conspiracy involving several people at a time for a period of more than two centuries, and this seems impossible. The data on the reporting of portents do not lend strong support to Bielensteins's hypothesis. Further, if the factor of indirect criticism did affect the reporting of portents, it was not a strong factor and other factors must have had a strong effect upon the reporting. Further, it is almost certain that a particular set of eclipse reports, about a fourth of the total, were calculated; I have called these the "impossible" eclipses. If Bielenstein is correct, the impossible eclipses should be correlated with the other portents. Instead, the correlation is negligible.

AAO = Newton [1970].

Acropolita, Georgius, Annales, ca. 1262. There is an edition
by Immanuel Bekker, with a parallel Latin translation by
Leo Allatius, in Corpus Scriptorum Historiae Byzantinae,
Weber's, Bonn, 1836.

Albericus, Chronicon, ca. 1241. There is an edition by MM.
Guigniaut and de Wailly in Recueil des Historiens des
Gaules et de la France, volume XXI, Imprimerie Royale,
Paris, 1855.

Alcobacense, Chronicon Alcobacense, ca. 1111. There is an
edition by Damião Peres in Revista Portuguesa de Historia, Tomo I, pp. 148-151, Coimbra, Portugal, 1940.

Allen, E.B., A Coptic solar eclipse record, Journal of the
American Oriental Society, 67, pp. 267-269, 1947. See
Djeme [ca. 601].

Alpertus (Monachus S. Symphoriani Metensis), Libello de
Diversitate Temporum, ca. 1020. Parts of this work are
printed in Recueil des Historiens des Gaules et de la
France, volume X, edited by the Benedictines of S. Maur,
Chez Martin, Guerin, Delatour, et Boudet, Paris, 1760.

Altahenses, Annales Altahenses Maiores, ca. 1073. There is
an edition by Wilhelm de Giesebrecht and Edmund L.B.
ab Oefele in Monumenta Germaniae Historica, Scriptores,
volume XX, G.H. Pertz, editor, Hahn's, Hannover, 1868.

Altahenses, Notae Altahenses, ca. 1585. This is actually a
collection of notes from documents kept at Nieder-
Alteich that was made by Philip Jaffé. It is found in
Monumenta Germaniae Historica, Scriptores, volume XVII,
G.H. Pertz, editor, Hahn's, Hannover, 1861.

American Ephemeris and Nautical Almanac, U.S. Government
Printing Office, Washington, published annually.

Andreas Bergomatis, Historia, ca. 877. There is an edition
by G. Waitz in Monumenta Germaniae Historica, Scriptores
Rerum Langobardicarum et Italicarum, Saec. VI-IX, L.
Bethmann and G. Waitz, editors, Hahn's, Hannover, 1878.

Angliae, Chronicon Angliae ab Anno Domini 1328 usque ad Annum
1388, ca. 1388. There is an edition by E.M. Thompson in
Rerum Britannicarum Medii Aevi Scriptores, no. 64,
Longman & Co., London, 1874.

APO = Newton [1976].

Augustani Minores, Annales Augustani Minores, ca. 1321. There
is an edition by G.H. Pertz in Monumenta Germaniae Historica, Scriptores, volume X, G.H. Pertz, editor, Hahn's,
Hannover, 1852.

Autissiodorense, Breve Chronicon Autissiodorense, ca. 1174.
    There is an edition by the Benedictines of S. Maur in
    Recueil des Historiens des Gaules et de la France,
    volumes X and XI. Volume X, Chez Martin, Guerin, Dela-
    tour, et Boudet, Paris, 1760; volume XI, Chez L.F.
    Delatour, Paris, 1767.

Balduinus, Balduini Ninovensis Chronicon, ca. 1294. There is
    an edition by O. Holder-Egger in Monumenta Germaniae
    Historica, Scriptores, volume XXV, G. Waitz, editor,
    Hahn's, Hannover, 1880.

Basileenses, Annales Basileenses, ca. 1277. There is an
    edition by Philip Jaffé in Monumenta Germaniae Historica,
    Scriptores, volume XVII, G.H. Pertz, editor, Hahn's,
    Hannover, 1861.

Beaver, D. deB, Bernard Walther: Innovator in astronomy,
    Journal for the History of Astronomy, 1, pp. 39-43,
    1970.

Benedict, Vita Henrici II Angliae Regis, ca. 1192. There is
    an edition in Recueil des Historiens des Gaules et de la
    France, volume 17, Michel-Jean-Joseph Brial, editor,
    Imprimerie Royale, Paris, 1818.

Benevenutus, Cronaca, ca. 1341. There is an edition by M.
    Faloci-Pulignani in Rerum Italicarum Scriptores, volume
    XXVI, part II, Nicola Zanichelli, Bologna, 1933.

Bergomates, Annales Bergomates, ca. 1241. There is an edition
    by Philip Jaffé in Monumenta Germaniae Historica, Scrip-
    tores, volume XVIII, G.H. Pertz, editor, Hahn's, Hannover,
    1863. There is a later edition by O. Holder-Egger in
    volume XXXI of the same series, edited by the Societas
    Aperiendis Fontibus, Hahn's, Hannover, 1903.

Bermondsey (Bermundeseia), Annales Monasterii de Bermundeseia,
    ca. 1433. There is an edition by H.R. Luard in Rerum
    Britannicarum Medii Aevi Scriptores, no. 36, volume 3,
    Longmans, Green, Reader, and Dyer, London, 1866.

Bielenstein, H., An interpretation of the portents in the
    Ts'ien-Han-Shu, Bulletin of the Museum of Far Eastern
    Antiquities (Stockholm), 22, pp. 127-143, 1956.

Bielenstein, H. and Sivin, N., Further comments on the use of
    statistics in the study of Han dynasty portents, Journal
    of the American Oriental Society, 97, pp. 185-187, 1977.

Bolognetti, Cronaca Detta Dei Bolognetti, ca. 1420. There is
    an undated edition by L.A. Muratori, revised and corrected
    by Albano Sorbelli, in Rerum Italicarum Scriptores,
    volume XVIII, in 3 parts, Stamperia di S. Lapi, Citta di
    Castello, 1906.

Britton, J.P., On the quality of solar and lunar observations
    and parameters in Ptolemy's Almagest, a dissertation
    presented to Yale University, 1967.

Brouwer, D., A study of the changes in the rate of rotation
of the earth, Astronomical Journal, 57, pp. 125-146,
1952.

Brown, E.W., with the assistance of H.B. Hedrick, Chief
Computer, Tables of the Motion of the Moon, in 3 volumes,
Yale University Press, New Haven, Connecticut, 1919.
The reader should also note Complement to the Tables of
the Moon, Containing the Remainder Terms for the Century
1800-1900, and Errata in the Tables, with the same
authors and publisher, 1926.

Brunwilarenses, Annales Brunwilarenses, ca. 1179. There is
an edition by G.H. Pertz, based upon information supplied
by L.C. Bethmann, in Monumenta Germaniae Historica,
Scriptores, volume XVI, G.H. Pertz, editor, Hahn's,
Hannover, 1859.

Burchardus and Cuonradus, Urspergensium Chronicon, ca. 1229.
There is an edition by Otto Abel and Ludwig Weiland in
Monumenta Germaniae Historica, volume XXIII, G.H. Pertz,
editor, Hahn's, Hannover, 1874.

Cantinellus, Petrus, Chronicon, ca. 1306. There is an edition
by Francesco Torraca in Rerum Italicarum Scriptores,
volume XXVIII, part II, Editore S. Lapi, Citta di Castello,
1902.

Capgrave, John, The Chronicle of England, ca. 1463. There is
an edition by F.C. Hingeston-Randolph in Rerum Britannicarum Medii Aevi Scriptores, no. 1, Longman, Brown,
Green, Longman, and Roberts, London, 1858.

Casinenses, Annales Casinenses ex Annalibus Montis Casini
Antiquis et Continuatis Excerpti, ca. 1098. There is
an edition by G. Smidt in Monumenta Germaniae Historica,
Scriptores, volume XXX, part II, section III, A. Hof-
meister, editor, K.W. Hiersemann, Leipzig, 1934.

Casinenses, Annales Casinenses, ca. 1212. There is an edition
by G.H. Pertz in Monumenta Germaniae Historica, Scriptores, volume XIX, G.H. Pertz, editor, Hahn's, Hannover,
1866.

Ceccanenses, Annales Ceccanenses, ca. 1217. There is an
edition by G.H. Pertz in Monumenta Germaniae Historica,
Scriptores, volume XIX, G.H. Pertz, editor, Hahn's,
Hannover, 1866.

Celoria, G., Sull' eclissi solare totale del 3 Giugno 1239,
Memorie del Reale Istituto Lombardo di Scienze e Letteri,
Classe di Scienze Matematiche e Naturali, 13, pp. 275-
300, 1877a.

Celoria, G., Sugle eclissi solari totali del 3 Giugno 1239
e del 6 Ottobre 1241, Memorie del Reale Istituto Lombardo
di Scienze e Letteri, Classe di Scienze Matematiche e
Naturali, 13, pp. 367-382, 1877b.

Chalipharum Liber, ca. 636. There is an edition by J.P.N. Land in Anecdota Syriaca, volume 1, J.P.N. Land, editor, E.J. Brill, Leiden, 1862.

Chounradus, Annales, ca. 1226. There is an edition by Philip Jaffé in Monumenta Germaniae Historica, Scriptores, volume XVII, G.H. Pertz, editor, Hahn's, Hannover, 1861.

Cicero, M. Tullius, De Oratore, -54. There is an edition by C.F.A. Nobbe in M. Tullii Ciceronis Opera Omnia, Tauchnitz, Leipzig, 1850.

Cicero, M. Tullius, De Re Publica, ca. -53. There is an edition by C.F.A. Nobbe in M. Tullii Ciceronis Opera Omnia, Tauchnitz, Leipzig, 1850.

Claustroneoburgensis, Continuatio Claustroneoburgensis, ca. 1383. There is an edition by W. Wattenbach in Monumenta Germaniae Historica, Scriptores, volume IX, G.H. Pertz, editor, Hahn's, Hannover, 1851.

Clavius, Christopher, In Sphaeram Ioannis de Sacro Bosco Commentarius, Sumptibus Fratrum de Gabiano, Lyons, 1593.

Clemens, S.L., A Connecticut Yankee in King Arthur's Court, Harper and Brothers, New York, 1889.

Cohen, A.P., Quick conversion of Chinese sexagenary year designations into Western dates, Journal of the Chinese Language Teachers Association, 11.3, pp. 194-198, 1976.

Cohen, A.P. and Newton, R.R., Solar eclipses recorded under the Tarng Dynasty, in preparation.

Colbazienses, Annales Colbazienses, ca. 1568. There is an edition by Wilhelm Arndt in Monumenta Germaniae Historica, Scriptores, volume XIX, G.H. Pertz, editor, Hahn's, Hannover, 1866.

Colmarienses, Annales Colmarienses Minores, ca. 1300. There is an edition by Philip Jaffé in Monumenta Germaniae Historica, Scriptores, volume XVII, G.H. Pertz, editor, Hahn's, Hannover, 1861.

Colonienses, Annales Colonienses Maximi, ca. 1238. There is an edition by K. Pertz in Monumenta Germaniae Historica, Scriptores, volume XVII, G.H. Pertz, editor, Hahn's, Hannover, 1861.

Coloniensis, Chronicae Regiae Coloniensis, ca. 1220. There is an edition by G. Waitz in Monumenta Germaniae Historica, Scriptores, volume XXIV, G. Waitz, editor, Hahn's, Hannover, 1879.

Confucius, Annals of Lu (sometimes called Ch'un Ch'iu), -479. There is an edition with an English translation by J. Legge, Chinese Classics, volume 5, Hong Kong University Press, Hong Kong, first edition, 1872, second edition, 1960. I assign the date -479 because the last event in the annals happened in -480. The ascription of these annals to Confucius is only traditional and they may have been compiled by someone else.

Copernicus, Nicolaus, De Revolutionibus Orbium Caelestium
Libri Sex, 1543. There is an edition, including a
facsimile of Copernicus's holograph, by Fritz Kubach,
Verlag von R. Oldenbourg, in 2 volumes: volume 1,
facsimile of the holograph, 1944; volume 2, a critical
edition of the text, 1949.

Cosmas (Cosmas Indicopleustes), Topographia Christiana, ca.
550. There is an edition with a parallel Latin translation by J.-P. Migne in Patrologiae Cursus Completus,
Series Graeca Prior, volume 88, J.-P. Migne, Paris, 1864.

Cotton, Bartholomaei de, Monachi Norwicensis, Historia Anglicana, ca. 1298. There is an edition by H.R. Luard in
Rerum Britannicarum Medii Aevi Scriptores, no. 16, Longman, Green, Longman, and Roberts, London, 1859.

Cracoviensis, Annales Capituli Cracoviensis, ca. 1331. There
is an edition by G.H. Pertz in Monumenta Germaniae Historica, Scriptores, volume XIX, G.H. Pertz, editor,
Hahn's, Hannover, 1866.

Cremonenses, Annales Cremonenses, ca. 1232. There is an edi-
tion by Philip Jaffé in Monumenta Germaniae Historica,
Scriptores, volume XVIII, G.H. Pertz, editor, Hahn's,
Hannover, 1863.

Curott, D.R., Earth deceleration from ancient solar eclipses,
Astronomical Journal, 71, pp. 264-269, 1966.

de Camp, L. Sprague, Lost Continents, revised edition, Dover
Publications, Inc., New York, 1970.

degli Unti, Pietruccio di Giacomo, Memoriale, 1440. There is
an edition by M. Faloci-Pulignani in Rerum Italicarum
Scriptores, volume XXVI, part II, Nicola Zanichelli,
Bologna, 1933.

Delambre, J.B.J., Histoire de l'Astronomie Ancienne, Chez Mme.
Veuve Courcier, Paris, in 2 volumes, 1817.

Delambre, J.B.J., Histoire de l'Astronomie du Moyen Âge, Chez
Mme. Veuve Courcier, Paris, 1819.

dello Mastro, Paolo di Benedetto di Cola, Memoriale, ca. 1484.
There is an edition by F. Isoldi in Rerum Italicarum
Scriptores, volume XXIV, part II, Editrice S. Lapi,
Citta di Castello, 1912.

de Sitter, W., On the secular accelerations and the fluctuations of the longitudes of the moon, the sun, Mercury,
and Venus, Bulletin of the Astronomical Institutes of
the Netherlands, IV, pp. 21-38, 1927.

Dicke, R.H., Average acceleration of the earth's rotation and
the viscosity of the deep mantle, Journal of Geophysical
Research, 74, pp. 5895-5902, 1969.

Diessenses, Notae Diessenses, ca. 1432. There is an edition
     by Philip Jaffé in Monumenta Germaniae Historica, Scrip-
     tores, volume XVII, G.H. Pertz, editor, Hahn's, Hannover,
     1861.

Dio Cassius, Romaikon, ca. 230. There is an edition with a
     parallel English translation by E. Cary in 9 volumes in
     Loeb's Classical Library, William Heinemann, London;
     different volumes were published in many different years.
     The author's name is sometimes given as Dion Cassius.

Djeme, Annals of Djeme, ca. 601. The solar eclipse record
     from this set of annals is discussed by E.B. Allen, A
     Coptic solar eclipse record, Journal of the American
     Oriental Society, 67, pp. 267-269, 1947. This is the
     same source as Allen [1947].

Dorenses, Annales Dorenses, ca. 1362. There is an edition by
     R. Paulus in Monumenta Germaniae Historica, Scriptores,
     volume XXVII, G. Waitz, editor, Hahn's, Hannover, 1885.

Dubs, H.H., Solar eclipses during the Former Han period,
     Osiris, 5, pp. 499-522, 1938.

Dubs, H.H., The History of the Former Han Dynasty, Waverly
     Press, Baltimore, in three volumes: vol. 1, 1938, vol.
     2, 1944, vol. 3, 1955. All volumes are cited here as
     1955.

Duncombe, R.L., Motion of Venus, 1750-1949, Astronomical
     Papers Prepared for the Use of the American Ephemeris
     and Nautical Almanac, XVI, Part 1, U.S. Government
     Printing Office, Washington, 1972.

Eckert, W.J., Jones, Rebecca, and Clark, H.K., Construction
     of the lunar ephemeris, in An Improved Lunar Ephemeris,
     1952-1959, issued as a Joint Supplement to the American
     Ephemeris and Nautical Almanac and the (British) Nautical
     Alamanac and Astronomical Ephemeris, U.S. Government
     Printing Office, Washington, 1954.

Elwacense, Chronicon Elwacense, ca. 1477. There is an edition
     by D.O. Abel in Monumenta Germaniae Historica, Scriptores,
     volume X, G.H. Pertz, editor, Hahn's, Hannover, 1852.

Ensdorfenses, Annales Ensdorfenses, ca. 1368. There is an
     edition by G.H. Pertz in Monumenta Germaniae Historica,
     Scriptores, volume X, G.H. Pertz, editor, Hahn's, Han-
     nover, 1852.

Eusebius (Pamphili), Chronicon, ca. 325. There is a collated
     edition of the principal texts edited by A. Schoene,
     Weidmann's, Berlin, 1866 and 1875.

Explanatory Supplement to The Astronomical Ephemeris and The
     American Ephemeris and Nautical Almanac, H.M. Stationery
     Office, London, 1961.

Feller, William, An Introduction to Probability Theory and Its Applications, Volume 1, Second Edition, John Wiley and Sons, New York, 1957.

Ferrarese, Diario Ferrarese, ca. 1502. There is an edition by G. Pardi in Rerum Italicarum Scriptores, volume XXIV, Part VII, Nicola Zanichelli, Bologna, 1933.

Flatøbogens, Flatøbogens Annaler, ca. 1394. There is an edition by Gustav Storm in Islandske Annaler indtil 1578, Grøndahl and Sons, Christiana, 1888.

Florence of Worcester (Florentius Wigorniensis), Chronicon ex Chronica, ca. 1118. There is an edition by B. Thorpe in Publications of the English Historical Society, London, 1848. There is a translation by T. Forester in Bohn's Antiquarian Library, Henry G. Bohn, London, 1854.

Flores Historiarum, ca. 1326. There is an edition by H.R. Luard in Rerum Britannicarum Medii Aevi Scriptores, no. 95, in 3 volumes, H.M. Stationery Office, London, 1890.

Florianensis, Continuatio Florianensis, ca. 1310. There is an edition by W. Wattenbach in Monumenta Germaniae Historica, Scriptores, volume IX, G.H. Pertz, editor, Hahn's, Hannover, 1851.

Forolivienses, Annales Forolivienses, ca. 1473. There is an edition by G. Mazzatinti in Rerum Italicarum Scriptores, volume XXII, part II, Editore S. Lapi, Citta di Castello, 1903.

Fossenses, Annales Fossenses, ca. 1384. There is an edition by G.H. Pertz in Monumenta Germaniae Historica, Scriptores, volume IV, G.H. Pertz, editor, Hahn's, Hannover, 1841.

Fotheringham, J.K., On the accuracy of the Alexandrian and Rhodian eclipse magnitudes, Monthly Notices of the Royal Astronomical Society, 69, pp. 666-668, 1909.

Fotheringham, J.K., (assisted by Gertrude Longbottom), The secular acceleration of the moon's mean motion as determined from the occultations in the Almagest, Monthly Notices of the Royal Astronomical Society, 75, pp. 377-394, 1915a.

Fotheringham, J.K., Note on some results of the new determination of the secular acceleration of the moon's mean motion, Monthly Notices of the Royal Astronomical Society, 75, pp. 395-396, 1915b.

Fotheringham, J.K., The secular acceleration of the sun as determined from Hipparchus' equinox observations; with a note on Ptolemy's false equinox, Monthly Notices of the Royal Astronomical Society, 78, pp. 406-423, 1918.

Fotheringham, J.K., A solution of ancient eclipses of the sun, Monthly Notices of the Royal Astronomical Society, 81, pp. 104-126, 1920.

Fotheringham, J.K., Two Babylonian eclipses, Monthly Notices of the Royal Astronomical Society, 95, pp. 719-723, 1935.

Frisacenses, Annales Frisacenses, ca. 1300. There is an edition by L. Weiland in Monumenta Germaniae Historica, Scriptores, volume XXIV, G. Waitz, editor, Hahn's, Hannover, 1879.

Frutolf, Chronicon Universale, 1103. There is an edition, in which the work is ascribed to Ekkehardus Uraugiensis, by G. Waitz in Monumenta Germaniae Historica, Scriptores, volume VI, G.H. Pertz, editor, Hahn's, Hannover, 1844.

Gardiner, Sir Alan, Egypt of the Pharaohs, Oxford University Press, Oxford, 1961.

Gaubil, A., Une Histoire de l'Astronomie Chinoise, which is volume 2 of Observations Mathématiques, Astronomiques, Géographiques, Chronométriques, et Physiques Tirées des Anciens Livres Chinois, E. Souciet, editor, Rollin Père, Paris, 1732a.

Gaubil, A., Un Traité de l'Astronomie Chinoise, which is volume 3 of the above, Paris, 1732b.

Gervase (of Canterbury), Chronica, ca. 1199. There is an edition with preface and notes by William Stubbs in Rerum Britannicarum Medii Aevi Scriptores, no. 73, volume 1, Longman and Co., London, 1879.

Ghirardacci, Cherubino, Historia di Bologna, ca. 1509. There is an edition by Albano Sorbelli in Rerum Italicarum Scriptores, volume XXXIII, Part I, Casa Editrice S. Lapi, Citta di Castello, 1932.

Ginzel, F.K., Astronomische Untersuchungen über Finsternisse, Part II, Sitzungberichte der Kaiserlichen Akademie des Wissenschaften, Wien, Math. - Naturwiss. Classe, 88, pp. 629-755, 1884.

Girardus de Fracheto, Continuatio Chronici, ca. 1328. There is an edition by MM. Guigniaut et de Wailly in Recueil des Historiens des Gaules et de la France, volume XXI, Imprimerie Royale, Paris, 1855.

Godel, Willelmus, Anonymous Continuation of the Chronicle of, ca. 1320. There is an edition by MM. Guigniaut et de Wailly in Recueil des Historiens des Gaules et de la France, volume XXI, Imprimerie Royale, Paris, 1855.

Gregoras, Nicephorus, Historia Byzantina, ca. 1359. There is an edition with a parallel Latin translation by Ludovicus Schopenus in Corpus Scriptorum Historiae Byzantinae, in 3 volumes, Weber's, Bonn, 1829.

Grierson, P., Les Annales de Saint-Pierre de Gand et de Saint-Amand, Commission Royale d'Histoire, Brussels, 1937.

Guerrierus, Chronicon Eugubinum, ca. 1472. There is an edition by G. Mazzatinti in Rerum Italicarum Scriptores, volume XXI, part IV, Editore S. Lapi, Citta di Castello, 1902.

Guidonis, Bernardus, Flores Historiarum et Chronicon Regum Francorum, ca. 1327. There is an edition by MM. Guigniaut et de Wailly in Recueil des Historiens des Gaules et de la France, volume XXI, Imprimerie Royal, Paris, 1855.

Guillelmus Armoricus, De Gestis Philippi Augusti, ca. 1224. There is an edition in Recueil des Historiens des Gaules et de la France, volume XVII, Michel-Jean-Joseph Brial, editor, Imprimerie Royale, Paris, 1818.

Guillelmus de Nangiaco, Chronicon, ca. 1368. There is an edition by Daunou and Naudet (other names or titles not given) in Recueil des Historiens des Gaules et de la France, volume XX, Imprimerie Royale, Paris, 1840.

Guillelmus de Podio-Laurentii, Historia Albigensium, ca. 1272. There is an edition of part of this work by Michel-Jean-Joseph Brial in Recueil des Historiens des Gaules et de la France, volume XIX, Imprimerie Royale, Paris, 1833. There is an edition of another part by Daunou et Naudet in volume XX, Imprimerie Royale, Paris, 1840.

Halley, Edmond, Some account of the ancient state of the city of Palmyra; with short remarks on the inscriptions found there, Philosphical Transactions of the Royal Society, 19, pp. 160-175, 1695.

Hamburgenses, Annales Hamburgenses, ca. 1265. There is an edition by I.M. Lappenberg in Monumenta Germaniae Historica, Scriptores, volume XVI, G.H. Pertz, editor, Hahn's, Hannover, 1859.

Heath, Sir Thomas, Aristarchus of Samos, Oxford University Press, Oxford, 1913.

Heimo, Chronographia, 1135. Parts of this work are edited by G.H. Pertz in Monumenta Germaniae Historica, Scriptores, volume X, G.H. Pertz, editor, Hahn's, Hannover, 1852.

Heinricus, Chronicon Lyvoniae, ca. 1227. There is an edition by Wilhelm Arndt in Monumenta Germaniae Historica, Scriptores, volume XXIII, G.H. Pertz, editor, Hahn's, Hannover, 1874.

Herbipolenses, Annales Herbipolenses Minores, ca. 1400. There is an edition by G. Waitz in Monumenta Germaniae Historica, Scriptores, volume XXIV, G. Waitz, editor, Hahn's, Hannover, 1879.

Hermannus, Altahenses Annales, 1273. There is an edition by Philip Jaffé in Monumenta Germaniae Historica, Scriptores, volume XVII, G.H. Pertz, editor, Hahn's, Hannover, 1861.

Hieronymus, Chronicon ab Anno MCCCXCVII usque ad Annum MCCCCXXXIII, ca. 1433. There is an edition by A. Pasini in Rerum Italicarum Scriptores, volume XIX, part V, Nicola Zanichelli, Bologna, 1931.

Hildesheimenses, Annales Hildesheimenses, ca. 1137. There is an edition by G.H. Pertz in Monumenta Germaniae Historica, Scriptores, volume III, G.H. Pertz, editor, Hahn's, Hannover, 1839.

Hoang, P., Catalogue des Éclipses de Soleil et de Lune (Variétés Sinologiques, No. 56), Imprimerie de la Mission Catholique, Shanghai, 1925.

Honorius (Honorius Augustodunensis), Summa Totius Mundi, ca. 1137. There is an edition by Roger Wilmans in Monumenta Germaniae Historica, Scriptores, volume X, G.H. Pertz, editor, Hahn's, Hannover, 1852.

Horner, Joseph, Biblical chronology, Society of Biblical Archaeology, Proceedings of June 7, 1898, p. 235, 1898.

Hyvanus, Anthonius, Gesta Unius Anni Memorabilia, ca. 1478. There is an edition by F.L. Mannucci in Rerum Italicarum Scriptores, volume XXIII, part IV, Editrice S. Lapi, Citta di Castello, 1913.

ibn Hayyan, Al-Muqtabis, ca. 1070. There is a translation of the last part of this work into Spanish by J.E. Guraieb, Cuadernos de Historia de España, XXXI-XXXII, pp. 316-321, 1960.

ibn Iyas, Chronique d'ibn Iyas, ca. 1522. There is a translation into French by Gaston Wiet, in 4 volumes, Imprimerie de l'Institut Francais d'Archéologie Orientale, Cairo, 1945.

ibn Yunis, Az-Zij al-Kabir al-Hakimi, 1008. There is an edition of part of this work, with a parallel translation into French, by J.J.A. Caussin de Perceval, under the title Le Livre de la Grande Table Hakémite, with the author's name spelled as Ebn Iounis, Imprimerie de la République, Paris, 1804. Page citations used here refer to this edition.

Ideler, Ludwig, Handbuch der Chronologie, volume 2, August Rücker, Berlin, 1826.

Igor, The Song of Igor's Campaign, ca. 1185. There is a translation into English, with critical notes, by Vladimir Nabokov, Vintage Books (Random House), New York, 1960.

Iohannes Longus, Chronica Monasterii Sancti Bertini, ca. 1365. There is an edition by O. Holder-Egger in Monumenta Germaniae Historica, Scriptores, volume XXV, G. Waitz, editor, Hahn's, Hannover, 1880.

Iterius, Bernardus, Chronicon, ca. 1225. There is an edition by H. Duplès-Agier in Chroniques de Saint-Martial de Limoges, Société de l'Histoire de la France, Publication no. 167, Paris, 1874.

Jao Tsung-i, <u>Oracle Bone Diviners of the Yin Dynasty</u>, v. 1, Hong Kong University Press, Hong Kong, 1959.

Jeffreys, Sir Harold, <u>The Earth</u>, Fifth Edition, Cambridge University Press, Cambridge, 1970.

Johannis à S. Victore, <u>Memorialiae Historiarum</u>, ca. 1322. There is an edition by MM. Guigniaut et de Wailly in <u>Recueil des Historiens des Gaules et de la France</u>, <u>Volume XXI</u>, Imprimerie Royale, Paris, 1855.

John of Reading, <u>Chronicon</u>, ca. 1367. There is an edition by James Tait in <u>Chronica Johannis de Reading et Anonymi Cantuariensis</u>, Manchester University Press, Manchester, 1914.

Johnson, S.J., <u>Eclipses, Past and Future</u>, Parker and Co., London, 1889.

Juvavenses, <u>Annales Juvavenses Maximi</u>, ca. 956. There is an edition by H. Bresslau in <u>Monumenta Germaniae Historica</u>, <u>Scriptores</u>, volume XXX, part II, K.W. Hiersemann, Leipzig, 1926.

Juvavenses, <u>Annales Juvavenses Maiores</u>, ca. 975. There is an edition by G.H. Pertz in <u>Monumenta Germaniae Historica</u>, <u>Scriptores</u>, <u>volume I</u>, G.H. Pertz, editor, Hahn's, Hannover, 1826.

Kalhana, <u>Rajatarangini</u>, 1150. There is a translation into English with notes, by M.A. Stein, with the title given as <u>A Chronicle of the Kings of Kasmir</u>, in 2 volumes, Archibald Constable and Co., Westminster, 1900.

Keightley, D.N., private communication, 31 October 1974.

Keightley, David N., <u>Sources of Shang History: The Oracle-Bone Inscriptions of Bronze Age China</u>, University of California Press, Berkeley, Los Angeles, and London, 1977.

Keightley, David N., On the misuse of ancient Chinese inscriptions: an astronomical fantasy, <u>History of Science</u>, in press. (Appeared in <u>v</u>. <u>xv</u>, pp. 267-272, 1977).

Kreyszig, Erwin, <u>Introductory Mathematical Statistics</u>, John Wiley and Sons, New York, 1970.

Kugler, F.X., <u>Sternkunde und Sterndienst in Babel</u>, <u>Ergänzungen zum Ersten und Zweiten Buch</u>, <u>II Teil</u>, Aschendorffsche Verlagsbuchhandlung, Münster, Westphalia, 1914.

Lambertus Waterlos, <u>Annales Cameracenses</u>, ca. 1170. There is an edition by G.H. Pertz in <u>Monumenta Germaniae Historica</u>, <u>Scriptores</u>, <u>volume XVI</u>, G.H. Pertz, editor, Hahn's, Hannover, 1859.

Legge, J., Introduction to <u>Chinese Classics</u>, <u>volume 5</u>, Hong Kong University Press, Hong Kong, first edition, 1872, second edition, 1960.

Legge, J., Confucius, Encyclopaedia Britannica, Eleventh
    Edition, volume VI, pp. 907-912, Encyclopaedia Britan-
    nica, Inc., New York, 1911.

Lemovicense, Majus Chronicon Lemovicense, ca. 1342. There is
    an edition of part of this work by MM. Guigniaut et de
    Wailly in Recueil des Historiens des Gaules et de la
    France, volume XXI, Imprimerie Royale, Paris, 1855.
    Other parts appear in volumes XII and XVIII of the same
    series.

Lemovicensia, Fragmenta Chronicorum Lemovicensia, ca. 1658.
    There is an edition by H. Duplès-Agier in Chroniques de
    Saint-Martial de Limoges, Société de l'Histoire de la
    France, publication no. 167, Paris, 1874.

Leo Diaconus, Historia, ca. 990. There is an edition by C.B.
    Hasius with a parallel Latin translation in Corpus Scrip-
    torum Historiae Byzantinae, B.G. Niebuhr, editor, Weber's,
    Bonn, 1828.

Livy (Titus Livius), Ab Urbe Condita Libri, ca. 0. There is
    an edition by Wilhelm Weissenborn, first appeared between
    1858 and 1862, revised by H.J. Müller, Weidmann's, Berlin,
    1900-1907.

Loch Cè, Annals of Loch Cè, ca. 1590. There is an edition by
    W.M. Hennessy in Rerum Britannicarum Medii Aevi Scriptores,
    no. 54, in 2 volumes, Longmann and Co., London, 1871.
    These have also been known at times as Annals of the Old
    Abbey of Inis-Macreen, an Island in Lough-Kea, as a Con-
    tinuation of the Annals of Tighernach, as the Book of the
    O'Duigenans of Kilronan, and as the Annals of Kilronan.
    The editor does not believe that any of these titles are
    justified.

Lögmanns, Lögmanns Annáll, ca. 1430. There is an edition by
    Gustav Storm in Islandske Annaler indtil 1578, Grøndahl
    and Sons, Christiana, 1888.

Londonienses, Annales Londonienses, ca. 1330. There is an
    edition by William Stubbs in Rerum Britannicarum Medii
    Aevi Scriptores, no. 76, part 1, Longman & Co., London,
    1882.

Lubenses, Annales Lubenses, ca. 1281. There is an edition
    by G.H. Pertz in Monumenta Germaniae Historica, Scrip-
    tores, volume XIX, G.H. Pertz, editor, Hahn's, Hannover,
    1866.

Lubicenses, Annales Lubicenses, ca. 1324. There is an edition
    by I.M. Lappenberg in Monumenta Germaniae Historica,
    Scriptores, volume XVI, G.H. Pertz, editor, Hahn's, Han-
    nover, 1859.

MacPike, E.F., Correspondence and Papers of Edmond Halley,
    Clarendon Press, Oxford, 1932.

M = Muller [1975].

Malalas, Ioannis, Chronographia, ca. 563. There is an edition with a parallel Latin translation in Corpus Scriptorum Historiae Byzantinae, Weber's, Bonn, 1831.

Malaterra, Gaufridus, Historia Sicula, ca. 1090. Excerpts
from this are printed in Recueil des Historiens des
Gaules et de la France, volume XIII, edited by the
Benedictines of S. Maur, Chez Vve. Desaint, Paris, 1786.

Mantuana, Chronica Pontificum et Imperatorum Mantuana, ca.
1250. There is an edition by G. Waitz in Monumenta
Germaniae Historica, Scriptores, volume XXIV, G. Waitz,
editor, Hahn's, Hannover, 1879.

Mantuani, Annales Mantuani, ca. 1299. There is an edition by
G.H. Pertz in Monumenta Germaniae Historica, Scriptores,
volume XIX, G.H. Pertz, editor, Hahn's, Hannover, 1866.

Maragone, Bernardo, Annales Pisani, 1182. There is an edition
by L.A. Muratori, revised by M.L. Gentile, in Rerum
Italicarum Scriptores, volume VI, part II, Nicola Zanichelli, Bologna, 1930.

Markowitz, W., Sudden changes in rotational acceleration of
the earth and secular motion of the pole, in Earthquake
Fields and the Rotation of the Earth, L. Mansinha, D.E.
Smylie, and A.E. Beck, editors, pp. 69-81, D. Reidel
Publishing Co., Dordrecht, Holland, 1970.

Martin, C.F., A study of the rate of rotation of the earth
from occultations of stars by the moon, 1627-1860, a
dissertation presented to Yale University, 1969. To be
published in the Astronomical Papers Prepared for the
Use of the American Ephemeris and Nautical Almanac.

Martin, C.F., and van Flandern, T.C., Secular changes in the
lunar elements, Science, 168, pp. 246-247, 1970.

Matthaeus (de Griffonibus), Memoriale Historicum, ca. 1472.
There is an edition by L. Frati and A. Sorbelli in
Rerum Italicarum Scriptores, volume XVIII, part II,
Editore S. Lapi, Citta di Castello, 1902.

Matthew Paris, Historia Anglorum (sometimes called Historia
Minor), ca. 1250. There is an edition by Sir Frederic
Madden in Rerum Britannicarum Medii Aevi Scriptores, no.
44, in 3 volumes, Longmans, Green, Reader, and Dyer,
London, 1866.

MCRE = Newton [1972].

Mechovienses, Annales Mechovienses, ca. 1434. There is an
edition by G.H. Pertz in Monumenta Germaniae Historica,
Scriptores, volume XIX, G.H. Pertz, editor, Hahn's,
Hannover, 1866.

Mediolanenses, Memoriae Mediolanenses, ca. 1251. There is an
edition by Philip Jaffé in Monumenta Germaniae Historica,
Scriptores, volume XVIII, G.H. Pertz, editor, Hahn's,
Hannover, 1863.

Mellicenses, Annales Mellicenses, ca. 1564.   There is an
   edition by W. Wattenbach in Monumenta Germaniae Historica, Scriptores, volume IX, G.H. Pertz, editor,
   Hahn's, Hannover, 1851.

Menkonis, Chronica Werumensium, ca. 1273.   There is an edi-
   tion by Ludwig Weiland in Monumenta Germaniae Historica,
   Scriptores, volume XXIII, G.H. Pertz, editor, Hahn's,
   Hannover, 1874.

Mercati, A., The new list of the popes Medieval Studies, An
   Annual Published by the Pontifical Institute of Mediaeval Studies of Toronto, IX, pp. 71-80, 1947.

Merton, Chronicon de Merton, ca. 1323.   There is an edition
   by H.R. Luard in Rerum Britannicarum Medii Aevi Scriptores, no. 95, volume 3, H.M. Stationery Office, London,
   1890.

Mettensis, Chronica Universalis Mettensis, ca. 1447. There is
   an edition by G. Waitz in Monumenta Germaniae Historica,
   Scriptores, volume XXIX, G. Waitz, editor, Hahn's, Hannover, 1879.

Michael I, Chronicle, ca. 1195.   There is an edition with a
   French translation by J.-B. Chabot, Imprimerie Camis et
   Cie., Paris, in 3 volumes. Volume 1, 1899, volume 2,
   1901, volume 3, 1905.

Michels, Agnes K., The Calendar of the Roman Republic,
   Princeton University Press, Princeton, N.J., 1967.

Miliolus, Albertus, Alberti Milioli Notarii Regini Liber de
   Temporibus et Aetatibus et Cronica Imperatorum, ca.
   1286.   There is an edition by O. Holder-Egger in Monumenta Germaniae Historica, Scriptores, volume XXXI,
   Hahn's, Hannover, 1903.

Moguntini, Annales Moguntini, ca. 1309. There is an edition
   by G.H. Pertz in Monumenta Germaniae Historica, Scriptores, volume XVII, G.H. Pertz, editor, Hahn's, Hannover, 1861.

Mommsen, Theodor, Römische Chronologie bis auf Caesar, Weid-
   mannsche Buchhandlung, Berlin, 1859.

Mommsen, Theodor, Chronica Minora, Saec. IV V VI VII, volume
   1 (of 3 volumes), which is volume 9 of Monumenta Germaniae Historica, Auctorum Antiquissimorum, Weidmann's,
   Berlin, 1892.

Morrison, L.V., Rotation of the earth from AD 1663-1972 and
   the constancy of G, Nature, 241, pp. 519-520, 1973.

Morrison, L.V. and Ward, C.G., An analysis of the transits
   of Mercury: 1677-1973, Monthly Notices of the Royal
   Astronomical Society, 173, 183-206, 1975.

MS = Muller and Stephenson [1975].

Muller, P.M., An analysis of the ancient astronomical observations with the implications for geophysics and cosmology, a dissertation presented to the School of Physics, University of Newcastle, Newcastle-upon-Tyne, 1975.

Muller, P.M. and Stephenson, F.R., The accelerations of the earth and moon from early astronomical observations, in Growth Rhythms and the History of the Earth's Rotation, G.D. Rosenberg and S.K. Runcorn, editors, pp. 459-533, John Wiley and Sons, London and New York, 1975.

Munk, W.H. and MacDonald, G.J.F., The Rotation of the Earth, Cambridge University Press, Cambridge, 1960.

Needham, J., Science and Civilization in China, volume 3, (with the collaboration of Wang Ling), Cambridge University Press, Cambridge, 1959.

Needham, J., Astronomy in ancient and medieval China, Philosophical Transactions of the Royal Society of London, 276A, pp. 67-82, 1974.

Neresheimenses, Annales Neresheimenses, ca. 1296. There is an edition by Otto Abel in Monumenta Germaniae Historica, Scriptores, volume X, G.H. Pertz, editor, Hahn's, Hannover, 1852.

Neugebauer, O., Astronomical Cuneiform Texts, in 3 volumes, Lund Humphries, London, 1955.

Neugebauer, O., The Exact Sciences in Antiquity, Second Edition, Brown University Press, Providence, Rhode Island, 1957.

Neugebauer, P.V. and Weidner, E.F., Ein astronomischer Beobachtungstext aus dem 37. Jahre Nebukadnezars II. (-567/66), Berichte über die Verhandlungen der Königlichen Sachsischen Akademie der Wissenschaften zu Leipzig, Philologie-Historie Klasse, Bd. 67, Heft 2, pp. 29-89, 1915.

Newcomb, S., Researches on the motion of the moon, Washington Observations, U.S. Naval Observatory, Washington, 1875.

Newcomb, S., Discussion of observed transits of Mercury, 1677-1881, Astronomical Papers Prepared for the Use of the American Ephemeris and Nautical Almanac, I, Part 6, U.S. Government Printing Office, Washington, 1882.

Newcomb, S., Tables of the motion of the earth on its axis and around the sun, Astronomical Papers Prepared for the Use of the American Ephemeris and Nautical Almanac, VI, Part 1, U.S. Government Printing Office, Washington, D.C., 1895.

Newcomb, S., Tables of Mercury, Astronomical Papers Prepared for the Use of the American Ephemeris and Nautical Almanac, VI, Part 2, U.S. Government Printing Office, Washington, D.C., 1895a.

Newcomb, S., Tables of Venus, Astronomical Papers Prepared
for the Use of the American Ephemeris and Nautical Almanac, VI, Part 3, U.S. Government Printing Office, Washington, D.C., 1895b.

Newton, R.R., Secular accelerations of the earth and moon,
Science, 166, pp. 825-831, 1969.

Newton, R.R., Ancient Astronomical Observations and the Accelerations of the Earth and Moon, Johns Hopkins Press,
Baltimore and London, 1970. This is usually designated
as AAO for brevity.

Newton, R.R., Medieval Chronicles and the Rotation of the
Earth, Johns Hopkins Press, Baltimore and London, 1972.
This is usually designated as MCRE for brevity.

Newton, R.R., Letter to the editor, Sky and Telescope, 44,
p. 303, 1972a.

Newton, R.R., The earth's acceleration as deduced from
al-Biruni's solar data, Memoirs of the Royal Astronomical
Society, 76, pp. 99-128, 1972b.

Newton, R.R., Astronomical evidence concerning non-gravitational
forces in the earth-moon system, Astrophysics and Space
Science, 16, 179-200, 1972c.

Newton, R.R., The authenticity of Ptolemy's parallax data -
Part I, Quarterly Journal of the Royal Astronomical
Society, 14, pp. 367-388, 1973.

Newton, R.R., The historical acceleration of the earth, Geophysical Surveys, 1, pp. 123-145, 1973a.

Newton, R.R., Two uses of ancient astronomy, Philosophical
Transactions of the Royal Society of London, 276A, pp.
99-116, 1974.

Newton, R.R., The authenticity of Ptolemy's parallax data -
Part II, Quarterly Journal of the Royal Astronomical
Society, 15, pp. 7-27, 1974a.

Newton, R.R., The authenticity of Ptolemy's eclipse and star
data, Quarterly Journal of the Royal Astronomical Society,
15, pp. 107-121, 1974b.

Newton, R.R., Ancient Planetary Observations and the Validity
of Ephemeris Time, Johns Hopkins Press, Baltimore and
London, 1976. This usually is designated as APO for
brevity.

Newton, R.R., The Crime of Claudius Ptolemy, Johns Hopkins
Press, Baltimore and London, 1977.

Newton, R.R., Correlations between culturally important quantities that depend upon variable time intervals, areas,
or populations, Journal of the American Oriental Society,
97, pp. 181-185, 1977a.

Nihongi, 720. There is a translation into English by W.G.
Aston originally published by the Japan Society in 2
volumes in 1896. This was reissued in 1 volume in 1924,
which was reprinted by Charles E. Tuttle Co., Rutland,
Vermont and Tokyo, 1972. The pagination is still that
of the edition in 2 volumes.

Novimontensis, Continuatio Novimontensis, ca. 1396. There is
an edition by W. Wattenbach in Monumenta Germaniae Historica, Scriptores, volume IX, G.H. Pertz, editor,
Hahn's, Hannover, 1851.

Oesterwinter, C. and Cohen, C.J., New orbital elements for
moon and planets, Celestial Mechanics, 5, pp. 317-395,
1972.

Oppert, Julius, Noli me tangere, a mathematical demonstration
of the exactness of Biblical chronology, Society of Biblical Archaeology, Proceedings of January 11, 1898,
pp. 24-44, 1898.

Oppolzer, T.R. von, Note über eine von Archilochos erwähnte
Sonnenfinsterniss, Sitzungsberichte der Königlichen
Akademie der Wissenschaften, Wien, 86, Band 2, pp. 790-
793, 1882.

Oppolzer, T.R. von, Canon der Finsternisse, Kaiserlich-König-
lichen Hof- und Staatsdruckerei, Wien, 1887. There is a
reprint, with the introduction translated into English
by O. Gingerich, by Dover Publications, New York, 1962.

Opus Chronicorum, ca. 1296. There is an edition by H.T. Riley
in Rerum Britannicarum Medii Aevi Scriptores, no. 28,
part 3, Longmans, Green, Reader, and Dyer, London, 1866.

Ordericus Vitalis, Historia Ecclesiastica, 1141. There is an
edition in Recueil des Historiens des Gaules et de la
France, volume XII, edited by the Benedictines of S.
Maur, Chez Vve. Desaint, Paris, 1781.

Ottenburani, Annales Ottenburani Minores, ca. 1298. There is
an edition by G.H. Pertz in Monumenta Germaniae Historica,
Scriptores, volume XVII, G.H. Pertz, editor, Hahn's,
Hannover, 1861.

Oxenedes, Johannis de, Chronica, ca. 1292. There is an edition
by Sir Henry Ellis in Rerum Britannicarum Medii Aevi
Scriptores, no. 13, Longman, Brown, Green, Longmans, and
Roberts, London, 1859.

Palmerius, Mattheus, Historia Florentina, ca. 1474. There is
an edition by G. Scaramella in Rerum Italicarum Scriptores, volume XXVI, part I, Editrice S. Lapi, Citta di
Castello, 1915.

Panaretos, Michael, Peri ton tes Trapezountos Basileon, ca.
1426. There is an edition, with a following German translation, by J.P. Fallmerayer in Original-Fragmente, Chron-
iken, Inschriften, und anderes Materiale zur Geschichte
des Kaiserthums Trapezunt, Abhandlungen der Historischen

Classe der Königlich Bayerischen Akademie der Wissenschaften, Band 4, Abteilung 2, Munich, 1846.

Parker, R.A., Ancient Egyptian astronomy, Philosophical
Transactions of the Royal Society of London, A. 276,
pp. 51-65, 1974.

Parker, R.A. and Dubberstein, W.H., Babylonian Chronology,
626 B.C. - A.D. 75, Brown University Press, Providence,
Rhode Island, 1956.

Parmenses, Annales Parmenses Maiores, ca. 1335. There is an
edition by Philip Jaffé in Monumenta Germaniae Historica,
Scriptores, volume XVIII, G.H. Pertz, editor, Hahn's,
Hannover, 1863. There is another edition by L.A. Muratori, revised by Giuliano Bonazzi, using the title
Chronicon Parmense ab Anno MXXXVIII usque ad Annum
MCCCXXXVIII, in Rerum Italicarum Scriptores, volume IX,
part IX, Coi Tipi dell'Editore S. Lapi, Citta di Castello, 1902.

Paschale, Chronicon, ca. 628. There is an edition by L.
Dindorf, with critical notes and a parallel Latin trans-
lation by C. DuCange, in Corpus Scriptorum Historiae
Byzantinae, B.G. Niebuhr, editor, Weber's, Bonn, 1832.

Paulini, Annales Paulini, ca. 1341. There is an edition by
William Stubbs in Rerum Britannicarum Medii Aevi Scriptores, no. 76, part 1, Longman & Co., London, 1882.

Petrus, Antonius, Diarium Romanum, ca. 1417. There is an
edition by F. Isoldi in Rerum Italicarum Scriptores,
volume XXIV, part V, Editrice S. Lapi, Citta di Castello,
1922.

Pisanum, Chronicon Pisanum seu Fragmentum Auctoris Incerti,
ca. 1136. There is an edition by L.A. Muratori, revised
by M.L. Gentile, in Rerum Italicarum Scriptores, volume
VI, part II, Nicola Zanichelli, Bologna, 1930.

Placentini, Annales Placentini Gibellini, ca. 1284. There is
an edition by G.H. Pertz in Monumenta Germaniae Historica,
Scriptores, volume XVIII, G.H. Pertz, editor, Hahn's,
Hannover, 1863.

Pliny (Plinius Secundus), Natural History, 77. There is an
edition with a parallel French translation by Jean
Beaujeu, Société d'Édition "Les Belles Lettres", Paris,
1950.

Plutarch, De Facie Quae in Orbe Lunae Apparet, ca. 90. There
is a translation by Harold Cherniss in Plutarch's Moralia,
v. 12, H. Cherniss and W.C. Helmbold, editors, Harvard
University Press, Cambridge, Mass., 1957.

Plutarch, Parallel Lives, ca. 100. Dryden's translation is
reprinted in part in Harvard Classics, volume 12, Col-
lier and Son, New York, 1909.

Povest' Vremennykh Let, ca. 1110. There is a translation into
    English by S.A. Cross and O.P. Sherbowitz-Wetzor, under
    the title The Russian Primary Chronicle, The Mediaeval
    Academy of America, Cambridge, Mass., 3rd printing, 1973.

Praedicatorum Vindobonensium, Continuatio Praedicatorum
    Vindobonensium, ca. 1283. There is an edition by W.
    Wattenbach in Monumenta Germaniae Historica, Scriptores,
    volume IX, G.H. Pertz, editor, Hahn's, Hannover, 1851.

Ptolemy, C., 'E Mathematike Syntaxis, ca. 142. There is an
    edition by J.L. Heiberg in C. Ptolemaei Opera Quae Ex-
    stant Omnia, B.G. Teubner, Leipzig, 1898. This edition
    is translated into German by K. Manitius, B.G. Teubner,
    Leipzig, 1913. There is also an edition with a parallel
    French translation by N.B. Halma, Henri Grand Libraire,
    Paris, 1813. The chapter numbering used here is that of
    Heiberg and Manitius, which differs occasionally from
    that used by Halma. This work is often designated as
    Syntaxis for brevity.

Rampona, Cronica Rampona, ca. 1425. There is an edition by
    L.A. Muratori, revised by Albano Sorbelli, in Rerum
    Italicarum Scriptores, volume XVIII, in 3 parts, Stamp-
    eria di S. Lapi, Citta di Castello, 1906.

Ratisbonenses, Annales Ratisbonenses, ca. 1201. There is an
    edition by W. Wattenbach in Monumenta Germaniae Historica,
    Scriptores, volume XVII, G.H. Pertz, editor, Hahn's,
    Hannover, 1861.

Remense, Chronicon Remense, ca. 1190. Some excerpts are
    printed in Recueil des Historiens des Gaules et de la
    France, volumes IX and X, edited by the Benedictines of
    S. Maur, Chez Martin, Guerin, Delatour, et Boudet, Paris,
    1760.

Revised Standard Version (of the Holy Bible), Thomas Nelson &
    Sons, New York, Toronto, and Edinburgh, 1952.

Riché, Pierre, Éducation et Culture dans l'Occident Barbare,
    Éditions du Seuil, 27, rue Jacob, Paris, 1962.

Rigordus, Gesta Philippi Augusti, Francorum Regis, ca. 1208.
    There is an edition by Michel-Jean-Joseph Brial, in
    Recueil des Historiens des Gaules et de la France, vol-
    ume XVII, Imprimerie Royale, Paris, 1818.

Robert of Gloucester, Metrical Chronicle, ca. 1300. There is
    an edition by W.A. Wright in Rerum Britannicarum Medii
    Aevi Scriptores, no. 86, in 2 volumes, H.M. Stationery
    Office, London, 1887.

Robertus de Monte, Cronica, ca. 1186. There is an edition
    in Recueil des Historiens des Gaules et de la France,
    volume XIII, edited by the Benedictines of S. Maur,
    Chez Vve. Desaint, Paris, 1786.

Rolandinus, Rolandini Patavini Chronica Facta, 1262. There is an edition by Philip Jaffé in Monumenta Germaniae Historica, Scriptores, volume XIX, G.H. Pertz, editor, Hahn's, Hannover, 1866.

Ryccardus (de Sancto Germano), Chronica, ca. 1243. There is an edition by L.A. Muratori, revised by C.A. Garufi, in Rerum Italicarum Scriptores, volume VII, part II, Nicola Zanichelli, Bologna, 1937.

S. Albani, Chronicon Rerum Gestarum in Monasterio S. Albani, ca. 1431. There is an edition by H.T. Riley in Rerum Britannicarum Medii Aevi Scriptores, no. 28, part 5, volume 1, Longmans, Green and Co., London, 1870.

S. Blasii, Annalium S. Blasii Brunsvicensium Maiorum Fragmenta, ca. 1173. There is an edition by O. Holder-Egger in Monumenta Germaniae Historica, Scriptores, volume XXX, part I, O. Holder-Egger, editor, Hahn's, Hannover, 1896.

S. Blasii, Annales S. Blasii Brunsvicenses, ca. 1314. There is an edition by G. Waitz in Monumenta Germaniae Historica, Scriptores, volume XXIV, G. Waitz, editor, Hahn's, Hannover, 1879.

S. Blasii, Notae S. Blasii Brunsvicensis, ca. 1482. There is an edition by G. Waitz in Monumenta Germaniae Historica, Scriptores, volume XXIV, G. Waitz, editor, Hahn's, Hannover, 1879.

S. Denis, Chroniques Francoises de S. Denis, ca. 1328. There is an edition by Daunou and Naudet in Recueil des Historiens des Gaules et de la France, volume XX, Imprimerie Royale, Paris, 1840.

S. Edmundi, Annales S. Edmundi, ca. 1212. There is an edition by Thomas Arnold in Rerum Britannicarum Medii Aevi Scriptores, no. 96, volume 2, H.M. Stationery Office, London, 1892.

S. Georgii, Notae S. Georgii Mediolanenses, ca. 1295. There is an edition by Philip Jaffé in Monumenta Germaniae Historica, Scriptores, volume XVIII, G.H. Pertz, editor, Hahn's, Hannover, 1863.

S. Georgii, Annales Montis S. Georgii, ca. 1415. There is an edition by O. Holder-Egger in Monumenta Germaniae Historica, Scriptores, volume XXX, part I, O. Holder-Egger, editor, Hahn's, Hannover, 1896.

S. Iacobi, Annales S. Iacobi Leodiensis, ca. 1174. There is an edition by G.H. Pertz in Monumenta Germaniae Historica, Scriptores, volume XVI, G.H. Pertz, editor, Hahn's, Hannover, 1859.

S. Iacobi, Annales S. Iacobi Leodiensis, ca. 1393. There is an edition by G.H. Pertz in Monumenta Germaniae Historica, Scriptores, volume XVI, G.H. Pertz, editor, Hahn's, Hannover, 1859.

S. Iustinae, Annales S. Iustinae Patavini, ca. 1270. There
is an edition by Philip Jaffé in Monumenta Germaniae
Historica, Scriptores, volume XIX, G.H. Pertz, editor,
Hahn's, Hannover, 1866.

S. Nicasii, Annales S. Nicasii Remenses, ca. 1309. There is
an edition by G. Waitz in Monumenta Germaniae Historica,
Scriptores, volume XIII, G.H. Pertz, editor, Hahn's,
Hannover, 1881.

S. Nicolai, Annales S. Nicolai Patavienses et Notae Wolfemi,
ca. 1287. There is an edition by G. Waitz in Monumenta
Germaniae Historica, Scriptores, volume XXIV, G. Waitz,
editor, Hahn's, Hannover, 1879.

S. Petri Erfordensis, Cronica S. Petri Erfordensis Moderna,
ca. 1353. There is an edition by O. Holder-Egger in
Monumenta Germaniae Historica, Scriptores, volume XXX,
Part I, O. Holder-Egger, editor, Hahn's, Hannover, 1896.

S. Rudberti, Continuatio Canonicorum S. Rudberti Salisburgensis,
ca. 1327. There is an edition by W. Wattenbach in Monumenta Germaniae Historica, Scriptores, volume IX, G.H.
Pertz, editor, Hahn's, Hannover, 1851.

S. Victoris, Annales Sancti Victoris Massilienses, ca. 1542.
There is an edition by G.H. Pertz in Monumenta Germaniae
Historica, Scriptores, volume XXIII, G.H. Pertz, editor,
Hahn's, Hannover, 1874.

Sambiensis, Canonici Sambiensis Annales, ca. 1352. There is
an edition by Wilhelm Arndt in Monumenta Germaniae Historica, Scriptores, volume XIX, G.H. Pertz, editor,
Hahn's, Hannover, 1866.

Sancrucensis, Continuatio Sancrucensis, ca. 1310. There is an
edition by W. Wattenbach in Monumenta Germaniae Historica,
Scriptores, volume IX, G.H. Pertz, editor, Hahn's,
Hannover, 1851.

Sawyer, J.F.A., Joshua 10:12-14 and the solar eclipse of 30
September 1131 B.C., Palestine Exploration Quarterly,
pp. 139-145, 1972a. (This does not carry a volume
number.)

Sawyer, J.F.A., Why is a solar eclipse mentioned in the Passion
narrative (Luke XXIII.44-45)?, Journal of Theological
Studies, New Series, 23, pp. 124-128, 1972b.

Sawyer, J.F.A. and Stephenson, F.R., Literary and astronomical
evidence for a total eclipse of the sun observed in
ancient Ugarit on 3 May 1375 B.C., Bulletin of the School
of Oriental and African Studies, 33, pp. 467-489, 1970.

Saxonici, Annales Saxonici, ca. 1271. There is an edition by
L.C. Bethmann in Monumenta Germaniae Historica, Scriptores,
volume XVI, G.H. Pertz, editor, Hahn's, Hannover, 1859.

Saxonicum, Chronicon Saxonicum, ca. 1139. The explanatory
material needed to understand this source is by M.
Bouquet in Recueil des Historiens des Gaules et de la
France, volume V, Chez Martin, Coignard, Mariette,
Guerin, et Guerin, Paris, 1744. The part of the text
that we need appears in Recueil des Historiens des Gaules
et de la France, volume X, edited by the Benedictines of
S. Maur, Chez Martin, Guerin, Delatour, et Boudet, Paris,
1760.

Senenses, Annales Senenses, ca. 1479. There is an edition
by J.F. Boehmer in Monumenta Germaniae Historica, Scriptores, volume XIX, G.H. Pertz, editor, Hahn's, Hannover,
1866.

Sifridus, Sifridi Presbyteri de Balnhusin Historia Universalis
et Compendium Historiarum, ca. 1307. There is an edition
by O. Holder-Egger in Monumenta Germaniae Historica,
Scriptores, volume XXV, G. Waitz, editor, Hahn's, Hannover, 1880.

Sigebertus (Sigebertus Gemblacensis), Chronica, ca. 1111.
There is an edition by L.C. Bethmann in Monumenta Germaniae Historica, Scriptores, volume VI, G.H. Pertz,
editor, Hahn's, Hannover, 1844.

Sivin, N., Cosmos and computation in early Chinese mathematical astronomy, T'oung Pao, 55, pp. 1-73, 1969.

Smith, George, Assyrian Discoveries; An Account of Explora-
tions and Discoveries on the Site of Nineveh, During
1873 and 1874, Sampson Low, Marston, Low and Searle,
London, 1875.

Smith, P.J., The intensity of the ancient geomagnetic field:
a review and analysis, Geophysical Journal of the Royal
Astronomical Society, 12, pp. 321-362, 1967.

Souciet, E., Editor, Observations Mathématiques, Astronomiques,
Géographiques, Chronométriques, et Physiques Tirées des
Anciens Livres Chinois, volume 1, Rollin Père, Paris,
1729.

Spencer Jones, H., The rotation of the earth, and the secular accelerations of the sun, moon, and planets, Monthly
Notices of the Royal Astronomical Society, 99, pp. 541-
558, 1939.

Stefani, Marchionne di Coppo, Cronaca Fiorentina, ca. 1386.
There is an edition by Niccolo Rodolico in Rerum Itali-
carum Scriptores, volume XXX, part I, Editore S. Lapi,
Citta di Castello, 1903.

Stephenson, F.R., The earliest known record of a solar eclipse,
Nature, 228, pp. 651-652, 1970.

Stirensis, Honorii Augustodunensis Continuatio Stirensis, ca.
1346. There is an edition by G. Waitz in Monumenta
Germaniae Historica, Scriptores, volume XXIV, G. Waitz,
editor, Hahn's, Hannover, 1879.

Stoyko, Anna, La variation seculaire de la rotation de la terre et les problemes connexes, Annales Guebhard, 46, pp. 293-316, 1970.

Syntaxis = Ptolemy [ca. 142].

Syriac Chronicle, Syriac Chronicle Known as That of Zachariah of Mitylene, 569. There is an English translation by F.J. Hamilton and E.W. Brooks, Methuen and Co., London, 1899.

Thompson, J.E.S., Maya astronomy, in The Place of Astronomy in the Ancient World, Philosophical Transactions of the Royal Society, 276A, pp. 83-98, 1974.

Thucydides, History of the Peloponnesian War, ca. -420. There is a translation into English by Richard Crawley, J.M. Dent and Sons, London, 1910.

Thuringici, Annales Thuringici Breves, ca. 1291. There is an edition by G. Waitz in Monumenta Germaniae Historica, Scriptores, volume XXIV, G. Waitz, editor, Hahn's, Hannover, 1879.

Times Atlas of the World, Mid-Century Edition, in 5 volumes, The Times Office, London, 1955.

Tinsley, B.M., The cosmological constant and cosmological change, Physics Today, 30, pp. 32-39, 1977.

Tolosanum, Chronicon Tolosanum, ca. 1271. This appears in two different volumes of Recueil des Historiens des Gaules et de la France. One part is in volume XII, edited by the Benedictines of S. Maur, Chez Vve. Desaint, Paris, 1781. The other part is in volume XIX, edited by Michel-Jean-Joseph Brial, Imprimerie Royale, Paris, 1833.

Treverorum, Gesta Treverorum Continuata, ca. 1242. There is an edition by G. Waitz in Monumenta Germaniae Historica, Scriptores, volume XXIV, G. Waitz, editor, Hahn's, Hannover, 1879.

Tso Ch'iu-Ming, Commentary on the Annals of Lu (see Confucius [-479]), ca. -450. There is an edition with a partial translation by J. Legge, Chinese Classics, volume 5, Hong Kong University Press, Hong Kong, first edition, 1872, second edition, 1960. Legge gives, for each year, the text of Confucius [-479], then the text of Tso [ca. -450], then a translation of Confucius, and finally a commentary on the events, including translations of part of Tso's comments. I believe that he does not give a full translation of Tso.

Ulster, Annals of Ulster, (sometimes known as the Annals of Senat; A Chronicle of Irish Affairs), ca. 1498. There is an edition by W.M. Hennessy, H.M. Stationery Office, London, in 4 volumes, from 1887-1901.

Urbevetani, Cronica Potestatum Urbevetani, ca. 1260. There
is an edition by L. Fumi in Rerum Italicarum Scriptores,
volume XV, part V, Volume I, Editore S. Lapi, Citta di
Castello, 1920.

Urbevetani, Cronica Potestatum Urbevetani, ca. 1276. There
is an edition by L. Fumi in Rerum Italicarum Scriptores,
volume XV, part V, Volume I, Editore S. Lapi, Citta di
Castello, 1920.

Urbevetani, Chronica Antiqua Urbevetani, ca. 1313. There is
an edition by L. Fumi in Rerum Italicarum Scriptores,
volume XV, part V, Volume I, Editore S. Lapi, Citta di
Castello, 1920.

van der Waerden, B.L., Secular terms and fluctuations in the
motions of the sun and moon, Astronomical Journal, 66,
pp. 138-147, 1961.

van Flandern, T.C., The secular acceleration of the moon,
Astronomical Journal, 75, pp. 657-658, 1970.

van Flandern, T.C., A determination of the rate of change of
G, paper presented to the American Geophysical Union,
Washington, D.C., on April 8, 1974.

van Flandern, T.C., Is gravity getting weaker?, Scientific
American, 234, no. 2, pp. 44-52, 1976.

Varignana, Cronaca Varignana, ca. 1425. There is an edition
by L.A. Muratori, revised by Albano Sorbelli, in Rerum
Italicarum Scriptores, volume XVIII, in 3 parts, Stamperia di S. Lapi, Citta di Castello, 1906.

Velikovsky, I., Worlds in Collision, Doubleday & Co., New
York, 1950. There is also an edition published by
Pocket Books, New York, 1977. Page references are to
the Pocket Book edition.

Veterocellenses, Annales Veterocellenses, ca. 1484. There is
an edition by G.H. Pertz in Monumenta Germaniae Historica,
Scriptores, volume XVI, G.H. Pertz, editor, Hahn's,
Hannover, 1859.

Villola, Pietro e Floriano da, Cronaca, ca. 1376. There is an
edition by L.A. Muratori, revised by Albano Sorbelli, in
Rerum Italicarum Scriptores, volume XVIII, in 3 parts,
Stamperia di S. Lapi, Citta di Castello, 1906.

Vincentina, Cronica Vincentina, ca. 1241. There is an edition
by G. Waitz in Monumenta Germaniae Historica, Scriptores,
volume XXIV, G. Waitz, editor, Hahn's, Hannover, 1879.

Vindobonensis, Continuatio Vindobonensis, ca. 1327. There is
an edition by W. Wattenbach in Monumenta Germaniae Historica, Scriptores, volume IX, G.H. Pertz, editor,
Hahn's, Hannover, 1851.

Walcher (of Malvern), Lunar Tables, ca. 1110. Part of this
is printed by C.H. Haskins in Studies in Mediaeval
Science, Harvard University Press, Cambridge, Mass.,
1924.

Walsingham, Thomae, Historica Anglicana, ca. 1422. There is
an edition by H.T. Riley in Rerum Britannicarum Medii
Aevi Scriptores, no. 28, in 2 volumes, Longman, Green,
Longman, Roberts, and Green, London, 1863.

Wendover, Roger of, Liber Qui Dicitur Flores Historiarum, ca.
1235. There is an edition by H.G. Hewlett in Rerum
Britannicarum Medii Aevi Scriptores, no. 84, in 3 volumes,
H.M. Stationery Office, London, 1886-1889.

Windbergenses, Annales Windbergenses, ca. 1407. There is an
edition by Philip Jaffé in Monumenta Germaniae Historica,
Scriptores, volume XVII, G.H. Pertz, editor, Hahn's,
Hannover, 1861.

Wormatienses, Annales Wormatienses Breves, ca. 1295. There
is an edition by G.H. Pertz in Monumenta Germaniae Historica, Scriptores, volume XVII, G.H. Pertz, editor,
Hahn's, Hannover, 1861.

Wykes, Thomas, Chronicon, ca. 1289. There is an edition with
notes in Annales Monastici, edited by H.R. Luard, which
is volume 4 of number 36 in the series Rerum Britannicarum
Medii Aevi Scriptores, Longmans, Green, Reader, and Dyer,
London, 1869.

Wylie, A., Chinese Researches, Shanghai, 1897. Reprinted by
Ch'eng Wen Publishing Co., Taipei, 1966.

Wyntershylle, William, Annales Henrici Quarti, Regis Angliae,
ca. 1406. There is an edition by H.T. Riley in Rerum
Britannicarum Medii Aevi Scriptores, no. 28, part 3,
Longmans, Green, Reader, and Dyer, London, 1866.

Zenkovsky, Serge A., Medieval Russia's Epics, Chronicles, and
Tales, E.P. Dutton and Co., New York, 1963.

Zosimus, Historias Neas, ca. 475. There is an edition with a
parallel Latin translation by Immanuel Bekker in Corpus
Scriptorum Historiae Byzantinae, B.G. Niebuhr, editor,
Weber's, Bonn, 1837.

Zwetlenses 285, Annales Zwetlenses Brevissimi, ca. 1281.
There is an edition by G. Waitz in Monumenta Germaniae
Historica, Scriptores, volume XXIV, G. Waitz, editor,
Hahn's, Hannover, 1879.

Zwetlensis, Continuatio Zwetlensis III, ca. 1329. There is an
edition by W. Wattenbach in Monumenta Germaniae Historica,
Scriptores, volume IX, G.H. Pertz, editor, Hahn's,
Hannover, 1851.

INDEX

A

Aachen, Germany, 281, 505

Abbeydore, Hereford, England, 239

acceleration
of earth's spin, 4, 10, 17, 55ff, 65, 105ff, Section XIV.9, 480ff
of the moon, 3ff, 10ff, 14ff, 19ff, 26ff, 55ff, 65, 103ff, Section IV.6, 222, Section XIV.4, Section XIV.9, 480
of the planets, 11, 19ff see: individual planets
of the sun, 11, 19ff, 25ff, 55ff, 104ff, 136ff, 222, Section XIV.3, 481

Acre, Palestine, 268

Acropolita, Georgius, 381, 393

Admont, Austria, 346, 347, 348, 505

Aegean Sea, 193, 229ff

Aegidius, S., 370

Aeneas, 183

Agathocles, 200, 218

Aijalon, Palestine, 39, 40

Albericus, 253, 257, 270, 273

Albi, France, 257

Alcala de Henares, Spain, 353

Alcobaca, Portugal, 353, 356, 372, 505

Alcobacense, Chronicon, 353, 356

Alexander III (pope), 322

Alexander (king of Scotland), 241

Alexandria, Egypt, 206, 218, 381

Alexius I Comnenus, 382, 395

Allen, E. B., 383, 388ff

Alpertus, 253, 261

Altahenses [ca. 1073] (Annales Altahenses Maiores), 279, 285, 288, 291, 292

Altahenses [ca. 1585](Notae Altahenses), 279, 298, 306, 309

American Ephemeris and Nautical Almanac, 4, 7, 402

Amida, Turkey, 385, 391

Ammersee, 284

Amorites, 39

Note: The word "Saint" is represented in this index, and throughout this work, by the abbreviation "S.", regardless of the language, gender, or number involved. In alphabetizing the name of a person who is a saint, I use the name of the person, followed by "S." to identify which person with this name is meant. Where "S." is a necessary part of a name, I alphabetize such a name by using "S." in front of the rest of the name. For example, the person known as S. Luke appears as "Luke, S." S. Paul's Cathedral in London, on the other hand, appears as "S. Paul's Cathedral". Names starting with "S." are alphabetized in a separate section that precedes the other S's.

Ampsanus, S., 318

Anastasius (Byzantine emperor), 386, 387

Anchin, France, 255
see: Douai, France

Andreas Bergomates, 319

Andronicus II (Byzantine emperor), 382, 394

Andronicus III (Byzantine emperor), 382

Angers, France, 255, 275, 505

Angliae (Chronicon Angliae ab Anno Domini 1328 usque ad Annum 1388), 238, 248, 249

Anglo-Saxon Chronicle, 238, 244

Angoulême, France, 254, 505

Anjou (France), 254, 504

Annales Maximi (Roman), 198

Antakya, Turkey, 42

Antioch, Turkey, 42, 208, 382, 384, 391ff, 468ff, 505

Antoninus Pius, 47

Anyang, China, 71ff, 465

Apollonian spectacles, 202

Archilochus (or "eclipse of"), 190ff, 193, 195, 210, 212, 218, 229ff, 468

Arezzo, Italy, 328

Aristarchus (of Samos), 46

Aristyllus, 46

Armenia, 187, 209, 210, 504

Arras, France, 254, 505

Arthur, 199

Ashbrook, J., 36

Asia Minor, 194, 218, 293, 469

Assur-nasir-abal, 189, 190

Assyria, 29, 49ff, 145, 190, 218, 219, 414

Aston, W. G., 379

Astun el Mata'na, Egypt, 383

Athens, 87, 187, 196, 197, 211, 218, 219, 376, 469, 493, 494, 505

Augsburg, Germany, 281, 282, 283, 303, 304, 305, 308, 505

Augustani Minores, Annales, 303, 304, 305

Augustus (Caesar), 199, 356, 484, 496, 497, 498ff

aurora, 264

Austria, 30, 345, 354, 363, 364ff, 370

Autissiodorense, Breve Chronicon, 256, 263

Autun, France, 255, 505

Auxerre, France, 256, 263, 275 505

Aventine Hill (Rome), 203

B

Babylon or Babylonia, 29, 47, 48, 106, 187, 188, 196, 205ff, 218, 220, 221, 413ff, 505

Baffin Island, Canada, 181

Baghdad, 390

Baily's beads, 319

139, 140, 160, 165, 171, 461, 506

Cherng-du, China, 81

China, 29, 70, 74ff, 84ff, 94ff, 126, Chapter V passim, 381, 413, 414, 462, 504
heartland China defined, 141

Chinq-yang, China 81

Chounradus, 283, 293, 295

chromosphere, 212ff, 262, 340, 467

Ch'u (Chinese state), 173ff

Chung K'ang (Chinese emperor), 142

Cicero, M. Tullius, 197, 198ff 467ff, 494

Cilicia, 266

Clarius of Sens, 128

Clark, H. K., 403

Claudius (Roman emperor), 209

Claustroneoburgensis, Continuatio, 349, 350, 365, 369

Clavius, Christopher, 212, 213, 313, 339ff, 353, 371, 374, 459, 470

Clemens, S. L., 192
see: Twain, Mark

Cluny, France, 254, 255, 262, 275, 459, 461, 462ff, 506

Coggeshall, England, 232, 233, 506

Cohen, A. P., 75ff, 139, 158, Section V.5 passim, 501ff

Cohen, C. J., 5, 453

Coimbra, Portugal, 371, 373, 374, 506

Colbaz, Poland, 351, 358, 363, 506, 510

Colbazienses, Annales, 351, 358, 363, 367

cold, during eclipse, 39

Collegio Romano (Rome), 340, 343, 506

Colmar, France, 256, 273, 276, 349, 506

Colmarienses (Annales Colmarienses Minores), 256, 272

Cologne, Germany, 281, 282, 284, 307, 506

Colonienses (Annales Colonienses Maximi), 283, 295, 296

Coloniensis, Chronicae Regiae, 283, 295

comet, 469

Confucius, 143ff, 173ff, 185

Constantine, 383, 384

Constantinople, 35, 37, 376, 382, 386, 387, 390, 393ff, 397, 469, 506
see: Istanbul

consuls, 49, 201ff

Copernicus, N., 43

Cordoba, Spain, 353, 355, 372, 506

core, earth's, Section XIV.9, 480

corona, 35, 212ff, 262, 357, 466ff, 469

565

Gubbio, Italy, 314, 337, 338, 343, 507

Guerrierus, 314, 337, 338

Guido de Coregia, 325

Guidonis, Bernardus, 256, 276

Guillelmus Armoricus, 257, 260

Guillelmus de Nangiaco, 253, 257, 258, 260, 265, 274

Guillelmus de Podio-Laurentii, 257, 265, 270, 271

H

Hadrian, 211

Halley, E., 222, 223

Hamburg, Germany, 65, 285, 302, 308, 507

Hamburgenses, Annales, 285, 302

Han dynasties
Former (Western), 74ff, 78, 83ff, 127, Section V.3 passim, 178, 527, 529ff
Later (Eastern), 77ff, 83ff, Section V.3 passim, 166ff

Harran, Turkey, 390

Harrison, Kenneth, 37

Harsefeld, Germany, 281, 507

Heath, Thomas, 208

Heiligenkreuz, Austria, 350, 351, 362, 363, 364, 365, 372, 373, 503, 507

Heilsbronn, Germany, 281, 282, 507

Heimo, 284

Heinricus, 352, 360

Hellespont, 87, 206, 218

Henry I (of England), 245

Henry II (of England), 244ff

Heraclius (Byzantine emperor), 384

Herbipolenses (Annales Herbipolenses Minores), 285, 300

Hermannus, 285, 289, 298

Herodotus, 194

Herzogenrath, Germany, 281, 507

Hieronymus, 314, 334, 335, 337

Hildesheim, Germany, 280, 281, 507

Hildesheimenses, Annales, 290

Hipparchus (or eclipse of), 22, 44, 46, 87, 200, 206ff, 216, 218, 220ff, 223, 225, 228, 383, 455, 456, 481

Hoang, P., 142

Holland, B. B., xvi

Holy Land, 30, 323, 375
see: Palestine

Honorius (pope), 248

Honorius Augustodunensis, 351

horizontal eclipse, 215

Horner, Joseph, 189

Howe, J. W., xvi

Hradisch, Czechoslovakia, 347, 507

Hubble, E. P., 476

Hugo of Flavigny, 128

Loch Cé, Annals of, 240, 250

Lögmanns Annáll, 367, 368

London, 232, 233, 238, 240ff, 247, 248, 251, 508, 512

Londonienses, Annales, 240ff, 248

Lorsch, Germany, 280, 289, 508

Lothar II (king of Italy), 244

Louis VIII (king of France), 260

Louis IX (king of France), 257

Lo-yang: see Luoh-yang

Lu (Chinese state), or Annals of Lu, 142, Section V.2 passim, 156, 161, Section V.6 passim, 179ff, 185, 461, 530ff

Lübeck, Germany, 285, 302, 308, 508

Lubenses, Annales, 352, 361

Lubicenses, Annales, 285, 302

Lubin, Poland, 352, 361, 372, 508

Luke, S., 41ff, 208

Lund, Sweden, 347, 508

Luoh-yang, China, 83, 86, 126ff, 139, 140, 164, 165, 176, 462

Lusitania, 371

Lynn, Norfolk, England, 238

Lyon, France, 254, 508

M

McConahy, R. J., 64

MacDonald, G. J. F., 10, 473

MacPike, E. F., 223

Maastricht, Netherlands, 37

Macrobius, 492, 493, 494ff

Madinat, Egypt, 383, 388ff, 397, 508, 510

Madrid, 353

Magdeburg, Germany, 281, 508

magnetic field (earth's), 137, Section XIV.9, 480

Mainz, Germany, 285, 307, 508

Malalas, Ioannis, 382, 385

Malaterra, Gaufridus, 315, 321

Malmesbury, England, 37, 232, 508

Man, Isle of, 29, 234, 504

Mantua, Italy, 315, 325, 341, 508

Mantuana, Chronica Pontificum et Imperatorem, 315, 325

Mantuani, Annales, 315, 325

Maragone, Bernardo, 315, 323

Maragone, Salem, 315

Marcian (Roman emperor), 386

Margan, Wales, 234, 508

Mark, S., 41ff

Market S. Florian, Austria, 349 see: S. Florian, monastery of

Markowitz, W., 16

Mars, 11, 66, 69, 70, Section V.8 passim, 205, 206

Marseilles, France, 260, 272, 276, 277, 508

Martin V (pope), 336, 337

Martin, C. F., 6ff, 16, 19, 89, 90, 91, 223, 404, 445, 456

Matthaeus de Griffonibus, 315, 325, 332

Matthew, S., 41ff

"Matthew of Westminster", 240

Matthew Paris, 238, 239, 242, 246

Mechovienses, Annales, 352, 367, 368

Mediolanenses, Memoriae, 315, 326

Mediterranean Sea, 29, 413ff

Melk, Austria, 346, 347, 349, 366, 370, 373, 508

Mellicenses, Annales, 349, 350, 351, 366, 370

Melrose, Scotland, 234, 251, 508

Memmingen, Germany, 285, 299, 307, 508

Menkonis, 345, 361

Mercati, A., 336

Mercedonius, 491

Mercury, 11, 14ff, 19ff, 25ff, 47, 205, 206

Merodachbaladan, 48

Merseburg, Germany, 307, 508

Merton, Chronicon de, 241, 247

Merton College, Oxford, 241

Meton, 44, 492, 493, 495

Mettensis, Chronica Universalis, 259, 269

Metz, France, 253, 255, 259, 261, 269, 275, 508

Michael I (patriarch of Antioch), 382, 384, 386ff, 390ff

Michels, Agnes K., 483, 485, 486, 488, 492, 493ff

Miechovia, monastery of (Krakow), 352

Milan, Italy, 312, 315, 316, 317, 323, 326, 341, 508

Miliolus, Albertus, 316, 326

Modena, Italy, 313

Moguntini, Annales, 285, 299

Mommsen, Theodor, 215, 483, 488ff, 491ff, 494, 496

Mongolia, 78, 168

Mons S. Georgii, Italy, 317

Monte Cassino, Italy, 312, 313, 320, 321, 322, 327, 341, 508 see: Cassino, Italy

Montfort, Simon de, 247

Mont S. Michel, 52, 255, 260, 265, 266, 267, 275, 508

moon
    acceleration of: see
    acceleration of the moon
    elongation of, 6ff, 55ff, 103ff,
    Section IV.6, 408ff, Section
    XIV.1, 479ff
    motion of, 14ff, 403ff
    node of orbit of, Section IV.6,
    Section XIV.2, 456, 481

Morrison, L. V., 5, 27, 132, 453, 457

Mt. Sumelas, Turkey, 395ff, 397, 508

Muirhead's Guides, 259

Numa (king of Rome), 484ff, 491

Nuremberg, Germany, 334

O

Ober-Alteich, Germany, 279,
288, 298, 307, 509

octaeteris, 492ff

Odoriscii, Thomas, 328

Oesterwinter, C., 5, 453

Olaf, S., 42, 355ff

O'Neill, M. J., xvi

Oppert, Julius, 189

Oppolzer, T. R. von, 48, 88,
173, 190, 203, 246, 247, 267,
274, 303, 321, 339, 359, 361,
380, 388, 389, 401, 466

Opus Chronicorum, 241ff, 247

oracle bones, 70ff

Ordericus Vitalis, 259, 264

Orléans, France, 254, 509

Orvieto, Italy, 318, 319, 327,
329, 341, 342, 509

Oslo, Norway, 347, 509

Osney, Oxfordshire, England,
232, 233, 505, 509

Ottenburani (Annales Ottenburani
Minores), 285, 299

Ottenburanus, monastery of
(Memmingen), 285

Oxenedes, Johannis de, 246

Oxford, England, 241, 243, 247

Oxnead, Norfolk, England, 36,
246, 251, 509

P

Paderborn, Germany, 281, 509

Padua, Italy, 317, 318, 324, 325,
329, 341, 342, 509

Palermo, Sicily, 145

Palestine, 41, 375
see: Holy Land

Palilia, festival of, 489

Palmerius, Mattheus, 316, 338

Pamplona, Spain, 271, 272, 361

Panaretos, Michael, 382, 394,
395ff

Pappus, 208

Paris, 248, 256, 258, 260, 274,
276, 509, 512

Parker, R. A., 51, 67, 181, 205

Parma, Italy, 316, 326, 331, 341,
342, 509

Parmenses (Annales Parmenses
Maiores), 316, 326, 331

Paros, 193, 194, 218, 230, 468

Paschale, Chronicon, 215

Passau, Germany, 286, 301, 308,
509

Paulini, Annales, 242, 248

Pegau, Germany, 281, 282, 509

Peking, China, 139, 140, 141

Peloponnesian War, 197

Peloponnesus, 197

Pericles, 197

Peterborough, England, 232, 238,
245, 246, 251, 509

Petrus, Antonius, 316, 335

Petrus Anibaldi, 327

Petrus Coral, 258

Philip II (king of France), 257, 260

Philip III (king of France), 257

Philip IV (king of France), 257

Phlegon, 208ff, 218

Piacenza, Italy, 317, 326, 341, 509

Pindar, 194, 195, 218

Pisa, Italy, 315, 317, 320, 323, 341, 509

Pisanum (Chronicon Pisanum seu Fragmentum Auctoris Incerti), 317, 320

Placentini (Annales Placentini Gibellini), 317, 325

Pliny (Plinius Secundus), 215

Plutarch (or "eclipse of"), 35, 197, 200, 202, 210ff, 218, 220, 466ff

Pohlde, Germany, 282, 509

Poland, 30, 345, 351ff, 366, 374

Pons Vicus, Italy, 313

Pont-à-Mousson, France, 254, 509

Portugal, 30, 313, 339, 345, 353ff, 374, 414

Povest' Vremennykh Let, 352, 356

Praedicatorum Vindobonensium, Continuatio, 350, 362

Prague, 346, 347, 348, 372, 509

Prum, Germany, 280, 509

Prussia, 353

Ptolemy, C., 11, 43ff, 47ff, 108, 147, 183, 188ff, 196, 221, 222, 223, 225, 228, 383, 384, 415, 447, 454, 480

Puylaurens, France, 257

Q

Quedlinburg, Germany, 280, 509

R

Raimond VII (count of Toulouse), 257

Rainerius, 318

Rampona, Chronica, 315, 329, 330, 331, 333

Ratisbonenses, Annales, 286, 291, 292

Ravenna, Italy, 312, 314, 509

Ravensburg, Germany, 36

Regensburg, Germany, 281, 286, 288, 289, 292, 301, 307, 308, 509

Reggio di Calabria, Italy, 316

Reggio nell'Emilia, Italy, 316, 326, 341, 509

Reichenau, Germany, 280, 509

Reichersberg, Austria, 347, 348, 509, 512

Reims, France, 259, 260, 261, 269, 270

Remense, Chronicon, 259, 261, 262, 263

Richard I (king of England), 238

Riché, Pierre, 36

S.[†]

[†]See the note at the beginning of the index.

Georgii Mediolanenses), 317, 323

S. Georgii [ca. 1415] (Annales Montis S. Georgii), 317, 329, 336

S. Giorgio d. E., Italy, 317, 329, 336, 342, 343, 510

S. Germain-des-Prés (Paris), 287

S. Germano, Italy, 317
see: Monte Cassino, Italy

S. Iacobi, Annales, 348, 365

S. Iustinae (Annales S. Iustinae Patavini), 318, 324, 329

S. Jacob, monastery of (Liège), 349

S. Mark's, church of (Venice), 352

S. Martial, monastery of (Limoges), 258

S. Michael, monastery of (Bamberg), 284

S. Nicasii (Annales S. Nicasii Remenses), 260, 269

S. Nicola, monastery of (Passau), 286

S. Nicolai (Annales S. Nicolai Patavienses et Notae Wolfemi), 286, 301

S. Omer, France, 254, 257, 268, 272, 275, 276, 361, 510

S. Paul's Cathedral (London), 242, 508

S. Peter's, abbey of (Salzburg), 350

S. Petri Erfordensis, Cronica Modena, 286, 293, 294, 295,

297, 302, 303, 305

S. Rudberti (Continuatio Canonicorum S. Rudberti Salisburgensis), 350, 366

S. Symphorianus, monastery of (Metz), 253

S. Trudperti, Germany, 282, 510

S. Victoris, monastery of (Marseilles), 260

S. Victoris (Annales Sancti Victoris Massilienses), 260, 272, 276

S[†]

Sachs, A. J., 204

Saladin, 323

Salerno, Italy, 312, 320, 510

Salzburg, Austria, 346ff, 348ff, 354, 366, 372, 373, 503, 510

Samarkand, 78, 168, 169

Sambiensis (Canonici Sambiensis Annales), 353, 363, 366, 367

Sancrucensis, Continuatio, 350, 362, 363, 364ff

Sandbach, S. H., 213

Santarem, Portugal, 353, 356, 372, 510

Sargon II, 189

saros, 157, 411

Saturn, 11, 206

Saumur, France, 255, 510

Sawyer, J. F. A., 38ff, 41ff, 66, 182, 208, 465, 468

---

†See the note at the beginning of the index.

Augustodunensis Continuatio
Stirensis), 351, 363

Stoyko, Anna, 65

Strasbourg, France, 253

Straubing, Germany, 288

Stuttgart, Germany, 256, 280,
510

Styria (Austria), 351

Sui dynasty: see Swei dynasty

Suiko (Japanese empress), 378ff

sun
    acceleration of: see
    acceleration of the sun
    motion of, 13ff, 403ff

Swei dynasty, 166ff

Switzerland, 30, 345, 354

Syntaxis: see Ptolemy, C.

Syria, 375, 376, 384ff, 386, 469,
504

Syriac Chronicle, 385, 387

T

T'ang dynasty: see Tarng dynasty

Tarng dynasty, 77ff, 127ff, 156,
158, 166, Section V.5 passim,
178, 180, 381

Tartars, 362

Temmu (Japanese emperor),
378ff

Terracina, Italy, 313, 322, 341, 510

Tewkesbury, England, 233, 510

Thales, 194, 218

Thasos, 87, 187, 193, 194, 196,
218, 229ff, 469, 510

Thebes, Egypt, 383, 388ff, 510

Thebes, Greece, 187, 195, 200,
218, 511

Theodosius, 386

Theon, 218, 219

Thomas, S., 248
    see: Becket, Thomas

Thompson, J. E. S., 487

Thucydides, 87, 196, 197, 218,
469

Thuringia , 288

Thuringici (Annales Thuringici
Breves), 288, 295, 300, 302

tidal friction, 10, 19, 137,
Section XIV.9 , 480

time
    atomic, 12, 453, 472ff
     defined, 2
    ephemeris, 403ff, 453, 471ff
     defined, 2
     validity of, 4, 10ff, 15ff
    sidereal, defined, 4
    solar, 403ff, 472
     defined, 1
    universal, 403
     defined, 3

Times Atlas, 143, 177, 279, 351,
383, 503, 507, 510, 512

Timocharis, 46

Tinsley, B. M., 476

Tolosanum , Chronicon, 261, 265,
269, 271

Toulouse, France, 257, 265, 269,
270, 271, 275, 511

Tournai, Belgium, 36, 258

Tours, France, 254, 511

Trabzon, Turkey, 382
see: Trebizond, Turkey

Trajan, 211

Trajectus, 36

Trebizond, Turkey, 382, 395, 397,
511

Trebizond, empire of, 382

Treverorum (Gesta Treverorum
Continuata), 288, 300

Trier, Germany, 280, 300, 308,
511

Trois-Fontaines, France, 253,
270, 273, 275, 276, 511

Trois-Fontaines (Sicily), 253

Trondheim, Norway, 355

Troy, 183

Ts'e (Chinese state), 175

Tso Ch'iu-Ming, 145ff, 150, 153,
178, 179

Twain, Mark, 192, 193, 212
see: Clemens, S. L.

U

Ugarit, 66ff, 147, 464ff, 479

Ukraine, 352

Ulm, Germany, 283

Ulster, Annals of, 240

unicorn, 151

Urbevetani [ca. 1260] (Cronica
Potestatum Urbevetani), 318,
327

Urbevetani [ca. 1276] (Cronica
Potestatum Urbevetani), 318,
327

Urbevetani [ca. 1313] (Chronica
Antiqua Urbevetani), 319, 329

Ursperg, monastery of (Germany),
283, 294, 295, 307, 511

Uzbek, S. S. R., 168

V

van der Waerden, B. L., 206, 225

van Flandern, T. C., 12, 453,
456, 472, 476, 477, 481

Varignana, Cronaca, 315, 330,
331

Vatican, 316

Velikovsky, I., Section V.8
passim

Vendôme, France, 255, 511

Venice, 312, 511

Venus, 5, 11, 14ff, 19ff, 26ff, 37,
47, 100, Section V.8 passim,
205, 206

Vespasian, 215

Vicenza, Italy, 319, 322, 324,
341, 511

Vienna, 286, 349, 350, 351, 362,
365, 372, 373, 511

Vigeois, France, 275, 511

Villola, Pietro e Floriano da,
315

Vincentina, Cronica, 319, 322,
323

Vindobonensis, Continuatio, 351,
365

Volterra, Italy, 315, 339, 343,
511

Vormezeele, Belgium, 346, 511

Library of Congress Cataloging in Publication Data

Newton, Robert R
    The moon's acceleration and its physical origins.

    Bibliography: v. 1, pp. 535-59
    Includes index.
    CONTENTS: v. 1. As deduced from solar eclipses.—

    1. Moon—Observations. 2. Acceleration (Mechanics)—
Observations. 3. Moon—Origin. I. Title.
QB581.N54   523.3'3   78-20529
ISBN 0-8018-2216-5 (v. 1)